Neurohormones and Neurohumors

Journal of Neuro-Visceral Relations

Supplementum IX

Neurohormones
and Neurohumors

Structure and Function of Regulatory Mechanisms

*Proceedings of the Symposium
of the International Society for Neurovegetative Research*

Amsterdam, July 25—28, 1967
Edited by J. Ariëns Kappers

Springer-Verlag
Wien GmbH 1969

With 142 Figures

ISBN 978-3-662-23465-5   ISBN 978-3-662-25519-3 (eBook)
DOI 10.1007/978-3-662-25519-3

© 1969 by Springer-Verlag Wien
Originally published by Springer-Verlag Wien New York in 1969
Softcover reprint of the hardcover 1st edition 1969

Library of Congress Catalog Card Number 71-90085

Title No. 9254

Introduction

Having accepted the task to organize, in Amsterdam, an international Congress in behalf of the International Society for Neurovegetative Research such as is held every second year, it occurred to me that it might be worthwhile to change its usual pattern somewhat. Instead of giving merely the opportunity for reading a number of rather short free papers, more or less grouped in many categories, we decided to start with a symposium on a special topic to be dealt with from several aspects by a number of invited speakers, the last day of the congress being only available for short free communications.

The central topic for the symposium finally chosen was "Neurohormones and Neurohumors" as this subject is of interest to many morphologists, physiologists, pharmacologists, neuro-endocrinologists, biochemists and clinicians, and belongs to a rapidly expanding field of research.

Moreover, it appeared to us that there is some confusion about the definition of these terms which are not seldom used indifferently. I hoped that such a symposium would contribute to a clearer apprehension of the terms based on morphological, physiological and biochemical criteria as well as to a clearer understanding of some problems involved which will have to be subjects of future investigations.

To obtain a maximum of coordination and integration of recent data it appeared useful to invite a number of prominent workers in several scientific fields which are related to this topic. It was, furthermore, decided not to publish the proceedings of the entire congress but only those of the symposium on Neurohormones and Neurohumors which developed into a kind of open table conference. The reader will see that most of the papers are more or less comprehensive and, in the opinion of the editor, very useful surveys based on data obtained by the authors as well as by others. Not a few communications contain also original work published here for the first time. Many authors, moreover, point to questions and problems which are still unsolved thus contributing to the future development of this special field which can be considered of fundamental importance, also for the clinician.

My sincere thanks are due to all of the invited speakers cooperating in this joint effort by contributing so many important papers and to them as well as to other participants for taking such a lively and fruitful part in the discussions. Quite a number of final papers published in this volume are more extensive and detailed than those originally read at the symposium. By this, the scientific value of the proceedings has certainly still been enhanced, but also a regrettable, however inevitable delay of the publication of this volume was caused. Being responsible for the reconstruction of the discussions, my apologies are offered in advance to anyone who will not be entirely satisfied with what has finally been published from his remarks and comments.

I feel most grateful to the staff members of the Netherlands Central Institute for Brain Research, especially to my secretary Miss *L. Kruse*, who helped in various ways in organizing the congress and in editing the proceedings of the symposium.

Financial help for the organization obtained from the Government of The Netherlands, Organon Ltd., CIBA A. G., Geigy A. G., Hoffmann-La Roche A. G., Sandoz A. G., and the Deutsche Gesellschaft für Neurovegetative Forschung is much appreciated as is the care bestowed by the Springer-Verlag, Vienna, on the publication of this volume.

Amsterdam, June 1969.   **J. Ariëns Kappers**

Contents

Contents

Neurohormonal and Neurohumoral Regulatory
Mechanisms

List of Contributors

Akert, K., Brain Research Institute, University of Zürich, Zürich, Switzerland.

Ariëns Kappers, J., The Netherlands Central Institute for Brain Research, Amsterdam, The Netherlands.

Bargmann, W., Department of Anatomy, University of Kiel, Kiel, Germany.

Birkmayer, W., Department of Neurology, Geriatric Hospital of the City of Vienna, Vienna-Lainz, Austria.

Carlsson, A., Department of Pharmacology, University of Göteborg, Göteborg, Sweden.

Csillik, B., Department of Anatomy, University Medical School, Szeged, Hungary.

De Robertis, E., Department of General Anatomy and Embryology, University of Buenos Aires, Buenos Aires, Argentina.

Feldberg, W., National Institute for Medical Research, London, Great Britain.

Halász, B., Department of Anatomy, University Medical School, Pécs, Hungary.

Knowles, Sir Francis, Department of Anatomy, University of London, London, Great Britain.

Lammers, H. J., Department of Anatomy and Embryology, University of Nijmegen, Nijmegen, The Netherlands.

Oksche, A., Department of Anatomy, University of Giessen, Giessen, Germany.

Picard, D., Department of Histology and Embryology, Faculty of Medicine, Marseille, France.

Quay, W. B., Department of Zoology, University of California, Berkely (Cal.), U.S.A.

Scharrer, B., Department of Anatomy, Albert Einstein College of Medicine, Bronx (N.Y.), U.S.A.

Smelik, P. G., Department of Pharmacology, Free University, Amsterdam, The Netherlands.

Stern, P., Department of Pharmacology, Faculty of Medicine, Sarajevo, Yugoslavia.

Stock, K., Institute of Pharmacology, University of Frankfurt, Frankfurt/Main, Germany.

Westermann, E., Institute of Pharmacology, University of Frankfurt, Frankfurt/Main, Germany.

Journal of Neuro-Visceral Relations. Suppl. IX, 1—20 (1969)

Neurohumors and Neurohormones:
Definitions and Terminology *

Berta Scharrer

Department of Anatomy
Albert Einstein College of Medicine
New York, U. S. A.

Summary

Neurohumors and neurohormones are physiologically active substances produced by the nervous systems of vertebrates and invertebrates. They have certain features in common and differ in others. Both are chemical messengers that are intermittently released from nerve cells to act on neuronal or non-neuronal effector cells. Significant characteristics of the regulatory mechanisms controlled by these two classes of neurochemical mediators concern the spatial relationship between site of origin of mediator and effector cell; the duration of the signal; the chemical nature of the active principles involved; and the ultrastructural features of their intracellular storage sites.

(1) *Neurohumors:* In "chemical synaptic transmission" the messenger substance elicits strictly localized postsynaptic responses of very short duration in effector cells that are contiguous with the respective presynaptic terminals. The active principles (*e. g.*, acetylcholine, noradrenaline) lack several essential attributes of endocrine substances and are, therefore, more appropriately classified as "chemical transmitters", "neurotransmitters", or "neurohumors" in contradistinction to "neurohormones". Electron micrographs of cholinergic neurons show small electron-lucent vesicles that are especially abundant in presynaptic areas. There is circumstantial evidence that these "synaptic vesicles" are intracellular storage sites of acetylcholine. Adrenergic axons and terminals (with deposits of noradrenaline as demonstrated by fluorescence microscopy) contain varying amounts of dense-core vesicles which seem to harbor some of this catecholamine.

(2) *Neurohormones:* Aside from synaptic transmission, neurons communicate with effector cells by means of hormonal mediators. However, this activity is restricted to specialized cell groups within the nervous system which possess glandular attributes above and beyond those of conventional neurons and which, because of their dual nature, are called "neurosecretory cells". Their product ("neurosecretory material") contains active principles capable of regulating

---

* Dedicated to the memory of Dr. *Tilly Edinger*, distinguished scholar, humanitarian, and friend.

multiple and diverse "target" cells (by bridging considerable distances via vascular channels) and of acting for sustained periods of time. These attributes parallel those of non-neuronal endocrine factors and serve to classify neurosecretory mediators as "neurohormones". Typical examples are polypeptides (e. g., vasopressin, oxytocin) bound to carrier proteins (neurophysins). Under the light microscope, the identification of classical neurosecretory neurons depends on selective staining and histochemical properties. Ultrastructurally, neurosecretory materials appear as membrane-bounded granules of several size ranges and of varying, but usually high, electron density.

That the borderline between these two classes of neural mediators is not as sharp as might be concluded from the preceding characterization, is illustrated by the following statements:

a) There is evidence for the existence of a class of neurohormones that differs from the classical by its non-proteinaceous, i. e., catecholamine nature and a somewhat different mode of operation.

b) In certain cases, where neurosecretory neurons appear to control endocrine effector cells by means of "neurosecretomotor junctions", the chemical mediator cannot be classified as a neurohormone.

At the outset of this symposium, at which participants representing several different disciplines will discuss mechanisms controlled by neurohumors and neurohormones, a brief general orientation seems in order. As in every rapidly expanding area of research, new information on the chemical mediators involved in the integrative activities of the nervous system requires continual modulation in interpretation and terminology. An attempt to do justice to the topic as it presents itself today turns out to be more difficult than anticipated, if our aim is conceptual synthesis rather than mere presentation of relevant definitions and currently accepted terms. Furthermore, the vast amount of literature on the subject does not permit extensive documentation of many of the statements made in this introductory discussion [*].

The formulation of the program of this symposium clearly indicates that we shall have to deal with two classes of active principles, neurohumors and neurohormones. Our central task will be to define the physiological mechanisms in which these mediators play a role, and to correlate distinctive structural and functional characteristics. Since neurohumoral agents are of primary concern in "chemical synaptic transmission" and neurohormones in neuroendocrine activities, there is merit in examining them separately. In the following presentation, covering information on vertebrates and invertebrates, the emphasis will be on typical manifestations of each of the two systems of neurochemical communication. At the same time, it will become apparent that both groups of mediators have certain features in common, and that a number of borderline cases more or less defy classification.

---

[*] In order to keep the list of references within manageable limits, it is largely restricted to publications that appeared in 1966—1967.

I. Neurohumors

Interest in the nature and mode of action of a group of biogenic amines, commonly called neurotransmitter substances, and classified as neurohumors, has remained vigorous ever since the classical discoveries of *Loewi*, *Dale*, and others which led to the concept of the chemical transmission of nervous impulses (see *Dale*, 1935). Aside from the more widespread representatives (acetylcholine, noradrenaline), we now know of other compounds with similar activities, among them 5-hydroxytryptamine (serotonin), dopamine, gamma-aminobutyric acid, glutamine, glycine and as yet unidentified neurotransmitters, particularly those occurring among invertebrates (*Gerschenfeld*, 1966; *Brown*, 1967).

It should perhaps be underlined here in passing that the chemical events during synaptic transmission are not the only ones for which these biogenic amines are responsible (see *Hinks*, 1967; *Kety* and *Samson*, 1967; and others). One need recall only the adrenal medulla with its large stores of catecholamines (*Douglas*, 1967), or the various neural and non-neural tissues rich in serotonin and other amines (see, for example, *Dixit* and *Buckley*, 1967; *Quay*, 1967). Some of these, belonging to the realm of endocrine organs, will be referred to in the second section of this discussion.

Characterization of neurotransmitters

To return to the special properties of biogenic amines involved in synaptic transfer of signals, we should remember that they pertain to practically all parts of the nervous system. The most important characteristics are precise localization, short time course of action, and (at least in the case of noradrenaline) "shuttle service" type of operation (see below). Since exceedingly brief signals must follow each other in rapid succession from neuron to neuron, two equally important requirements are instant availability of the transmitter substances involved, and effective means for their speedy inactivation after the completion of the signal.

It is generally believed (see *De Robertis*, 1964, 1967) that part of the neurotransmitter material present is stored in inactive form within small, membrane-limited vesicles in the preterminal area of the synapse. In response to nervous stimulation, release from these "synaptic vesicles" occurs into the synaptic cleft (about 100—200 Å in width), the distance traversed by the chemical mediator to reach the postsynaptic receptor thus being negligible. Subsequent abolishment of transmitter action occurs *in loco* and may be the result of enzymatic action, desensitization of the receptor, diffusion, or re-uptake by storage vesicles on the presynaptic side. From such a pool of sequestered transmitter, packaged in quantal units, the same molecule may be called into service again and again (shuttle service type of action). Experiments to determine the localization of infused labelled noradrenaline by means of density gradient centrifugation have revealed that lysosomes may serve as additional sites for storage and inactivation of (perhaps only excessive amounts of) noradrenaline (*Laduron* and coll., 1966).

To return briefly to the chain of events by which neurotransmitters are released from axon terminals (or perhaps periterminal swellings; *Barer,* 1967), the same factors are under consideration as in the case of neurohormones (p. 10). Much information on this subject is contained in a volume edited by *von Euler* and coll. (1966). An increase in the neuron's calcium uptake associated with impulse conduction (*Hodgkin* and *Keynes,* 1957; *Douglas* and *Poisner,* 1964; *Douglas,* 1966) seems to be one of the decisive factors in the liberation of chemical mediators such as noradrenaline.

According to *Burn* (1967), the prime mover in adrenergic neurons is acetylcholine which in this case acts as a "pretransmitter" or "modulator substance" (*Csillik* and *Kása,* 1967) to facilitate the activity of the actual transmitter. If this should turn out to be so, actylcholine would in this instance play a role equivalent to that attributable to it in other cases of its presumed intracellular action (see *Koelle,* 1961), and different from that in cholinergic synaptic transmission.

Typically, the neurotransmitters, after being released from their cells of origin, neither enter the body's circulation to remain active for some time, nor do they affect multiple and distant target cells. Thus they lack essential criteria for classification as hormones. Instead, neurotransmitters, being confined to the synaptic site in conformity with the special requirements of chemical transmission, represent a separate class of agents, generally called neurohumors.

Experimental and ultrastructural analysis of neurohumoral activities

Our understanding of the distinctive nature of neurohumoral transmitters derives from a combination of investigative methods that belong to the realms of neurophysiology, biochemical pharmacology, and ultracytochemistry.

Pharmacological tests can be carried out with small pieces of nervous tissue; and thus the existence of adrenergic neurons within certain regions of the central nervous system has been demonstrated (*Vogt,* 1954). This result is supported and extended by subsequent histo- and cytochemical studies, in particular those with highly sensitive and specific fluorescence methods (see *Hillarp* and coll., 1966; *Fuxe* and coll., 1966; *Eneström* and *Svalander,* 1967), which permit the differentiation between noradrenaline, serotonin, and dopamine, and their localization in axon terminals or other parts of the neuron. Another important contribution, especially towards the understanding of events at the synaptic level, comes from the electron microscope. The results from these and various other approaches in invertebrates and vertebrates are remarkably similar. The aim here is not merely the identification of subcellular storage sites for one or another type of transmitter, but the interpretation of alterations paralleling known physiological or experimentally induced changes in the synaptic apparatus. Among the latter are attempts at changing the amount and distribution of transmitters by section or constriction of axons (*Steg,* 1966; *Kapeller* and *Mayor,* 1967). Sympathectomy may be employed to cause depletion of biologically active amines in sympathetic neurons. *In vivo* and *in vitro* application of drugs (such as reserpine) with known effects on the elaboration or storage of neurohumors permit comparisons with untreated controls, or with sympathectomized animals. Autoradiography, both at the light and electron microscopic levels, yields information on the sites of synthesis and activity of neurotransmitters. An example is the selective uptake of label in terminals with granular vesicles after the injection of tritiated noradrenaline (see *Aghajanian* and *Bloom,* 1966; *Ishii* and *Friede,* 1967; *Kapeller* and *Mayor,* 1967; *Lenn,* 1967). Another approach is subcellular fractionation and density gradient centrifugation (see, for example, *Austin* and coll., 1967) which, in the case of cholinergic neurons, yields "synaptosomes" (detached nerve terminals, vesicular fraction) rich in acetylcholine, etc. The same procedure, carried out with posterior pituitaries (*Bindler* and coll., 1967), results in "neurosecretosomes" with high contents of neurohormones (see second section).

The results of many of these experimental procedures would be difficult to interpret without concurrent examinations of the micromorphology of intracellular binding sites of various neural mediators. In this connection the existence of more than one type of ultrastructurally distinct vesicle is of particular interest and will, therefore, be discussed here in some detail.

These vesicles may be classified according to size, shape, and content (see *Halaris* and coll., 1967; *Palay,* 1967). There is a whole spectrum of vesicular cores from highly opaque to electron-translucent. However, it has to be emphasized that the degree of opacity depends, in part, on the method of fixation (osmium tetroxide, glutaraldehyde, or potassium permanganate; see *Tranzer* and *Thoenen,* 1967; *Hökfelt,* 1967). Other differences concern the extent to which the vesicles appear to be filled with electron-opaque material, but this criterion, too, has to be judged in terms of possible fixation artifacts.

As to the interpretation of "clear vesicles", it is important to consider that some of them may have lost their contents during tissue preparation, that others may appear translucent because the active principle they contain does not lend itself to visualization with electron-microscopic "staining" procedures currently in use. Or the lack of a visible content may mean that, at the time of fixation, the agranular vesicle had been unoccupied only temporarily and was thus ready to reaccept its specific quantum of transmitter substance. This interpretation is particularly applicable in the case of neurons with mixed populations of electron-lucent and dense-core vesicles.

Agranular vesicles of different sizes and shapes may be correlated with different functional types of synapses (excitatory, inhibitory; see *Bodian,* 1966; *Pappas* and *Bennett,* 1966; *Larramendi* and coll., 1967). Finally, there are species differences with respect to the micromorphology and diameter of various storage vesicles. Therefore, the results and conclusions obtained in one species do not necessarily apply to another.

At present, the assignment of specific types of chemical mediators to morphologically distinct classes of vesicles is not yet possible with confidence. However, certain relationships are becoming apparent.

The accumulation in recognized cholinergic terminals of uniform populations of small (200—500 Å) clear vesicles has led to the postulate that these are storage sites for acetylcholine. They have been accordingly termed "synaptic vesicles", but it should be borne in mind that vesicles of this kind, especially when found at different sites, may fulfill functions other than those connected with synaptic transmission. Thus, the role of morphologically similar clear vesicles in adrenergic terminals, where they occur intermingled with dense-core vesicles in various proportions (see, for example, *Bondareff*, 1965; *Bak*, 1967), is more difficult to interpret. Acetylcholine as well as noradrenaline have been considered as possible contents of these electron-lucent vesicles.

According to *Fuxe* and coll. (1965, 1966), a combined fluorescence and electron microscopic study of appropriate areas of the central nervous system shows a good correlation between the localization of agranular vesicles and catecholamine. However, other evidence favors the dense-core vesicles (especially those of small size; see *Austin* and coll., 1967; *Hökfelt*, 1967) as storage sites for noradrenaline, in which case electron-lucent vesicles of the same diameter may be considered as temporarily depleted but ready for renewed uptake of noradrenaline transmitter (see *Ferry*, 1967). This view is supported by the following observations. Adrenomedullary cells (*Douglas*, 1967; *Elfvin*, 1967) whose noradrenaline content has been identified histochemically, show strongly electron-dense membrane-bounded inclusions under the electron microscope, especially after glutaraldehyde fixation (*Tranzer* and *Thoenen*, 1967). Purified fractions of catecholamine-depleted adrenomedullary granules have a high proportion of electron-lucent vesicles as contrasted with normal controls (*Malamed* and coll., 1967). Dense granules with a comparable morphology in parts of the nervous system known to be rich in noradrenaline by fluorescence microscopy (see *Eränkö*, 1967; *Sano* and coll., 1967) or other histochemical procedures (*Wood*, 1966), therefore suggest that at least part of this transmitter resides in dense-core vesicles.

An even more convincing argument for the concept that electron-dense vesicles contain catecholamine is the close correlation between the increased amounts of these vesicles and of fluorescent material specifically attributable to noradrenaline in the proximal portions of constricted sympathetic fibers (*Kapeller* and *Mayor*, 1967). Furthermore, the degree of fluorescence and the amount of noradrenaline are reduced after the administration of reserpine (*Elofsson* and coll., 1966). The experiments involving constriction of axons also illustrate the concept that noradrenaline (or its precursor), along with other neuroplasmic components, reaches the terminal of the postganglionic fiber by proximo-distal transport (*Steg*, 1966).

In general terms, dense-core vesicles characteristic of adrenergic neurons differ from electron-opaque inclusions of "typical" neurosecretory neurons (see second section) by their lower frequency, their smaller size, and the varying, often incomplete degree to which the vesicles are filled.

The presence of only small electron-lucent vesicles has been mentioned as a characteristic of cholinergic neurons. In other neurons, microvesicles of the same ("synaptic") type are intermingled, in varying proportions, not only with noradrenergic, but with "peptidergic" (*Bargmann* and coll., 1967), *i. e.*, "classical" neurosecretory, granules as well (see below). In the latter two cases, the role of acetylcholine, if indeed supplied by these clear vesicles, must differ from that of a neurotransmitter in the strict sense of the word. It is, therefore, not desirable to classify neurosecretory neurons as "cholinergic", as proposed, for example, by *Matsui* and *Kobayashi* (1965). Presumably, there are comparable neurohumoral activities in the nervous system about which even less is known, both regarding the chemical identity of the active principles involved and their mode of operation. Cases in point are stores of serotonin, dopamine, and related substances in distinct locations within the nervous systems of vertebrates and invertebrates. Only those among them for which *in loco* activity can be demonstrated qualify as neurohumors.

Some of the preceding statements descriptive of neurotransmitter activities do not strictly apply to the sympathetic innervation of smooth muscle. Here, true neuromuscular junctions are missing. Instead, beaded adrenergic fibers terminate at distances of several thousand Å or more from their multiple effector cells (*Lever* and coll., 1967; *Verity* and *Bevan*, 1967). The extracellular stromal material filling this wide neuromuscular gap seems to serve as a suitable channel for a relatively slow propagation of transmitter substance. Release into this compartment may occur from several sites (preterminal swellings as well as actual fiber terminations; *Barer*, 1967). The action of this "freely stored" neurohumor must be more sustained and more widespread than in the case of "typical" synaptic processes.

II. Neurohormones

Characterization of neurohormones

Nervous tissue is endowed with the capacity to use chemical mediators in ways and for purposes quite different from those of neurotransmitter substances. This capacity, unlike that involved in synaptic transmission, is restricted to specialized neurons. Typical messenger substances belonging to the second class, to be discussed now, are polypeptides. They can reach distant non-neural "targets" by way of the circulation and are not as speedily inactivated as the neurohumors making up the first class. There is no evidence that, after their release, these polypeptides are used by the organism more than once for the same purpose. In short, they act in the manner of endocrine messengers and can therefore be classified as "neurohormones". Many, though by no means all, of the cells controlled by neurohormones are endocrine elements. Thus a major task of these active principles is that of providing channels of communication between nervous

and non-nervous integrative centers. Detailed information on these well
established phenomena is available in several comprehensive publications
(see, for example, *Scharrer* and *Scharrer*, 1963; *Bern* and *Knowles*, 1966 a;
Gabe, 1966).

As has been mentioned before and will be further discussed, there are
certain deviations from this general scheme, so that the distinction between
neurohumoral and neurohormonal activities is not always easy. These
borderline areas concern the chemical nature of the mediators as well as
the manner in which the signals are dispatched.

There are in addition to "classical" neurosecretory centers, whose polypeptide products are known to act in the manner of hormones, certain
others which apparently discharge materials of non-peptide character into
circulatory channels. Among the latter are biogenic amines corresponding
chemically, but not functionally, to the neurotransmitters discussed in the
first section. The neurons in question differ from ordinary nerve cells in
that their content of active amines seems to be higher, that the vesicles
indicative of the presence of amines populate all parts of the neuron, and
that their axons may terminate at neurohemal areas. Indications for such
intraneuronal stores of monoamines (*e. g.*, serotonin) exist in the central
nervous systems of many animals, *i. e.*, coelenterates and planarians (see
Lentz, 1967), annelids (*Bianchi*, 1967; see also *Oosaki*, 1966), insects
(*Frontali* and *Norberg*, 1966; *Hinks*, 1967), as well as vertebrates from
fishes (*Knowles* and coll., 1967) to mammals (*Knowles*, 1965; *Knowles* and
Bern, 1966; *Sano* and coll., 1967).

Ultrastructurally, catecholamine-rich neurosecretory neurons show small
dense-core and clear vesicles comparable to those of adrenergic terminals.
Other "neurosecretory neurons" are characterized by the presence of closely
packed agranular vesicles of the kind that in ordinary neurons are found
in such concentration only in preterminal areas. Such neurons occur, for
example, in the diencephalic floor of the hagfish, *Polistotrema stoutii*
(*Nishioka* and *Bern*, 1966), the median eminence of the frog, *Hyla regilla*
(*Smoller*, 1966), or the anterior median eminence of the bird, *Zonotrichia
leucophrys gambelii* (*Bern* and coll., 1966 b). These too appear to furnish
a non-proteinaceous product.

Thus we have to consider "peptidergic" A fibers and "aminergic"
B fibers * as sources of neurohormones. However, since our knowledge
about the latter type of fibers is still fragmentary, the products of the
"classical" A fibers will be in the center of interest in the following discussion.

Cells specializing in the production of neurohormones occur in specific
locations in the nervous systems of invertebrates and vertebrates. Since
they display cytological signs of glandular activity much more extensive

---

* The terminology proposed by *Knowles* (1965) for two different classes
of neurosecretory axons as A and B fibers is used here with reservation, because
the established usage in the neurophysiological literature of the same letters
for functionally distinctive types of conventional neurons is somewhat confusing.

than those of "ordinary" neurons they have been given the designation of "neurosecretory cells". Even though it may be argued that all neurons are capable of secreting certain substances (*De Robertis*, 1964), the term "neurosecretion", which is now firmly established, should remain restricted to the activity of those neural elements in which the production of chemical mediators has become a dominant feature, and which provide all of the existing neurohormones (see *Welsh*, 1959; *E. Scharrer*, 1965 b; *Rinne* and *Arstila*, 1965—66; *Bern*, 1966; *Bern* and *Hagadorn*, 1965; *Bern* and *Knowles*, 1966 a; *Gabe*, 1966; *Green*, 1966; *Knowles* and *Bern*, 1966; *Bajusz*, 1967; *Bargmann*, 1966; *Bargmann* and coll., 1967; *B. Scharrer*, 1967).

Ultrastructurally, "classical" neurosecretory neurons may, as a rule, be distinguished from ordinary nerve cells by their prominent content of membrane-bounded granules of varying, but frequently pronounced, electron opacity. Their appearance may be homogeneous or finely granulated. Several size categories of granules are observed within the same animal, and there are variations from species to species. Generally speaking, for polypeptide elements the range is of the order of 1000—3000 Å, *i. e.*, above the average for "catecholamine granules" (500—1400 Å). Intermingled with typical proteinaceous granules are smaller electron-lucent structures that cannot be distinguished from synaptic vesicles of cholinergic neurons. In this respect, classical neurosecretory cells resemble neurons (or terminals) rich in monoamines.

Preterminal areas of "peptidergic" or "adrenergic" neurosecretory neurons may contain such small vesicles in dense aggregates (see, for example, *Oota*, 1963 a; *B. Scharrer*, 1963, 1968). Since the plasma membrane on which these vesicles abut may be thickened and electron-opaque, such contact areas are similar to synaptic structures of non-neurosecretory neurons, and may, therefore, be called "synaptoids" [*].

Demonstrations of the functional significance of neurosecretory neurons in a great variety of animal groups have been numerous and conclusive. The existing information has been extensively reviewed in the past (for references see, for example, *Gabe*, 1966). Therefore, it will suffice here to state that classical methods of endocrinology, such as extirpation and implantation, also apply to the search for neurohormonal activities. Furthermore, the insights gained from a combination of structural and functional approaches, during the last decade or two, have led the way to the exploration of the relationships between the nervous and endocrine systems on a broad front, and thus to the establishment of a new discipline, Neuroendocrinology (see *Scharrer* and *Scharrer*, 1963; *E. Scharrer*, 1966; *Bajusz* and *Jasmin*, 1964; *Klotz*, 1966; *Martini* and *Ganong*, 1966; *Weitzman*, 1964—1969). The role of the neurosecretory neuron as an essential link in the communication between the nervous and endocrine systems has

---

[*] This term, for lack of a better one, is suggested to designate ultrastructural specializations which are characteristic of chemical synapses, but which occur at sites not, or not necessarily, involved in regular synaptic transmission.

been discussed elsewhere (*E. Scharrer*, 1952, 1965 a; *B. Scharrer*, 1967).
In the present contribution, attention will be focused on the mechanisms
by which neurosecretory substances are dispatched to various effector cells.

Release of neurohormones

Neurosecretory granules reach the fiber terminal by proximo-distal
axonal transport and accumulate there in a bulbous enlargement, an
important site of release of neurosecretory hormones.

The mode of their liberation from the cell is not entirely understood
(see *B. Scharrer*, 1967, 1968). However, the process seems to have certain
features in common with the release of similar secretory products from non-
neuronal gland cells. Intact (membrane-bounded) polypeptide granules have
as a rule not been demonstrated outside of the neurons in with they are produced and stored. Therefore, the active components must be assumed to
leave the neurosecretory vesicles before being secreted from the cells. They
may leave the vesicles either still bound to protein carriers (neurophysins,
see *LaBella* and coll., 1967), or, more likely, separated from them, before
passing through the plasma membrane. In either case the material becomes
sufficiently dispersed to be no longer detectable as a morphological entity
under the electron microscope. Several investigators (among them *Sachs*,
1966, 1967) are of the opinion that heterogeneous pools of polypeptide
hormones exist side by side in at least some neurosecretory neurons, one
membrane-bounded, the other not.

Empty membranous sacs (with the same diameters as those of the intact
neurosecretory vesicles with their "dense" content), frequently observed in
"depleted" neurosecretory terminals, suggest intracellular dispersal of
granule content before release. In some cases, but not in others, omega-
shaped membranous structures in contact with the plasma membrane speak
for temporary fusion between it and the vesicular wall which effects liberation of the granular content by exocytosis (*Normann*, 1965; *Smith* and
Smith, 1966).

The small agranular vesicles (of the "synaptic" type), often concentrated
near the contact area of the plasma membrane, have been interpreted in
several ways. Perhaps it is helpful to recall that not all clear vesicles need
serve the same purpose. Some may be derived from fragmented larger
vesicles whose task of harboring neurosecretory material had been terminated and which had, therefore, broken down and then become re-
arranged (see, for example, *Streefkerk*, 1967). Small vesicles of this type
perhaps participate in the discharge of secretory polypeptide products. Or
small vesicles may contain another type of active principle, most likely
acetylcholine. This is suggested by biochemical indications for the presence
of this substance in neurosecretory neurons of invertebrates (see *Beaulaton*,
1967), as well as vertebrates (*De Robertis*, 1967; *Duffy* and coll., 1967;
and others). If, indeed, acetylcholine should be the content of some of
these synaptic-type vesicles, it might facilitate a calcium-activated release
mechanism for neurosecretory polypeptides as well as for catecholamines

in neurosecretory B fibers, or postsynaptic sympathetic fibers. However, depolarization of the plasma membrane during impulse conduction seems to suffice for the entry of Ca-ions into the terminal and subsequent hormone release (*Douglas* and *Poisner*, 1964; *Douglas*, 1966; *Haller* and coll., 1965).

Several observations, among them the absence of increased numbers of "synaptic vesicles" in neural lobes depleted of neurosecretory material by thirsting experiments (*Oota* and *Kobayashi*, 1966), speak against the possibility that these vesicles are derived from neurosecretory elements.

Aside from serving in an "auxiliary" capacity within the same neurosecretory neuron, accumulations of small vesicles may be involved in hormone release mechanisms on a cell-to-cell basis. This is suggested by the observation of "synaptic" contacts between peptidergic neurosecretory axons (A fibers) and either ordinary neurons (see, for example, *Oota*, 1963 b), or catecholamine containing neurosecretory B fibers. Synaptoids of the latter kind were observed in "neurohemal" organs of insects (*B. Scharrer*, 1963, 1968; *Normann*, 1965; *Bowers* and *Johnson*, 1966; *Brady* and *Maddrell*, 1967) and in vertebrates (*Nishioka* and *Bern*, 1966). More specifically, these morphological data suggest an indirect (auxiliary) role of contiguous B fibers in the functional management of certain A fibers, for example those furnishing the adenohypophysis with special polypeptides, the "regulating factors" (also called "releasing" or "hypophysiotropic" factors).

However, an alternative interpretation of the selective accumulation of biogenic amines in certain neurohemal areas cannot be excluded. It will be discussed in conjunction with the question to be taken up next, *i. e.*, how neurosecretory substances, once they have left their intracellular storage sites, reach their destinations.

Pathways between neurosecretory neurons and effector cells

Active substances released from neurosecretory neurons may reach their "targets" by one of several possible routes.

1. The most important pathway is the vascular system. In higher vertebrates this involves the close spatial affiliation between granule-laden terminals and capillary walls giving access to either the general circulation (neural lobe) or a special portal system (median eminence).

The median eminence and its equivalent structures in lower vertebrates (see *Nishioka* and *Bern*, 1966; *Smoller*, 1966; and others) are distinctive neurohemal organs not only in that they provide pathways circumventing the general systemic circulation, but also because they are sites of release of non-peptide mediators from B fibers. Some of these may represent a distinctive class of hypophysiotropic factors (see *Guillemin* and coll., 1966) whose role and possible relationship with the peptidergic regulating factors mentioned earlier are still largely unknown.

In invertebrates lacking a capillary system, such as insects, secretory terminals line up near a sheath (or its processes) which separates the neurohemal organ from the surrounding hemolymph. This extracellular stromal "barrier" corresponds in its properties to the boundary membrane of the capillary wall.

It is of considerable interest that a structural and functional dualism such as that exemplified by the posterior lobe and the median eminence of higher vertebrates, also occurs among neurohemal organs of insects (*Raabe*, 1966; *Brady* and *Maddrell*, 1967). Non-proteinaceous neurohormones of vertebrates seem to have a counterpart among invertebrates. An example is the demonstration by *Hinks* (1967) that serotonin, furnished by a special type of neurosecretory cells in the pars intercerebralis of the nocturnal moth, *Noctua*, apparently acts as a hormone in the control of "dark-activation" in the circadian rhythm of this insect.

Another parallelism concerns the fact that invertebrate as well as vertebrate effector cells reached via the circulation may be either "terminal targets" (*e. g.*, kidney cells; so-called first-order neuroendocrine mechanisms) or endocrine way stations (*e. g.*, insect molting gland; second-order mechanisms).

A word might be added about the possible role of the cerebrospinal fluid as a selective circulatory channel for the transport of neurosecretory material in vertebrates. This proposition is based on the observation of neuronal processes protruding into the ventricular space with bulbous endings that contain peptidergic, adrenergic, and cholinergic vesicles in varying proportions (see *Fridberg* and *Nishioka*, 1966; *Sterba* and *Weiss*, 1967; *Takeichi*, 1967). Release of neural mediators into the cerebrospinal fluid has been demonstrated (see, for example, *Baumgarten* and *Braak*, 1967). Perhaps in this case neural elements located at some distance from the site of hormone release might be their effector cells.

2. Aside from the hypophysial portal system, there exist several other possibilities by which neurosecretory mediators reach effector organs (especially endocrine cells) in a "directed" fashion. Granule-laden terminals may penetrate pericapillary spaces within the adenohypophysis, as demonstrated, for example, in the teleost fishes *Anguilla* and *Conger* (see *Vollrath*, 1967), and thus considerably shorten the distance of travel for the released mediator. The vascular link may even be abandoned altogether. In certain other teleosts, such as *Tinca*, the pituitaries show neurosecretory terminals to be separated from the endocrine gland cells by nothing more than a "basement membrane", and again in others (*Hippocampus*) even this last intervening structure has disappeared and the terminals of A and B fibers are contiguous with the effector cells. This is an interesting case of direct dual neurosecretory innervation (*Knowles* and coll., 1967). Furthermore, *Bargmann* and coll. (1967) have demonstrated peptidergic, side by side with aminergic, terminals in direct contact with the glandular cells of the pars intermedia of a mammal, the cat.

Morphological relationships similar to those described for vertebrates occur in endocrine organs of insects, the corpus allatum and the prothoracic gland (*B. Scharrer*, 1964 a, b; and others). Neurosecretory fibers of insects may make contact also with non-endocrine cells, such as muscular elements (*Bowers* and *Johnson*, 1966), and finally there are "synaptoid" relationships between peptidergic and presumed aminergic fibers mentioned earlier (p. 11).

It is apparent that the close relationships between certain neurosecretory terminals and effector cells parallel the spatial arrangements at the terminations of ordinary neurons. This is true not only for cases of contiguity, but even for those of "near contact". The latter situation is exemplified by special cases of muscular innervation. In the intrafusal fiber a "basement membrane" layer intervenes between motor ending and effector cell (*Landon*, 1966). As to the innervation of smooth muscle discussed earlier (p. 7), "diffusional" distances of 4000 Å and above have been observed, exceeding considerably the usual intersynaptic space of about 150 Å. In these special situations among vertebrates, the mucopolysaccharide layer facing the axon terminal may play a distinctive role in the storage and/or transport of active substances (*Barer*, 1967), as it seems to do in insects (*B. Scharrer*, 1965).

In brief, wherever neurosecretory neurons display ultrastructural features reminiscent of synaptic contacts (close spatial relationships, "presynaptic vesicles", electron-opaque membrane thickenings), it is difficult to decide whether they signify "transmitting synapses" or "specialized sites of hormone release" (see *Vollrath*, 1967). All we can state is that close contacts between neurosecretory fibers and cells of internal secretion strongly suggest functional interaction by means of chemical messengers. The latter may differ (in quantitative, if not always in qualitative terms) from ordinary neurotransmitters, and the similarity of the contact areas in both cases of chemical mediation may merely be the expression of a parallelism in requirements for structural specialization (vesicles, adhesion devices; *Kelly*, 1967).

Active substances released from neurosecretory terminals for strictly localized action should not be ranked among neurohormones for the same reasons that disqualify neurotransmitters.

Conclusions

Our current views on neurohumors and neurohormones are closely linked with the phenomenon of neurosecretion. The interpretation of its relationship with humoral activities in synaptic transmission has undergone stepwise modulations in the past years, and it appears that the end point has not yet been reached. To put it succinctly, ordinary and neurosecretory neurons overlap in more respects than was originally anticipated. Some of the features thought at first to be the exclusive property of neurosecretory elements are shared by non-neurosecretory neurons. Conversely, properties of ordinary neurons, formerly considered as absent in neurosecretory cells, are observed in both categories. Among these are ultrastructural correlates of the secretory processes by which their respective active principles are manufactured, transported along axons, stored, and liberated from the cells.

Another point is that "non-synaptic terminations", a characteristic of the prototypes of neurosecretory neurons, also seem to occur among ordinary nerve cells sending processes into perimuscular or periglandular areas (*Robertson*, 1967), or into the ventricular space (*Leonhardt* and *Lindner*, 1967; *Takeichi*, 1967). On the other hand "synaptic endings", the hallmark of regular neurons, are paralleled by "synaptic" vesicles and membrane specializations in neurosecretory neurons that are in direct contact with effector cells.

Observations of such structural and functional similarities perhaps call for a re-evaluation of the term "synapse". In order to avoid using it too loosely, thickened, electron-opaque plasma membranes affiliated with clusters of small clear vesicles at neurosecretory terminals (including "neurosecretomotor junctions") should perhaps be referred to as "synaptoid" configurations. The same holds for endings and perhaps other axonal specializations (*Robertson*, 1967) of sympathetic fibers releasing chemical mediators into extracellular stromal areas. In both instances, time course and physiological effect differ from those of regular synaptic activity. Particularly in the case of neurosecretory "innervation" the effects are sustained; inactivation of neurosecretory substances presumably occurs at a much slower rate than that of neurotransmitters. On the other hand, it must be emphasized that those neurosecretory mediators, acting *in loco*, have in a sense more in common with neurohumors than with neurohormones. Therefore, the terms "neurohormones" and "neurosecretory substances", although closely related to each other, are not synonymous.

According to this view, active principles furnished by neurosecretory neurons include all neurohormones, plus certain special non-hormonal mediators that bridge the gap between neurotransmitters and neurohormones. From the time of its introduction the term "neurosecretory" was intended to leave room for all possible functional aspects (*E. Scharrer*, 1965 a). It was not restricted to neurons with hormonogenic activity (contrary to the statement by *Bern*, 1966), and was therefore always considered as preferable to "neuroendocrine" or "neurocrine" for the designation of nerve cells in which secretory activity is a predominant feature. Even for those neurosecretory neurons whose products qualify as hormones the term "neuroendocrine" would no longer be useful, since it has acquired a broader meaning referring to all interactions between the nervous and the endocrine systems.

Another feature by which neurohormones can no longer be categorically separated from neurohumors is their chemical nature. Originally the polypeptide character of neurosecretory products was considered a valid characteristic in contradistinction to the biogenic amines serving as neurotransmitters. More recent information indicates, however, that some of the existing neurosecretory elements (B fibers) make use of substances other than the "customary" octapeptides, among them catecholamines, for purposes of special forms of endocrine mediation. Perhaps the effects of these non-peptide neurohormones are of shorter duration than those of the products of classical A fibers (*Knowles*, 1965).

These new insights have not detracted from the distinctiveness of the neurosecretory neuron as a special cell type, a view held by the great majority of investigators in this field. Neither is there a valid reason for abandoning the separation of chemical mediators of the nervous system into neurohumors and neurohormones. Even if the boundary line is somewhat blurred, the end points are clear. In summary, neurohumors are biogenic amines generally present in neurons to serve the chemical transmission of impulses, and are characterized by high speed and very short range of action. Neurohormones, predominantly furnished by peptidergic special neurons (neurosecretory cells), exert sustained actions on multiple distant effector cells and serve neuroendocrine integrative functions. However, not all (peptide or non-peptide) neurosecretory substances need to act in the manner of blood-borne hormones, and conversely aminergic (neurohumoral type) substances may enter the circulation to function as endocrine mediators.

Acknowledgements

The preparation of this article was aided by U. S. P. H. S. grant NB-00840, part of which supports the compilation of Bibliographia Neuroendocrinologica.

I owe profound thanks to Dr. *Mary Weitzman* for her substantial bibliographic assistance.

References

Aghajanian, G. K. and *F. E. Bloom:* Electron-microscopic autoradiography of rat hypothalamus after intraventricular H³-norepinephrine. Science *153,* 308-310 (1966).

Austin, L., I. W. Chubb, and *B. G. Livett:* The subcellular localization of catecholamines in nerve terminals in smooth muscle tissue. J. Neurochem. *14,* 473—478 (1967).

Bajusz, E., ed.: An Introduction to Clinical Neuroendocrinology. 573 pp. Basel and New York: Karger, 1967.

Bajusz, E., and *G. Jasmin,* eds.: Major Problems in Neuroendocrinology. Basel and New York: Karger, 1964.

Bak, I. J.: The ultrastructure of the substantia nigra and caudate nucleus of the mouse and the cellular localization of catecholamines. Exp. Brain Res. *3,* 40—57 (1967).

Barer, R: Speculations on the storage and release of hormones and transmitter substances. Symp. electr. Activ. Innerv. Blood Vessels, Cambridge 1966. Bibl. anat. *8,* 72—75. Basel and New York: Karger, 1967.

Bargmann, W.: Neurosecretion. International Review of Cytology *19,* 183—201 (1966).

Bargmann, W., E. Lindner, and *K. H. Andres:* Über Synapsen an endokrinen Epithelzellen und die Definition sekretorischer Neurone. Untersuchungen am Zwischenlappen der Katzenhypophyse. Zschr. Zellforsch. *77,* 282—298 (1967).

Baumgarten, H. G., and *H. Braak:* Catecholamine im Hypothalamus vom Goldfisch *(Carassius auratus).* Zschr. Zellforsch. *80,* 246—263 (1967).

Beaulaton, J.: Sur la localisation ultrastructurale d'une activité cholinestérasique dans le corps cardiaque de *Rhodnius prolixus* Stal. (Hétéroptère, Reduvidae) aux quatrième et cinquième stades larvaires. J. Microscopie *6,* 65—80 (1967).

Bern, H. A.: On the production of hormones by neurones and the role of neurosecretion in neuroendocrine mechanisms. Symp. Soc. Exp. Biol. *20,* 325—344 (1966).

16   Berta Scharrer:

Bern, H. A., and *I. R. Hagadorn:* Neurosecretion. In: Structure and Function in the Nervous Systems of Invertebrates, by *T. H. Bullock* and *G. A. Horridge.* pp. 353—429. San Francisco and London: W. H. Freeman and Co. 1965.

Bern, H. A., and *F. G. W. Knowles:* Neurosecretion. In: Neuroendocrinology, *Martini, L.,* and *W. F. Ganong,* eds. Vol. I, pp. 139—186. New York and London: Academic Press, 1966a.

Bern, H. A., R. S. Nishioka, L. R. Mewaldt, and *D. S. Farner:* Photoperiodic and osmotic influences on the ultrastructure of the hypothalamic neurosecretory system of the white-crowned sparrow, *Zonotrichia leucophrys gambelii.* Zschr. Zellforsch. *69,* 198—227 (1966b).

Bianchi, S.: The amine secreting neurons in the central nervous system of the earthworm *(Octolasium complanatum)* and their possible neurosecretory role. Gen. comp. Endocrinol. *9,* 343—348 (1967).

Bindler, E., F. S. LaBella, and *M. Sanwal:* Isolated nerve endings (neurosecretosomes) from the posterior pituitary. Partial separation of vasopressin and oxytocin and the isolation of microvesicles. J. Cell Biol. *34,* 185—205 (1967).

Bodian, D.: Electron microscopy: two major synaptic types on spinal motoneurons. Science *151,* 1093—1094 (1966).

Bondareff, W.: Submicroscopic morphology of granular vesicles in sympathetic nerves of rat pineal body. Zschr. Zellforsch. *67,* 211—218 (1965).

Bowers, B., and *B. Johnson:* An electron microscope study of the corpora cardiaca and secretory neurons in the aphid, *Myzus persicae* Sulz. Gen. comp. Endocrinol. *6,* 213—230 (1966).

Brady, J., and *S. H. P. Maddrell:* Neurohaemal organs in the medial nervous system of insects. Zschr. Zellforsch. *76,* 389-404 (1967).

Brown, B. E.: Neuromuscular transmitter substance in insect visceral muscle. Science *155,* 595—597 (1967).

Burn, J. H.: Release of noradrenaline from the sympathetic postganglionic fibre. Brit. med. J. *1967,* 2, 197—201 (1967).

Csillik, B., and *P. Kása:* Localization of acetylcholinesterase in the guinea pig cerebellar cortex. Acta neuroveget. Wien *29,* 289—296 (1967).

Dale, H.: Pharmacology and nerve-endings. Proc. Roy. Soc. Med., London, *28,* 319—332 (1935).

De Robertis, E.: Histophysiology of Synapses and Neurosecretion. pp. 1—244, Oxford: Pergamon Press, 1964.

De Robertis, E.: Ultrastructure and cytochemistry of the synaptic region. Science *156,* 907—914 (1967).

Dixit, B. N., and *J. P. Buckley:* Circadian changes in brain 5-hydroxytryptamine and plasma corticosterone in the rat. Life Sci. *6* (No. 7), 755—758 (1967).

Douglas, W. W.: Calcium-dependent links in stimulus-secretion coupling in the adrenal medulla and neurohypophysis. In: Mechanisms of release of biogenic amines. *U. S. von Euler, S. Rosell and B. Uvnäs,* eds. 267—290, Oxford: Pergamon Press, 1966.

Douglas, W. W.: Mechanism of release of catecholamines in adrenal medulla. Neurosciences Res. Prog. Bull. *5,* 45—47 (1967).

Douglas, W. W., and *A. M. Poisner:* Stimulus-secretion coupling in a neurosecretory organ: the role of calcium in the release of vasopressin from the neurohypophysis. J. Physiol. *172,* 1—18 (1964).

Duffy, P. E., V. M. Tennyson, and *M. Brzin:* Cholinesterase in adult and embryonic hypothalamus. A combined cytochemical electron microscopic study. Arch. Neurol. *16,* 385—403 (1967).

Elfvin, L. G.: The development of the secretory granules in the rat adrenal medulla. J. Ultrastr. Res. *17,* 45—62 (1967).

Elofsson, R., T. Kauri, S. O. Nielsen, and *J. O. Strömberg:* Localization of monoaminergic neurons in the central nervous system of *Astacus astacus* Linné (Crustacea). Zschr. Zellforsch. *74,* 464—473 (1966).

Eneström, S., and *C. Svalander:* Liquid formaldehyde in catecholamine studies. A new approach to the morphological localization of monoamines in the adrenal medulla and the supraoptic nucleus of the rat. Histochemie *8,* 155—163 (1967).

Eränkö, O.: Histochemistry of nervous tissues: catecholamines and cholinesterases. Ann. Rev. Pharmacol. *7,* 203—222 (1967).

Euler, U. S. von, S. Rosell, and *B. Uvnäs,* eds.: Mechanisms of Release of Biogenic Amines. Wenner-Gren Center International Symposia Series. Vol. 5, 482 pp. Oxford: Pergamon Press, 1966.

Ferry, C. B.: The autonomic nervous system. Ann. Rev. Pharmacol. *7,* 185—202 (1967).

Fridberg, G., and *R. S. Nishioka:* Secretion into the cerebrospinal fluid by caudal neurosecretory neurons. Science *152,* 90—91 (1966).

Frontali, N., and *K.-A. Norberg:* Catecholamine containing neurons in the cockroach brain. Acta physiol. Scand. *66,* 243—244 (1966).

Fuxe, K., T. Hökfelt, and *O. Nilsson:* A fluorescence and electron microscopic study on certain brain regions rich in monoamine terminals. Amer. J. Anat. *117,* 33—46 (1965).

Fuxe, K., T. Hökfelt, O. Nilsson, and *S. Reinius:* A fluorescence and electron microscopic study on central monoamine nerve cells. Anat. Rec. *155,* 33—40 (1966).

Gabe, M.: Neurosecretion. Internat. Ser. Monogr. Biol. Vol. 28, 872 pp. Oxford, London, New York: Pergamon Press, 1966.

Gerschenfeld, H. M.: Chemical transmitters in invertebrate nervous systems. In: Nervous and Hormonal Mechanisms of Integration. Symp. Soc. Exp. Biol. *20,* 299—323 (1966).

Green, J. D.: Microanatomical aspects of the formation of neurohypophysial hormones and neurosecretion. In: The Pituitary Gland. *G. W. Harris* and *B. T. Donovan,* eds. Vol. 3, 240—268. Berkeley and Los Angeles: University of California Press, 1966.

Guillemin, R., R. Burgus, E. Sakiz, and *D. N. Ward:* Nouvelles données sur la purification de l'hormone hypothalamique TSH-hypophysiotrope, TRF. C. R. Acad. Sc. Paris, Série D, *262,* 2278—2280 (1966).

Halaris, A., E. Rüther, and *N. Matussek:* Effect of a benzoquinolizine (RO4-1284) on granulated vesicles of the rat brain. Zschr. Zellforsch. *76,* 100—107 (1967).

Haller, E. W., H. Sachs, N. Sperelakis, and *L. Share:* Release of vasopressin from isolated guinea pig posterior pituitaries. Amer. J. Physiol. *209,* 79—83 (1965).

Hillarp, N. A., K. Fuxe, and *A. Dahlström:* Central monoamine neurons. In: Mechanisms of release of biogenic amines. *U. S. von Euler, S. Rosell* and *B. Uvnäs,* eds. 31—57, Oxford: Pergamon Press, 1966.

Hinks, C. F.: Relationship between serotonin and the circadian rhythm in some nocturnal moths. Nature *214,* 386—387 (1967).

Hodgkin, A. L., and *R. D. Keynes:* Movements of labelled calcium in squid giant axons. J. Physiol. *138,* 253—281 (1957).

Hökfelt, T.: On the ultrastructural localization of noradrenaline in the central nervous system of the rat. Zschr. Zellforsch. *79,* 110—117 (1967).

Ishii, T. and *R. L. Friede:* Distribution of a catecholamine-binding mechanism in rat brain. Histochemie *9,* 126—135 (1967).

Kapeller, K., and *D. Mayor:* The accumulation of noradrenaline in constricted sympathetic nerves as studied by fluorescence and electron microscopy. Proc. Roy. Soc. London, Biol. Sc. *167,* 282—292 (1967).

Kelly, D. E.: Fine structure of cell contact and the synapse. Anesthesiology *28,* 6—30 (1967).

Kety, S. S., and *F. E. Samson,* eds.: Neural properties of the biogenic amines. Neurosciences Res. Prog. Bull. 5, No. 1, 1967.

Klotz, H. P., ed.: Symposium international sur la neuroendocrinologie. Probl. act. d'endocrinol. et de nutrition. Expansion Scient. Française, Paris. 1966.

Knowles, F:. Neuroendocrine correlations at the level of ultrastructure. Arch. d'Anat. Micr. *54,* 343—357 (1965).

Knowles, F., and *H. A. Bern:* The function of neurosecretion in endocrine regulation. Nature *210,* 271—272 (1966).

Knowles, F., L. Vollrath and *R. S. Nishioka:* Dual neurosecretory innervation of the adenohypophysis of *Hippocampus,* the sea-horse. Nature *214,* 309 (1967).

Koelle, G. B.: A proposed dual neurohumoral role of acetylcholine: its functions at the pre- and post-synaptic sites. Nature *190,* 208—211 (1961).

LaBella, F. S., S. Vivian, and *E. Bindler:* Amino acid composition of neuro-hypophysial secretory granules and van Dyke protein. Biochem. Pharmacol. *16,* 1126—1130 (1967).

Laduron, P., W. De Potter, and *F. Belpaire:* Storage of labeled noradrenaline in lysosomes. Life Sci. *5,* 2085—2094 (1966).

Landon, D. N.: Electron microscopy of muscle spindles. In: Control and Inner-vation of Skeletal Muscle. *B. L. Andrew,* ed., 91—111, Dundee, Scotland: D. C. Thomson, Ltd., 1966.

Larramendi, L. M. H., L. Fickenscher, and *N. Lemkey-Johnston:* Synaptic vesicles of inhibitory and excitatory terminals in the cerebellum. Science *156,* 967—969 (1967).

Lenn, N. J.: Localization of uptake of tritiated norepinephrine by rat brain *in vivo* and *in vitro* using electron microscopic autoradiography. Amer. J. Anat. *120,* 377—389 (1967).

Lentz, T. L.: Fine structure of nerve cells in a planarian. J. Morph. *121,* 323—338 (1967).

Leonhardt, H., and *E. Lindner:* Marklose Nervenfasern im III. und IV. Ventrikel des Kaninchen- und Katzengehirns. Zschr. Zellforsch. *78,* 1—18 (1967).

Lever, J. D., J. D. P. Graham, and *T. L. B. Spriggs:* Electron microscopy of nerves in relation to the arteriolar wall. Symp. electr. Activ. Innerv. Blood Vessels, Cambridge 1966; Bibl. anat. *8,* 51—55. Basel and New York: Karger, 1967.

Malamed, S., A. M. Poisner, and *M. Trifaro:* Recovery of electron-translucent granules from homogenates of catecholamine-depleted adrenal medullae of cats. Anat. Rec. *157,* 282—283 (1967).

Malamed, S., A. M. Poisner, J. M. Trifaro, and *W. W. Douglas:* The fate of the chromaffin granule during catecholamine release from the adrenal medulla-III. Recovery of a purified fraction of electron-translucent structures. Biochem. Phar-macol. *17,* 241—246 (1968).

Martini, L., and *W. F. Ganong,* eds.: Neuroendocrinology. Vol. 1. New York and London: Academic Press, 1966.

Matsui, T., and *H. Kobayashi:* Histochemical demonstration of monoamine oxidase in the hypothalamo-hypophysial system of the tree sparrow and the rat. Zschr. Zellforsch. *68,* 172—182 (1965).

Nishioka, R. S., and *H. A. Bern:* Fine structure of the neurohemal areas associated with the hypophysis in the hagfish, *Polistotrema stoutii.* Gen. comp. Endocrinol. *7,* 457—462 (1966).

Normann, T. C.: The neurosecretory system of the adult *Calliphora erythrocephala.* I. The fine structure of he corpus cardiacum with some observations on adjacent organs. Zschr. Zellforsch. *67,* 461—501 (1965).

Oosaki, T.: Observations on the ultrastructure of nerve cells in the brain of the earthworm, *Eisenia foetida,* with special reference to neurosecretion. Zschr. Zellforsch. *72,* 534—542 (1966).

Oota, Y.: On the synaptic vesicles in the neurosecretory organs of the carp, bullfrog, pigeon and mouse. Annotnes. zool. jap. *36,* 167—172 (1963a).

Oota, Y.: Fine structure of the median eminence and the pars nervosa of the turtle, *Clemmys japonica.* J. Fac. Sci. Tokyo Univ., Sec. IV, *10,* 170-179 (1963b).

Oota, Y., and *H. Kobayashi:* On the synaptic vesicle-like structures in the neurosecretory axon of the mouse neural lobe. Annotnes. zool. jap. 39, 193—201 (1966).

Palay, S.: Classification of vesicles according to size and stored chemical. Neurosciences Res. Prog. Bull. 5, 9—10 (1967).

Pappas, G. D., and *M. V. L. Bennett:* The fine structure of vesicles associated with excitatory and inhibitory junctions. Biol. Bull. *131,* 381 (1966).

Quay, W. B.: Twenty-four-hour rhythms in cerebral and brainstem contents of 5-hydroxytryptamine in a turtle, *Pseudemys scripta elegans.* Comp. Biochem. Physiol. *20,* 217—221 (1967).

Raabe, M.: Etude des phénomènes de neurosécrétion au niveau de la chaîne nerveuse ventrale des phasmides. Bull. Soc. zool. France *90,* 631—654 (1966).

Rinne, U. K., and *A. U. Arstila:* Ultrastructure of the neurovascular link between the hypothalamus and anterior pituitary gland in the median eminence of the rat. Neuroendocrinol. *1,* 214—227 (1965—66).

Robertson, D. R.: The ultimobranchial body in *Rana pipiens.* III. Sympathetic innervation of the secretory parenchyma. Zschr. Zellforsch. *78,* 328—340 (1967).

Sachs, H.: Neurosecretion in the mammalian hypothalamo-neurohypophysial complex. In: Protides of the Biological Fluids. H. Peeters ed., 181—192. Amsterdam: Elsevier Publishing Company, 1966.

Sachs, H.: Biosynthesis and release of vasopressin. Amer. J. Med. *42,* 687—700 (1967).

Sano, Y., G. Odake, and *S. Taketomo:* Fluorescence microscopic and electron microscopic observations on the tuberohypophyseal tract. Neuroendocrinology 2, 30—42 (1967).

Scharrer, B.: Neurosecretion. XIII. The ultrastructure of the corpus cardiacum of the insect *Leucophaea maderae.* Zschr. Zellforsch. *60,* 761—796 (1963).

Scharrer, B.: Histophysiological studies on the corpus allatum of *Leucophaea maderae.* IV. Ultrastructure during normal activity cycle. Zschr. Zellforsch. *62,* 125—148 (1964a).

Scharrer, B.: The fine structure of the blattarian prothoracic glands. Zschr. Zellforsch. *64,* 301—326 (1964b).

Scharrer, B.: The stromal element in endocrine organs of insects. Proc. VIIIth Internat. Congr. Anat., Wiesbaden, p. 107. Stuttgart: Thieme, 1965.

Scharrer, B.: The neurosecretory neuron in neuroendocrine regulatory mechanisms. Amer Zool. *7,* 161—169 (1967).

Scharrer, B.: Neurosecretion. XIV. Ultrastructural study of sites of release of neurosecretory material in blattarian insects. Zschr. Zellforsch. *89,* 1—16 (1968).

Scharrer, E.: The general significance of the neurosecretory cell. Scientia *46,* 177—183 (1952).

Scharrer, E.: The final common path in neuroendocrine integration. Arch. d'Anat. micr. *54,* 359—370 (1965a).

Scharrer, E.: On Terminology. Bibliogr. Neuroendocrinol. 2, Nr. 2, IV—VI (1965b).

Scharrer, E.: Principles of neuroendocrine integration. In: Endocrines and the Central Nervous System. Res. Publ. A. Nerv. & Ment. Dis. *43,* 1—35 (1966).

Scharrer, E., and *B. Scharrer:* Neuroendocrinology. 289 pp. New York: Columbia University Press, 1963.

Smith, U., and *D. S. Smith:* Observations on the secretory processes in the corpus cardiacum of the stick insect, *Carausius morosus.* J. Cell Sci. *1,* 59—66 (1966).

Smoller, C. G.: Ultrastructural studies on the developing neurohypophysis of the Pacific tree frog, *Hyla regilla.* Gen. comp. Endocrinol. *7,* 44—73 (1966).

Steg, G.: Effects on α- and γ-efferents of drugs influencing neostriatal monoaminergic and acetylcholinergic transmission. In: Control and Innervation of Skeletal Muscle. *B. L. Andrew,* ed., 139—149. Dundee, Scotland: D. C. Thomson, Ltd., 1966.

Sterba, G., and *J. Weiss:* Beiträge zur Hydrencephalokrinie: I. Hypothalamische Hydrencephalokrinie der Bachforelle *(Salmo trutta fario).* Journ. Hirnforsch. *9,* 359—371 (1967).

Streefkerk, J. G.: Functional changes in the morphological appearance of the hypothalamo-hypophyseal neurosecretory and catecholaminergic neural system, and in the adenhypophysis of the rat. A light, fluorescence and electron microscopic study. Med. Thesis, 110 pp., Amsterdam: G. Van Soest N. V., 1967.

Takeichi, M.: The fine structure of ependymal cells. Part II: An electron microscopic study of the soft-shelled turtle paraventricular organ, with special reference to the fine structure of ependymal cells and so-called albuminous substance. Zschr. Zellforsch. *76,* 471—485 (1967).

Tranzer, J. P., and *H. Thoenen:* Significance of "empty vesicles" in postganglionic sympathetic nerve terminals. Experientia *23,* 123—124 (1967).

Verity, M. A., and *J. A. Bevan:* A morphopharmacologic study of vascular smooth muscle innervation. Symp. electr. Activ. Innerv. Blood Vessels, Cambridge 1966. Bibl. anat. *8,* 60—65. Basel and New York: Karger, 1967.

Vogt, M.: The concentration of sympathin in different parts of the central nervous system under normal conditions and after the administration of drugs. J. Physiol. *123,* 451—481 (1954).

Vollrath, L.: Über die neurosekretorische Innervation der Adenohypophyse von Teleostiern, insbesondere von *Hippocampus cuda* und *Tinca tinca.* Zschr. Zellforsch. *78,* 234—260 (1967).

Weitzman, M., ed.: Bibliographia Neuroendocrinologica. Vols. 1—6. Albert Einstein College of Medicine, New York, 1964—69.

Welsh, J. H.: Neuroendocrine substances. In: Comparative Endocrinology. A. *Gorbman* ed., 121—133. New York: John Wiley & Sons, 1959.

Wood, J. G.: Electron microscopic localization of amines in central nervous tissue. Nature *209,* 1131—1133 (1966).

Author's address: Prof. *Berta Scharrer,* Department of Anatomy, Albert Einstein College of Medicine, Eastchester Road and Morris Park Ave., Bronx, New York 10461, U.S.A.

Neuronal Secretion

Journal of Neuro-Visceral Relations, Suppl. IX, 23—62 (1969)

The Neurosecretory Cell of Vertebrates

D. Picard

Department of Histology, Faculty of Medicine, Marseille, France

With 11 Figures

Summary

Being a double specialized cell, the neurosecretory cell exhibits the fundamental features of a neuron including bioelectrical activity. It receives afferent connections by way of synaptic junctions, both cholinergic and monoaminergic, and gives responses in the shape of action potentials. It shows also the character of a glandular cell which secretes granules of varying shape originating from the Golgi apparatus. The granules are driven along the processes, mainly the axon, down to their endings, and bear hormonal substances. The hormones are released either in the vicinity of vessels in neurohaemal storage organs, or in contact with glandular cells (neurosecretory innervation).

The synthesis of secreted material might partly occur progressively along the axon. It appears likely that hormones might be released not only at the ending, but also along the course of the axon. The nerve impulse conducted by the neurosecretory neuron itself is likely to be responsible for hormone release. The morphological and biochemical aspects of the mechanism of hormone release, for instance the part possibly played by acetylcholine, remain uncertain. There seems to be a conspicuous depletion of the dense content of granules. Attention is drawn to two different kinds of neurosecretory fibers and granules respectively bearing peptidic hormones or other substances which might be monoamines in some neurosecretory systems. The comparison between the monoaminergic and neurosecretory neurons leads to the necessity of a physiological definition of a neurosecretory cell: the product of a neurosecretory cell is of a hormonal nature.

Résumé

La cellule neurosécrétrice, doublement spécialisée, possède les caractères fondamentaux d'un neurone, y compris son comportement bioélectrique; elle reçoit des afférences de type synaptique, cholinergiques et monoaminergiques, et fournit des réponses sous forme de potentiels d'action. Elle a aussi les caractères d'une cellule glandulaire, sécrétant à partir de l'appareil de Golgi des granules protéiques acheminés le long des prolongements et notamment de l'axone, jusqu'à leur extrémité; ils sont porteurs de substances hormonales libérées soit au contact de vaisseaux dans des organes neuro-hémaux de stockage, soit au contact de cellules glandulaires (innervation neuro-sécrétoire).

La question est aujourd'hui sérieusement envisagée d'une synthèse partielle et progressive du matériel sécrété se faisant dans l'axone; il est aussi envisagé que les hormones soient excrétées non seulement à l'extrémité, mais le long du trajet de l'axone. L'influx nerveux émis et conduit dans le neurone neurosécréteur intervient probablement pour la libération des hormones; sauf pour la déplétion du contenu dense des granules, les aspects morphologiques et biochimiques de l'excrétion, et notamment le rôle éventuel de l'acétylcholine, restent incertains.

Une grande importance est aujourd'hui accordée à l'existence de deux catégories de fibres neurosécrétrices et de granules, porteurs les uns d'hormones peptidiques, les autres de substances différentes qui pourraient être des monoamines dans certains systèmes neurosécréteurs. La comparaison entre les neurones monoaminergiques et les neurones neurosécréteurs fait ressortir la nécessité de définir ces derniers par le critère physiologique de la nature hormonale de leurs produits.

Introduction

Although neurosecretory cells * are widespread in the whole animal kingdom and much of our knowledge results from studies on Invertebrates, the present paper is supposed to provide a review of data and problems related to the limited field of the Vertebrate neurosecretory cell. Some aspects will be completely omitted with the aim to present a selection adapted to the framework of this symposion while other points will be dealt with to serve as an introduction to subsequent papers.

The vertebrate neurosecretory cell will be considered mainly from the cytophysiological point of view.

Data useful for this purpose will be taken from:

a) the hypothalamo-neurohypophyseal and hypothalamo-infundibular systems from all vertebrate classes,

b) the hypothalamo-adenohypophyseal system in fishes,

c) the caudal neurosecretory system in fishes.

An analytical study concerning the functional significance of the mentioned neurosecretory systems was not attempted for reasons outlined above.

In this review we consider the neurosecretory cell to be a neuron which has acquired a glandular function, in other words as a cell with a double specialization. The features it has in common with nerve cells will be considered first, then its glandular characteristics will follow. From a functional point of view it is not quite possible to separate both aspects of cellular activity. This integration of secretory and nervous functions in a single unit represents the special character of the neurosecretory cell.

In the bibliographical part of this review a list of general references on cytophysiology of neurosecretion is included providing means for a more extensive investigation of the literature.

I. The Neurosecretory Cell as a Nerve Cell

As to their general morphological features, the cells of neurosecretory nuclei have been but little investigated by conventional neurohistological

---

* As a general introduction to the problems regarding terminology the reader is referred to the paper of Dr. *Berta Scharrer* in this volume.

methods with the prospect of getting information on interconnections and afferences such as would provide morphological support to physiological data. The morphology of these cells is fairly similar to that of ordinary neurons. They are unipolar in lower and multipolar in higher Vertebrates. Attention has been paid to dendrites only when these appeared to be loaded with stainable neurosecretory material and to have a particular destination such as terminating in the ependymal lining. This phenomenon does not occur frequently and has been neglected by most authors, who considered the axons the only pathway for transport of secretory material.

The cytological characteristics of neurosecretory cells are also, on the whole, those of other neurons: a large vesicular nucleus with broken-up chromatin containing, generally, a single nucleolus, Nissl-bodies, situated as a rule in a peripheral position, neurofibrils, small-sized mitochondria, and finally, at the ultrastructural level, neurofilaments and neurotubules.

The teleostean neurosecretory cell body is covered with many synaptic terminals (*Ishibashi*, 1962; *Follenius*, 1965; *Fridberg* and coll., 1966 a). Numerous physiological data indicate that mammalian neurosecretory cells are submitted to adrenergic as well as cholinergic influences (*Pickford* and *Watt*, 1951; *Abrahams* and *Pickford*, 1956; *Soulairac* and *Soulairac*, 1964). The dual innervation of the neurosecretory neuron is visualized by histo-chemical observations on the localization of acetylcholinesterase and cate-cholamines. *Rodriguez* (1965), using a modification of Koelle's technique, showed that acetylcholinesterase is localized at the surface of mammalian neurosecretory cell bodies suggesting the presence of a great number of cholinergic synaptic boutons. On the other hand, observations by *Carlsson* and coll. (1962) and by *Fuxe* and *Ljunggren* (1965), using the catecholamine fluorescence technique of *Falck* (1962), are in favour of numerous noradren-ergic terminals being in contact with the surface of supraoptical and para-ventricular neurons. The physiological and histochemical evidence in favour of aminergic and cholinergic axo-somatic synapses is supported by electron microscopical observations on rat supraoptic nucleus (Fig. 1). Axo-axonic synapses on neurosecretory axons have also been described (*Oota*, 1963 a—d; *Kobayashi* and coll., 1965; *Bern* and coll., 1965).

Neurosecretory neurons are influenced by many pharmacological agents known to have a direct action upon nerve cells in general. Some of these substances show a selective effect upon neuronal *metabolism*. In this respect, reference may be made to the effects of tricyano-aminopropene *(Triap)* on the Nissl-substance of hypothalamic neurosecretory neurons of rat and cat as shown by *Seïte* (see: *Picard* and coll., 1965), similar to the effects of *Triap* on other neurons such as Deiters' cells (*Hydén* and *Egyhazi*, 1962). Other substances that stimulate or inhibit *nervous activity* of neurons are able to change the secretory activity of neurosecretory neurons (*Seïte* and coll., 1964; review *Barry*, 1966). As the same can be said for *Triap*, these findings favour the view that nervous and glandular functions of these partic-ular neurons are closely linked and not merely coexistent in the same cell unit.

The most important data on the nervous character of neurosecretory cells are electrophysiological. Research in this field, however, is extremely difficult due to the complexity of the systems investigated. In many cases it is impossible to decide whether the results are due to neurosecretory elements or to intermingled ordinary neurons.

The first unit recordings of action potentials from neurosecretory neurons — low frequency discharges — were obtained in the rabbit by *Cross* and *Green* (1959). Further electrophysiological investigations were not only car-

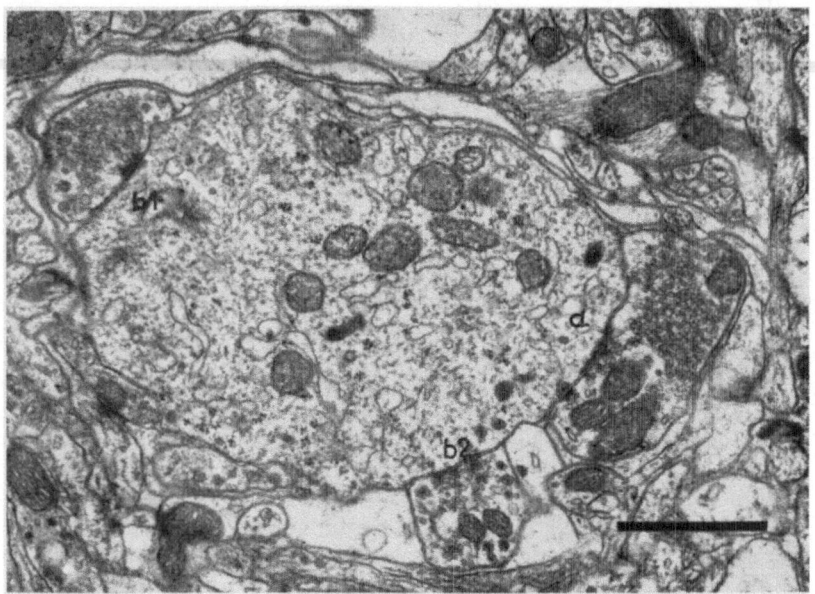

Fig. 1. Axo-somatic synapses. Transverse section through an elongated part of a neurosecretory cell body (rat — Nucleus supraopticus). Possibly two kinds of presynaptic fibers are shown, cholinergic (a) and monoaminergic (b) respectively. This part of the neurosecretory cell is devoid of granules. All electron micrographs by Dr. *Cotte*, with technical assistance of *A. M. Haon* and S. *Rua-Durand*. Rat hypothalamo-hypophyseal system; fixation with glutaraldehyde and osmium tetroxyde; embedding in Epon. Scale: 1 μ.

ried out on the mammalian hypothalamus, but also on the caudal neurosecretory system of teleost because of its large and fairly accessible neurons (*Morita* and coll., 1961; *Ishibashi*, 1962; *Bennett* and *Fox*, 1962; *Kinosita* and coll., 1962; *Yagi* and *Bern*, 1963, 1965; *Fridberg* and coll., 1966 b). *Ishibashi* obtained evoked potentials by antidromic stimulation. The particularly long lasting action potentials in the teleost system do not seem to be a general characteristic of all neurosecretory neurons (*Bern* and *Yagi*, 1965).

The hypothalamic systems have been mainly investigated in mammals, although recordings obtained by *Kandel* (1962, 1964) in the preoptic nucleus of a fish should be mentioned. After some first results obtained by *Cross* and *Green* (1959) in the rabbit, action potentials were clearly demonstrated by *Brooks* and coll. (1962) and *Suda* and coll. (1963) in the cat, and by *Barraclough* and *Cross* (1963) in the rat. Frequency and amplitude of these action potentials may be influenced not only by non-specific stimuli, but also by specific stimulation of neurosecretory "hormonogenic" activity (*Brooks* and coll., 1962; *Koizumi* and coll., 1964; *Suda* and coll., 1963). Recent findings by *Brooks* and coll. (1966) even established an accurate correlation between unitary activity of neurons in the paraventricular nucleus and oxytocin secretion induced by a peripheral stimulus.

Some of the above mentioned findings also point to another aspect of neurosecretory neurons as nerve cells, namely that they might act as osmoreceptors. The existence of such receptors was already surmised by *Verney* (1947) and has been supported by the important experimental work of *Jewell* and *Verney* (1957). The opinion has come forth that these receptors could be the preoptic or supraoptic neurons themselves, the dendrites of which may extend as far as the ependymal lining (*Dierickx*, 1962; *Cross*, 1964; *Smoller*, 1965). On the other hand, the neurons of the caudal neurosecretory system of teleosts do not seem to play a role in osmoreception (*Yagi* and *Bern*, 1965). For more complete information we refer to reviews by *Bern* and *Yagi* (1965) and by *Cross* and *Silver* (1966).

It might be concluded that neurosecretory cells have morphological and structural features, as well as an electrical behaviour, essentially similar to those of all neurons. They show, however, one morphological characteristic not present in ordinary neurons: the axon terminals do not form a synapse with either another neuron or an effector cell. This particular phenomenon, which, from the first, has been clearly apparent in the hypothalamo-neurohypophyseal system, seemed to be basic for their definition (*Knowles* and *Carlisle*, 1956). Though, presently, it should be questioned whether this assertion is not too rigorous, it is in line with the generally accepted definition of a neurosecretory system which includes a junction between free axon endings and blood vessels, situated in a neurohaemal organ. Models of such systems would be the neurohypophysis, the median eminence and the urophysis of fishes (*E.* and *B. Scharrer*, 1954; *Bern*, 1962) (Fig. 2).

On the other hand, *Barry* (1956) suggested and repeatedly maintained that axons bearing stainable neurosecretory material outside the hypothalamo-neurohypophyseal system ("extra-hypophyseal neurosecretory pathways": *E.* and *H. Legait*, 1956, 1957, 1958; *Grignon* and *Lamarche*, 1959; *Grignon*, 1961) terminate in contact with perikarya of various neurons in several mammalian species, thus forming interneuronal neurosecretory synapses. The exact meaning of such contacts, however, awaits further investigation. Moreover, contacts between axons loaded with neurosecretory material and gland cells have been described in detail at the ultrastructural level, especially in the intermediate lobe of fishes (*Bargmann* and *Knoop*, 1960; *Knowles*, 1965 b; *Knowles* and *Vollrath*, 1966 a, b), of amphibians (*Cohen*, 1964) and in neurosecretory systems of Invertebrates. Such data have brought about an evolution of the concept of a neurosecretory system *Bern* (1963), *Knowles* (1965 a), *Knowles* and *Bern* (1966).

One has to take into account a true morphologically and physiologically defined neurosecretory innervation therefore including close contacts between neurosecretory axon terminals and glandular effector cells, beside the original concept of neurosecretory cells acting at a distance by way of a humoral pathway (blood or cerebrospinal fluid). Careful re-examination of these neuroglandular contacts, however, seems to be essential. Thus it appeared that in the adenohypophysis of the perch (*Follenius* and *Porte*, 1961, 1962 a) and in the intermediate lobe of the eel (*Knowles* and *Vollrath*, 1966 a) a basal membrane, or rather a narrow stretch of perivascular space, is interposed between nerve endings and glandular elements. Recent observations in the adenohypophysis of *Hippocampus cuda* (*Vollrath*, 1967) have shown that such neuroglandular contacts exist without interposition of anything but a narrow space like a synaptic cleft (Fig. 2). However, in a strict neurophysiological sense, the synaptic nature of this contact is not likely.

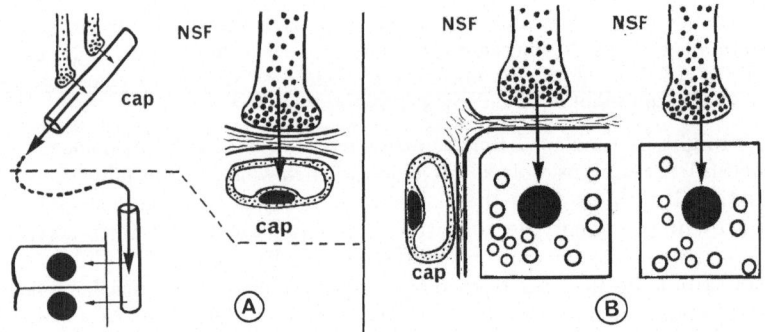

Fig. 2. Different possibilities of neurohormone transfer to reactive cells. — A. Neurohaemal "storage and release" organ: hormone transported to reactive cells by the blood stream; no synaptic junction at axon endings. — B. Neurosecretory innervation, with (left) or without (right) interposition of an extension of the perivascular space; no transport by the blood stream. — N. B. More complex pathways, such as involving the cerebrospinal fluid, are not figured here.

Consequently, the question about the significance of a neuronal electrical activity in neurosecretory cells has to be raised; this problem will be discussed below, together with mechanisms of excretion and transport of secretory material.

II. The Neurosecretory Cell as a Glandular Cell

A. General cytological features

The glandular characteristics of neurosecretory cells are apparent from observations of the cells during spontaneous or experimentally induced variations in activity. Several cell organelles display changes in relation to such variations.

The nucleus is frequently described as having an irregular outline, sometimes showing deep invaginations of the nuclear membrane resulting in a particularly distorted shape which is specially conspicuous in the caudal neurosecretory system of fishes. The increase in size of more ellipsoid nuclei

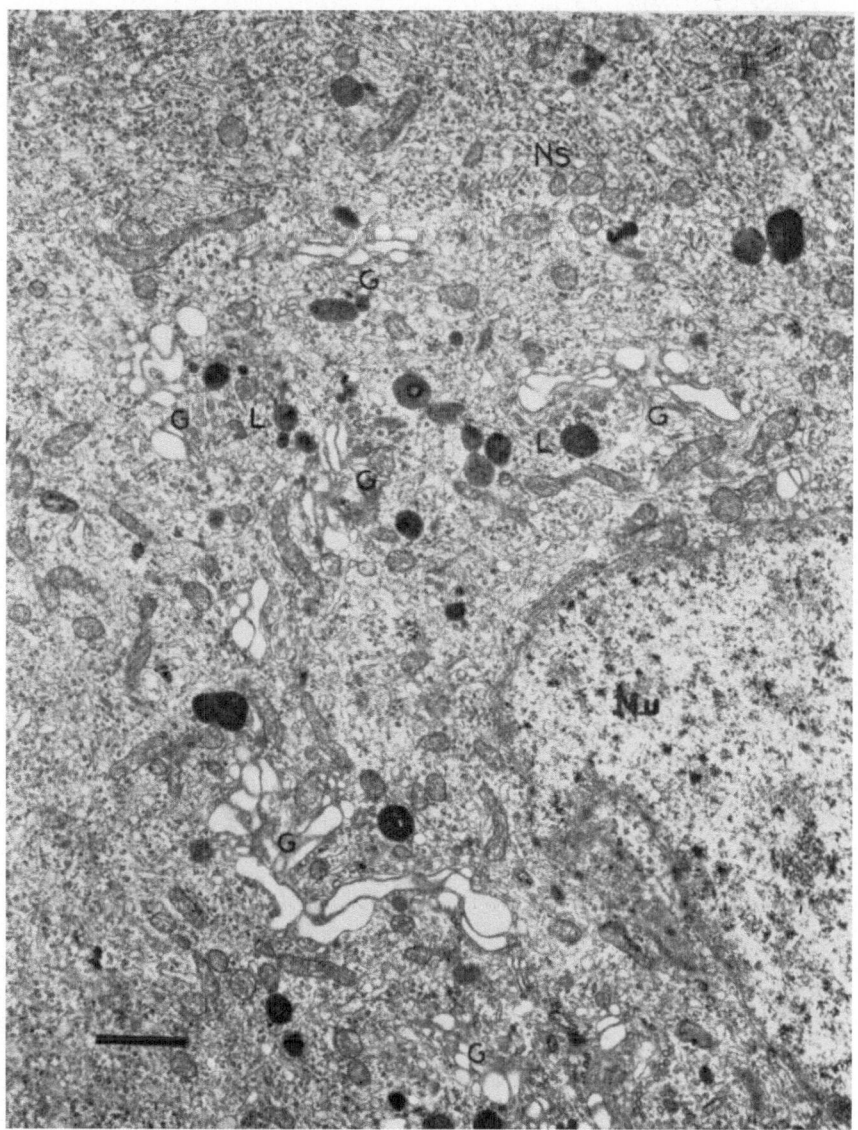

Fig. 3. Juxta-nuclear crescent-shaped area with numerous Golgi elements (G) and lysosomes (L). Nu, nucleus; NS, Nissl-substance in a compact state with narrow ergastoplasmic cisternae and numerous polysomes.

together with the increase of nucleolar size is often considered a good criterion for active secretion (*Stahl* and coll., 1955). This parameter has been
frequently used in recent research (*e. g. Soulairac* and *Soulairac*, 1962;
Uemura and *Kobayashi*, 1963; *Ifft*, 1964; *Ifft* and *Berkowitz*, 1965; *Zambrano*
and *Mordoh*, 1966; *Talanti*, 1967).

After preliminary observations (*Romieu* and coll., 1953), variations of the
Golgi apparatus were demonstrated in this laboratory with standard light
microscopical techniques (*Stahl* and coll., 1955). The increase of both the
surface of this area and the quantity of Golgi material after stimulation obviously reminds of what happens in glandular cells. It will be shown below
that Golgi structures indeed do play a part in the formation of neurosecretory
material. Electron microscopy confirms the presence of a large crescent
shaped juxta-nuclear area, with few ribosomes, where most Golgi structures
are located (Fig. 3). According to functional activity, these are mainly composed either of elongated flattened sacs, or of vesicles due to the dilatation
and budding of the sacs.

Ribosomes are always present in a great number, either along ergastoplasmic membranes at the cell periphery, or non-bound as isolated ribosomes
and polysomal formations. Consequent on ribosomal abundance, clear or
dark cellular aspects result which are related to cell maturation or functional
activity (*Murakami*, 1964; *Nemetschek-Gansler*, 1965; *Zambrano* and *Mordoh*,
1966). Although long ago a relation had been postulated between Nissl-
bodies and elaboration of secretory material, the formation of large granules
inside dilated endoplasmic reticulum cisternae is but seldom observed, for
instance in the gecko and the toad (*Murakami*, 1963, 1964). This, however,
might represent the ultrastructural aspect of "colloid-like" droplets observed
in lower Vertebrates with the light microscope. Usually the relation between
Nissl-substance and production of neurosecretory material is considered to
be a more indirect one reflecting the general role of ergastoplasm in protein
synthesis. Such a relation between Nissl-substance and variations in neurosecretory material was emphasized by recent work on the increased synthetic
activity of hypothalamic neurosecretory cells due to the administration of
tricyano-aminopropene in rat (*Seïte* and *Monneron*, 1964) and cat (*Luciani*
and *Seïte*, 1965). This phenomenon is probably related to the enhanced protein synthesis in neurons due to tricyanoaminopropene (*Hydén* and *Egyhazi*,
1962). Other pharmacodynamic agents generally known to influence nerve
cells (see review by *Barry*, 1966) show also numerous effects on amount and
distribution of neurosecretory material. Experiments with thyroxin and
thiouracil (*Talanti*, 1967) also agree with the opinion that a non-specific action is exerted on neurosecretion by neurotropic drugs.

Little will be said here about the cytoplasmic matrix which contains
neurofilaments, neurotubules and smooth endoplasmic reticulum mainly
located in the Nissl-free perinuclear area. As in other neurons, mitochondria
are small, frequently showing longitudinal cristae.

The lysosomes (Fig. 3) appear mainly in the shape of large dense granules
(*Palay*, 1960; *Murakami*, 1962, 1963; *Follenius* and *Porte*, 1962 b; *Nemet-*

schek-Gansler, 1965). The lysosomal nature of these large granules in the rat, often visible with the light microscope and closely related to pigment granules, was stressed by *Stutinsky* and *Porte* (1965). Another aspect of the lysosomes in the rat is that they stain with aldehydefuchsin. Much light microscopical work demonstrating differences in amount and distribution of aldehydefuchsin-positive material should be seriously re-examined. At the ultrastructural level no significant changes of lysosomes are noticed during marked variations of secretory activity (*Zambrano* and *De Robertis*, 1966).

B. Enzymatic activities

Numerous enzymes are present in the vertebrate neurosecretory cells. A few of them only will be discussed here in relation to their possible functional importance for the neurosecretory process.

On the basis of extensive histochemical investigations *Arvy* (1962) is of the opinion that the respiration in neurosecretory cells might be due to other enzymes than succino-dehydrogenase and cytochrome-oxydase, for instance to amino-oxydase. Referring to *Hydén's* (1960) findings about interrelations between neurons and glial cells, and to his own observations on the caudal system of fishes, *Peyrot* (1964, 1965) suggests that neurosecretory cells might draw energy for their metabolic processes not from oxydation, but from anaerobic glycolysis. Recent findings of *Iijima* and coll. (1967) are in agreement with this hypothesis making it likely that glial cells might act as energy donors for neurosecretory cells.

Phosphatases, especially acid phosphatases, are present in neurosecretory cells and have been frequently used as a significant test of secretory activity (*Wislocki* and *Dempsey*, 1948; *Eränkö*, 1951; *Sloper*, 1955; *Kivalo* and coll., 1958; *Holmes*, 1961; *Kobayashi* and coll., 1962; *Wolfson* and *Kobayashi*, 1962; *Uemura*, 1964; *Kobayashi* and *Farner*, 1966). A relation between acid phosphatase and formation of neurosecretory products may be present and is supported by histochemical observations at the ultrastructural level (*Osinchak*, 1964).

As to cholinesterase activity, statements are somewhat conflicting. In the caudal system of fishes, positive results (*Kobayashi* and coll., 1963) are in contradiction with negative findings (*Peyrot*, 1964). Specific acetylcholinesterase activity was observed in the hypothalamo-hypophyseal system by numerous authors (*Abrahams* and coll., 1957; *Pepler* and *Pearse*, 1957; *Pearse*, 1958; *Kivalo* and coll., 1958; *Bloom*, 1960; *Holmes*, 1961; *Arvy*, 1962; *Kobayashi* and *Farner*, 1964; *Rodriguez*, 1965; *Iijima* and coll., 1967); this activity, however, appears to be moderate or low, or absent even in the same species. This phenomenon can be related with the very low choline-acetylase activity in the posterior lobe as compared to other areas in the central nervous system (*Feldberg* and *Vogt*, 1948). Concerning the acetylcholinesterase present in the neurosecretory neurons two hypotheses are put forward: a) acetyl cholinesterase is synthetized in the perikaryon and transported along the axons toward their endings, or, b) the acetylcholinesterase is related to acetylcholine metabolism in presynaptic terminals at the surface of neurosecretory perikarya (*Kobayashi* and *Farner*, 1964). Observations by *Rodriguez* (1965) and *Iijima* and coll. (1967) favour the second possibility and these authors consider the hypothalamic neurosecretory neurons as non-cholinergic. In the neurophysiological sense this seems more and more likely, though it does not rule out the possibility of acetylcholine production by these neurons and the presence of this transmitter in the neurosecretory axon endings. This point will be discussed below in connection with the mechanism of hormone release.

C. Origin and formation of secreted substances

This aspect of the cytophysiology of neurosecretion is closely related to methods used for demonstrating these substances. A glandular activity of hypothalamic neurons was suggested on the basis of ordinary histological techniques which enabled *E.* and *B. Scharrer* and their group to state that neurosecretory material is formed in the perikaryon and transported along the axon. Since chromalum-haematoxylin (*Bargmann*, 1949) and paraldehyde-fuchsin were applied for staining the neurosecretory material, a decisive impulse was given to research in that field. From that time on it became possible to visualize the hypothalamic cells diversely loaded with stainable granules, the fibers being marked by droplets or granules (Fig. 4), and finally

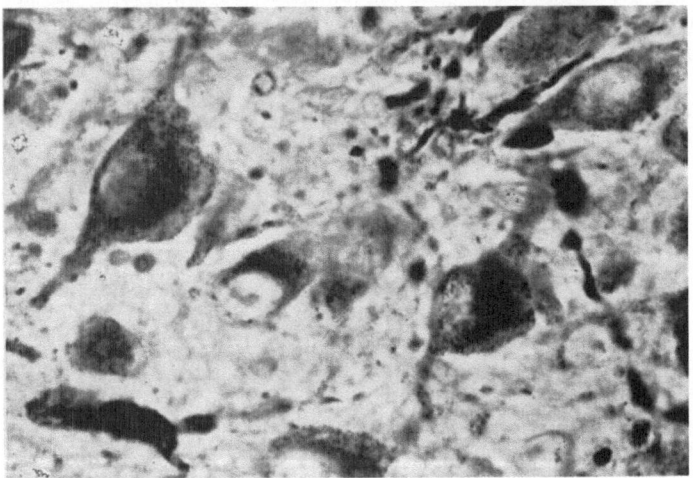

Fig. 4. Nucleus supraopticus of cat stained with aldehydefuchsin. Granular appearance of neurosecretory material in cell bodies; small or large droplets along axons (courtesy of Prof. *Seïte*).

the accumulation of stainable material in areas such as the posterior lobe where neurosecretory fibers make contact with blood vessels. After that it became feasible to determine semiquantitatively spontaneous or experimentally induced variations in amount of neurosecretory material in cells, fibers and posterior lobe, *e. g.* after increased antidiuretic stimulation.

The concept of neurosecretion was in accordance with the notion of proximo-distal axoplasmic flow, postulated by *Weiss* (1944) and *Weiss* and *Hiscoe* (1948) which was fully confirmed afterwards. Experimental section of the pituitary stalk and hypophysectomy (*Stutinsky*, 1951 a, b, 1953; *Hild*, 1951; *Scharrer* and *Wittenstein*, 1952; *Mazzi*, 1953; *Benoit* and *Assenmacher*, 1952, 1953; *Sloper* and *Adams*, 1956) provided evidence for the direction of the movement of stainable neurosecretory material, especially when the neurohypophysis had been experimentally depleted prior to operation (*Hild* and *Zetler*, 1953). Hypophyseal stalk sections performed in various species of Vertebrates established the hypothalamic origin of both visible neurosecretory material and the "neurohypophyseal" hormones. The demonstration of the hypothalamic origin of releasing factors influencing the adenohypophysis (*Benoit* and *Assenmacher*) followed.

Bodian (1951) and *Mosinger* (1951), however, suggested a distal origin for neurosecretory material. *Diepen*, *Christ* and their group (since 1954; see *Christ*, 1962) were of the opinion that stainable material results from a degenerative process, either a physiological degeneration of neurosecretory axon endings, or a post-traumatic degeneration of fibers after section of the stalk. They supported this view by making reference to the fact than in the developing animal, at least in some species, the neurosecretory material will appear first in the posterior lobe and only later in hypothalamic cells. This fact has been furthermore confirmed (*Wurster* and *Benirschke*, 1964). One has to keep in mind, however, that a sufficient amount of a stainable substance should be present in order to visualize this with the light microscope. Therefore, it is not surprising that such a substance can first be demonstrated at sites where it accumulates at the time of onset of hypothalamic function not being enough concentrated to visualize it at the sites where it is, in fact, produced. Consequently, the theory of a distal origin of neurosecretory products has been severely criticized by most investigators during the Bristol symposion (1961).

Moreover, the administration of radio-active substances like ^{35}S-cysteine (*Sloper*, 1958; *Flament-Durand*, 1961, 1966; *Leray*, 1963) provided additional evidence for the hypothalamic origin and proximo-distal migration of neurosecretory material towards the neurohypophysis (Fig. 5).

In the meantime, electron microscopy had revealed dense "elementary" granules (*Palay*, 1955; *Green* and *Van Breemen*, 1955; *Brettschneider*, 1956; *Duncan*, 1965; *Barry*, 1957; *Bargmann* and *Knoop*, 1957, etc.), which have now been observed in all animal groups. It is a rather remarkable finding that granules of identical appearance are not only present in the well known neurosecretory systems showing chromalum-haematoxylin and aldehyde-fuchsin stainable material in both, all higher Vertebrates and in Invertebrates, but also in "Gömöri-negative" systems like the caudal system of fishes and the latero-dorsal interstitial hypothalamic nucleus of the Guinea pig (see review in: *Barry*, 1966). Thus, despite differences in staining properties a morphological uniformity of the neurosecretory process seems to be present.

Neurosecretory granules are generally 1000 to 2000 Å, seldom 3000 Å, in diameter and they are surrounded by a single membrane. Their content, which is of variable electron density, may even completely disappear morphologically. Then the granules present the aspect of empty vesicles. The electron-dense content is often surrounded by a narrow, clear halo separating it from the membrane and suggesting the possibility of intermediate stages between "full" and "empty" granules. These different aspects and their interpretation in view of functional problems related to the secretory process will be discussed below (Fig. 6).

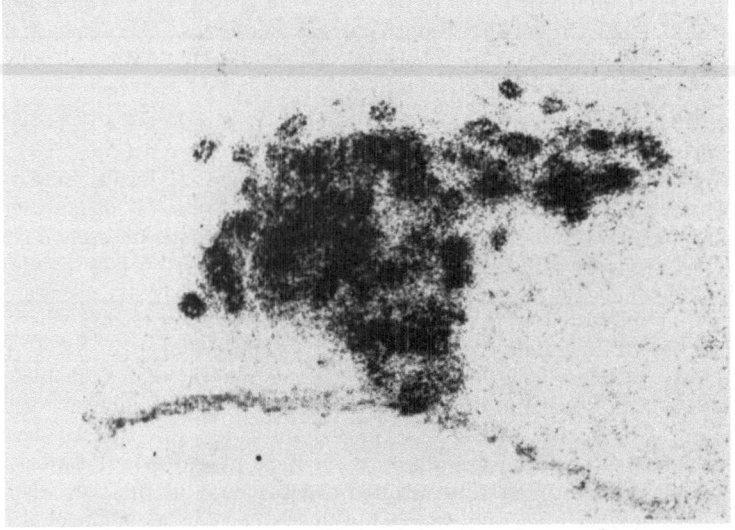

Fig. 5. Historadio-autograph of nucleus supraopticus of rat; selective incorporation of ³⁵S, 1 hour after intra-cisternal injection of labelled cysteine (courtesy of Prof. *Seïte*).

The uniformity of the neurosecretory process was made even more conspicuous when it was shown that "elementary" granules (or vesicles) are formed in the cell body by vesiculation of the Golgi cisternae (*Sano* and *Knoop*, 1959; *Palay*, 1960; *Bern* and coll., 1961). After this process had been very accurately described in *Lumbricus* by *Scharrer* and *Brown* (1961), numerous other observations pointed to the Golgi origin of dense neurosecretory granules (Fig. 7) thus providing a confirmation of previous findings with the light microscope (see above: *Stahl* and coll., 1955). At the ultrastructural level changes in appearance of ergastoplasmic cisternae and ribosomes (Fig. 7) after activation of the neurosecretory process could be influenced by protein synthesis inhibitors (*Zambrano* and *De Robertis*, 1966, 1967).

Regarding the simultaneous changes in the Golgi structures one can say that the neurosecretory neuron behaves rather like a protein secreting gland cell: protein, synthetized in ergastoplasmic structures, is transferred to Golgi cisternae where the product acquires the shape of dense granules and is transported towards the "excretory pole" of the cell. In this particular case, the "excretory pole" seems to be located at the axon ending. Most often, the accumulation of secretory granules takes place at the ending, but may also be observed along the axon (Herring bodies as described with the light microscope).

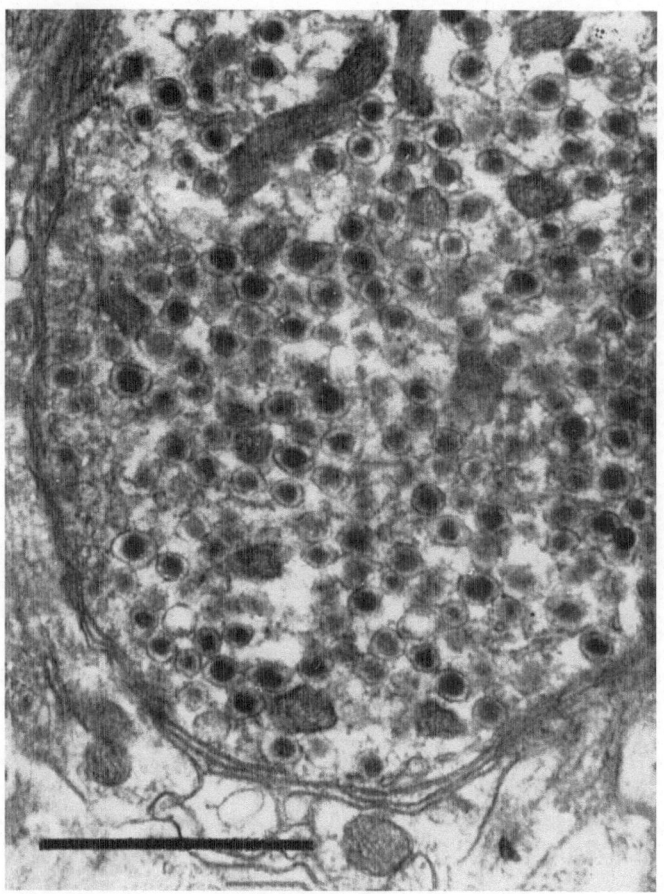

Fig. 6. Dilated segment of a neurosecretory axon in the proximal part of the supraoptico-hypophyseal tract. Axoplasm crowded with elementary neurosecretory granules surrounded by a membrane; notice the variable density of the granules and the clear halo, frequently present between membrane and dense material.

By several authors this clear and attractive outline is not accepted in such a simple form. Indeed, more than one point has presently to be reconsidered. Before we look into these objections it seems necessary to take seriously into account the fact that correspondence between data obtained from different techniques is frequently somewhat uncertain and too easily admitted without sufficient criticism.

a) Morphologically we observe substances stainable by "non-specific" stains (chromalum-haematoxylin, aldehydefuchsin, etc.) or by means of histochemical reactions such as those revealing either sulfhydril or disulfide

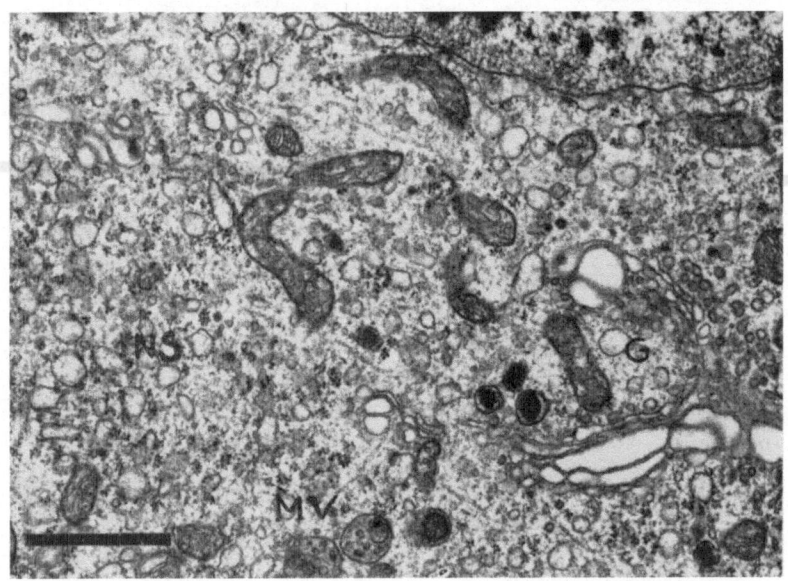

Fig. 7. Granule formation in Golgi structures (G) which consist of dilated cavities and flattened sacs giving rise to numerous vesicles which, by an increase of their size and electron density, reach the aspect of large elementary granules, here 2000 Å in diameter. NS, Nissl-substance in an "active" state with dilated ergastoplasmic cisternae (compare with Fig. 3). MV, multivesicular bodies.

groups (*Barrnett*, 1954; *Sloper*, 1955; *Adams* and *Sloper*, 1955, 1956; *Howe* and *Pearse*, 1956; *Gabe*, 1955, 1960), or glucids in very variable amounts, or perhaps lipids as stated by *Schiebler* (1951, 1952) and *Bachrach* and coll. (1953), but rather convincingly denied by *Sloper* (1955, 1966), and finally in any case proteins.

b) Radio-isotopic labelling reveals incorporation of sulfur and migration of labelled neurosecretory material.

c) With the electron microscope, we observe vesicles loaded with dense material (elementary granules).

d) By means of *physiological* and *biochemical* investigations, we know that the hypothalamo-neurohypophyseal system does contain peptide hormones, probably bound to an inactive protein such as Acher's neurophysin.

Several facts established during the last few years prompted many investigators to doubt an exact correlation between these different sets of observations. Until a recent past there seemed to exist, however, an unanimous agreement about the following scheme:

stainable material = *dense granules* = *presence of hormones*
(light microscopy)   (electron microscopy) (biochemistry, physiology)

According to some authors stainability or electron density should be due to a complex carrier protein and not to the hormones themselves (*Schiebler*, 1952; *Hild* and *Zetler*, 1953), whereas, according to others, the material stained either non specifically or histochemically (cystine) is actually identical with hormones (*Sloper, Adams* and *Sloper; Rodeck;* see: *Sloper*, 1966), even if these are bound to a carrier substance.

Against this view, which should now be regarded as too simple, several arguments may be retained.

A. Numerous consistent observations, mainly in Invertebrates, show the presence of electron dense granules morphologically identical to neurosecretory elementary granules, but located in different kinds of nerve cells, in some species even in all nerve cells. These observations in Invertebrates reduce the morphological specificity of the ultrastructural criterion. In higher Vertebrates this criterion still keeps its significance.

B. In the pars intermedia of an amphibian (*Iturriza*, 1964) the fibers are loaded with stainable neurosecretory material, although no elementary granules are present.

C. In the posterior median eminence of a bird fibers are present which are filled with granules, although these are not stainable with aldehyde-fuchsin (*Bern* and coll., 1966). Consequently, aldehydefuchsin might stain substances different from granular material (*Bern* and coll.), and all aldehyde-fuchsin-positive material might not necessarily be related to granules.

D. *Moses* and coll. (1963) and *Lederis* (1964) pointed to cases in which stainability of neurosecretory material and electron density of granules obviously do not parallel the presence or the amount of hormones. In conclusion, the absence of elementary granules or stainable material in a fiber is no proof for the assumption that the fiber is not of a neurosecretory nature and does not actually contain a hormonal substance. This appears consistent with critical remarks by *Jörgensen* (1965) about the consideration of morphological criteria regarding neurosecretion. Moreover, the opinion has already been put forward that hypothalamic hormones might, at least partly, be present in the axoplasm in a dispersed state (*Heller* and *Ginsburg*, 1966).

On the whole, we have now rather substantial reasons for being more cautious than we used to be a short time ago in interpreting experimental results obtained by a single technical approach. It may be suitable to investigate from this viewpoint problems like, for instance, that of mammalian supraoptical and paraventricular dendrites which are often loaded with stainable neurosecretory material, as has been mentioned previously (*Seïte* and coll., 1964; *Palay*, 1953; and others).

These problems had to be considered before returning to the question of the formation of neurosecretory substances. It appeared necessary to reconsider the simple scheme of a neurosecretory neuron manufacturing hormonal substances in the perikaryon and delivering the substances to the "milieu intérieur" at the excretory pole.

The question arises, whether secreted substances may originate from elsewhere than the perikaryon.

Green and *Maxwell* (1959) calculated that, if all elementary granules were formed in the perikaryon and driven all along the axon, this would imply both an inadmissible rate of granule production and a transport of up to 520 granules per minute in a single neuron. In their opinion, granules are formed along the axon.

Investigation of the vasopressin/oxytocin ratio in mammals (*Vogt*, 1953; *Lederis*, 1962), as well as radio-isotopic labelling of vasopressin (*Sachs*, 1959, 1961) leads to the possibility that, at least in the dog, oxytocin is synthetized for the major part at a site distal to the tuber (*Van Dyke* and coll., 1957), probably in the neural lobe.

Formation of vesicles and granules at the distal end of axons was strongly suggested on morphological grounds (*Knowles*, 1962, 1964; *Holmes* and *Kiernan*, 1964; *Lederis*, 1964, 1965). Neurosecretory fiber terminals showing a tubular reticulum which is likely to produce granules are present in a bird (*Bern* and coll., 1966). Similar observations were reported in the caudal neurosecretory system of fishes in consequence of osmotic changes (*Fridberg* and coll., 1966 a) and during regeneration (*Fridberg* and coll., 1966 c). In consequence of these findings, the possibility of a distal site of origin of neurosecretory granules in the axon endings has now to be seriously taken into account (*Bern*, 1966).

Some authors express a more discriminative opinion, thinking that secreted substances are indeed formed in the perikaryon by means of the well established mechanism of granule genesis from Golgi structures, but that this is not the end of the process. This agrees with the progressive synthesis hypothesis of *De Robertis*, based on the increase in volume of neurosecretory granules during their transport along the axon as observed in the toad by *Gerschenfeld* and coll. (1960), in the rat by *De Robertis* and *Pellegrino de Iraldi* (1961), and again in several groups of Vertebrates by *Oota* (1962, 1963 a—d), *Oota* and *Kobayashi* (1962, 1963), *Kobayashi* and coll. (1966). Moreover, while in some species such as fishes perikarya are filled with a large amount of elementary granules, these granules are obviously scarce in mammals, in contrast with their large amount in axons. This discrepancy might be explained by a very slow intra-axonal transport, allowing granules to accumulate. Various arguments, however, support the idea of these granules being formed progressively on their way along the axon such as for instance the above mentioned changes in vasopressin/oxytocin ratio indicating synthesis of oxytocin only in the distal segment of axons in the dog neurohypophysis. It is possible that the neurosecretory material undergoes considerable changes during its axonal transit, such as an increase due to addition of locally formed components. This view is supported by ultrastructural observations of a variable electron density of the granules in the cell body as compared with the proximal part of the axon. Sometimes the granular vesicles are mixed with clear vesicles and some authors are inclined to consider these empty granules as immature ones. In the opinion of *Nemetschek-Gansler* (1965) one cannot avoid taking into account at least a complementary synthesis in the axons. It remains, however, to be decided which additional components might be synthetized in the axon. *Sachs* and *Takabatake* (1964) concluded that the hormone is synthetized in the perikaryon in a bound, biologically inactive form, and progressively activated along the axon, perhaps by means of degradation of a large protein molecule. In addition, *Takabatake* and *Sachs* (1964), in *in vitro* experiments, showed that, in the median eminence, labelled amino-acids are incorporated into vasopressin. From dehydration and rehydration experiments in the rat, *Zambrano* and *De Robertis* (1966) visualized the role played by the ergastoplasm and Golgi apparatus in the formation of neurosecretory material, and they confirmed the increase in volume of granules during their transport in the axon. They assume that this increase is likely to be due to an increase of their hormonal content, since it has been shown that in a micro-organism a peptide synthesis up to 10 amino-acids may take place independently from RNA (*Mach* and coll., 1963). Considering these findings the following succession of events would seem to occur:

1. a perikaryonal synthesis of a granule-shaped inactive carrier substance;
2. an intra-axonal synthesis of active hormonal peptides which would be bound to previously formed granules.

It should be noticed that granules are always bounded by a membrane originating from Golgi structures, and therefore important chemical reactions can take place inside the granules. One of these reactions might be the synthesis of peptide hormones, but this attractive hypothesis still requires more direct evidence.

D. Excretion of elaborated substances

From what is known about various protein secreting glandular cells, the excretion might occur by coalescence of granules with the cell membrane (*i. e.* the axolemma) followed by extrusion of their dense material. However, direct confirmation of this theory has not yet been brought forward. The only exceptions occur in the caudal neurosecretory system of fishes. Here, granules having left axon terminals are sometimes still visible underneath the basal membrane limiting the pericapillary space (*Bern* and coll., 1965). Under normal conditions and more conspicuously during and after regeneration following section of axons in the caudal system, an output of neurosecretory granules through the ependyma into the cerebrospinal fluid has also been reported (*Fridberg* and coll., 1966 c; *Fridberg* and *Nishioka*, 1966).

Except for these rare findings, the excretion of substances by neurosecretory neurons does not seem to be visible in the form of extrusion of granules. Stimulation of hormone release results, in the posterior lobe of the hypophysis, in a decrease of electron density of the "elementary" granules (*Palay*, 1955, 1957; *Hartmann*, 1958; *Gerschenfeld* and coll., 1960). It is, therefore, possible that granules undergo physico-chemical alterations at the site of excretion resulting in a splitting of the hormone-carrier complex. It remains to be decided whether this process occurs inside the granule. The carrier substance itself may penetrate the granule membrane reaching the axolemma. Anyhow, the morphological consequence of every stimulation of hormone delivery (for instance antidiuretic regulation) is the appearance of numerous clear vesicles of variable electron density in the axon terminals (Fig. 8). This phenomenon takes place in the neurohypophysis as well as in the caudal neurosecretory system of fishes (*Oota* and *Kobayashi*, 1966; *Fridberg* and coll., 1966 a, b). The change in morphological appearance of the granules after stimulation does introduce new questions about membrane permeability and the transport mechanism which, at the present time, have only been dealt with in a speculative way.

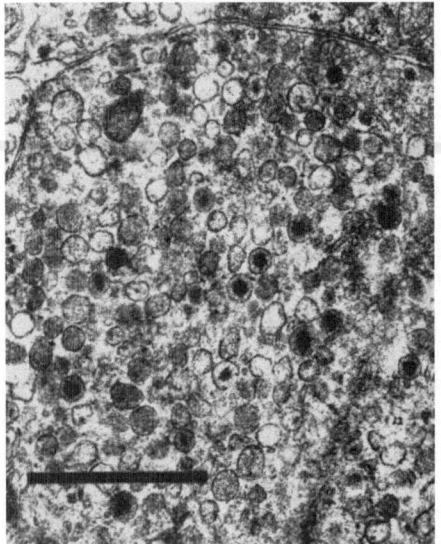

Fig. 8. Preterminal segment of axon in neurohypophysis: numerous granules of low density, some of them depleted and appearing as empty vesicles (stage of high level of hormone release).

The question of a possible neuro-humoral agent operating at the time of excretion can also be raised.

In the neurosecretory axon terminal the presence of clear, electron-lucent microvesicles showing a morphological appearance identical to that of microvesicles in pre-synaptic nerve endings was demonstrated in the rat (*Palay*, 1955), in several fishes (*Palay*, 1960; *Bargmann* and *Knoop*, 1960; *Follenius* and *Porte*, 1962 a, b), in the toad (*Gerschenfeld* and coll., 1960), and in the Guinea pig (*Barry* and *Cotte*, 1961). Microvesicles appear to be concentrated in clusters close to the axolemmal membrane, either at the axon endings or in its preterminal segment. Although the similarity to an ordinary presynaptic arrangement is striking, the post-synaptic element defining a synaptic junction is lacking. In the present case, the cluster of vesicles is facing a pericapillary space with interposition of a basal membrane, while no change in thickness or electron density of the axolemmal membrane is observed (Fig. 9). Recently, special attention has been paid to synaptic-like arrangements with corresponding membrane modifications at points of contact between neurosecretory axons and non-nervous elements (Fig. 9), such as glial cells and their processes (*Matsui*, 1966 a, b), and ependymal cells in birds (*Nishioka* and coll., 1964; *Matsui*, 1966 a), fishes (*Knowles* and *Vollrath*, 1966 a) or mammals (*Matsui*, 1966 b). Possibly, either the secretory activity of glial and ependymal cells themselves is stimulated by mear.s of "transient" synapses (*Knowles* and *Vollrath*, 1966 a), or the hormones produced by neurosecretory neurons are transported by glial or ependymal cells to the cerebrospinal fluid (*Nishioka* and coll., 1964). In the latter view, the synaptic-like contact areas, with accumulations of microvesicles, would represent the "points of output" where the excretion of hormonal substances

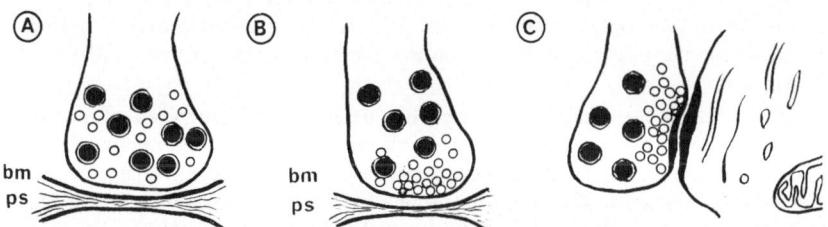

Fig. 9. Microvesicles in neurosecretory endings. — A. Microvesicles scattered between elementary granules. — B. Cluster of microvesicles in contact with the perivascular space (ps) limited by the basal membrane (bm) in a neurohaemal area. — C. Synaptic-like contact of a neurosecretory fiber with a glial or ependymal cell.

takes place. This interpretation parallels the hypothesis of *Gerschenfeld* and coll. (1960) following which the microvesicles might represent acetylcholine containing organelles. Acetylcholine would not act as a neurotransmitter but its release from microvesicles might bring about changes in permeability of the granular membrane and the axolemma. In this way acetylcholine might be responsible for the excretion of active substances, and, moreover, it would perhaps be of some functional significance for the dissociation of the carrier-hormone complex. This explanation finds an interesting analogy in the mechanism of excretion of catecholamines in the adrenal medulla which is caused by acetylcholine released from the axon terminals of the splanchnic nerve. The theory mentioned has been strongly supported by *De Robertis* (1964), *Kobayashi* and *Oota* (1964), *Kobayashi* and coll. (1965), and others. It is based particularly on the correlation between the amount of synaptic microvesicles and the decrease of the dense content of neurosecretory granules although other observations point to the fact that this correlation is not constantly present. *Oota* and *Kobayashi* (1966) for instance stated that in dehydrated mice synaptic microvesicles were unchanged although the axons were depleted of neurosecretory material. However, this finding is not necessarily in contradiction with the possible role of acetylcholine-containing microvesicles in the process of hormone excretion.

The *Gerschenfeld* and *De Robertis* hypothesis has been severely criticized. Several authors stressed the approximate similarity in morphological appearance of the microvesicles in neurosecretory and in true pre-synaptic terminals. This, however, does not say anything about the identity of their chemical content so that the microvesicles might as well result from fragmentation of depleted neurosecretory granules. When depleted, these granules become clear vesicles of variable size, down to 200—600 Å, being similar in all morphological aspects to synaptic vesicles. This is one of the reasons why *Holmes* and *Knowles* (1960), *Lederis* (1963), *Bern* (1963), *Knowles* (1965 a), *Bern* and coll. (1964), *La Bella* and *Sanwal* (1965) put forward the idea that the microvesicles, even grouped in clusters, might result from fragmentation of large vesicles after depletion of their dense content. In that case they would represent an intermediary stage of granules disappearing during excretion. However, *Knowles* (1964, 1965 a) and *Follenius* (1965) suggested this interpretation only cautiously, and it is also the present author's opinion that it is not yet possible to decide on this particular point (*Picard* and *Stahl*, 1966).

As has been mentioned, evidence has been brought forward of neurosecretory pre-synaptic arrangements with corresponding alterations of the plasma membrane in areas of contact between neurosecretory axons and non-nervous elements. The demonstration of similar arrangements facing the pericapillary spaces may not be far away. In the literature of the two last years one generally meets the opinion that, while some small vesicles actually result from fragmentation of depleted neurosecretory granules, others are identical to synaptic microvesicles being probably of the same nature.

This, however, does not mean that these microvesicles would contain acetylcholine. Hypothalamic neurosecretory neurons are probably not cholinergic as indicated by the low choline-acetylase activity in the pituitary stalk and the posterior lobe (*Feldberg* and *Vogt*, 1948); the cholinesterase activity, which is either low (*Koelle* and *Geesey*, 1961) or non-specific (*Parmar* and coll., 1961), is not conclusive. The same can be said of cholinesterase activity of fish urophysis (*Peyrot*, 1964, 1965). In addition, whenever this activity is present, it cannot be assigned with certainty to neurosecretory fibers, since these are intermingled with other ones. Finally, while the vasopressin output from neurohypophyseal neurosecretory endings is induced *in vivo* and *in vitro* by acetylcholine, this response fails to occur *in vitro* in the isolated neural lobe, disconnected from the hypothalamus (*Daniel* and *Lederis*, 1963, 1966). Therefore, it seems that if acetylcholine is involved in the mechanism of hormone excretion it is likely to act at a central level, on the perikaryon, and not at the axon endings. At present microvesicles may be considered to be related to the excretory process and similar to the microvesicles present in the synaptic junctions of the central or peripheral nervous system, but their acetylcholine content still remains a matter of speculation. The results of recent investigations on adrenergic, or monoaminergic neurons in general indicate that this line of research is likely to provide a successful approach to this problem.

The discussion of this hypothesis has drawn attention towards two fundamental aspects: the speed of hormone release and the possible role of the electrical activity of neurosecretory cells in the production of hormones. *Harris* (1948) already mentioned the direct response due to antidiuretic stimulation. Such a rapid response indicates that the axon ending receives a fast signal whereupon hormone excretion takes place. Since this response is abolished by stalk section, the signal reaches the ending by way of the axon proper. The velocity of the response indicates that the phenomenon cannot be any but electrical. The role played by nerve impulses on the hormone output from axon endings in neurosecretory systems has been brought forward by *Enami* (1957) and was supported by *Welsh* (1959) and *B. Scharrer* (1959). The question may arise whether the impulse is transmitted by the neurosecretory neuron itself, or that non-neurosecretory neurons are acting as secretory excitators (*Bern*, 1962). In the neurosecretory systems investigated fibers are present which are devoid of neurosecretory material and located between fibers loaded with granules. These "empty" fibers terminate together with the loaded ones in the immediate vicinity of blood vessels in the corresponding neurohaemal area. Synaptic junctions between such fibers and neurosecretory axons have been observed in the hypothalamic system of various Vertebrates (*Oota*, 1963 a—d; *Kobayashi* and coll., 1965) and in the caudal neurosecretory system of fishes (*Bern* and coll., 1965). These junctions do not occur in the terminal segment of neurosecretory axons, but more proximally which, however, does not necessarily rule out their significance in the control of neurohormone excretion. It should, however, be stressed that a careful screening of the literature brings out their scarcity rather than their abundance, so that this question ought to the submitted to supplementary investigation. On the other hand, considering our present knowledge of electrical activity of neurosecretory neurons and the absence of true synapses at their axon ending, the hypothesis is justified that the impulse conducted along these neurons plays a part in the excretory process.

This interpretation would provide an explanation for the quick hormonal response to physiological stimulation of antidiuresis (*Harris*, 1948) or to direct electrical stimulation of hypothalamic nuclei (*Harris*, 1960; *Fang* and coll., 1962; *Bissett* and coll., 1963; *Hayward* and *Smith*, 1964; *Stutinsky* and *Befort*, 1964; *D'Angelo* and coll., 1964; *Stutinsky* and *Guerne*, 1965). It has already been mentioned that acetylcholine cannot release vasopressin in the isolated neurohypophysis, but, on the other hand, direct electrical stimulation of the isolated neurohypophysis results in vasopressin release (*Haller* and coll., 1965). In addition, electrical stimulation *in vivo* as well as *in vitro* with optimal intensities of the fish caudal neurosecretory system produces depletion and even complete disappearing of elementary granules (*Fridberg* and coll., 1966 b). The ultrastructural observations pointed to variable effects of stimulation on individual axons. It, thus, appears that each neurosecretory neuron reacts as a single unit. This is in agreement with findings of *Brooks* and coll. (1966) who demonstrated a well defined correlation between oxytocin release induced by a peripheral stimulus and unitary electrical activity recorded in neurons of the paraventricular nucleus.

These consistent findings favour the opinion, which is now prevailing, that the nerve impulse from the neurosecretory neuron is at least partly responsible for hormone release, although they do not rule out a possible role played by non-neurosecretory fibers in controlling excretion.

It is not proven that acetylcholine or any other neurohumoral substance of the group of neurotransmitters acts in the electrophysiological release mechanism. It is known that a nerve impulse induces in the axon important and reversible physico-chemical alterations, especially of some kinds of proteins. The electrical phenomenon itself might therefore induce changes of membrane permeability and splitting of the hormone-carrier complex, resulting in depletion of granules and diffusion of active substances out of the axon.

It is not certain that in the hypothalamo-hypophyseal system excretion does occur only at the axon endings. *Lederis* (1964) discussed this problem in connection with his experiences on trout brought into sea water; all along the course of the axons the anatomical pattern is fairly similar to that existing at the endings. Another interesting point, stated above, is that along the axons the neurosecretory granules may show differences in electron density. This has already been interpreted as a granule maturation during their transport, but, on the other hand, it is also possible that the mature granules are depleted during this transport. The question remains whether a decrease in electron density is actually a morphological parameter for a decrease of the hormone content of the granules. This correlation still remains controversial, though it is likely from what is known about depletion of monoamine granules in adrenal medulla and monoaminergic nerve fibers, as well as of observations in the newborn rabbit (*Heller* and *Lederis*, 1962), that a very low amount of hormones correlates well with a low density of granules. Consequently, the question can be raised about a possible permanent or occasional output of hormones along neurosecretory axons. Considerations in this line cannot be developed here.

III. The Different Kinds of Neurosecretory Fibers and the Question of Monoaminergic Fibers

Several authors tried to elucidate the significance of the variability in size of the elementary granules. One explanation, which could be applied to some cases, would be that of a progressive enlargement of granules during their intra-axonal transport. *Mellinger* (1963) and *Follenius* (1963), however, showed that two distinct categories of fibers in the hypothalamo-hypophyseal tract of some fishes contain larger and smaller granules, respectively, and that they seem to originate either from the preoptic nucleus or from the lateral nucleus of the tuber. Numerous investigations have been performed to assign a specific origin and a specific hormone content to granules of different sizes and to the fibers transporting them. Interesting results have already been obtained in this line by means of ultracentrifugation.

Knowles (1964, 1965 a) and *Knowles* and *Vollrath* (1966 a, b) put forward the concept of two fundamentally different kinds of fibers, both observed in Invertebrates and in Vertebrates: A-fibers bearing granules of a relatively large size (more than 1000 Å diameter) and of high electron density, and B-fibers bearing smaller granules (less than 1000 Å in Vertebrates) the content of which is of lower density and which are surrounded by a wide, clear halo. According to *Knowles*, fibers with A-granules would be related to peptidic hormones, while B-fibers would be of a different significance. A single target element (like the intermediate lobe in fish adenohypophysis) may receive a double neurosecretory innervation from both kinds of fibers. A-fibers might carry a peptidic substance particularly influencing the synthesis of products secreted by the gland cells, whereas B-fibers might bring a non-peptidic substance fit to control the output of these products. It seems indeed established that both kinds of fibers may be found in a single neurosecretory system, since "small granular fibers" were described in the posterior lobe of various mammals (*Holmes*, 1964, 1966) and of man (*Lederis*, 1965).

Special attention has been drawn to these two kinds of fibers in studying the extremely complex structure of the median eminence. The A- and B-fibers of *Knowles* apparently correspond with type 1 and type 2 axon endings of *Mazzuca* (1965), and also with type 1 and type 2 endings of *Matsui* (1966 a, b). In the observations of the latter author large and small granules are found together with non-granulated clear vesicles similar to synaptic microvesicles (Fig. 10). According to *Matsui*, additional fibers of a type 3 contain only clear microvesicles and no dense granules. The small granules, or small dense-core vesicles were supposed to be characteristic of adrenergic terminals (*Pellegrino de Iraldi* and coll., 1963; *Richardson*, 1964). It was considered to be likely that among fibers encountered in the median eminence which frequently end on vessels of the

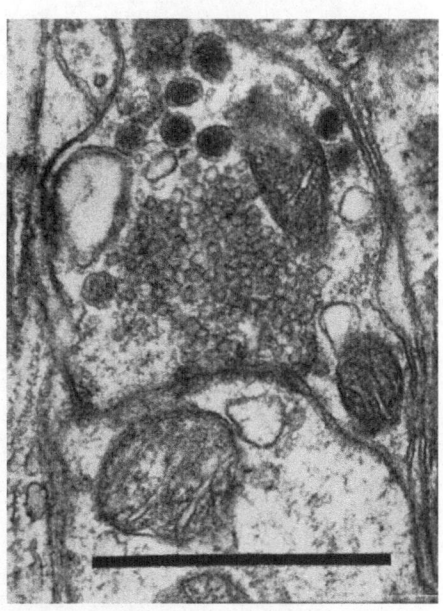

Fig. 10. Neurosecretory fiber sectioned between the nucleus supraopticus and the median eminence. Large neurosecretory granules (1000—1500 Å) are associated with microvesicles of regular shape and size (300—400 Å).

primary capillary plexus (*Barry* and *Cotte*, 1961; *Mazzuca*, 1965), some fibers bear monoamine granules (*Röhlich* and call., 1965; *Bern* and *Nishioka*, 1965; *Kobayashi* and coll., 1966; *Matsui*, 1966 a, b, 1967).

After the basic findings of *Benoit* and *Assenmacher* (1952, 1953), the morphological analysis of the median eminence has made considerable progress. Morphological research (the "parvicellular neurosecretory system" of *Szentágothai*, 1964), the experimental approach ("hypophysiotropic area", resp. "medial basal hypothalamus" in the rat: *Halász* and coll., 1962; *Halász* and coll., 1967; ventromedial area of tuber in the frog: *Dierickx*, 1963, 1965, 1967) and histochemical investigations (*Carlsson* and coll., 1962; *Falck*, 1964;

Dahlström and *Fuxe*, 1964 a, b; *Fuxe* and *Hökfelt*, 1966) are now leading to the concept of a tubero-infundibular system, responsible for the influence exerted by the hypothalamus on the anterior hypophyseal lobe. The question remains whether neurons of this system produce one or several "releasing factors" and whether the monoamines present in neurons of this system act as neurotransmitters or as hormones. This physiological and pharmacological question will be discussed later during this symposion.

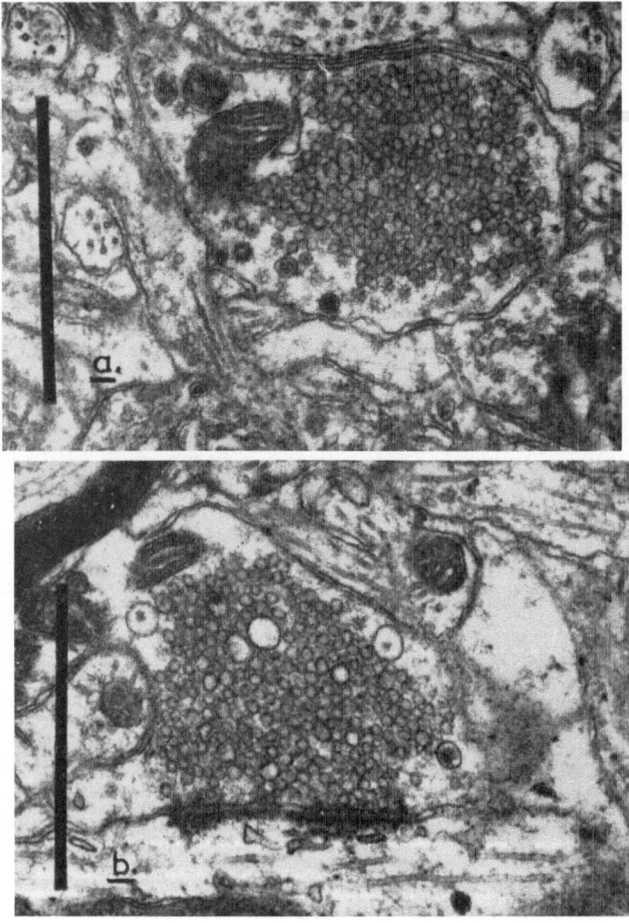

Fig. 11. Fibers in the anterior part of median eminence. Clusters of microvesicles of unequal diameter (300—1000 Å) and density, some of them showing the well known aspect of "dense-core vesicles" (a), others with a content of lesser density surrounded by a large, clear halo reminding of aspects described by *Knowles* for his "B-fibers" (b). The fiber shown at *b* makes synaptic contact with a nerve element of undetermined nature (possibly an axo-axonal synapse?). These fibers may be considered to be monoaminergic.

When we consider these particular neurons from histochemical and ultra-structural points of view they will bring us back to problems regarding neuro-secretory cells. Up to 1966 it was impossible to distinguish between electron-lucent microvesicles and empty granulated vesicles deprived of their content either by reserpine or spontaneously (*Sano* and coll., 1967). Consequently, it was impossible to establish a relation between synaptic-like microvesicles and acetylcholine. Some authors are inclined to admit that the small electron-lucent vesicles are, in fact, the normal aspect of monoamine containing vesicles (Fig. 1 b and 11) (*Fuxe* and coll., 1965). This idea is in agreement with findings by *Richardson* (1964, 1966), *Tranzer* and *Thoenen* (1967 a, b), *Thoenen* and coll. (1966), who showed that changes in the fixation procedure of adrenergic endings of the peripheral nervous system results in a marked increase of the proportion of dense-core vesicles. Recent results obtained by *Taxi* (1967) demonstrated that peripheral fibers containing clear vesicles exclusively are able to incorporate labelled noradrenaline. In rat hypothalamus and caudate nucleus which were incubated with α-methyl-noradrenaline and fixed in permanganate according to *Richardson*'s technique (1966), *Hökfelt* (1966) showed that a large number of dense-core vesicles admittedly has stored the amine. Both the *in vitro* experiments of *Hökfelt* and the *in vivo* experiments of *Taxi* are in agreement with each other. Finally, *Hökfelt* (1967) studied with the same fixation technique the locus coeruleus of rat. In this area, which contains noradrenaline-rich neurons (*Dahlström* and *Fuxe*, 1964 b), terminal boutons were shown to be present which are identical to peripheral adrenergic endings. Moreover, these neurons contain noradrenaline in all their parts, including the perikaryon in which the small granulated vesicles may be formed at the expense of the Golgi apparatus [*].

The vesicles once depleted of their amine content in the nerve endings are replaced by means of axonal transport of newly formed vesicles originating in the perikaryon. *Hökfelt* suggests that this would explain the fact that, while depletion by reserpine of granulated vesicles in the hypothalamus is very marked after 3 hours, it is much less conspicuous after 24 hours (*Shimizu* and *Ishii*, 1964). This time range is consistent with the speed of axoplasmic flow and very similar to the speed of transport of the peptidic neurosecretory granules (*Sloper*, 1958; *Flament-Durand*, 1966).

The preceding considerations may appear somewhat far from our subject, the neurosecretory cell. We, on the contrary, are inclined to think that all these data can be integrated in a general comprehensive view of the cyto-physiology of this particular element. In our opinion it is possible to draw synthetic conclusions although some of them will still be of a speculative character.

---

[*] The larger dense granules observed together with granulated microvesicles are considered by *Hökfelt* to contain noradrenaline. These findings might throw light on the still unknown significance of the large "neurosecretory-like" granules described by *Taxi* in peripheral fibers.

General Conclusions

1. The neurosecretory cell is a nerve cell showing the morphological and electrophysiological features of neurons. From various afferent pathways it receives signals transmitted by axo-somatic synaptic junctions which, at least in neurosecretory hypothalamic nuclei of Vertebrates, may be either cholinergic or monoaminergic, or both.

2. Various neurons of the central nervous system give rise to vesicles being either clear or showing an electron-dense content. These vesicles are formed in the perikaryon and transported along the axon as far as the presynaptic endings. These neurons are monoaminergic and the transported substance is a neurotransmitter acting at the synaptic junctions.

3. Among the monoaminergic neurons, some have axons with synaptic-like endings located either in close contact with or in close vicinity of epithelial endocrine glandular cells (secreto-motor innervation in fish adenohypophysis), or in the vicinity of vascular walls and pericapillary spaces (tubero-infundibular system of various lower as well as higher Vertebrates). In the latter case the question is raised, whether monoamines released in the portal vessels of the median eminence act as true hormones comparable to releasing factors, or if they act more indirectly as neurotransmitters influencing the secretory activity of the adenohypophysis by means of another substance of true hormonal nature, thus controlling the release of hormones from those reactive cells.

4. From the cytological point of view, the formation of these monoamine-containing vesicles is strikingly similar to the genesis of catecholamine-bearing granules in the adrenal medullary cell.

5. By means of a cytologically identical process, groups of neurons composing the neurosecretory nuclei give rise to large granules containing peptide hormones which possibly reach their final chemical composition only during their transport along neuronal processes, as is true in the hypothalamo-neurohypophyseal system. These cells are doubly specialized neurons which have acquired the character of nerve cells as well as that of gland cells. The hormonal content of these granules might perhaps be partly released during their transport along the axon, although the excretory process resulting in hormone release takes place mainly in the neurosecretory fiber endings. These storage sites are located either in a neurohaemal area (axon endings in the neurohypophysis) or in close relation to the ependyma.

6. The intra-axonal transport of the granules depends on the speed of the axoplasmic flow. The velocity of this flow can be estimated with staining methods, but more accurately by means of radio-isotopic labelling. By means of enhanced (or reduced) synthesis in the cell bodies this transport results in a possible increase (or decrease) of material available at these sites. This process, with its time of latency, accounts for the rather slow and long lasting *adaptive response* characteristic of such a neurosecretory system in neuroendocrine regulations. This is the first aspect of the response of a neurosecretory cell to physiological stimulation.

7. Release of hormones, that is the *excretory response* of these glandular neurons, may start quickly following a physiological or pharmacological stimulation. This is the second aspect of their responses as secretory cells. This response, induced by a nervous impulse, may be experimentally obtained by electrical stimulation of the neurosecretory nuclei. Though it is possible that this impulse might reach the neurosecretory endings along other non-neurosecretory axons, many observations point to the fact that this specific nervous impulse originates from and is conducted by the neurosecretory neurons themselves. The effect of the impulse might be either exerted upon neurosecretory granules, or alternatively mediated by a substance of another category borne by the same neurosecretory axon. Acetylcholine has been taken into account, but at present this does not appear very likely so that this question should be re-investigated on the basis of new technical approaches suggested by recent data.

Probably, some neurosecretory neurons act as receptor cells themselves, not receiving information by the indirect way of synaptic contacts with fibers reaching their perikaryon. It appears, therefore, that in some cases a neurosecretory neuron might represent a very complete system, integrating an entire neuroendocrine mechanism in a single cell unit. This would undoubtedly be the most peculiar feature of the neurosecretory cells. Neurosecretory cells are found throughout the whole animal kingdom apparently representing one of the most fundamental forms of correlative systems.

8. Some years ago, while discussing another aspect of the secretory activity of nerve cells (*Picard* and *Stahl*, 1956; *Picard* and coll., 1957), we suggested that the neurosecretory phenomenon in some groups of neurons should be considered to be a particular and specialized adaptation of a cytophysiological behaviour common to various nerve elements. This idea was in agreement with requirements of *E.* and *B. Scharrer* for an essentially physiological criterion of neurosecretory phenomena. It was, however, in contrast with the tendency of some authors to regard the whole nervous system as one wide assemblage of neurosecretory elements. Our opinion is now supported by the recent development of knowledge in this particular field. During the symposion on neurosecretion in Strasbourg (1966), the criterion of a neurohaemal organ as being characteristic of neurosecretory systems was abandoned for a more physiological basis in agreement with *Knowles* and *Bern* (1966). This seems even more justified because of the recent findings on non-neurosecretory monoaminergic neurons in the peripheral as well as in the central nervous system. The special character of a neurosecretory cell is that it provides a reactive organ or tissue with a hormonal substance. A multitude of pathways is present such as the blood circulation, a specialized vascular apparatus (*e. g.* the portal system of the median eminence), or the cerebrospinal fluid. It is also possible that the hormone reaches the target organ more directly either by means of a very narrow pericapillary space or of a true close contact (neurosecretory innervation). Consequently, the criterion defining a neurosecretory neuron seems to be the hormonal nature of the secreted substance which is: 1) a substance probably having a rather

slow and long lasting action being different in this respect from a neuro-transmitter showing a rapid action of the quantic type, 2) a substance acting on a specific effector or target element. Even if the target organ is remote and has to pick up the hormonal substance out of all the substances carried by the blood stream, the receptor cell chooses the hormone which is rec-ognized by its biochemical code fitting into the code of the receptor. In contrast to a too strict opinion which I wrongly worded not long ago, I am now convinced that this comprehension of the significance of neurosecretory cells applies to the case of a now admitted neurosecretory innervation, as well as to the hormonal influence exerted at a distance by way of the blood stream. As often happens in biological research, the story of the neuro-secretory cell has been carried through various complicated and winding ways to a more and more simplified and clear understanding which is likely to be reliable.

Acknowledgements

We are greatly indebted to Professors A. *Stahl* and R. *Seïte* for personal data provided by their previous research in this laboratory, and to Dr. *Cotte* for electron microscopical investigations as well as for providing illustrations of yet unpublished documents related to his present research.

We also want to thank Mrs. J. *Bottini* for invaluable help in bibliograph-ical work and preparation of the manuscript and Dr. *J. F. Jongkind* for revision of the English text.

References

General references

Neurosecretion (Proc. 3rd Intern. Sympos. on Neurosecretion, Bristol 1961; ed. H. *Heller* and R. B. *Clark*). London and New York: Academic Press, 1962.

Neurosécrétion (4e Sympos. Intern. sur la Neurosécrétion, Strasbourg 1966; ed. F. *Stutinsky*). Berlin: Springer-Verl. 1967.

Travaux du 4e Sympos. Intern. Endocrin. Comp., Paris 1964. In: Arch. Anat. micr., *54*, 1—658 (1965).

Szentágothai, J., B. Flerkó, B. Mess, and *B. Halász:* Hypothalamic control of the anterior pituitary. Budapest: Akademiai Kiado, 1962.

Diepen, R.: Der Hypothalamus. V. Möllendorff's Handb. d. mikr. Anat., IV/7. Berlin: Springer, 1962.

Gabe, M.: Neurosecretion. Oxford: Pergamon Press, 1966.

Picard, D., and A. *Stahl:* La cellule neurosécrétrice chez les Vertébrés. Bull. Assoc. Anat., *130 bis*, 1—75 (1966).

Barry, J.: Les neurosécrétats hypothalamiques. Etude histologique. In: Sympos. Intern. Neuro-Endocrinol., Paris 1966, 5—63. Paris: l'Expansion, 1966.

Recent studies on the hypothalamus, ed. by K. *Brown-Grant* and B. A. *Cross*, Brit. Med. Bull. *22/3*, 195—277 (1966).

Special references

Abrahams, V. C., and M. *Pickford:* The effects of anticholinesterase injected into the supraoptic nuclei of chloralosed dogs on the release of the oxytocic factor of the posterior pituitary. J. Physiol., London, *133*, 330—333 (1956).

Abrahams, V. C., G. B. Koelle, and *P. Smart:* Histochemical demonstration of cholinesterase in the hypothalamus of the dog. J. Physiol., London, *139,* 137—144 (1957).

Adams, C. W. M., and *J. C. Sloper:* Preuve de l'existence d'une substance neurosécrétoire riche en cystine dans l'hypothalamus de l'Homme, du Rat, du Chien et du Phoque, à l'aide d'une méthode à l'acide performique-bleu alcian. VIe Congr. Féd. Intern. Anat., 1—2. Paris: Masson édit., 1955.

Adams, C. W. M., and *J. C. Sloper:* The hypothalamic elaboration of posterior pituitary principles in man, rat and dog. Histochemical evidence derived from a performic acid-alcian blue reaction for cystine. J. Endocr. *31,* 221—228 (1956).

Arvy, L.: Histochemical demonstration of enzymatic activities in neurosecretory centres of some homoiothermic animals. In: Neurosecretion (Proceedings of the Third Intern. Sympos. on Neurosecretion, Bristol 1961), 215—225. London and New York: Academic Press, 1962.

Bachrach, D., K. Kovacs, F. Olah, and *V. Varro:* Histochemical examination of the colloids of the hypothalamus-hypophysis system. Acta Morph. Acad. Sci. Hung. *3,* 169—182 (1953).

Bargmann, W.: Über die neurosekretorische Verknüpfung von Hypothalamus und Neurohypophyse. Zschr. Zellforsch. *34,* 610—634 (1949).

Bargmann, W., and *A. Knoop:* Elektronenmikroskopische Beobachtungen an der Neurohypophyse. Zschr. Zellforsch. *46,* 242—251 (1957).

Bargmann, W., and *A. Knoop:* Über die morphologischen Beziehungen des neurosekretorischen Zwischenhirnsystems zum Zwischenlappen der Hypophyse. Zschr. Zellforsch. *52,* 256—277 (1960).

Barraclough, C. A., and *B. A. Cross:* Unit activity in the hypothalamus of the cyclic female rat: effect of genital stimuli and progesterone. J. Endocr. *26,* 339—359 (1963).

Barrnett, R. J.: Histochemical demonstration of disulfide groups in the neurohypophysis under normal and experimental conditions. Endocrinology 55, 484—501 (1954).

Barry, J.: Les voies extra-hypophysaires de la neurosécrétion diencéphalique (C. R. Assoc. Anat., 42e Réun., Paris 1955). In: Bull. Assoc. Anat. *89,* 464—476 (1956).

Barry, J.: Etude au microscope électronique de certains éléments de la voie neurosécrétoire hypothalamo-hypophysaire. C. R. Soc. Biol. *151,* 156—158 (1957).

Barry, J., and *G. Cotte:* Etude préliminaire, au microscope électronique, de l'éminence médiane du Cobaye. Zschr. Zellforsch. *53,* 714—724 (1961).

Bennett, M. V. L., and *S. Fox:* Electrophysiology of caudal neurosecretory cells in the skate and fluke. Gen. Comp. Endocrin. *2,* 77—95 (1962).

Benoit, J., and *I. Assenmacher:* Influence de lésions hautes et basses de l'infundibulum sur la gonadostimulation chez le Canard domestique. C. R. Acad. Sci. *235,* 1547—1549 (1952).

Benoit, J., and *I. Assenmacher:* Rapport entre la stimulation sexuelle préhypophysaire et la neurosécrétion chez l'Oiseau. Arch. Anat. micr. *42,* 334—386 (1953).

Bern, H. A.: The properties of neurosecretory cells. In: Progress in Comparative Endocrinology (Proc. 3rd Int. Sympos. Compar. Endocrin., Oiso 1961). Gen. Compar. Endocrin. *suppl. 1,* 117—132 (1962).

Bern, H. A.: The secretory neuron as a doubly specialized cell. In: The General Physiology of Cell Specialization, 349—366. New York: McGraw-Hill, 1963.

Bern, H. A.: On the production of hormones by neurones and the role of neurosecretion in neuro-endocrine mechanisms. Symp. Soc. exp. Biol. *20*, 325—344 (1966).

Bern, H. A., and *R. S. Nishioka:* Fine structure of the median eminence of some passerine birds. Proc. zool. Soc. Calcutta *18*, 107—119 (1965).

Bern, H. A., and *K. Yagi:* Electrophysiology of neurosecretory systems. In: Proc. 2nd Int. Congr. Endocrin., London 1964, 577—583. Amsterdam: Excerpta Medica Foundation, 1965.

Bern, H. A., R. S. Nishioka, and *I. R. Hagadorn:* Association of elementary neurosecretory granules with the Golgi complex. J. Ultrastruc. Res. *5*, 311—320 (1961).

Bern, H. A., R. S. Nishioka, L. R. Mewaldt, and *D. S. Farner:* Photoperiodic and osmotic influences on the ultrastructure of the hypothalamic neurosecretory system of the white-crowned sparrow *Zonotrichia leucophrys gambelii.* Zschr. Zellforsch. *69*, 198—227 (1966).

Bern, H. A., K. Yagi, and *R. S. Nishioka:* Structure and function of the caudal neurosecretory system of fishes. (Proc. IV Int. Sympos. on Comparative Endocrinology, Paris 1964). In: Arch. Anat. micr. *54*, 217—238 (1965).

Bisset, G. W., S. M. Hilton, and *A. M. Poisner:* Parallel assays of vasopressin and oxytocin in blood on localized electrical stimulation of the hypothalamus. J. Physiol., London, *169*, 40 p. (1963).

Bloom, R. S.: Thesis, Birmingham 1960. Quoted by *Arvy, L.:* Histoenzymologie des glandes endocrines. Paris: Gauthiers-Villars, 1963.

Bodian, D.: Nerve endings, neurosecretory substance and lobular organization of the neurohypophysis. Bull. Johns Hopkins Hosp. *89*, 354—376 (1951).

Brettschneider, H.: Die Feinstruktur des nervösen Parenchyms des Infundibulum. I. Faserstrukturen. Zschr. mikrosk.-anat. Forsch. *62*, 247—266 (1956).

Brooks, C. Mc C., J. Ushiyama, and *G. Lange:* Reactions of neurons in or near the supraoptic nuclei. Amer. J. Physiol. *202*, 487—490 (1962).

Brooks, C. Mc C., T. Ishikawa, K. Koizumi, and *H. H. Lu:* Activity of neurons in the paraventricular nucleus of the hypothalamus and its control. J. Physiol., London, *182*, 217—231 (1966).

Carlsson, A., B. Falck, and *N. A. Hillarp:* Cellular localization of brain monoamines. Acta Physiol. Scand., *Suppl. 196* (1962).

Christ, J. F.: The early changes in the hypophysial neurosecretory fibres after coagulation. In: Neurosecretion (Proceedings of the Third Intern. Sympos. on Neurosecretion, Bristol 1961), 125—142. London and New York: Academic Press, 1962.

Cohen, A. G.: B. Sc. Thesis, Birmingham 1964. Quoted by *F. Knowles* and *L. Vollrath.*

Cross, B. A.: The hypothalamus in mammalian homeostasis. Sympos. Soc. Exp. Biol. *18*, 157—193 (1964).

Cross, B. A., and *J. D. Green:* Activity of single neurons in the hypothalamus: effect of osmotic and other stimuli. J. Physiol., London, *148*, 554—569 (1959).

Cross, B. A., and *I. A. Silver:* Electrophysiological studies on the hypothalamus. Brit. Med. Bull. *22*, 254—260 (1966).

Dahlström, A., and *K. Fuxe:* A method for the demonstration of monoamine-containing nerve fibres in the central nervous system. Acta Physiol. Scand. *60*, 293—294 (1964 a).

Dahlström, A., and *K. Fuxe:* Evidence for the existence of monoamine-containing neurons in the central nervous system. I. Demonstration of monoamines in the cell bodies of brain stem neurons. Acta Physiol. Scand. *62*, suppl. 232, 1—55 (1964 b).

D'Angelo, S. A., J. Snyder, and *J. M. Grodin:* Electrical stimulation of the hypothalamus: simultaneous effects on the pituitary-adrenal and thyroid systems of the rat. Endocrinology *75*, 417—427 (1964).

Daniel, A. R., and *K. Lederis:* Hormone release from the neurohypophysis *in vitro.* Gen. Comp. Endocrin. *3*, 693—694 (1963).

Daniel, A. R., and *K. Lederis:* Effects of acetylcholine on the release of neurohypophysial hormones *in vitro.* Proc. Soc. Endocrin. In: J. Endocr. *34*, X—XI (1966).

De Robertis, E.: Histophysiologie des synapses et neurosécrétion. Paris: Gauthiers-Villars, 1964.

De Robertis, E.: and *A. Pellegrino de Iraldi:* Plurivesicular secretory processes and nerve endings in the pineal gland of the rat. J. Biophys. Biochem. Cytol. *10*, 361—372 (1961).

Dierickx, K.: The dendrites of the preoptic neurosecretory nucleus of *Rana temporaria* and the osmoreceptors. Arch. Int. Pharmacodyn. *140*, 708—725 (1962).

Dierickx, K.: The extirpation of the neurosecretory preoptic nucleus and the reproduction of *Rana temporaria.* Arch. Int. Pharmacodyn. *145*, 580—589 (1963).

Dierickx, K.: The origin of the aldehyde-fuchsin-negative nerve fibres of the median eminence of the hypophysis: a gonadotropic centre. Zschr. Zellforsch. *66*, 504—518 (1965).

Dierickx, K.: The gonadotropic centre of the tuber cinereum hypothalami and ovulation. Zschr. Zellforsch. *77*, 188—203 (1967).

Duncan, D.: An electron microscope study of the neurohypophysis of a bird. Anat. Rec. *125*, 457—472 (1956).

Enami, M.: 1957 (quoted by *Ishibashi*).

Eränkö, O.: Histochemical evidence of intense phosphatase activity in the hypothalamic magnocellular nuclei of the rat. Acta Physiol. Scand. *24*, 1—6 (1951).

Falck, B.: Observations on the possibilities of the cellular localization of monoamines by a fluorescence method. Acta Physiol. Scand. *suppl. 197* (1962).

Falck, B.: Cellular localization of monoamines. Progress in Brain Research *8*, 28—44, Amsterdam: Elsevier Publ. 1964.

Fang, H. S., H. M. Liu, and *S. C. Wang:* Liberation of antidiuretic hormone following hypothalamic stimulation in the dog. Amer. J. Physiol. *202*, 212—216 (1962).

Feldberg, W., and *M. Vogt:* Acetylcholine synthesis in different regions of the central nervous system. J. Physiol., London, *107*, 372—381 (1948).

Flament-Durand, J.: Etude des relations hypothalamo-hypophysaires à l'aide de radioisotopes marqués au soufre 35. C. R. Acad. Sci. *252*, 3487—3500 (1961).

Flament-Durand, J.: Contribution à l'étude des relations hypothalamo-hypophysaires. Ann. Soc. Roy. Sci. Med. et Natur. Bruxelles *19*, 1—119 (1966).

Follenius, E.: Etude comparative de la cytologie fine du noyau préoptique (NPO) et du noyau latéral du tuber (NLT) chez la Truite *(Salmo irideus Gibb)* et chez la Perche *(Perca fluviatilis).* Comparaison des 2 types de neurosécrétion. Gen. Comp. Endocrin. *3*, 66—85 (1963).

Follenius, E.: Bases structurales et ultrastructurales des corrélations hypothalamo-hypophysaires chez quelques espèces de Poissons téléostéens. Ann. Sci. Nat. Zool. (12e série) *VII*, 1—150 (1965).

54   D. Picard:

Follenius, E., and *A. Porte:* Etude des différents lobes de l'hypophyse de la Perche *(Perca fluviatilis L.)* au microscope électronique. C. R. Soc. Biol. *155,* 128—131 (1961).

Follenius, E., and *A. Porte:* Appearance, ultrastructure and distribution of the neurosecretory material in the pituitary gland of two teleost fishes, *Lebistes reticulatus* and *Perca fluviatilis.* In: Neurosecretion (Proc. 3rd Intern. Sympos. on Neurosecretion, Bristol 1961) 51—69. London and New York: Academic Press 1962 a.

Follenius, E., and *A. Porte:* Etude du noyau préoptique de la Perche *(Perca fluviatilis L.)* au microscope électronique. C. R. Acad. Sci. *254,* 930—932 (1962 b).

Fridberg, G., and *R. S. Nishioka:* Secretion into the cerebrospinal fluid by caudal neurosecretory neurons. Science *152,* 90—91 (1966).

Fridberg, G., H. A. Bern, and *R. S. Nishioka:* The caudal neurosecretory system of the isospondylous teleost *Albula vulpes,* from different habitats. Gen. Comp. Endocrin. *6,* 195—212 (1966 a).

Fridberg, G., S. Iwasaki, K. Yagi, H. A. Bern, D. M. Wilson, and *R. S. Nishioka:* Relation of impulse conduction to electrically induced release of neurosecretory material from the urophysis of the teleost fish *Tilapia mossambica.* J. Exper. Zool., *161,* 137—142 (1966 b).

Fridberg, G. R. S. Nishioka, H. A. Bern, and *W. R. Fleming:* Regeneration of the caudal neurosecretory system in the cichlid teleost *Tilapia mossambica.* J. Exper. Zool., *162,* 311—336 (1966 c).

Fuxe, K.: Cellular localization of monoamines in the median eminence and the infundibular system of some mammals. Zschr. Zellforsch. *61,* 710—724 (1964).

Fuxe, K., and *T. Hökfelt:* Further evidence for the existence of tuberoinfundibular dopamine neurons. Acta Physiol. Scand. *66,* 245—246 (1966).

Fuxe, K., and *L. Ljunggren:* Cellular localization of the monoamines in the upper brain stem of the pigeon. J. Comp. Neurol., *125,* 355—388 (1965).

Fuxe, K., T. Hökfelt, and *O. Nilsson:* A fluorescence and electron microscopic study on certain brain regions rich in monoamine terminals. Amer. J. Anat. *117,* 33—45 (1965).

Gabe, M.: Signification histochimique de certaines affinités tinctoriales du produit de neurosécrétion hypothalamique. C. R. Soc. Biol. *149,* 462—464 (1955).

Gabe, M.: Présence de composés décelables par la réaction à l'acide periodique-Schiff dans le produit de neurosécrétion hypothalamique chez quelques Vertébrés. C. R. Acad. Sci. *250,* 937—939 (1960).

Gerschenfeld, H. M., J. Tramezzani, and *E. de Robertis:* Ultrastructure and function in neurohypophysis of the toad. Endocrinology *66,* 741—762 (1960).

Green, J. D., and *D. S. Maxwell:* Comparative anatomy of the hypophysis and observations on the mechanism of neurosecretion. In: Comparative Endocrinology, 368—392. New York: J. Wiley, 1959.

Green, J. D., and *V. L. van Breemen:* Electron microscopy of the pituitary and observations on neurosecretion. Amer. J. Anat. *97,* 177—227 (1955).

Grignon, G.: Mise en évidence d'une activité antidiurétique d'extraits de quelques régions de l'encéphale chez la Tortue terrestre *(T. mauritanica Dumer.).* C. R. Soc. Biol. *155,* 1523—1526 (1961).

Grignon, G., and *M. Lamarche:* Etude de l'activité ocytocique d'extraits de différentes parties du cerveau chez la Tortue terrestre *(Testudo mauritanica).* C. R. Soc. Biol. *153,* 2030—2032 (1959).

Halász, B., L. Pupp, and *S. Uhlarik:* Hypophysiotrophic area in the hypothalamus. J. Endocr. *25,* 147—154 (1962).

Halász, B., M. A. Slusher, and *R. A. Gorski:* Adrenocorticotrophic hormone secretion in Rats after partial or total deafferentation of the medial basal hypothalamus. Neuroendocrin. **2,** 43—55 (1967).

Haller, E. W., H. Sachs, N. Sperelakis, and *L. Share:* Release of vasopressin from isolated guinea pig posterior pituitaries. Amer. J. Physiol. **209,** 79—83 (1965).

Harris, G. W.: Stimulation of the supraoptico-hypophyseal tract in the conscious rabbit with currents of different wave forms. J. Physiol., London, **107,** 412—448 (1948).

Harris, G. W.: Central control of pituitary secretion. In: Neurophysiology 2 (*Field, J., H. W. Magoun,* and *V. E. Hall,* eds.), 1007—1038. Baltimore: Waverly Press, 1960.

Hartmann, J. F.: Electron microscopy of the neurohypophysis in normal and histamine-treated rats. Zschr. Zellforsch. **48,** 291—308 (1958).

Hayward, J. N., and *W. K. Smith:* Antidiuretic response to electrical stimulation in brain stem of the monkey. Amer. J. Physiol. **206,** 15—20 (1964).

Heller, H., and *M. Ginsburg:* in *Harris, G. W.,* and *B. T. Donovan:* The pituitary Gland. London: Butterworths, 1966.

Heller, H., and *K. Lederis:* Characteristics of isolated neurosecretory vesicles from mammalian neural lobes. In: Neurosecretion (*Heller, H.,* and *R. B. Clark,* eds), 35—46. London: Academic Press, 1962.

Hild, W.: Experimentell-morphologische Untersuchungen über das Verhalten der "Neurosekretorischen Bahn" nach Hypophysenstieldurchtrennungen, Eingriffen in den Wasserhaushalt und Belastung der Osmoregulation. Virchows Arch. path. Anat. **319,** 526—546 (1951).

Hild, W., and *G. Zetler:* Über die Funktion des Neurosekrets im Zwischenhirn-Neurohypophysensystem als Trägersubstanz für Vasopressin, Adiuretin und Ocytocin. Zschr. exper. Med. **120,** 236—243 (1953).

Hökfelt, T.: Electron microscopic studies on brain slices from regions rich in catecholamine nerve terminals. (Quoted by *Hökfelt,* 1967.) Acta physiol. Scand. **69,** 119—120 (1967).

Hökfelt, T.: On the ultrastructural localization of noradrenaline in the central nervous system of the rat. Zschr. Zellforsch. **79,** 110—117 (1967).

Holmes, R. L.: Esterases of the hypothalamo-hypophysial system of the monkey. J. Endocr. **23,** 63—67 (1961).

Holmes, R. L.: Comparative observations on inclusions in nerve fibers of the mammalian neurohypophysis. Zschr. Zellforsch. **64,** 474—492 (1964).

Holmes, R. L.: The neurohypophysis of the foetal monkey. Zschr. Zellforsch. **69,** 288—295 (1966).

Holmes, R. L., and *J. A. Kiernan:* The fine structure of the infundibular process of the hedgehog. Zschr. Zellforsch. **61,** 894—912 (1964).

Holmes, R. L., and *F. G. Knowles:* "Synaptic vesicles" in the neurohypophysis. Nature **185,** 710—711 (1960).

Howe, A., and *A. G. E. Pearse:* A histochemical investigation of neurosecretory substance in the rat. J. Histochem. Cytochem. **4,** 561—569 (1956).

Hydén, H.: The neuron. In: The cell (*J. Brachet* and *A. E. Mirsky,* eds) IV, 216—232. New York: Academic Press, 1960.

Hydén, H., and *E. Egyhazi:* Changes in the base compositions of nuclear ribonucleic acid of neurons during a short period of enhanced protein production. J. Cell Biol. **15,** 37—44 (1962).

Ifft, J. D.: The effect of endocrine gland extirpation on the size of nucleoli in rat hypothalamic neurons. Anat. Rec. **148,** 599—603 (1964).

Ifft, J. D., and *W. Berkowitz:* A comparison of selected morphological and chemical methods for measuring neuron activity in the supraoptic nucleus of dehydrated rats. Anat. Rec. *152,* 231—234 (1965).

IIjima, K., T. R. Shantha, and *G. H. Bourne:* Enzyme histochemical studies on the hypothalamus, with special reference to supraoptic and paraventricular nuclei of the squirrel monkey *(Saimiri sciurens).* Zschr. Zellforsch. *79,* 76—91 (1967).

Ishibashi, T.: Electrical activity of the caudal neurosecretory cells in the eel *Anguilla japonica* with special reference to synaptic transmission. Gen. Comp. Endocrin. *2,* 415—424 (1962).

Iturriza, F. C.: Electron microscopic study of the pars intermedia of the pituitary of the toad, *Bufo arenarum.* Gen. Comp. Endocrin. *4,* 492—502 (1964).

Jewell, P. A., and *E. B. Verney:* An experimental attempt to determine the site of the neurohypophysial osmoreceptors in the dog. Philos. Transact. Roy. Soc., London, Biol. Sc., Sér. B, *240,* 197—324 (1957).

Jørgensen, C. B.: Brain-pituitary relationships in amphibians, birds and mammals: on the origin and nature of the neurons by which hypothalamic control of pars distalis functions are mediated. (Trav. 4e Sympos. Endocrin. Comp., Paris 1964). In: Arch. Anat. microsc. *54,* 261—276 (1965).

Kandel, E. R.: Spike and synaptic potentials in hypothalamic neuroendocrine cells. Fed. Proc. *21,* 361 (1962).

Kandel, E. R.: Electrical properties of hypothalamic neuroendocrine cells. J. gen. Physiol. *47,* 691—718 (1964).

Kinosita, H., K. Yagi, and *M. Yasuda:* Electrophysiological studies on the caudal neurosecretory system of fish. Dobutsu Zasshi (Zool. Mag.) *71,* 371 (1962).

Kivalo, E., U. K. Rinne, and *S. Mäkelä:* Acetylcholinesterase, acid phosphatase and succinic dehydrogenase in the hypothalamic magnocellular nuclei after chlorpromazine administration. Experientia *14,* 293—296 (1958).

Knowles, F.: The ultrastructure of a crustacean neurohaemal organ. In: Neurosecretion (Proc. 3rd Intern. Sympos. on Neurosecretion, Bristol 1961), 71—88. London and New York: Academic Press 1962.

Knowles, F.: Vesicle formation in the distal part of a neurosecretory system. Proc. Roy. Soc., London, Ser. B, *160,* 360—372 (1964).

Knowles, F.: Neuroendocrine correlations at the level of ultrastructure. (Trav. 4e Sympos. Endocrin. Comp., Paris 1964). In: Arch. Anat. micr. *54,* 343—358 (1965 a).

Knowles, F.: Evidence for a dual control, by neurosecretion, of hormone synthesis and hormone release in the pituitary of the dogfish *Scylliorhinus stellaris.* Phil. Transact. Roy. Soc., Ser. B, Biol. Sc., *249,* 435—456 (1965 b).

Knowles, F., and *H. A. Bern:* The function of neurosecretion in endocrine regulation. Nature *210,* 271—272 (1966).

Knowles, F., and *D. B. Carlisle:* Endocrine control in the crustacea. Biol. Rev. *31,* 396—473 (1956).

Knowles, F., and *L. Vollrath:* Neurosecretory innervation of the pituitary of the eels *Anguilla* and *Conger.* I. The structure and ultrastructure of the neuro-intermediate lobe under normal and experimental conditions. Phil. Transact. Roy. Soc., Ser. B, Biol. Sc., *250,* 311—327 (1966 a).

Knowles, F., and *L. Vollrath:* Neurosecretory innervation of the pituitary of the eels *Anguilla* and *Conger.* II. The structure and innervation of the pars distalis at different stages of the life cycle. Phil. Transact. Roy. Soc., Ser. B, Biol. Sc., *250,* 329—341 (1966 b).

Kobayashi, H., and *D. S. Farner:* Cholinesterase in the hypothalamo-hypophysial neurosecretory system of the white-crowned sparrow, *Zonotrichia Leucophrys Gambelii.* Zschr. Zellforsch. *63,* 965—973 (1964).

Kobayashi, H., and *D. S. Farner:* Evidence of a negative feed-back on photoperiodically induced gonadal development in the white-crowned sparrow, *Zonotrichia leucophrys Gambelii.* Gen. Comp. Endocrin. *6,* 443—452 (1966).

Kobayashi, H., and *Y. Oota:* Functional electron microscopy of the vertebrate neurosecretory storage-release organs. Gunma Sympos. Endocrinol. *1,* 63—79 (1964).

Kobayashi, H., T. Hirano, and *Y. Oota:* Electron microscopic and pharmacological studies on the median eminence and pars nervosa. (Trav. 4e Sympos. Endocrin. Comp., Paris 1964.) In: Arch. Anat. micr. *54,* 277—294 (1965).

Kobayashi, H., Y. Oota, and *T. Hirano:* Acid phosphatase activity of the hypothalamo-hypophysial system of dehydrated rats and pigeons in relation to neurosecretion. Gen. Comp. Endocrinol. *2,* 495—498 (1962).

Kobayashi, H., Y. Oota, and *T. Hirano:* Electron microscopic and pharmacological studies on the rat median eminence. Zschr. Zellforsch. *71,* 387—404 (1966).

Kobayashi, H., H. Uemura, Y. Oota, and *S. Ishii:* Cholinergic substance in the caudal neurosecretory storage organ of fish. Science *141,* 714—716 (1963).

Koelle, G. B., and *C. N. Geesey:* Localization of acetylcholinesterase in the neurohypophysis and its functional implications. Proc. Soc. Exper. Biol. Med. *106,* 625—628 (1961).

Koizumi, K., T. Ishikawa, and *C. Mc C. Brooks:* Control of activity of neurons in the supraoptic nucleus. J. Neurophysiol. *27,* 878—892 (1964).

La Bella, F. S., and *M. Sanwal:* Isolation of nerve endings from the posterior pituitary gland. Electron microscopy of fractions isolated by centrifugation. J. Cell Biol. *25,* 179—193 (1965).

Lederis, K.: The distribution of vasopressin and oxytocin in hypothalamic nuclei. In: Neurosecretion (Proc. 3rd Intern. Sympos. on Neurosecretion, Bristol 1961), 227—239. London and New York: Academic Press 1962.

Lederis, K.: A preliminary report on the ultrastructure of the human neurohypophysis. J. Endocr. *27,* 133—135 (1963).

Lederis, K.: Fine structure and hormone content of the hypothalamo-neurohypophysial system of the rainbow trout *(Salmo irideus)* exposed to sea water. Gen. Comp. Endocrin. *4,* 638—661 (1964).

Lederis, K.: An electron microscopical study of the human neurohypophysis. Zschr. Zellforsch. *65,* 847—868 (1965).

Legait, H., and *E. Legait:* Mise en évidence de voies neurosécrétoires extrahypophysaires chez quelques Batraciens et Reptiles. C. R. Soc. Biol. *150,* 1429—1431 (1956).

Legait, H., and *E. Legait:* Les voies extra-hypophysaires des noyaux neurosécrétoires hypothalamiques chez les Batraciens et les Reptiles. Acta Anat. *30,* 429—433 (1957).

Legait, H., and *E. Legait:* Présence d'une voie neurosécrétoire hypothalamohabénulaire et mise en évidence d'une activité antidiurétique au niveau des ganglions de l'habénula chez la Poule. C. R. Soc. Biol. *152,* 828—830 (1958).

Leray, C.: Etude de l'incorporation de cystéine marquée au soufre 35 dans le système hypothalamo-hypophysaire et plus spécialement dans l'adénohypophyse chez un téléostéen: *Mugil cephalus L.* C. R. Acad. Sci. *256,* 795—798 (1963).

Luciani, J., and *R. Seïte:* Sur une différence de comportement des cellules neurosécrétrices des noyaux supra-optique et paraventriculaire du Chat, après injection de tricyano-amino-propène (TRIAP). C. R. Soc. Biol. *159,* 422—424 (1965).

Mach, B., E. Rich, and *E. L. Tatum:* Separation of the biosynthesis of the antibiotic polypeptide tyrocidine from protein biosynthesis. Proc. nat. Acad. Sci. (Wash.) *50,* 175—181 (1963).

Matsui, T.: Fine structure of the posterior median eminence of the pigeon, *Columba livia domestica.* J. Fac. Sci. Tokyo, sect. IV, *11,* 49—70 (1966 a).

Matsui, T.: Fine structure of the median eminence of the rat. J. Fac. Sc. Tokyo, sect. IV, *11,* 71—96 (1966 b).

Matsui, T.: Effect of reserpine on the distribution of granulated vesicles in the mouse median eminence. Neuroendocrinol. *2,* 99—106 (1967).

Mazzi V.: I fenomeni neurosecretori nella femmina del tritone crestato in condizioni sperimentali. Zschr. Zellforsch. *39,* 298—317 (1953).

Mazzuca, M.: Structure fine de l'éminence médiane du cobaye. J. Microscopie *4,* 225—238 (1965).

Mellinger, J.: Les relations neuro-vasculo-glandulaires dans l'appareil hypophysaire de la Roussette, *Scylliorhinus caniculus L.* Thesis Sci., Strasbourg 1963.

Morita, H., T. Ishibashi, and *S. Yamashita:* Synaptic transmission in neurosecretory cells. Nature *191,* 183 (1961).

Moses, A. M., T. F. Leveque, M. Giambatista, and *C. W. Lloyd:* Dissociation between the content of vasopressin and neurosecretory material in the rat neurohypophysis. J. Endocr. *26,* 273—278 (1963).

Mosinger, M.: Bases d'une médecine et d'une biologie intégratives. Diencéphale, neuro-endocrinologie et neuro-ergonologie. Coimbra: Coimbra Editora, 1951.

Murakami, M.: Elektronenmikroskopische Untersuchung der neurosekretorischen Zellen im Hypothalamus der Maus. Zschr. Zellforsch. *56,* 277—299 (1962).

Murakami, M.: Weitere Untersuchungen über die Feinstruktur der neurosekretorischen Zellen im Nucleus Supra-opticus von *Gecko japonicus.* Zschr. Zellforsch. *59,* 684—699 (1963).

Murakami, M.: Elektronenmikroskopische Untersuchungen am Nucleus praeopticus der Kröte *(Bufo vulgaris formosus).* Zschr. Zellforsch. *63,* 208—225 (1964).

Nemetschek-Gansler, H.: Zur Ultrastruktur des Hypophysen-Zwischenhirnsystems der Ratte. Zschr. Zellforsch. *67,* 844—862 (1965).

Nishioka, R. S., H. A. Bern, and *L. R. Mewaldt:* Ultrastructural aspects of the neurohypophysis of the white-crowned sparrow, *Zonotrichia leucophrys Gambelii,* with special reference to the relation of neurosecretory axons to ependyma in the pars nervosa. Gen. Comp. Endocrin. *4,* 304—313 (1964).

Oota, Y.: Effect of dehydration on the fine structures of pars nervosa of pigeon. Dobutsu. Zasshi (Zool. Mag.) *71,* 235—242 (1962).

Oota, Y.: Fine structure of the caudal neurosecretory system of the carp, *Cyprinus carpio.* J. Fac. Sc. Tokyo Univ., sect. IV, *10,* 129—141 (1963 a).

Oota, Y.: Electron microscopic studies on the region of the hypothalamus contiguous to the hypophysis and the neurohypophysis of the fish, *Oryzias latipes.* J. Fac. Sc. Tokyo Univ., sect. IV, *10,* 143—153 (1963 b).

Oota, Y.: Fine structure of the median eminence and the pars nervosa of the mouse. J. Fac. Sc. Tokyo Univ., sect. IV, *10,* 155—168 (1963 c).

Oota, Y.: Fine structure of the median eminence and the pars nervosa of the turtle, *Clemmys japonica.* J. Fac. Sc. Tokyo Univ., sect. IV, *10,* 169—179 (1963 d).

Oota, Y., and *H. Kobayashi:* Fine structure of the median eminence and pars nervosa of the pigeon. Annot. Zool. Japon. *35,* 128—138 (1962).

Oota, Y., and *H. Kobayashi:* Fine structure of the median eminence and the pars nervosa of the bullfrog, *Rana catesbeiana.* Zschr. Zellforsch. *60,* 667—687 (1963).

Oota, Y., and *H. Kobayashi:* On the synaptic vesicle-like structures in the neurosecretory axon of the mouse neural lobe. Annot. Zool. Japon. *39,* 193—201 (1966).

Osinchak, J.: Electron microscopic localization of acid phosphatase and thiamine pyrophosphatase activity in hypothalamic neurosecretory cells of the rat. J. Cell Biol. *21,* 35—48 (1964).

Palay, S. L.: Neurosecretory phenomena in the hypothalamo-hypophysial system of man and monkey. Amer. J. Anat. *93,* 107—141 (1953).

Palay, S. L.: An electron microscope study of the neurohypophysis in normal, hydrated and dehydrated rats. Anat. Rec. *121,* 348 (1955).

Palay, S. L.: The fine structure of the neurohypophysis. In: Progress in Neurobiology, II. Ultrastructure and cellular chemistry of neural tissue, 31—49. New York: Hoeber, 1957.

Palay, S. L.: The fine structure of secretory neurone in the preoptic nucleus of the goldfish. Anat. Rec. *138,* 417—444 (1960).

Parmar, S. S. M. C. Sutter, and *M. Nickerson:* Localization and characterization of cholinesterase in subcellular fractions of rat brain and beef pituitary. Can. J. Physiol. and Pharmacol. *39,* 1335 (1961).

Pearse, A. G.: Esterases of the hypothalamus and neurohypophysis and their functional significance. In: Pathophysiologia diencephalica, 329—335. Wien: Springer, 1958.

Pellegrino de Iraldi, A., H. F. Duggan, and *E. de Robertis:* Adrenergic synaptic vesicles in the anterior hypothalamus of the rat. Anat. Rec. *145,* 521—531 (1963).

Pepler, W. J., and *A. G. Pearse:* The histochemistry of esterases of rat brain with special reference to those of the hypothalamic nuclei. J. Neurochem. *1,* 193—202 (1957).

Peyrot, A.: Recherches histoenzymologiques sur le système neurosécréteur caudal de quelques téléostéens (*Cyprinus carpio L., Tinca tinca L.* et *Salmo fario L.*). Gen. Comp. Endocrin. *4,* 320—330 (1964).

Peyrot, A.: Il sistema neurosecernente caudale dei pesci. Ric. Sci. *35* (I), 115—140 (1965).

Picard, D., and *A. Stahl:* Signification fondamentale du certaines activités élaboratrices des cellules nerveuses. Etude critique de la notion actuelle de neurosécrétion. J. de Physiol. *48,* 73—95 (1956).

Picard, D., A. Stahl, and *R. Seïte:* Elaborations neuronales dans des territoires ganglionnaires et encéphaliques du système végétatif. Acta neuroveget. *16,* 110—129 (1957).

Picard, D., R. Seïte, A. Monneron, and *J. Luciani:* Influence du tricyanoaminopropène sur les noyaux neurosécrétoires hypothalamiques du Rat et du Chat. J. de Physiol. *57,* 270—271 (1965).

Pickford, M., and *J. A. Watt:* A comparison of the effect of intravenous and intracarotid injection of acetylcholine in the dog. J. Physiol., London, *114,* 333—335 (1951).

Richardson, K. C.: The fine structure of the albino rabbit iris with special reference to the identification of adrenergic and cholinergic nerve endings in its intrinsic muscles. Amer. J. Anat. *114,* 173—206 (1964).

Richardson, K. C.: Electron microscopic identification of autonomic nerve endings. Nature *210,* 756 (1966).

Rodriguez, R. M.: Etude comparative de l'activité acétylcholinestérasique dans les noyaux hypothalamiques chez plusieurs mammifères. Trab. Inst. Cajal Invest. Biol. 57, 29—44 (1965).

Röhlich, P., B. Vigh, I. Teichmann, and *B. Aros:* Electron microscopy of the median eminence of the rat. Acta Biol. Acad. Sc. Hung. *15,* 431—457 (1965).

Romieu, M., A. Stahl, and *G. Cotte:* Cytologie des cellules nerveuses de l'hypothalamus. Acta anat. *18,* 74—79 (1953).

Sachs, H.: Vasopressin biosynthesis. Biochem. Biophys. Acta *34,* 572—573 (1959).

Sachs, H.: Studies concerned with vasopressin biosynthesis. In: Regional neurochemistry, 265—273. London and New York: Pergamon Press, 1961.

Sachs, H., and *Y. Takabatake:* Evidence for a precursor in vasopressin biosynthesis. Endocrinology 75, 943—948 (1964).

Sano, Y., and *A. Knoop:* Elektronenmikroskopische Untersuchungen am kaudalen neurosekretorischen System von *Tinca vulgaris.* Zschr. Zellforsch. *49,* 464—492 (1959).

Sano, Y., G. Odake, and *S. Taketomo:* Fluorescence microscopic and electron microscopic observations on the tuberohypophyseal tract. Neuroendocrinology *2,* 30—42 (1967).

Scharrer, B.: The role of neurosecretion in endocrine integration. In: Comparative endocrinology, 134—148. New York: J. Wiley, 1959.

Scharrer, E., and *S. Brown:* Neurosecretion. XII. The formation of neurosecretory granules in the earthworm, *Lumbricus terrestris L.* Zschr. Zellforsch. *54,* 530—540 (1961).

Scharrer, E., and *B. Scharrer:* Neurosekretion. In: V. Möllendorff's Handb. d. mikr. Anat. VI/5. Berlin: Springer, 1954.

Scharrer, E., and *G. Wittenstein:* The effect of the interruption of the hypothalamo-hypophyseal neurosecretory pathway in the dog. Anat. Rec. *112,* 387 (1952).

Schiebler, T. H.: Zur Histochemie des neurosekretorischen hypothalamisch-neurohypophysären Systems. Acta anat. *13,* 233—255 (1951).

Schiebler, T. H.: Zur Histochemie des neurosekretorischen hypothalamisch-neurohypophysären Systems. Acta anat. *15,* 393—416 (1952).

Seïte, R., and *A. Monneron:* Effets du tricyanoaminopropène (TRIAP) sur les signes histologiques de l'activité noosécrétrice de l'hypothalamus du Rat. C. R. Acad. Sci. *258,* 6527—6529 (1964).

Seïte, R., D. Picard, and *J. Luciani:* Effets précoces du choc cardiazolique sur les noyaux neurosécréteurs hypothalamiques chez le Chat. In: Progress in Brain Research 5, 171—190. Amsterdam: Elsevier Publ., 1964.

Shimizu, N., and *S. Ishii:* Electron microscopic observation of catecholamine-containing granules in the hypothalamus and area postrema and their changes following reserpine injection. Arch. histol. japon. *24,* 489—497 (1964).

Sloper, J. C.: Hypothalamic neurosecretion in the dog and cat, with particular reference to the identification of neurosecretory material with posterior lobe hormone. J. Anat. *89,* 301—316 (1955).

Sloper, J. C.: Hypothalamo-neurohypophysial neurosecretion. Int. Rev. Cytol. 7, 337—389 (1958).

Sloper, J.C.: Hypothalamic neurosecretion. The validity of the concept of neurosecretion and its physiological and pathological implications. Brit. Med. Bull. *22,* 209—215 (1966).

Sloper, J.C., and *C.W. M. Adams:* The hypothalamic elaboration of posterior pituitary principle in man. Evidence derived from hypophysectomy. J. Path. Bact. *72,* 587—602 (1956).

Smoller, C.G.: Neurosecretory processes extending into third ventricle: secretory or sensory? Science *147,* 882—884 (1965).

Soulairac, A., and *M.L. Soulairac:* Effet de l'acétazolamide sur la faim et la soif du rat. Modifications concomitantes des noyaux hypothalamiques et du cortex surrénal. Ann. endocr., Paris, *23,* 723—731 (1962).

Soulairac, A., and *M.L. Soulairac:* Action de certains inhibiteurs de la L-DOPA/5 HTP décarboxylase sur la neurosécrétion hypothalamo-hypophysaire chez le Rat. C. R. Soc. Biol. *158,* 1445—1447 (1964).

Stahl. A., G. Cotte, and *R. Seïte:* Modifications cytologiques et cytochimiques des cellules neurosécrétoires de l'hypothalamus après perturbation expérimentale du métabolisme hydrique. (C. R. Assoc. Anat. 41e Réun., Gênes 1954.) In: Bull. Assoc. Anat. *85,* 455—464 (1955).

Stutinsky, F.: Sur l'origine de la substance Gömöri-positive du complexe hypothalamo-hypophysaire. C. R. Soc. Biol. *145,* 367—370 (1951 a).

Stutinsky, F.: Sur la substance Gömöri-positive du complexe hypothalamo-hypophysaire du rat. (C. R. Assoc. Anat., 38e Réun., Nancy 1951 b.) In: Bull. Assoc. Anat. *70,* 942—950 (1952).

Stutinsky, F.: La neurosécrétion chez l'Anguille normale et hypophysectomisée. Zschr. Zellforsch. *39,* 276—297 (1953).

Stutinsky, F., and *J.J. Befort:* Effets des stimulations électriques du diencéphale de *Rana esculenta* mâle. Gen. Comp. Endocrinol. *4,* 370—379 (1964).

Stutinsky, F., and *Y. Guerne:* Effets des stimulations électriques hypothalamiques sur la pression artérielle du Rat. C. R. Soc. Biol. *159,* 1420—1422 (1965).

Stutinsky, F., and *A. Porte:* Sur certaines inclusions neuronales chez le Rat blanc. (C. R. Assoc. Anat., 49e Réun., Madrid 1964.) In: Bull. Assoc. Anat. *129,* 1647—1652 (1965).

Suda, I., K. Koizumi, and *C. Mc C. Brooks:* Study of unitary activity in the supraoptic nucleus of the hypothalamus. Jap. J. Physiol. *13,* 374—385 (1963).

Szentágothai, J.: The parvicellular neurosecretory system. In: Progress in Brain Research 5, 135—146. Amsterdam: Elsevier Publ., 1964.

Takabatake, Y., and *H. Sachs:* Vasopressin biosynthesis. III. *In vitro* studies. Endocrinology 75, 934—941 (1964).

Talanti, S.: The effect of thiouracil and excess thyroxine on the hypothalamus of the rat, with special reference to neurosecretory phenomena. Zschr. Zellforsch. *79,* 92—109 (1967).

Taxi, J.: Identification des fibres nerveuses adrénergiques par la méthode radioautographique dans quelques muscles lisses de Mammifères. (C. R. Assoc. Anat., 52e Réun. Paris-Orsay 1967.) In: Bull. Assoc. Anat. *139,* 1132—1139 (1968).

Thoenen, H., J.P. Tranzer, A. Hürlimann, and *W. Haefely:* Untersuchungen zur Frage eines cholinergischen Gliedes in der postganglionären sympathischen Transmission. Helv. Physiol. Acta *24,* 229—246 (1966).

Tranzer, J.P., und *H. Thoenen:* Elektronenmikroskopische Untersuchungen am peripheren sympathischen Nervensystem der Katze: physiologische und phar-

D. Picard:

makologische Aspekte. (30. Tag. Dtsch. Pharmak. Gesellsch., Kiel 1966.) In: Naunyn-Schmiedebergs Arch. exper. Path. *257*, 73 (1967 a).

Tranzer, J. .P, and *H. Thoenen:* Significance of "empty vesicles" in post-ganglionic sympathetic nerve terminals. Experientia, Basel, *23*, 123—124 (1967 b).

Uemura, H.: Effects of water deprivation on the hypothalamo-hypophysial neurosecretory system of the Grass Parakeet *(Melopsittacus undulatus),* Gen. Comp. Endocrin. *4*, 193—198 (1964).

Uemura, H., and *H. Kobayashi:* Effects of prolonged daily photoperiods and estrogen on the hypothalamic neurosecretory system of the passerine bird, *Zosterops palpebrosa japonica.* Gen. Comp. Endocrin. *3*, 253—264 (1963).

Van Dyke, H. B., K. Adamsons, and *S. L. Engel:* in: *Heller, H.:* The neuro-hypophysis. London: Butterworths 1957.

Verney, E. B.: The antidiuretic hormone and the factors which determine its release. Proc. Roy. Soc., London, Ser. B, *135*, 25—106 (1947).

Vogt, M.: Vasopressor, antidiuretic and oxytocic activities of extracts of the dog's hypothalamus. Brit. J. Pharmacol. exp. Chemother. *8*, 193—196 (1953).

Vollrath, L.: Über die neurosekretorische Innervation der Adenohypophyse von Teleostiern, insbesondere von *Hippocampus cuda* und *Tinca tinca.* Zschr. Zellforsch. *78*, 234—260 (1967).

Weiss, P.: Damming of axoplasm in constricted nerve: a sign of perpetual growth in nerve fibers. Anat. Rec. *88*, 464 (1944).

Weiss, P., and *H. B. Hiscoe:* Experiments on the mechanism of nerve growth. J. Exper. Zool. *107*, 315—395 (1948).

Welsh, J. H.: Neuroendocrine substances. In: Comparative Endocrinology, 121—133. New York: J. Wiley, 1959.

Wislocki, G. B., and *E. W. Dempsey:* The chemical cytology of the choroid plexus and blood brain barrier of the rhesus monkey *(Macaca mulatta).* J. Comp. Neurol. *88*, 319—345 (1948).

Wolfson, A., and *H. Kobayashi:* Phosphatase activity and neurosecretion in the hypothalamo-hypophyseal system in relation to the photoperiodic gonadal response in *Zonotrichia albicollis.* Gen. Comp. Endocrin. *suppl. 1*, 168—179 (1962).

Wurster, D. H., and *K. Benirschke:* Development of the hypothalamo-hypophyseal neurosecretory system in the fetal Armadillo *(Dasypus novemcinctus),* with notes on rabbit, cat and dog. Gen. Comp. Endocr. *4*, 433—441 (1964).

Yagi, K., and *H. A. Bern:* Evidence for two kinds of neurosecretory fibers in the caudal neurosecretory system of *Tilapia mossambica.* Amer. Zool. *3*, 508—509 (1963).

Yagi, K., and *H. A. Bern:* Electrophysiologic analysis of the response of the caudal neurosecretory system of *Tilapia mossambica* to osmotic manipulations. Gen. Comp. Endocr. *5*, 509—526 (1965).

Zambrano, D., and *E. de Robertis:* The secretory cycle of supraoptic neurons in the rat. A structural-functional correlation. Zschr. Zellforsch. *73*, 414—431 (1966).

Zambrano, D., and *E. de Robertis:* Ultrastructural aspects of the inhibition of neurosecretion by puromycin. Zschr. Zellforsch. *76*, 458—470 (1967).

Zambrano, D., and *J. Mordoh:* Neurosecretory activity in supraoptic nucleus of normal rats. Zschr. Zellforsch. *73*, 405—413 (1966).

Author's address: Prof. *D. Picard*, Laboratoire d'Histologie I, Faculté de Médecine, Boulevard Jean-Moulin, 13-Marseille (5e), France.

Discussion

Issidorides: Are the two methods for demonstration of neurosecretory substance, aldehyde fuchsin and Gomori's chromalum haematoxylin method, interchangable or do they stain different granules? In human nucleus supraopticus cells aldehyde fuchsin demonstrates granules in the perinuclear central area of the cells while Gomori's chromalum haematoxylin stains granules in the periphery of the cell bodies.

Picard: It may be that aldehyde fuchsin and chromalum haematoxylin, which are known to stain other components than neurosecretory material, *e. g.* lysosomes, stain these components to a different degree. It is likely that aldehyde fuchsin strongly stains lysosomes, which have a tendency to be gathered in the Golgi area.

Bargmann: It is my opinion, too, that aldehyde fuchsin stains lysosomes more intensively than chromalum does.

Journal of Neuro-Visceral Relations, Suppl. IX, 64—77 (1969)

Das neurosekretorische Zwischenhirn-Hypophysensystem und seine synaptischen Verknüpfungen*

W. Bargmann**

Anatomisches Institut der Universität Kiel, Deutschland

Mit 6 Abbildungen

Summary

The secretory diencephalo-hypophyseal system is constituted by peptidergic and aminergic nerve cells of the hypothalamus. The terminals of their axons reach the walls of the capillaries in the median eminence and in the neurohypophysis. Part of the peptidergic and aminergic fibers end by means of synapses on the epithelial cells of the pars intermedia and on the pituicytes of the posterior hypophyseal lobe. Small aminergic nerve fibers of diencephalic origin penetrate into the interstitium of the pars tuberalis of the hypophysis but are not involved in the formation of synapses. Evidently, these are perivascular nerve fibers. In the remaining parts of the mammalian adenohypophysis we have not been able to demonstrate synaptic endings ultramicroscopically. However, such negative findings should not be generalized.

Furthermore, the question has been dealt with whether the secretory nerve cells connecting the hypothalamus and the hypophysis are innervated. It is pointed out that the pericarya and axons, and possibly also the dendrites, of neurosecretory cells in diencephalic nuclei are covered by synapses. It is still not clear in which nuclei the non-peptidergic fibers originate which are synaptically connected with the peptidergic nerve cells. Concerning this matter reference is made to the investigations by *Rothballer* (1966).

Bevor erörtert wird, durch welche synaptischen Verbindungen das neurosekretorische Zwischenhirn-Hypophysensystem einerseits mit anderen Neuronensystemen, anderseits mit nichtneuronalen Elementen verknüpft ist, muß definiert werden, was als neurosekretorisches Zwischenhirnsystem verstanden werden soll. Dabei wird die Frage auftauchen, ob die bisherige Definition dieses Systems aufrecht erhalten werden kann.

Nach der bisher üblichen Auffassung besteht das neurosekretorische Zwischenhirn-Hypophysensystem aus Neuronen, die elektiv angefärbt werden können, in deren Perikaryen hormonal aktive Oktapeptide produzier·

---

* Die Untersuchungen wurden mit dankenswerter Hilfe der Deutschen Forschungsgemeinschaft durchgeführt.

** Herrn Prof. Dr. med. *Kurt Goerttler*, Freiburg i. Br., zum 70. Geburtstag gewidmet.

und in deren Axonen sie — an eine „carrier substance" gebunden — zur Neurohypophyse transportiert werden. Hier, im Stapelorgan, werden die Hormone bei Bedarf von ihrer Trägersubstanz abgetrennt und an den Kreislauf abgegeben. Elektronenmikroskopische Untersuchungen haben gezeigt, daß das lichtmikroskopisch-färberisch darstellbare Neurosekret aus Elementargranula besteht, deren Durchmesser je nach Spezies zwischen 800 und 3000 Å schwankt. Bei dem „release" der Wirkstoffe aus den Nervenendigungen innerhalb des Hinterlappens der Hypophyse verwandeln sich die Elementargranula in elektronenoptisch leere Bläschen, d. h. der massendichte Inhalt der Körnchen schwindet bei der Hormonabgabe. Die sekretorischen Zwischenhirnneurone gehören bei niederen Wirbeltieren dem Nucleus praeopticus, bei den Sauropsiden und Mammaliern den Nuclei supraopticus und paraventricularis an. *Bargmann, Lindner* und *Andres* (1967) haben im Anschluß an Überlegungen von *Knowles* (1965) vorgeschlagen, diese Neurone als *peptiderge Neurone* zu bezeichnen.

In diese Kategorie gehören vermutlich auch jene hypothalamischen Nervenzellen, die „releasing and inhibitory factors" produzieren, also Oktapeptide, welche die Adenohypophyse auf dem Wege des hypophysären Pfortaderkreislaufes erreichen und sie zur Ausschüttung von Vorderlappenhormonen veranlassen bzw. diese Ausschüttung blockieren. Da neurosekretorische Fasern, die sich lichtmikroskopisch nicht von den zum Hinterlappen ziehenden Axonen unterscheiden, an den Spezialgefäßen des Infundibulum enden, ist es denkbar, daß die klassischen neurosekretorischen Kerne im Hypothalamus auch Zellen einschließen, die „releasing" oder „inhibitory factors" bilden, ohne sich morphologisch-färberisch von den Produzenten des Vasopressins und Oxytocins bzw. ihrer Äquivalente abgrenzen zu lassen. Denkbar ist aber auch, daß Neurone anderer Kerngebiete als die oben genannten die hypothalamischen „factors" hervorbringen, bisher jedoch durch elektive Färbeverfahren nicht dargestellt werden können. Dennoch besteht meines Erachtens kein Grund, derartige peptidergen Neurone nicht in die Gruppe der neurosekretorischen Systeme einzubeziehen.

Schwieriger als die Frage der Zusammenfassung von Nervenzellen, die teils Hinterlappenhormone, teils „releasing and inhibitory factors" bilden, zur Gruppe der peptidergen, d. h. neurosekretorischen Elemente, erscheint die Frage, ob jene die Eminentia mediana und den Hinterlappen, d. h. die Neurohypophyse im weiteren Sinne erreichenden Neurone gleichfalls als neurosekretorische Zellen bezeichnet werden können, die keine peptidergen Nervenzellen sind. Ich meine damit jene Elemente, die sich zwar nicht durch Spezialfärbungen wie die Färbung mit Chromalaunhämatoxylin, Aldehydfuchsin, Alcianblau, Pseudoisocyanin usw., wohl aber fluorescenzmikroskopisch sichtbar machen lassen und *aminerge Neurone* verkörpern (*Dahlström* und *Fuxe*, 1964). Endigungen von dopaminhaltigen Neuronen bilden an dem primären Kapillarplexus in der Eminentia mediana und im Infundibularstamm einiger Säuger dichte Geflechte (*Fuxe*, 1964). Als das elektronenmikroskopische Äquivalent der entsprechenden Nervenendigungen sind boutons mit Vesikeln anzusehen, die einen massendichten Inhalt aufweisen und einen Durchmesser von rund 500 Å besitzen (vgl. *Hökfelt*, 1967, Lit.). Da sich diese Granulärvesikel im elektronenmikroskopischen Bild von den neurosekretorischen Elementargranula — jedenfalls soweit bisher bekannt — lediglich durch ihre geringere Größe unterscheiden, da ferner eine Reihe von Beobachtungen dafür spricht, daß diese Partikel in den Perikaryen der aminergen Neurone entstehen und durch einen axonalen Strom in die Nervenendigungen transportiert werden (*Dahlström*, 1966; *Norberg*, 1967), wie er für das klassische neurosekretorische System beschrieben wurde, sehe ich kein zytologisches Argument, das gegen die Bezeichnung der aminergen Neurone als neurosekretorische Neurone spricht. Umso weniger, als eine Reihe dieser Zellen mit ihren Axonen die Kapillaren des Portalsystems erreicht und sich somit wie die peptidergen, d. h. neurosekretorischen Neurone im herkömmlichen Sinne verhält. Auch dann, wenn nicht die „small dense core granules", sondern „small clear vesicles" die Träger von Monoaminen sein sollten (vgl. *B. Scharrer*, 1967), scheint mir kein zwingender Einwand gegen diese Auffassung zu bestehen.

Da es wenig sinnvoll sein dürfte, peptiderge und aminerge Neurone als neurosekretorische Elemente den cholinergen Neuronen als nichtsekretorischen gegenüberzustellen, muß man sich fragen, ob die Bezeichnung „neurosekretorisch" in näherer oder weiterer Zukunft durch eine biochemisch begründete Charakterisierung der jeweils in Rede stehenden Neurone ersetzt werden muß, die preliminar als sekretorisch gekennzeichnet wurden. Damit nähern wir uns der Ansicht von *De Robertis* (1964), „synaptic and neuroeffector processes" seien „manifestations of a more general neurosecretory function".

Wie dem auch sei, unter dem neurosekretorischen Zwischenhirn-Hypophysensystem möge im folgenden 1. das *peptiderge* hypothalamisch-hypophysäre System, und 2. ein *aminerges* System verstanden werden, das den Hypothalamus mit der Neurohypophyse und dem Portalkreislauf verbindet. Es fragt sich nunmehr, ob die komplex aufgebaute hypothalamisch-hypophysäre Bahn a) mit synaptischen Endigungen ausgestattet ist und b) ob sie selbst durch Synapsen oder ihnen vergleichbare, sog. synaptoide Kontaktstrukturen, mit anderen Neurosenensystemen verknüpft ist.

Für das sekretorische Neuron galt lange Zeit der Satz, es handle sich um „a nervous cell, containing granules, having an axon which does not present any synaptic contact either with other neurones or with other target organs, and secreting directly into the bloodstream" (*Knowles* und *Bern*, 1966; siehe auch *Bern*, 1962, 1963, 1966). Schon frühere lichtmikroskopische Beobachtungen ließen Zweifel daran aufkommen, ob diese Definition der Wirklichkeit voll entspreche, soweit es sich um die Beziehungen der neurosekretorischen Zellen zu Erfolgsorganen („target organs") handelt. In Schnittpräparaten, die mit Chromalaunhämatoxylin-Phloxin gefärbt wurden, ließen sich nämlich feinste, mit Neurosekret beladene Perlschnurfasern inmitten des Gewebes der *Pars intermedia* der Adenohypophyse von Säugetieren und Fischen nachweisen (*Bargmann*, 1949, 1953, 1965, 1966; *Bargmann* und *Knoop*, 1960; *E. Scharrer*, 1952). Die Abbildung 1 gibt das lichtmikroskopische Bild neurosekretorischer Endigungen in der Pars intermedia eines Primaten (*Cebus* spec.) wieder. Die Auffassung, es handle sich hier möglicherweise um aberrierende Fasern, die im Verlauf von Regenerationsvorgängen in die Pars intermedia eindringen, ist unter dem Eindruck neuerer elektronenmikroskopischer Beobachtungen nicht mehr zu vertreten. Unter den verschiedenartigen Nervenendigungen, die mit der Oberfläche der *Epithelzellen* in der Pars intermedia in enge Kontaktbeziehungen treten, befinden sich nämlich solche, die neurosekretorische Elementargranula ent-

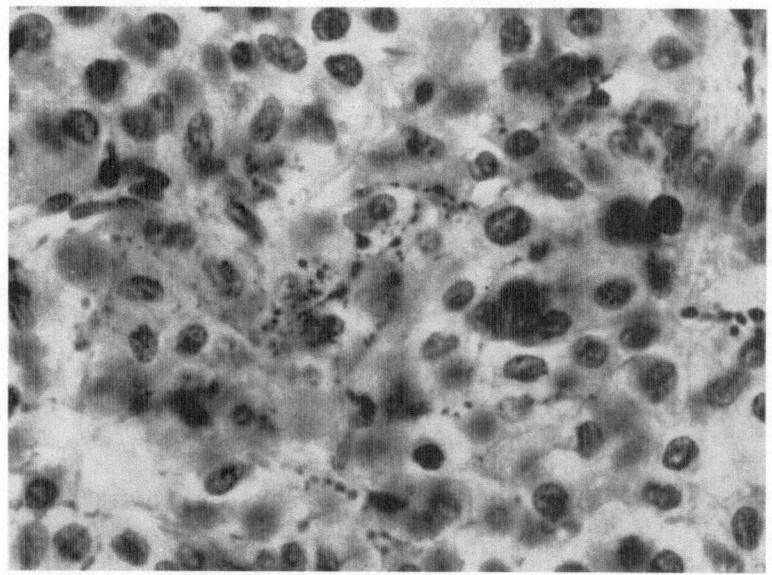

Abb. 1. Pars intermedia der Hypophyse von *Cebus* spec. Färbung mit Chromalaunhämatoxylin-Phloxin. Beachtet die zarten, mit Neurosekret gefüllten Perlschnurfasern zwischen den Epithelzellen. × 580.

halten, wie u. a. durch *Legait* und *Legait* (1957, 1958), *Ziegler* (1963), *Knowles* und *Vollrath* (1966), *Bargmann, Lindner* und *Andres* (1967) sowie *Dent* und *Gupta* (1967) an verschiedenen Spezies gezeigt wurde. *Kobayashi* (1956) spricht zu Recht von „neuroglandulären Synapsen". Wie *Knowles* und *Vollrath* (1966) an der Hypophyse von Fischen feststellen konnten, bestehen ferner echte synaptische Kontakte zwischen *Pituizyten* und neurosekretorischen Fasern. Neuerdings berichtet *Wittkowski* (1967 b) über synaptische Endigungen von Fasern mit kleinen oder großen Elementargranula an den Pituizyten des Meerschweinchens.

Aminerge Fasern kommen offenbar reichlicher als peptiderge Nervenfasern zwischen den Epithelzellen der Pars intermedia vor. *Dahlström* und *Fuxe* (1966) finden die Intermediazellen der Ratte von einem reich entwickelten Plexus umgeben, der auf Grund fluoreszenzmikroskopischer Untersuchungen als katecholaminhaltig anzusehen ist. Das elektronenmikroskopische Bild steht mit dieser Beobachtung im Einklang. Es enthüllt typische Synapsen an den Epithelzellen, die mit kleinen Granula vom Typus der Katecholaminkörnchen und mit synaptic vesicles ausgestattet sind (Abb. 2), wie aus den Untersuchungen von *Bargmann, Lindner* und *Andres* (1967) an der Katze sowie von *Wittkowski* (1967 a) am Meer-

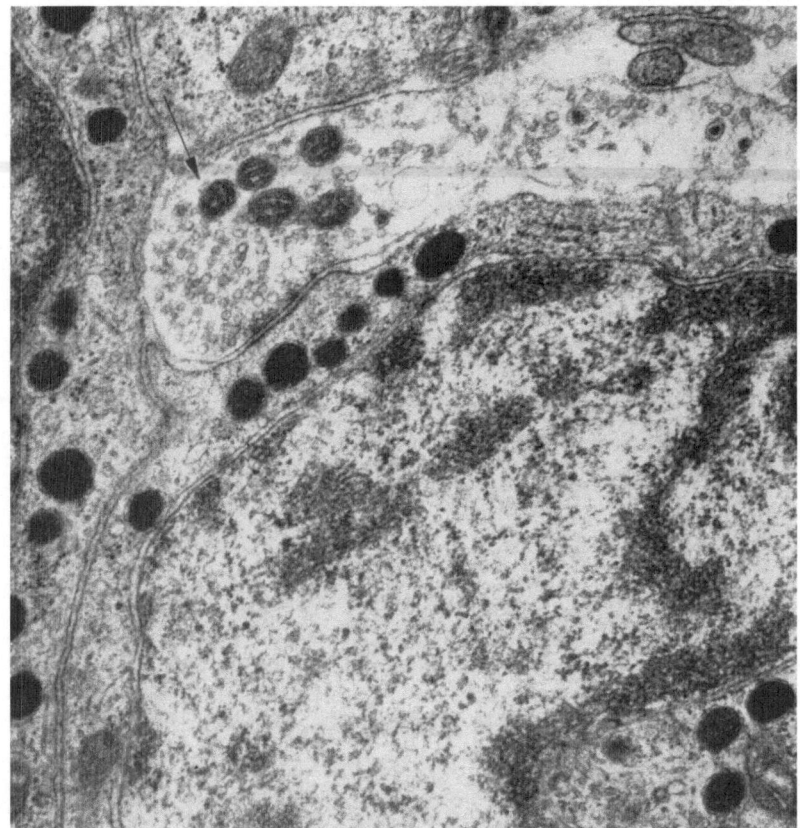

Abb. 2. Synapse einer vermutlich aminergen Nervenfaser (Pfeil) an der Oberfläche einer Epithelzelle der Pars intermedia der Hundehypophyse. Elektronenmikroskopische Aufnahme von Doz. Dr. *K. H. Andres*, × 25.000.

schweinchen hervorgeht. Ob die von *Bargmann* und Mitarbeitern sowie von *Wittkowski* beschriebenen Synapsen, die nur „synaptic vesicles" enthalten, ausschließlich *cholinerge Endigungen* sind oder ob es sich um das Äquivalent von Funktionsstadien aminerger oder peptiderger Endigungen handelt, sei dahingestellt.

Diese Hinweise mögen genügen, um klarzumachen, daß die konventionelle Definition neurosekretorischer Neurone insofern nicht mehr gelten kann, als gesagt wurde, es handle sich um sezernierende Nervenzellen, die keine synaptischen Kontakte mit Erfolgsorganen eingehen, sondern lediglich neurokapilläre Kontakte bilden. Welche funktionelle Bedeutung diese Synapsen besitzen, wissen wir noch nicht. Zunächst wird man experimentell untersuchen müssen, ob und wie neurohypophysäre Oktapeptide einerseits, Monoamine anderseits die Zellen der Pars intermedia beeinflussen. *Smoller* (1966) hält es für möglich, daß die neurosekretorischen Fasern, die bei *Hyla regilla* in die Pars intermedia einwandern, die Sekretion von Intermedin regulieren. Weiter muß man sich mit der Frage befassen, ob der Mechanismus der Wirkung neurosekretorischer Nervenendigungen demjenigen zu vergleichen ist, der von den klassischen Transmitter-Substanzen auf „target neurons" ausgeht.

Die Tatsache, daß die synaptische Verbindung diencephaler sekretorischer, d. h. peptiderger und aminerger Neurone, mit der Pars intermedia der Hypophyse bisher in den Vordergrund der Betrachtung gestellt wurde, darf nicht zur Vernachlässigung der Frage führen, ob hypothalamische Neuronensysteme auch zu anderen Abschnitten der Adenohypophyse in Beziehung stehen.

Eine Reihe von lichtmikroskopischen Untersuchungen von *Stutinsky* (1948), *Oberti* (1957), *Metuzals* (1954, 1955, 1959) u. a., über die *Diepen* (1962) zusammenfassend berichtet, sagt aus, daß in die *Pars infundibularis* (tuberalis) der Hypophyse hypothalamische Nervenfasern eindringen; sie sollen sich von dort in den Vorderlappen fortsetzen. *Diepen* meint, es könne sich um die Fehlinterpretation von Glia- bzw. Ependymfasern handeln, welche die Basis des Hypothalamus verlassen und dem perivasculären Bindegewebe im Bereich der Portalgefäße zustreben. Die Entscheidung über die Natur der sehr zarten Faserstrukturen kann durch die elektronenmikroskopische Untersuchung erbracht werden. Nach jüngsten Beobachtungen von *Wittkowski* (1967 b) kommen zwischen den Drüsenzellen der Pars infundibularis der Adenohypophyse des Meerschweinchens in der Tat Axone vor, die teils als vegetative Faserbündel ohne Granula, teils als Nervenfasern mit Granula beschrieben werden, deren Durchmesser 800 Å beträgt. Diese Körnchen entsprechen somit den Granulärvesikeln, die für die Axone des Nucleus infundibularis tuberalis bezeichnend sind.

Nach gemeinsamen Beobachtungen von *K. H. Andres* und *W. Bargmann* liegen beim Hund folgende Verhältnisse vor. Sehr dünne markhaltige Nervenfäserchen durchbrechen die Oberfläche des Tuber cinereum in lockeren Bündelchen und verlaufen in den Bindegewebsräumen zwischen den Gefäßen, die zur Pars tuberalis ziehen (Abb. 3). Zum Teil handelt es sich um Fasern ohne Elementargranula, zum kleineren Teil um solche, die Körnchen vom Typus der Katecholamin-Granula führen. Synapsen an Tuberaliszellen wurden jedoch nicht festgestellt. Diese Befunde stimmen mit der Mitteilung von *Dahlström* und *Fuxe* (1966) überein, die katecholaminhaltige Fäserchen in der Pars tuberalis der Hypophyse nur in Begleitung von Blutgefäßen fanden. Die Frage, ob es sich, wie *Stutinsky* (1948) meint, um vasomotorische und sensitive Fasern zugleich handelt, bedarf weiterer Prüfung. Synapsen bzw. Faserendigungen im *Vorderlappen* von Säugern haben wir im Gegensatz zu den Angaben von *Vazquez-Lopez* (1949) über

„knob-like nerve endings in the pars distalis" elektronenmikroskopisch bisher nicht nachweisen können. Aus Untersuchungen von *Vollrath* (1967) geht jedoch hervor, daß es bei Teleostiern neurosekretorische synaptische Kontakte an Vorderlappenzellen gibt. Frühere lichtmikroskopische Angaben über eine reiche sekretorische Innervation der Pars distalis der Vogelhypophyse (*Metuzals*, 1955) sollten elektronenmikroskopisch nachgeprüft werden.

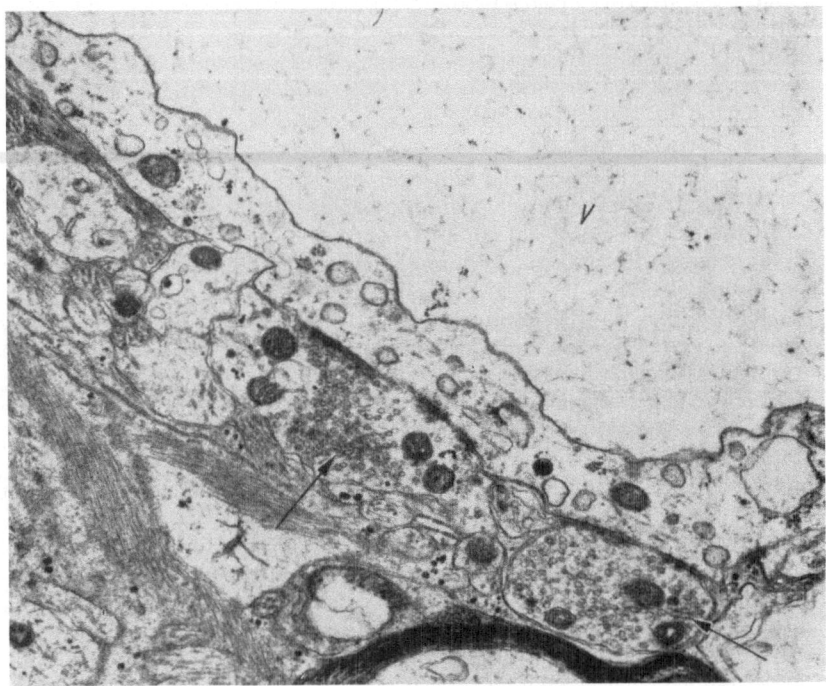

Abb. 3. Synapsen (Pfeile) an der Oberfläche einer vakuolisierten Nervenzelle des Nucleus supraopticus eines älteren Hundes. V=Vakuole. Elektronenmikroskopische Aufnahme von Doz. Dr. *K. H. Andres*, × 22.000.

Zusammenfassend können wir also feststellen:

1. Peptiderge und aminerge Nervenzellen des Hypothalamus bilden das sekretorische Zwischenhirn-Hypophysensystem. Beide erreichen mit ihren Endigungen die Wandung der Blutkapillaren in der Eminentia mediana und in der Neurohypophyse.

2. Ein Teil der peptidergen und aminergen Neurone endet mit Synapsen an den Epithelzellen der Pars intermedia bzw. an den Pituizyten.

3. Diencephale aminerge Fäserchen dringen in das Interstitium der Pars tuberalis ein, ohne hier Synapsen zu bilden; es handelt sich offenbar um perivasculäre Fasern.

4. In den übrigen Teilen der Adenohypophyse von Säugern wurden synaptische Endigungen elektronenmikroskopisch von uns nicht festgestellt.

Die Beobachtungen von *Vollrath* (1967) an Teleostiern zeigen jedoch, daß derartige negative Befunde nicht verallgemeinert werden dürfen.

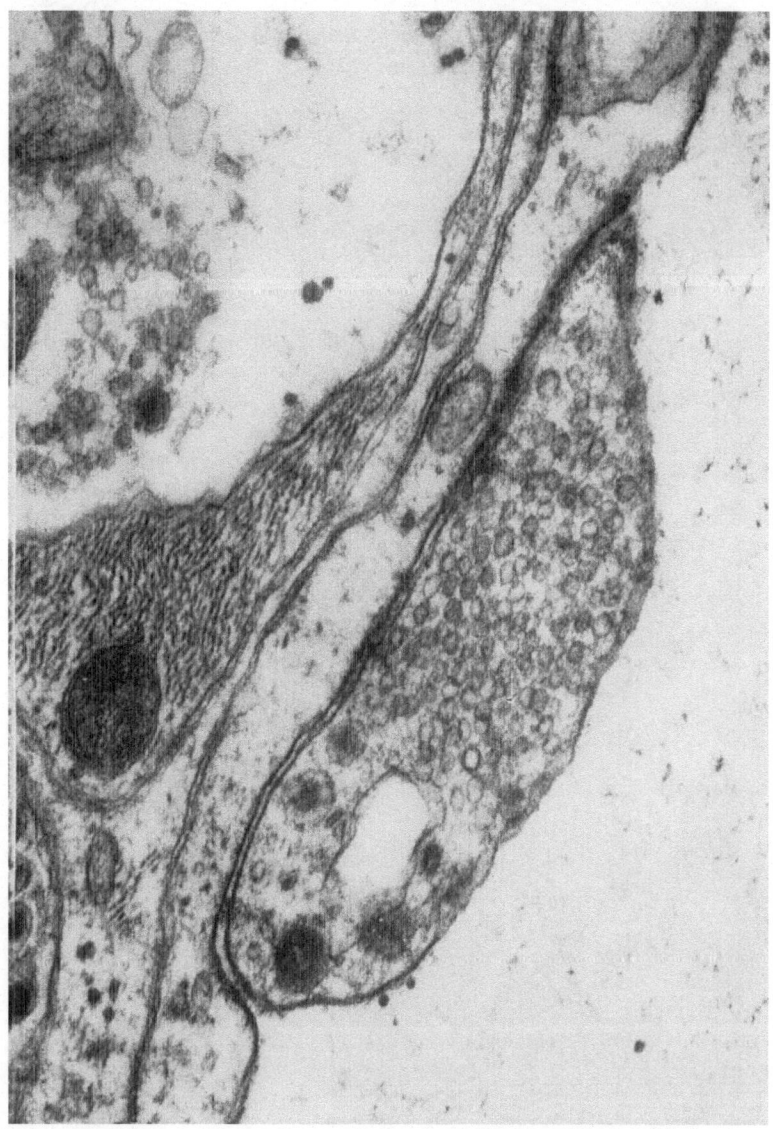

Abb. 4. Synapse im Inneren der Vakuole einer Nervenzelle des Nucleus supraopticus eines älteren Hundes. Elektronenmikroskopische Aufnahme von Doz. Dr. *K. H. Andres,* × 50.000.

Wir wenden uns nun der Frage zu, ob die sekretorischen Nervenzellen, welche den Hypothalamus mit der Hypophyse verbinden, innerviert

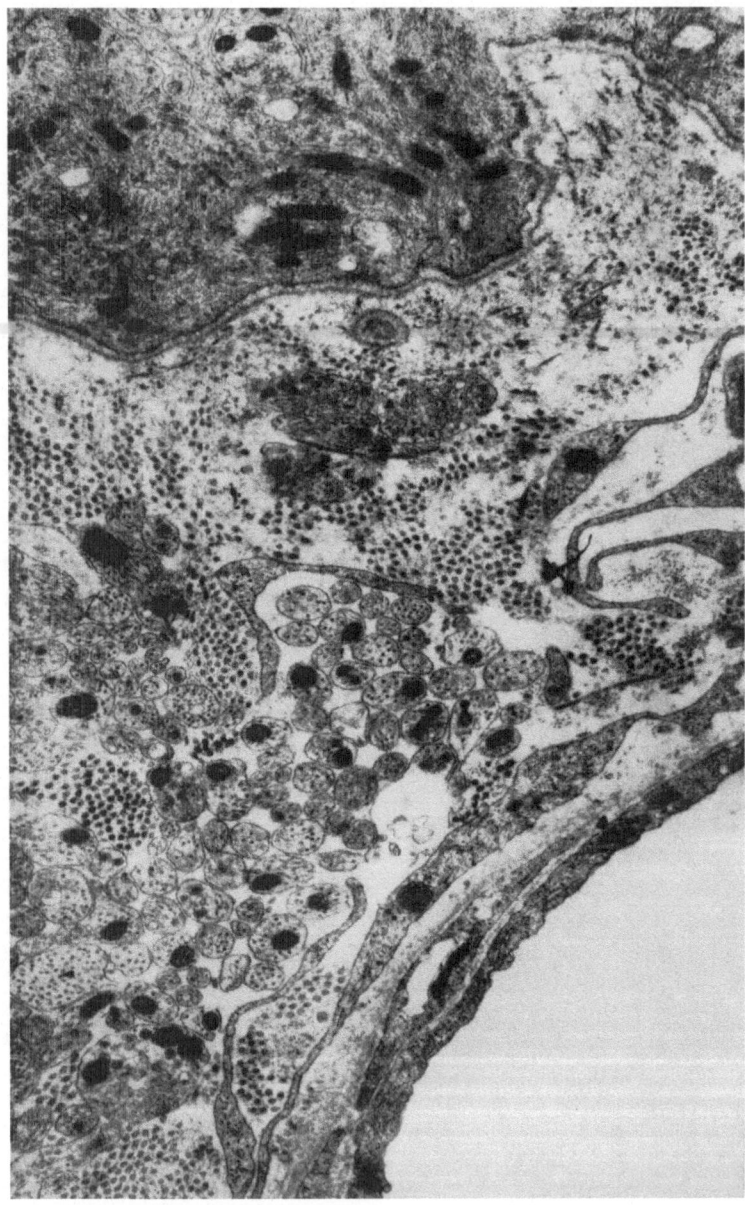

Abb. 5. Zahlreiche marklose Axone zwischen der Basis des Tuber cinereum und den hypophysären Portalgefäßen eines jungen Hundes. Lichtung eines Gefäßes rechts unten, Basis des Tuber cinereum oben. Elektronenmikroskopische Aufnahme von Doz. Dr. *K. H. Andres*, × 18.000.

werden. Zunächst ist auf die bereits bekannte Tatsache hinzuweisen, daß an den *Perikaryen* der Zellen des Nucleus supraopticus und paraventricularis von Säugern feinste Nervenendigungen fluoreszenzmikroskopisch nachgewiesen wurden (*Falck*, 1964). Elektronenmikroskopisch wurden axosomatische Synapsen mit „dense core" granula an Zellen des Nucleus supraopticus von Katze und Ratte dargestellt. Wie aus neueren Untersuchungen von *Andres* (im Druck) hervorgeht, zeichnen sich die mit großen Blasen ausgestatteten Ganglienzellen des Nucleus supraopticus des Hundes, die schon *Verney* (1947) beschrieb, durch eine besondere Art der Innervation ihres Soma aus. Die Abbildung 4 zeigt an der Außenfläche der Perikaryen ansetzende Synapsen, die Abbildung 5 eine Synapse vom aminergen Typus im Inneren der Flüssigkeitsblase, die das Perikaryon weitgehend erfüllt. Möglicherweise handelt es sich um eine abnorme Topik dieser Endformation, die auf die degenerativen Vorgänge zurückzuführen sein dürfte, denen die blasigen Ganglienzellen nach unserer Auffassung ihre Entstehung verdanken. Vergleichbare Befunde erhob *Unsicker* (1967) an sekretorischen Nervenzellen des Nebennierenmarkes.

Außer axosomatischen gibt es *axo-axonale Synapsen* an peptidergen Neuronen des Zwischenhirn-Hypophysensystems. *Stutinsky* (1967) hat auf dem Straßburger Symposion eine derartige Synapse an einem Herringkörper demonstriert, *Bargmann* und *Lindner* (*Bargmann*, 1966) zeigen eine Synapse vom cholinergen Typus an einer neurosekrethaltigen Faserverdickung in der Neurohypophyse der Ratte. Die funktionelle Bedeutung dieser Kontakte kennen wir nicht. *Oota* und *Kobayashi* (1963) fanden im Hilus der Neurohypophyse eines Anuren (*Rana catesbeiana*) häufig Synapsen zwischen neurosekretorischen Fasern mit entsprechenden Elementargranula und Fasern ohne Elementargranula. Die Autoren halten es für möglich, daß es sich hier um eine Verbindung zwischen neurosekretorischen Axonen und den Dendriten nichtsekretorischer Neurone handelt, doch bedarf es weiterer Bemühungen, das Vorkommen auch *axo-dendritischer Synapsen* im neurosekretorischen Zwischenhirnsystem von Säugern zu sichern. Im Diencephalon des Teleostiers *Carassius auratus* haben *Baumgarten* und *Braak* (1967) kräftig fluoreszierende katecholaminhaltige Fasern festgestellt, deren Endigungen mit den Dendriten des Nucleus praeopticus in Verbindung treten sollen.

Es ist im einzelnen unklar, welchen Kerngebieten die nicht-peptidergen Neurone angehören, die mit den Perikaryen der peptidergen Zellen und ihren Fortsätzen in synaptische Verbindungen treten, wenn man von dem Hinweis von *Baumgarten* und *Braak* (1967) absieht, wonach es sich bei *Carassius auratus* um den Nucleus recessus posterioris und lateralis handelt. Die Aufdeckung der Zusammenhänge zwischen anderen Kerngebieten, aber auch corticalen Arealen und den neurosekretorischen Zwischenhirnneuronen, würde uns einem Verständnis der Kontrolle der neurohormonal aktiven Systeme näherbringen, die Vasopressin und Oxytocin beziehungsweise bei niederen Wirbeltieren analoge Oktapeptide produzieren, transportieren und abgeben. In diesem Zusammenhang sind die Resultate neurophysiologischer Experimente von *Rothballer* (1966) an der Ratte von besonderem Interesse. Ihm gelang der Nachweis, daß die Formatio reticularis, der Lemniscus medialis und das limbische System zu den afferenten Kontrollsystemen (vgl. hiezu Abb. 6) der sekretorischen Zwischenhirnneurone und damit des Hinterlappens gehören, da sich durch ihre Stimulation die

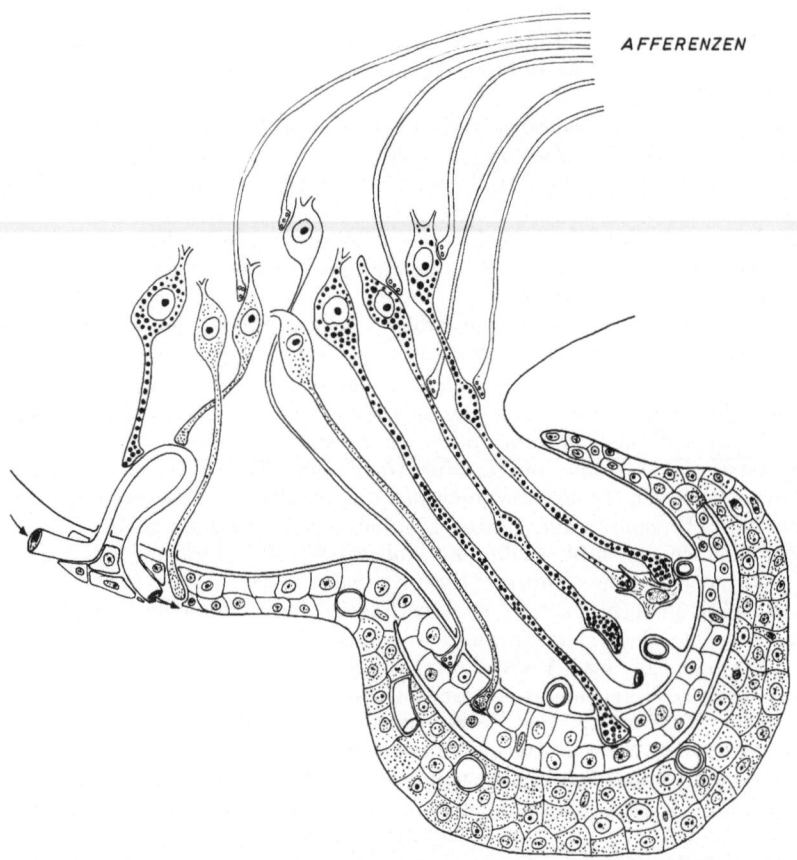

Abb. 6. Synaptische Verbindung der Hypophyse eines Säugers mit verschiedenartigen hypothalamischen Neuronen und deren Afferenzen, Schema. Grobe Punktierung: Peptiderge Neurone; zarte Punktierung: Aminerge Neurone; ohne Punktierung: Mutmaßlich cholinerge Neurone. Im Hinterlappen ein innervierter Pituizyt.

Abgabe von neurohypophysären Hormonen hervorrufen läßt. Die bereits erwähnten axo-somatischen Synapsen am Nucleus supraopticus der Ratte und der Katze sind möglicherweise zum Teil die Endigungen des medialen Vorderhirnbündels, das *Rothballer* in einem Schema wiedergibt. Da es sich bei diesem Bündel wie bei zahlreichen anderen Afferenzen der sekretorischen Kerne im Diencephalon um dünne marklose oder markarme Fasern handelt, wird es voraussichtlich schwierig sein, ein zuverlässiges Bild der

Zusammenhänge der neurosekretorischen Zentren und ihrer Axone und Dendriten mit den cholinergen und aminergen Systemen zu gewinnen, die auf sie einwirken, sei es excitatorisch, sei es inhibitorisch (vgl. *Shute* und *Lewis*, 1966). Daß die Analyse dieser Verknüpfungen des hypothalamisch-hypophysären Komplexes, d. h. einer Endstrecke für viele zentralnervöse Systeme, auch für das Begreifen krankhafter Prozesse Bedeutung erlangen kann, hat *Ernst Scharrer* (1966) in einem weitgespannten Überblick dargelegt.

Literatur

Bargmann, W.: Über die neurosekretorische Verknüpfung von Hypothalamus und Neurohypophyse. Z. Zellf. *34*, 610—634 (1949).

Bargmann, W.: Über das Zwischenhirn-Hypophysensystem von Fischen. Z. Zellf. *38*, 275—298 (1953).

Bargmann, W.: Über Synapsen im endokrinen System. Nova Acta Leopoldina N. F. *30*, 199—206 (1965).

Bargmann, W.: Neurosecretion. Internat. Rev. Cytol. *19*, 183—201 (1966).

Bargmann, W., und *A. Knoop*: Über die morphologischen Beziehungen des neurosekretorischen Zwischenhirnsystems zum Zwischenlappen der Hypophyse (licht- und elektronenmikroskopische Untersuchungen). Z. Zellf. *52*, 256—277 (1960).

Bargmann, W., E. Lindner, und *K. H. Andres*: Über Synapsen an endokrinen Epithelzellen und die Definition sekretorischer Neurone. Untersuchungen am Zwischenlappen der Katzenhypophyse. Z. Zellf. *77*, 282—298 (1967).

Baumgarten, H. G., und *H. Braak*: Catecholamine im Hypothalamus vom Goldfisch (Carassius auratus). Z. Zellf. *80*, 246—263 (1967).

Bern, Howard A.: The Properties of Neurosecretory Cells. Gen. and compar. Endocrinol. Suppl. *1*, 117—132 (1962).

Bern, Howard A.: The secretory neuron as a doubly specialized cell. In: General physiology of cell specialization, 350—366. D. Mazia and A. Tylor, eds, London, McGran-Hill, 1963.

Bern, Howard A.: On the production of hormones by neurons and the rôle of neurosecretion in neuroendocrine mechanism. Symp. Soc. exp. Biol. *20*, 325—344 (1966).

Dahlström, A.: The intraneuronal distribution of noradrenaline and the transport and life-span of amine storage granules in the sympathetic adrenergic neuron. A histochemical and biochemical study. Stockholm, 1966.

Dahlström, A., und *Kj. Fuxe*: Evidence for the existence of monoamine-containing neurons in the central nervous system. Acta Physiol. Scand. *62*, Suppl. 232, 1—55 (1964).

Dahlström, A., und *Kj. Fuxe*: Monoamines and the pituitary gland. Acta Endocrinol. *51*, 301—314 (1966).

Dent, I. N., und *B. L. Gupta*: Ultrastructural observations on the developmental cytology of the pituitary gland in the spotted newt. Gen. Compar. Endocrinol. *8*, 273—288 (1967).

De Robertis, E.: Histophysiology of synapses and neurosecretion. Oxford-London-New York, Pergamon-Press, 1964.

Diepen, R.: Der Hypothalamus. In: Handb. der mikrosk. Anat. des Menschen, herausgegeben von W. *Bargmann*, IV/7, Berlin-Göttingen-Heidelberg, Springer-Verlag, 1962.

Falck, B.: Cellular localization of monoamines. Progr. in Brain Res. *8*, 28—44 (1964).

Fuxe, K.: Cellular localization of monoamines in the median eminence and the infundibular stem of some mammals. Z. Zellf. *61*, 710—724 (1964).

Hökfelt, T.: On the ultrastructural localization of noradrenaline in the central nervous system of the rat. Z. Zellf. *79*, 110—117 (1967).

Knowles, F.: Neuroendocrine correlations at the level of ultrastructure. Arch. Anat. Anat. mier. Morph. exper. *54*, 343—358 (1965).

Knowles, F., und *H. A. Bern:* The function of neurosecretion in endocrine regulation. Nature. (Lond.), *210*, 271—272 (1966).

Knowles, F., und *L. Vollrath:* Neurosecretory innervation of the pituitary of the eels, *Anguilla* and *Conger.* Proc. roy. Soc. *250*, 311—342 (1966).

Kobayashi, Y.: Functional morphology of the pars intermedia of the rat hypophysis as revealed with the electron microscope. II. Z. Zellf. *68*, 155—171 (1965).

Legait, E., und *H. Legait:* Recherches sur l'ultrastructure de l'hypophyse de quelques téléostéens. C. R. Soc. Biol., Paris, *152*, 130—133 (1958).

Legait, H., und *E. Legait:* Terminaisons neurosécrétoires au niveau de l'adénohypophyse chez quelques téléostéens. Etude au microscope électronique. C. R. Soc. Biol., Paris, *151*, 1943—1946 (1957).

Metuzals, J.: Neurohistologische Studien über die nervöse Verbindung der Pars distalis der Hypophyse mit dem Hypothalamus auf dem Wege des Hypophysenstiels. Acta anat., Basel, *20*, 258—285 (1954).

Metuzals, J.: Die Innervation der Drüsenzellen der Pars distalis der Hypophyse bei der Ente. Z. Zellf. *43*, 319—334 (1955).

Metuzals, J.: Hypothalamic nerve fibres in the pars tuberalis and pia-arachnoid tissue of the cat and their degeneration pattern after a lesion in the hypothalamus. Experientia, Basel, *15*, 36 (1959).

Norberg, K.-A.: Transmitter histochemistry of the sympathetic adrenergic nervous system. Brain Res. *5*, 125—170 (1967).

Oberti, C.: Zur Kenntnis der Glia und Nerven der Adenohypophyse. Z. Zellf. *46*, 252—258 (1957).

Oota, Y., und *H. Kobayashi:* Fine structure of the median eminence and the pars nervosa of the bullfrog, *Rana catesbeiana.* Z. Zellf. *60*, 667—687 (1963).

Rothballer, A. B.: Pathways of secretion and regulation of posterior pituitary factors. Res. Publ. Ass. nerv. ment. Dis. *43*, 86—131 (1966).

Scharrer, B.: The neurosecretory neuron in neuroendocrine regulatory mechanisms. Am. Zool *7*, 161—169 (1967).

Scharrer, E.: Das Hypophysen-Zwischenhirnsystem von *Scyllium stellare.* Z. Zellf. *37*, 196—204 (1952).

Scharrer, E.: Principles of neuroendocrine integration. In: Endocrines and the central nervous system. Ass. Res. in Nerv. and Mental Dis. *43*, 1—35 (1966).

Smoller, C. G.: Ultrastructural studies on the developing neurohypophysis of the Pacific treefrog, *Hyla regilla.* Gen. Compar. Endocrinol. *7*, 44—73 (1966).

Shute, C. C. D., und *P. R. Lewis:* Cholinergic and Monoaminergic pathways in the hypothalamus. Brit. Med. Bull. *22*, No. 3, 221—226 (1966).

Stutinsky, F.: Sur l'innervation de la pars tuberalis de quelques mammifères. C. r. l'Assoc. Anat. Strasbourg 1948, 1—9.

Stutinsky, F.: (1967, im Druck).

Unsicker, K.: Über die Ganglienzellen im Nebennierenmark des Goldhamsters *(Mesocricetus auratus).* Z. Zellf. *76*, 187—219 (1967).

Vazquez-Lopez, E.: Innervation of the rabbit adenohypophysis. J. Endocrinol. *6*, 158—168 (1949).

Verney, E. B.: The antidiuretic hormone and the factors which determine its release. Proc. roy. Soc., London, *135,* 25—106 (1947).

Vollrath, L.: Über die neurosekretorische Innervation der Adenohypophyse von Teleostiern, insbesondere von *Hippocampus cuda* und *Tinca tinca.* Z. Zellf. *78,* 234—260 (1967).

Wittkowski, H.: Kapillaren und perikapilläre Räume im Hypothalamus-Hypophysensystem und ihre Beziehungen zum Nervengewebe: eine elektronenmikroskopische Studie am Meerschweinchen. Z. Zellf. *81,* 344—364 (1967 a).

Wittkowski, H.: Synaptische Strukturen und Elementargranula in der Neurohypophyse des Meerschweinchens. Z. Zellf. *82,* 434—458 (1967 b).

Ziegler, B.: Licht- und elektronenmikroskopische Untersuchungen an Pars intermedia und Neurohypophyse der Ratte. Z. Zellf. *59,* 486—506 (1963).

Anschrift des Verfassers: Prof. Dr. W. *Bargmann,* Anatomisches Institut der Universität, 23 Kiel, Neue Universität, Eingang F 1.

Discussion

Knowles: In your diagram of the pars intermedia of the cat you label some nerve endings as cholinergic. Could they not be peptidergic fibers without electron-dense vesicles or even aminergic ones with the amine contained in small clear vesicles?

Bargmann: Yes, as I mentioned previously the pictures of the so-called cholinergic fibers are possibly equivalents of peptidergic or aminergic fibers with emptied granules.

Mohri: You showed us a fluorescent micrograph of Fuxe. What is your opinion about the specifictiy of Falck's fluorescence-microscopical method?

Bargmann: According to my experience the specificity of this technique is high. However, investigations into this problem will be useful.

Journal of Neuro-Visceral Relations, Suppl. IX, 78—93 (1969)

The Mammalian Subfornical Organ*

K. Akert

Brain Research Institute, University of Zürich, Switzerland

With 9 Figures

Summary

Current investigations on the ultrastructure and function of the mammalian subfornical organ have been reviewed. Almost no information exists on the physiological significance of this structure. The results derived from anatomical research point towards a secretory function of its nerve cells. One type of secretory granules originates at least in part within the SFO itself and seems to be released within the pericapillary spaces. This material is Gomori-positive and may be considered "peptidergic". *(Bargmann)*. The second type of secretory material arises from "dark neurons" and accumulates within its fiber terminals in the form of large vacuoles whence it is released into the ventricular fluid. Functions of the SFO need also be considered in the light of its location outside the blood-brain barrier and the brain-cerebrospinal fluid barrier. Finally, the presence of a richly developed cholinergic plexus of extramural origin and the formation of peculiar "double plug synapses" with SFO neurons is emphasized. Relationships with monoaminergic mechanisms seem less obvious at the present time.

Several studies on the subfornical organ (SFO) have been carried out since the last reviews (*Akert, Potter* and *Anderson,* 1961; *Hofer,* 1965) have appeared. While little progress can be reported with respect to functional properties there is a considerable body of information about the structure of the organ which shall be summarized briefly in the subsequent sections of this article.

1. Blood Supply and Vascular Fine Structure

Spoerri (1963) has analysed the SFO vascular supply in rats by india ink injection techniques. Confirming earlier attempts by *Cohrs* and *Knobloch* (1936) and *Hofer* (1958) he established a remarkably dense capillary network with characteristic sinusoid and glomerular loops, and discovered its three-fold arterial supply by branches of the anterior cerebral, anterior choroid and posterior choroid arteries. The connections with the vascular bed of the choroid plexus of the third ventricle were also documented. *Duvernoy* and *Koritké* (1965) working with similar techniques in cats and several

---

° This work was supported by UPSH grant Nr. NB-03644 and by the Swiss National Foundation of Scientific Research (Nr. 4356).

other species arrived at similar conclusions (Fig. 1). It seems that the next step would consist in the intravital study of blood flow with special attention to the question of a portal circulation in the direction of certain hypothalamic and septal areas. Also the vascular connection with the choroid plexus of the third ventricle is in need of further clarification.

The fine structure of SFO capillaries is of considerable interest in view of the fact that the so-called blood-brain barrier is notably reduced or lacking in this area (*Wislocki* and *Leduc*, 1952). A thorough analysis of *Rohr* (1966 a) has revealed a number of properties of the capillary endo-

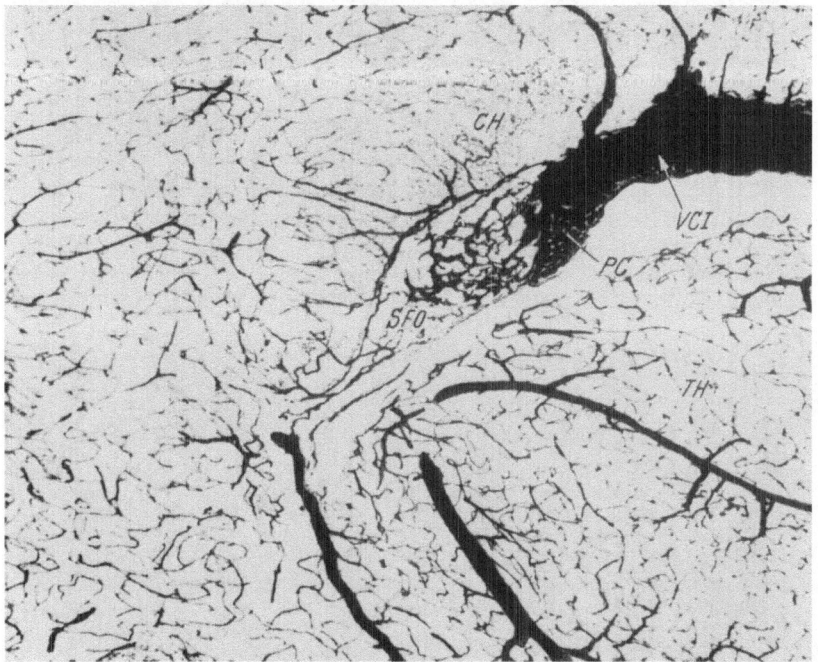

Fig. 1. Vascularisation of the rat SFO. India ink injection. Sagittal section. *TH* thalamus, *CH* commissura hippocampi, *PC* plexus chorioideus, *VCI* vena cerebri interna, (Courtesy O. Spoerri)

thelium which seem consistent with the notion of increased transport activities, i. e. the highly developed pinocytotic activity, the fenestrations and the expansion of luminal surface by means of microvilli and folds. Moreover, the adventitial space is greatly enlarged and its outer surface is formed by a palisade of densely packed cell processes which converge radially to the vascular wall and seem to derive from neural, glial and ependymal elements (Fig. 2). While many processes contain secretory material and extend into the adventitial space, the majority of very fine fibers arise from ependymal or hypendymal cells. These processes contain numerous fibrils and seem to impinge upon the outer basement membrane which

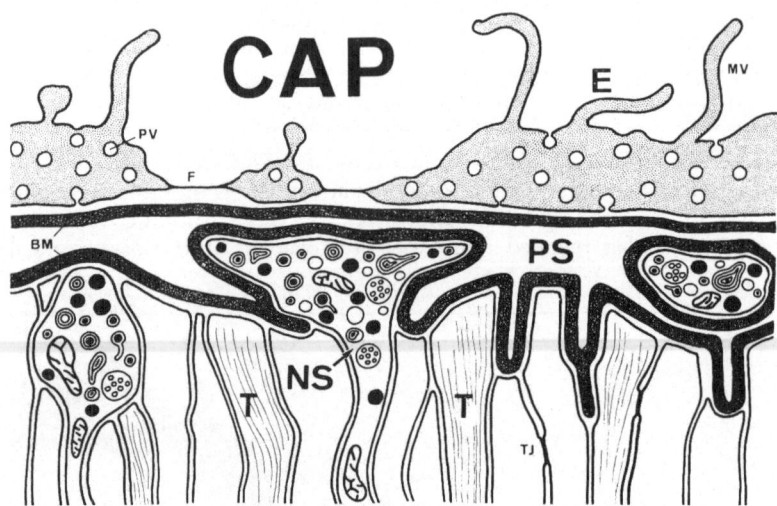

Fig. 2. SFO capillary reconstructed from electron-micrographs. *CAP* capillary lumen, *E* endothelium, *F* fenestration *BM* basement membrane, *NS* secretory nerve endings, *PV* pinocytotic vesicle, *T* fibrillary process of ependymal tanycyte, *TJ* tight junction.

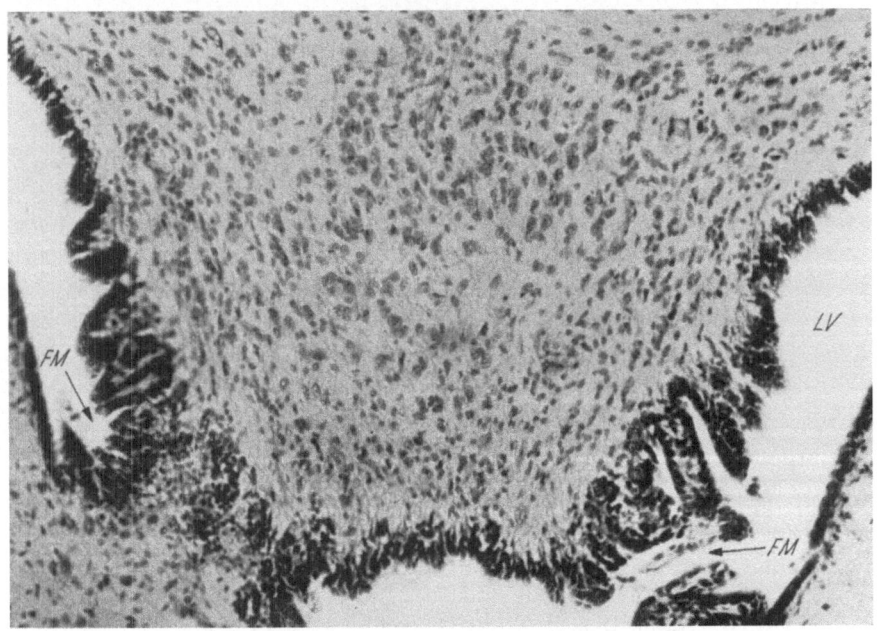

Fig. 3. SFO histology in frontal section (squirrel monkey). Note the ependymal folds. Parenchyma contains smaller cells in the peripheral zone, larger ones in the center. *FM* foramen Monroi, *LV* lateral ventricle. PAS stain.

forms a complex labyrinth by lining the distal ends of pericapillary processes.

Dempsey and *Wislocki* (1955) have shown that intravenously injected silver nitrate particles are primarily deposited within the complex basement membrane system of the SFO while *Marchesi* and *Barrnett* (1964) have demonstrated the significance of this structure in enzyme activity. It seems therefore justified to consider the enlarged and highly complex adventitial space of SFO capillaries to be an additional correlate of intensive transport mechanisms that may be going on between SFO parenchyma and its blood supply system. Further studies will have to elucidate more specifically its role and direction. Presently, the content of monoamines in areas outside the blood-brain barrier is of considerable interest (*Weil-Malherbe, Whitby* and *Axelrod*, 1961). *Lichtensteiger* (1967) has investigated this aspect more specifically using the histochemical fluorescence method of *Falck* and coll. (1962) in mice. In the normal resting state, the SFO content of fluorescing monoamines was rather minimal. However, under specific experimental conditions which favor the accumulation of endogenous serotonine, he found a fairly large number of nerve fibers containing this substance. Furthermore, many cell bodies had taken up exogenous L-Dopa, 5-hydroxytryptophan and, rarely, noradrenaline. The significance of these substances in sleep mechanisms have recently been demonstrated by *Jouvet* (1965) and by *Koella* and coll. (1965) and *Koella* (1966) while evidence in support of their role in temperature regulation has been accumulated by *Feldberg* and *Myers* (1964). However, the role of the SFO with respect to these vital regulatory functions remains yet to be elucidated.

2. Relations to the Cerebrospinal Fluid System

The SFO and the area postrema are glomus-like structures which seem to have similar relations to the cerebrospinal fluid, both regions being located at opposite ends of the intracerebral cavities (*Watermann* and *Abdel-Messeih*, 1957). Both areas have vascular connections and are in contiguity with the choroid plexus of the third and fourth ventricle respectively (*Duvernoy* and *Koritké*, 1964). *Akert, Potter* and *Anderson* (1961) have shown that the mammalian SFO is consistently found at the junction between tela choroidea of the third ventricle and the lamina terminalis. *Sprankel* (1960) pointed out that the SFO is not only located at the transition of the lateral and third ventricles, but also between the intracerebral and extracerebral (cisterna ambiens) fluid spaces as well. Unfortunately, the significance of this strategic location is not understood, although there is no lack of speculation on this point (*Rabl*, 1966).

An additional problem is concerned with the barrier between brain parenchyma and cerebrospinal fluid. Recently, *Fleischhauer* (1964) using fluorescent dyes noted the differential uptake of intrathecally injected materials by various cerebral areas forming the linings of the ventricular system. The subfornical organ was among the regions with increased permeability. Two structural findings may be relevant with respect to this observation. One is concerned with the expanded interphase between the

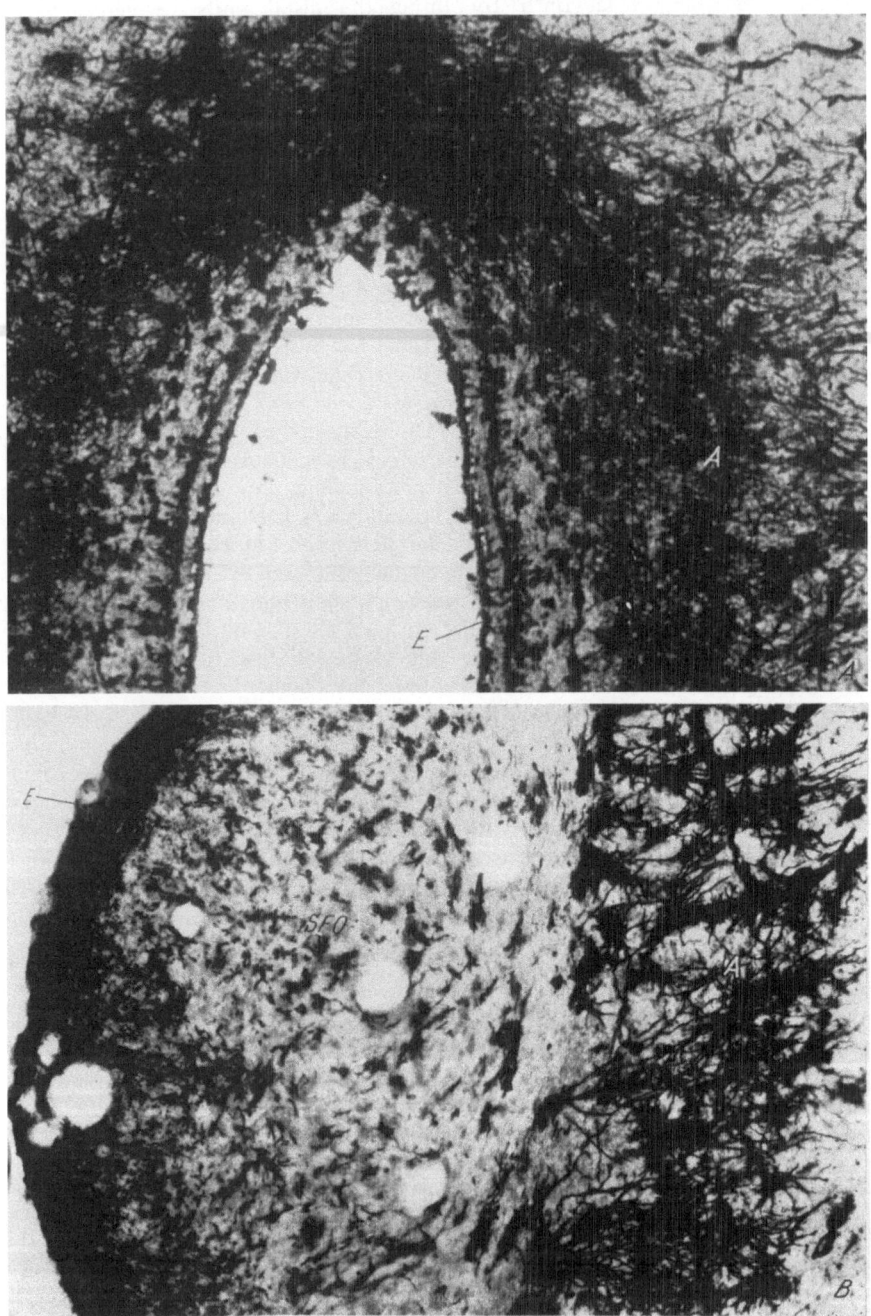

SFO and cerebrospinal fluid. Figure 3 shows the corrugated appearance of the ependymal lining as it characterizes the SFO surface in the squirrel monkey. On the other hand, *Watermann* (1965) and *Andres* (1965 a) have demonstrated the existence of ependymal canaliculi in dogs which seem to penetrate the depth of the SFO from its crenated surface, and seem to provide an increased area of interaction. The second finding is related to the histological property of the ventricular wall. Figure 4 a shows a segment of the cat's third ventricle in its posterior and dorsal extent. A dense feltwork of astrocytes is situated under the ependyma leaving only a very narrow band of tissue in between. It seems remarkable that no astroglial barrier separates the SFO from the ventricular border. Rather, the astrocytes form the boundary between the SFO parenchyma and the adjacent hippocampal commissure (Fig. 4 b). Thus the SFO is clearly outside the astroglial zone and by virtue of ependymal canaliculi the most intimate relations with the cerebrospinal fluid seem to be maintained.

3. Neurosecretory Activity

Legait and *Legait* (1956) were among the first to draw attention to the presence of neurosecretory material in the SFO of lower vertebrates. They identified the typically beaded nerve fibers with the aid of the Gomori method and thought that these fibers originated from the region of the supraoptic nuclei. Several other investigators (e. g. *Creswell, Reis* and *MacLean*, 1964) have also reported the finding of Gomori- and aldehyde-fuchsin positive material in the SFO of several species.

Studies with the aid of the electro microscope (*Rohr*, 1966 b) have confirmed the presence of neurosecretory granules in the SFO. Two types of secretory systems were found: the first type consists of nerve fibers containing electron dense 1000—2000 Å granules similar to the "peptidergic" vesicles as defined by *Bargmann* (1966) in the hypothalamo-neurohypo-physeal system. This material corresponds to the Gomori-positive material mentioned above. The origin of these nerve fibers could not be established with certainty. Contrary to the views of previous authors, it is conceivable that these elements arise to some extent from within the SFO. Figure 5 illustrates an SFO cell containing dark secretory granules in the soma and in the axon.

More definite information was provided by *Rohr* (1966 b) on the site of release. Club-like endings containing secretory granules, mitochondria and lamellated bodies were found in the pericapillary spaces from which the material may be discharged into the blood stream. These endings bear close similarity to those described in the palisade of the simian neurohypophysis by *Bodian* (1966) (Fig. 2 and 5).

Fig. 4. *A:* Subependymal astrocytes in the wall of the third ventricle (dorsocaudal moiety) in cat. Note the small zone between ependyma and astrocyte (A) barrier. Golgi-Bubenaite stain. x 100. *B:* SFO and subependymal astrocyte barrier (A) in cat. Golgi-Bubenaite stain. Same magnification as in A.

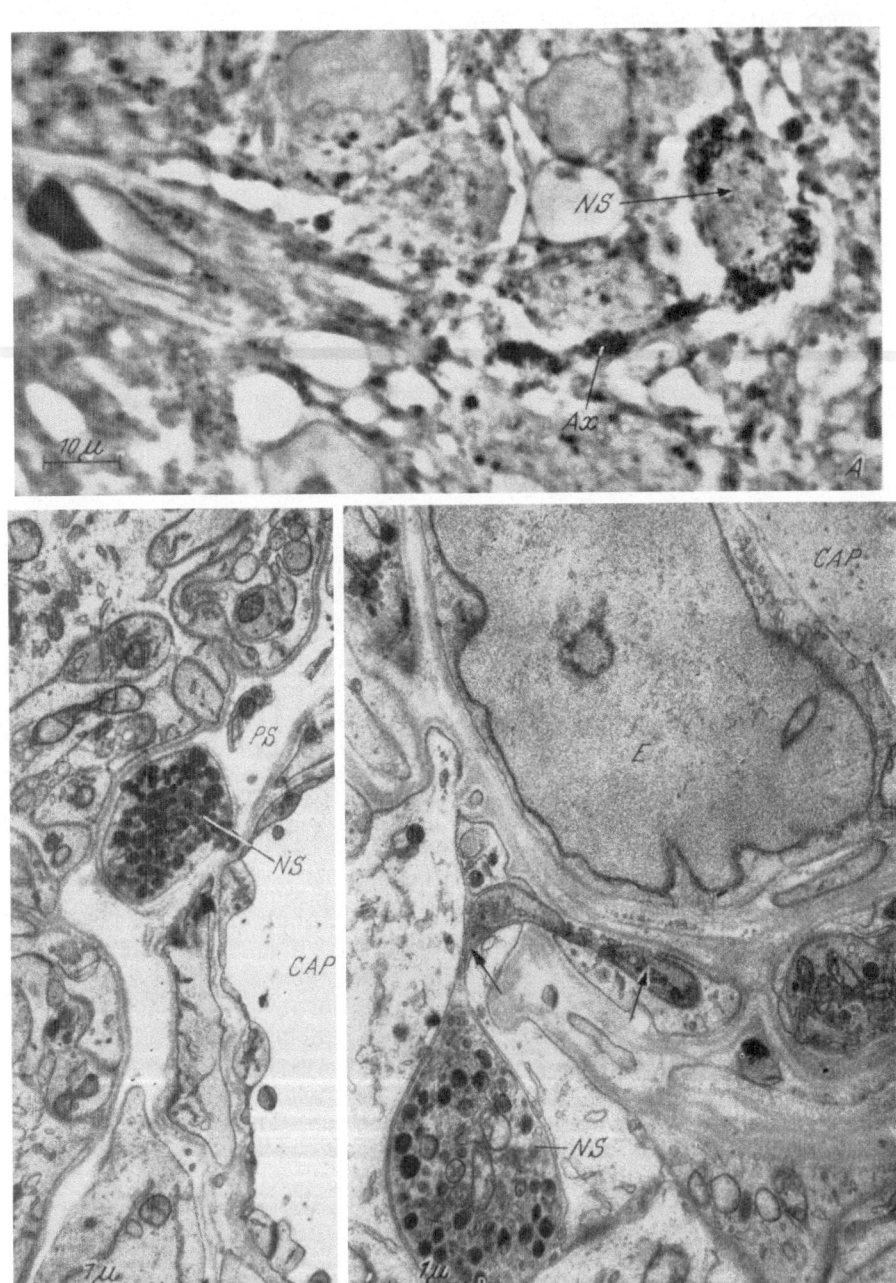

The function of this secretory system is unknown and the conditions which favor its activation are still uncertain. *Pachomov* (1963) reported an increase of secretory activity after intraperitoneal injection of alcohol and intramuscular application of estrogens in rats, while *Dierickx* (1962) and *Rudert* (1965) claim that osmotic stress may activate this system in

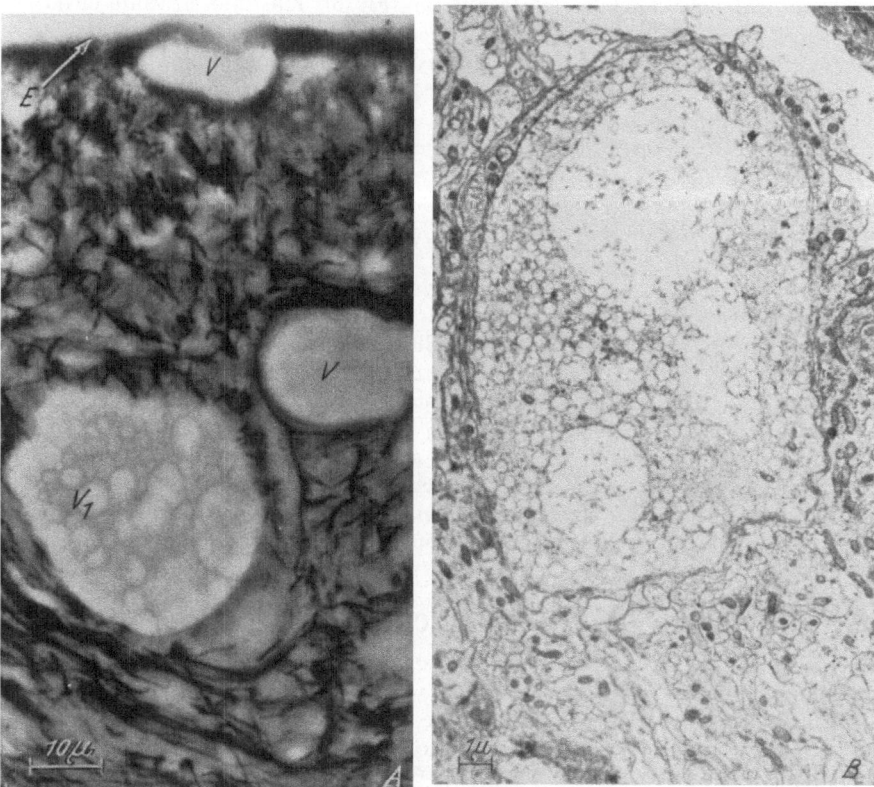

Fig. 6. Neurosecretory activity in cat SFO: colloid-like secretion. A: Vacuoles in the ependymal (E) and subependymal zone. Note that two vacuoles (V) seem empty, while one (V₁) contains vesicles of varying size. Phase contrast. x 850. B: Vacuole similar to that seen in A in low power electronmicrograph (OsO₄ Fixation); it seems about to discharge into a larger space situated in the ependymal zone. x 4,750.

the frog, particularly in the absence of the neurohypophysis. In our own material it was found that the Gomori-positive material was consistently increased in monkeys which had sustained prolonged anesthesia and cerebral experimentation.

Fig. 5. Neurosecretory activity in cat SFO: *A:* Neuron containing secretory granules in soma (NS) and axon (Ax). Phase contrast. x 1,000. *B:* Neurosecretory fiber ending in pericapillary space (PS) near capillary lumen (Cap) x 16,000. *C:* Similar neurosecretory fiber with protrusion of ending (arrow). E endothelial cell. KMnO₄ Fixation. x 12,800.

The second type of secretory activity discovered by *Rohr* (1966 b) has to do with the "Saftlücken" and the interstitial vacuoles described repeatedly by previous investigators (*Watermann*, 1956; *Hofer*, 1965; *J. Ariëns Kappers*, personal communication). These vacuoles were shown to be identical with expanded nerve processes which had accumulated an enormous amount of colloid-like secretory material (Fig. 6). The fine structure of this material reveals a multitude of vesicles of varying size. These vesicles are coated by a unit membrane and contain a finely granulated substance which is irregularly dispersed within a matrix of extremely low electron density (Fig. 7 a).

The vesicles seem to arise within the endoplasmic reticulum of "dark neurons" (Fig. 7 b). They can be followed in myelinated and unmyelinated processes of these cells which seem to be oriented towards the subependymal region. The vesicles have a strong tendency to coalesce within the enlarged nerve endings; eventually their content seems to be discharged into the ventricular cavity. This would explain the fact that vacuoles encountered within the ependymal area are often empty. Apparently, the ependymal canaliculi in the deeper layers of the SFO are also frequently the recipients of secretory material which may be released from more centrally located "vacuoles". The intracellular origin of SFO vacuoles and their secretory nature has been suggested by *Weindl* (1965) on the basis of light microscopic investigations. *Rohr's* (1966 b) analysis based on alternating phase contrast and electron microscopic sections has provided the final proof. The chemical nature of this secretion is unknown and little more than speculations can be offered with respect to the mode of its activation and function.

4. Nerve Cells and Their Cholinergic Innervation

Spiegel (1918) was among the first to consider the cells of the SFO (which he called "ganglion psalterii") to be neurons. Many of the subsequent investigators were less certain and the term "parenchymal cell" was more commonly used. Several light microscopic descriptions of SFO cells are attempts to classify the parenchyma (*Cohrs* and *Knobloch*, 1936; *Brizzee*, 1954; *Hofer*, 1957; *Sprankel*, 1957). The most modern analysis is the one made with the aid of the electron microscope by *Andres* (1965 b). This author was able to differentiate three main cell types in the dog: i) type I neurons of small size (8—12μ); ii) type II neurons of larger size (14—20μ); and iii) neuroglial satellite cells. His findings could be confirmed in cats and monkeys by similar investigations in our laboratory. However, a "dark neuron" should be added to this list, which is mainly concerned with secretory functions. This neuron has been briefly mentioned in this article ("dark neuron", Fig. 7 b) and described in more detail elsewhere (*Pfenninger, Akert, Sandri, Bruppacher*, 1967).

Andres (1965b) also described ample contacts between the neural elements in the SFO. Axo-dendritic synapses of intramural and extramural afferent fibers were mainly differentiated on the basis of mitochondrial form and size.

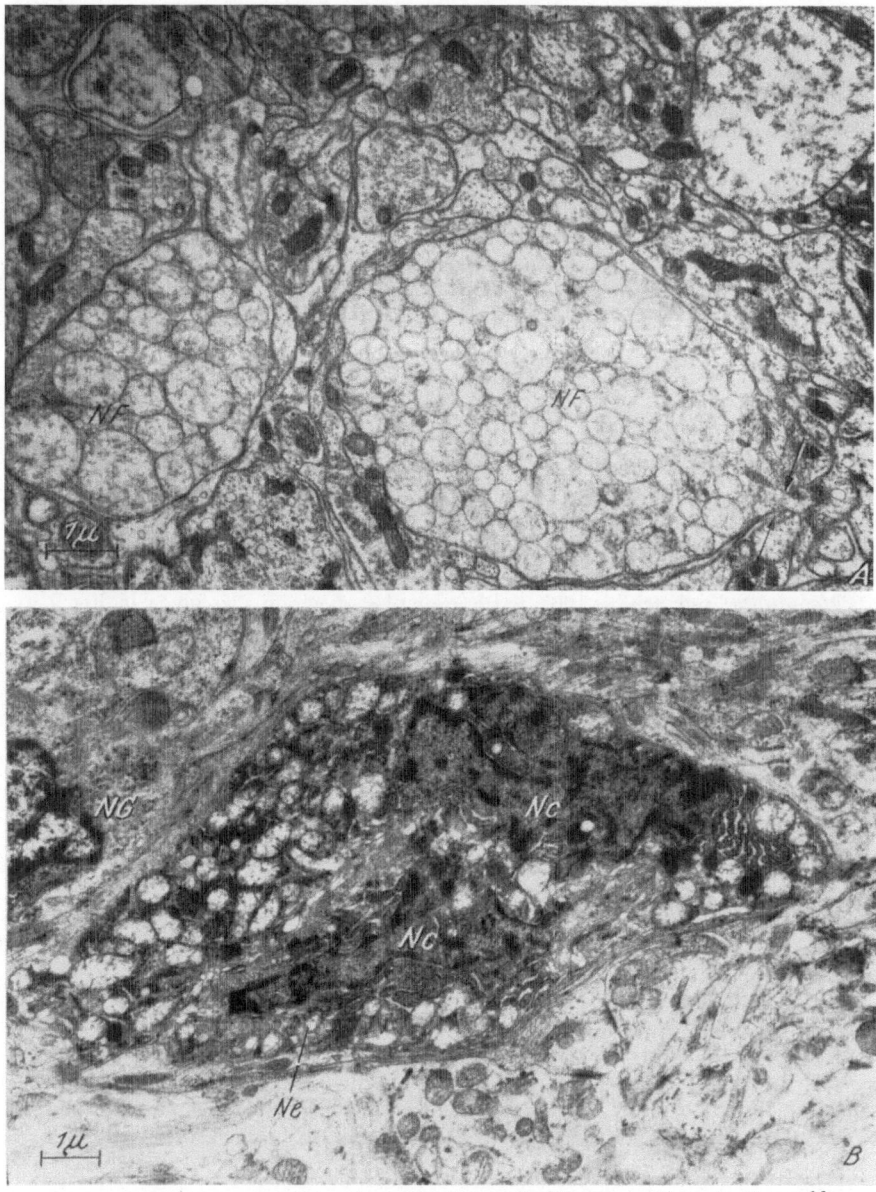

Fig. 7. Neurosecretory activity in cat SFO: colloidlike secretion. A: Nerve fiber varicosities (NF) containing vesicles with very fine granular material. Normal fiber size is resumed at arrows. x 9,000. B: "Dark neuron" (Ne). The endoplasmic reticulum contains the same vesicles as seen in (A). Nc nucleus, NG neuroglial satellite cell. Evidence of the neuronal nature of this cell rests on the presence of synapses (*Pfenninger* et al., 1967). x 7,500.

Evidence in support of an extramural innervation has been brought forward by *Shute* and *Lewis* (1966) who offered a new approach to the problem. These authors, using acetylcholinesterase staining techniques, have mapped the cholinergic pathways within the rat forebrain and found heavy deposits of this enzyme within the SFO (*Shute* and *Lewis*, 1963). This was confirmed in rat (Fig. 8 a), cat and monkey by similar studies in our laboratory. *Shute* and *Lewis* (1966) have also been able to trace the origin of SFO cholinesterase activity to cells in the raphe of the septal area and to fibers of the dorsal fornix.

The presence of a dense nerve plexus in the SFO was recently established in our laboratory by yet another procedure. Using the osmium-zincjodide stain of *Maillet* (1962) nerve fibers and terminals were found to be localized throughout the SFO parenchyma with the exception of the ependymal layer (Fig. 8 b). Typical varicosities are frequently seen in close proximity to nerve cell bodies and proximal dendrites. The capillary walls and the secretory "vacuoles" seem to have no relations with this innervation. In a parallel study, the electron microscopic aspect of this innervation was examined. A vast number of classical Gray 1 and 2 synapses was found between axons of unknown origin and the somatic and dendritic membranes of SFO nerve cells. Of particular interest is a peculiar type of junction called the "double-plug crest synapse with subjunctional bodies" (*Akert, Pfenninger* and *Sandri*, 1967 a, b), examples of which are given in Fig. 9.

Similar junctions have been found by *Milhaud* and *Pappas* (1965) in the medial habenular and the interpenduncular nuclei, of which the content in cholinesterase is considerable (*Friede*, 1966).

The vast majority of synaptic vesicles to be found in these junctions is of the clear type, 200—500 Å in size. In contrast, typical catecholamine granules (*Wolfe, Axelrod, Potter* and *Richardson*, 1962) are lacking, and the few large (1000—1500 Å) granulated vesicles proved resistant to reserpine. Additional evidence in support of the notion that the SFO contains a dense cholinergic innervation of extramural origin is rapidly accumulating in our laboratory. Of special interest was the observation of SFO neurons which responded with an increased rate of discharge when acetylcholine was applied locally by microelectrophoresis (*Steiner, Akert, Pfenninger* and *Villoz*, unpublished).

The function of SFO cholinergic innervation is not clear. *Shute* and *Lewis* (1966) consider the cholinergic pathways of the forebrain as part of the "ascending reticular activating system" of neurophysiologists. Neither is the role of SFO target neurons fully known. It seems doubtful whether the secretory activity which we have mentioned before represents their exclusive function. Many authors have suggested some sort of sensory

---

Fig. 8. A: Cholinesterase activity in the cat SFO. Note the prominent reaction with Koelle's thiocholine method. No inhibitor for pseudocholinesterase was used in this preparation. *FX* fornix, *HC* hippocampal commissure, *KP* choroid plexus. B: Terminal reticulum of an extramural nerve plexus in the cat SFO. Note the sparing of the ependymal cell layer (E). This plexus is believed to be cholinergic (see text). *Nc* nerve cells. Maillet stain. x 750.

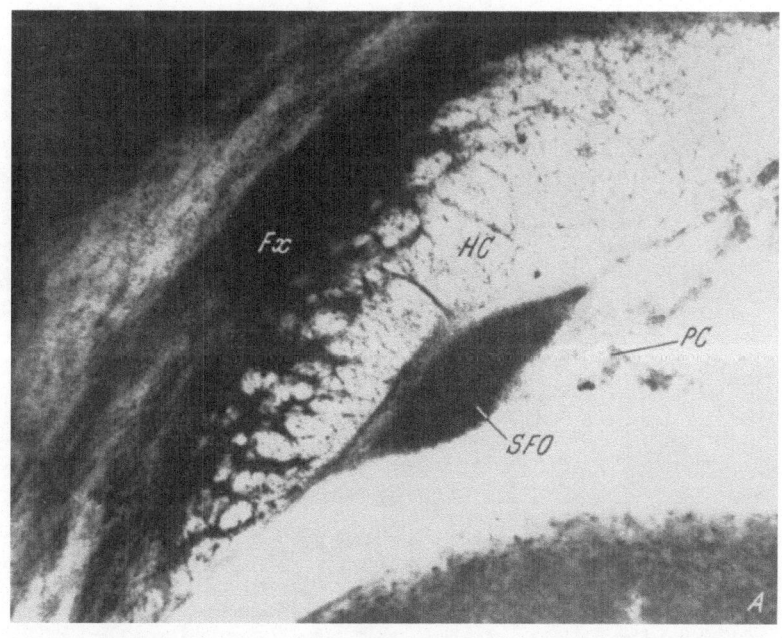

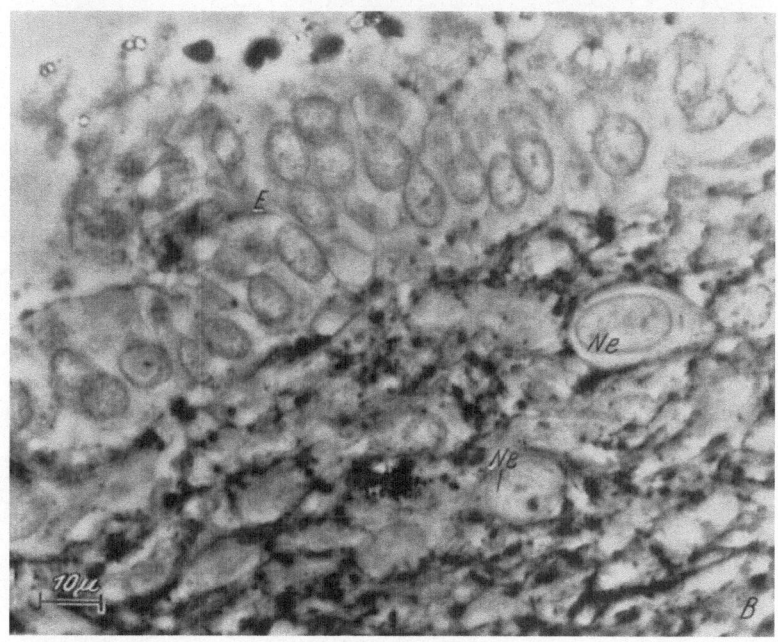

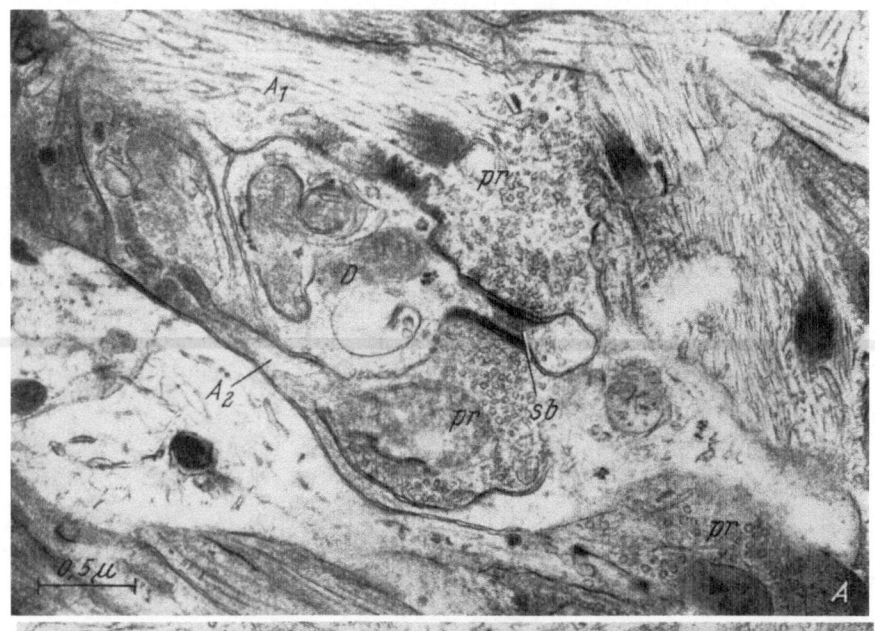

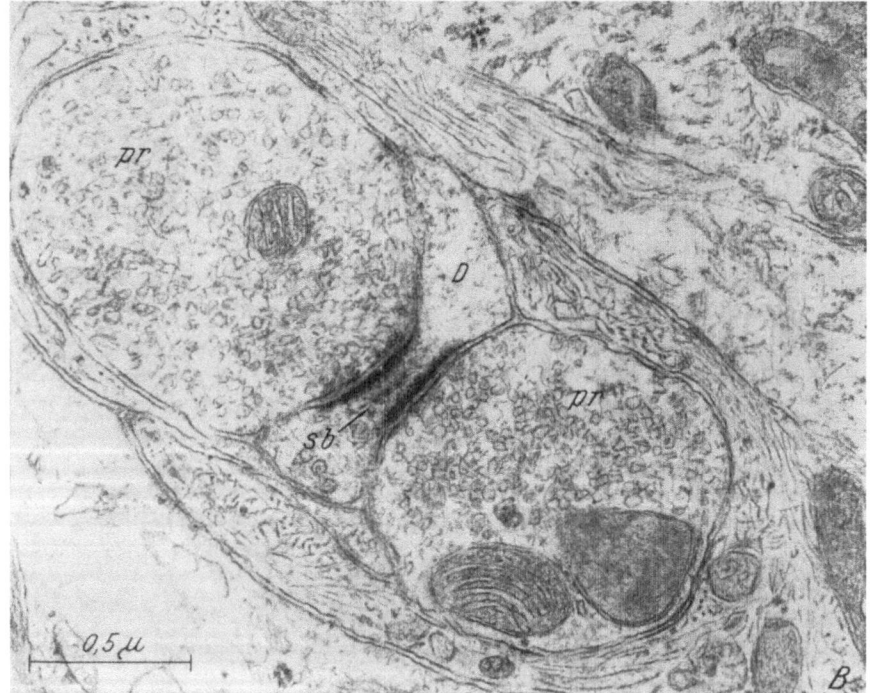

activity to be present within the SFO (*Andres*, 1965 b; *Weindl*, 1965). Attractive as such an hypothesis may appear, we have to admit that little supportive physiological or pharmacological evidence exists at the present time. But even the anatomist's task seems far from being solved. The neuropil of the SFO — dense and inextricable as it is — and its interrelations with ependyma and capillary walls offers many puzzling problems.

References

Akert, K., H. D. Potter, and *J. W. Anderson:* The subfornical organ in Mammals I. Comparative and topographical anatomy. J. Comp. Neurol., Philadelphia *116,* 1—14 (1961).

Akert, K., K. Pfenninger, and *C. Sandri:* Crest synapses with subjunctional bodies in the subfornical organ. Brain Res. *5,* 118—121 (1967a).

Akert, K., K. Pfenninger, and *C. Sandri:* The fine structure of synapses in the subfornical organ of the cat. Zschr. Zellforsch. *81,* 537—556 (1967).

Andres K. H.: Ependymkanälchen im Subfornikalorgan vom Hund. Naturwissenschaften *52,* 433 (1965 a).

Andres K. H.: Der Feinbau des Subfornikalorganes vom Hund. Zschr. Zellforsch. *68,* 445—473 (1965b).

Bargmann, W.: Neurosecretion. Int. Rev. cytol. *19,* (1966).

Bodian, D: Herring bodies and neuro-apocrine secretion in the monkey. An electron microscopic study of the fate of the neurosecretory product. Bull. Johns Hopk. Hosp. *118,* 282—326 (1966).

Brizze, K. R.: A comparison of cell structure in the area postrema, supraoptic crest, and intercolumnar tubercle with notes on the neurohypophysis and pineal body in the cat. J. Comp. Neurol., *100,* 699—716, (1954).

Cohrs, P., and *D. v. Knobloch:* Das subfornikale Organ des dritten Ventrikels. Zschr. Anat. Entw. gesch., *105,* 491—518 (1936).

Creswell, G. F., D. J. Reis, and *P. D. MacLean:* Aldehyde-fuchsin positive material in brain of squirrel monkey. *(Saimiri sciureus).* Amer. J. Anat. *115,* 543—557 (1964).

Dempsey, E. W., and *G. B. Wislocki:* An electron microscopic study of the blood-brain barrier in the rat employing silver nitrate as vital stain. J. biophys. biochem. Cytol. *1,* 245—256 (1955).

Dierickx,: K. The dendrites of the preoptic neurosecretory nucleus of *Rana temporaria* and the osmoreceptors. Arch. int. Pharmacodyn. *140,* 708—725 (1962).

Duvernoy, H., and *J. G. Koritké:* Contribution à l'étude de l'angioarchitectonie des organes circum-ventriculaires. Arch. biol., Paris *75,* 693—748 (1964).

Duvernoy, H., and *J. G. Koritké:* Recherches sur la vascularisation de l'organe subfornical. J. Méd. Besançon *1,* 115—130 (1965).

Fig. 9. Crest synapses with subjunctional bodies in cat SFO. Double plug synapses (*Akert, Pfenninger* and *Sandri,* 1967 a, b) with predominantly nongranulated vesicles. *A:* Two axon endings from presynaptic end knobs at dendritic crest. x 8,000. (prim. magn.) *B:* Similar synapse sectioned at a different angle. *A* axon, *D* dendrite, *pr* presynaptic ending, *sb* subjunctional bodies. x 20,000 (prim. magn.)

Falck, B., N.-A. Hillarp, G. Thieme, and *A. Torp:* Fluorescence of catecholamines and related compounds condensed with formaldehyde. J. Histochem. Cytochem. *10,* 348—354 (1962).

Feldberg, W., and *R. D. Myers:* Effects on temperature of amines injected into the cerebral ventricles. A new concept of temperature regulation. J. Physiol. (Lond.) *173,* 226—237 (1964).

Fleischhauer, K.: Fluoreszenzmikroskopische Untersuchungen über den Stofftransport zwischen Ventrikelliquor und Gehirn. Zschr. Zellforsch. *62,* 639—654 (1964).

Friede, R. L.: Topographic Brain Chemistry. New York and London: Academic Press 1966.

Hofer, H.: Beobachtungen an der Glia des Subfornicalorgans von *Galago crassicaudatus* Geoffroy 1812. Zschr. Anat. Entw.-gesch., *120,* 1—14 (1957).

Hofer, H.: Zur Morphologie der circumventriculären Organe des Zwischenhirns der Säugetiere. Verh. Dtsch. Zool. Ges. Fankfurt a. M., 202—251, 1958.

Hofer, H.: Circumventrikuläre Organe des Zwischenhirns. In: Primatologia (H. Hofer, A. H. Schultz, D. Starck, eds.) II/2, 1—104, Basel and New York: Karger 1965.

Jouvet, M: Etude électrophysiologique et neuropharmacologique des états de sommeil. Acta Pharmacol. *18,* 109—173, (1965).

Koella W. P.: The mode and locus of action of serotonin in its effects on the recruiting responses and the EEG of cats. In: Molecular Basis of some Aspects of Mental Activity. (O. Walaas, ed.), *1,* 431—442, London: Academic Press 1966.

Koella, W. P., C. M. Trunca, and *J. S. Czicman:* Serotonin: effect on recruiting responses of the cat. Life Sci. *4,* 173—181 (1965).

Legait, H., and *E. Legait:* Mise en évidence de voies neurosécrétoires extrahypothalamo-hypophysaires chez quelques batraciens et reptiles. C. R. Soc. Biol. (Paris) *150,* 1429—1431 (1956).

Lichtensteiger W.: Monoamines in the subfornical organ. Brain Res. *4,* 52—59 (1967).

Maillet, M.: La technique de Champy à l'osmium iodure de potassium et la modification de Maillet à l'osmium-iodure de zinc. Trab. Inst. Cajal sec. fisiol, Madrid, *54,* 1—36 (1962).

Marchesi, V. T., and *R. J. Barrnett:* Localization of nucleosidephosphatase activity in different types of small blood vessels. J. Ultrastruct. Res. *10,* 103—115 (1964).

Milhaud, M., and *G. Pappas:* Postsynaptic bodies in the habenula and interpeduncular nuclei of the cat. J. Cell. Biol. *30,* 437—441 (1966).

Pachomov, N.: Morphologische Untersuchungen zur Frage der Funktion des subfornikalen Organs der Ratte. Dtsch. Zschr. Nervenh. *185,* 13—19 (1963).

Pfenninger, K., K. Akert, C. Sandri, and *H. Bruppacher:* Die Feinstruktur der Parenchymzellen im Subfornikalorgan der Katze. Schweiz. Arch. Neurol. Neurochir. Psychiat. *100,* 232—254 (1967).

Rabl, R.: Das Subfornikalorgan des Menschen. J. Hirnforsch. *8,* 529—545 (1966).

Rohr, V. U.: Zum Feinbau des Subfornikal-Organs der Katze. I. Der Gefäß-Apparat. Zschr. Zellforsch. *73,* 246—271 (1966a).

Rohr, V. U.: Zum Feinbau des Subfornikal-Organs der Katze. II. Neurosekretorische Aktivität. Z. Zellforsch. *75,* 11—34 (1966b).

Rudert, H.: Das Subfornicalorgan und seine Beziehungen zu dem neurosekretorischen System im Zwischenhirn des Frosches. Z. Zellforsch. *65,* 799—804 (1965).

Shute, C. C. D., and *P. R. Lewis:* The subfornical organ (intercolumnar tubercle) of the rat. J. Anat. (Lond.) *27,* 301 (1963).

Shute, C. C. D., and *P. R. Lewis:* Cholinergic and monoaminergic pathways in the hypothalamus. Brit. med. Bull. *22,* 221—226 (1966).

Spiegel, E. A.: Das Ganglion psalterii. Anat. Anz., Jena, *51,* 454—462 (1918).

Spoerri, O.: Über die Gefäßversorgung des Subfornikalorganes der Ratte. Acta anat., Basel, *54,* 333—348 (1963).

Sprankel, H.: Zur Zytologie des subfornikalen Organes bei Affen. Verh. dtsch. zool. Ges., Graz. Zool. Anz. (Suppl.) *21,* 444—451 (1957).

Sprankel, H.: Über die Beziehungen des Plexus des dritten Ventrikels zum subfornikalen Organ bei den Primaten. Naturwissenschaften *47,* 383—384 (1960).

Watermann, R.: Über das Vorkommen interstitieller Vakuolen im Subfornikalorgan. Dtsch. Zbl. Nervenhk. *174,* 593—596 (1956).

Watermann, R.: Interventrikularorgan und Trigonum supracommissurale. Anat. Anz., Jena, *117,* 261—279 (1965).

Watermann, R., and *G. Abdel-Messeih:* Ein Vergleich von Subfornikalorgan und Area postrema. Zschr. Morph. Tiere, Berlin, *45,* 603—615 (1957).

Weil-Malherbe, H., L. G. Whitby, and *J. Axelrod:* The bloodbrain barrier for catecholamines in different regions of the brain. In Regional Neurochemistry, S. S. Kety and J. Elkes, eds., 284—292. Oxford: Pergamon Press 1961.

Weindl, A.: Zur Morphologie und Histochemie von Subfornicalorgan, Organum vasculosum laminae terminalis und Area postrema bei Kaninchen und Ratte. Zschr. Zellforsch. *67,* 740—755 (1965).

Wislocki, G. B., and *E. H. Leduc:* Vital staining of the hematoencephalic barrier by silver nitrate and trypanblue, and cytological comparisons of the neurohypophysis, pineal body, area postrema, intercolumnar tubercle and supraoptic crest. J. Comp. Neurol., Philadelphia, *96,* 371—413 (1952).

Wolfe, D. E., J. Axelrod, L. T. Potter, and *K. C. Richardson:* Localization of norepinephrine in adrenergic axons by light- and electron-microscopic autoradiography. In: Fifth Int. Congr. Electron Microscopy, S. Breese, Jr., ed., L-12, Philadelphia: Academ. Press 1962.

Author's address: Prof. *K. Akert,* Brain Research Institute, University of Zürich, August Forelstraße 1, 8008 Zürich, Switzerland.

Discussion

Bargmann: Dr. *Akert,* you mentioned a "double" innervation. Would it be possible that your opinion about this kind of innervation is based on sagittal sections of invaginated synapses, that is by the direction of cutting?

Akert: The synapses which I have shown consist for the most part of two separate presynaptic endings. They attach on opposite sides, *i. e.* vis-à-vis, on a dendritic crest. The evidence was based on serial sections as well as on a large material of individual sections with different orientation. The situation of invaginated single synapses may also occur, but rather rarely.

Oksche: I am wondering whether the aldehyde fuchsin positive fibers you observed within the subfornical organ are fluorescent with the pseudoisocyanine method of Sterba. This method is more specific for neurosecretory material originating in the supraoptic and paraventricular nuclei than the aldehyde fuchsin stain. It must be excluded that some of the monoamine-rich endings produce a similar aldehyde fuchsin positive picture as do the peptidergic fiber elements. For this reason a relatively wide spectrum of methods including the monoamine-fluorescence techniques of the Swedish school is necessary.

Akert: Thus far we have not made any fluorescent study with the pseudo-isocyanine method of Sterba. This would be certainly most interesting and necessary to do. On the other hand, the monoamine fluorescence technique of *Falck* and *Hillarp* has been used by *Lichtensteiger* (1967) both in the normal subfornical organ of cat and mouse and in experimental material in which catecholamines and indoleamines were allowed to accumulate. It was shown that a number of serotonin containing nerve fibers are present in this organ. Their origin and exact destination, however, remain to be further investigated.

Watermann: Using the silver technique of *Van Campenhout* we demonstrated spherical and polygonal granules in the large vacuoles of the cat and rabbit subfornical organ. Two types of these granules could be distinguished, a larger and denser and a smaller and clearer type. Fiber-like structures were also seen using *Van Campenhout's* method at the border of the vacuoles. Could these possibly represent the partly myelinated sheaths shown by *Akert* and *Rohr* to be present at the border of the neuronal large vacuoles?

Akert: No comment.

Neuroglial Secretion

Journal of Neuro-Visceral Relations, Suppl. IX, 97 110 (1060)

Ependymal Secretion, especially in the Hypothalamic Region

Sir **Francis Knowles**

Department of Anatomy, The Medical School, Birmingham 15, England

With 3 Figures

Summary

Certain circumscribed areas of ependyma, lining the ventricles of the brain, appear to be specialized in structure and activity. In contrast to "normal" or "classical" ependymal cells, which are ciliated and apparently engaged in some synthesis, secretion and adsorption, though not very actively, specialised ependymal cells have morphological characteristics generally associated with considerable secretory and/or absorptive function.

One area of specialized ependyma in the ventro-lateral walls of the III. ventricle lies near the so-called hypophysiotropic area, believed to be concerned in regulation of pituitary function; it is therefore of special interest to endocrinologists. Recently, correlations between the structure of these ependymal cells lining the infundibular recess and pituitary functions related to colour-change and reproductive cycles have been described, and are here considered in relation to hypothalamic control of the pituitary.

I. Introduction

Within recent years attention has been drawn to a possibility that certain ependymal cells may be concerned in neuroendocrine control. Although the ventricles of the brain are classically described as having a simple cuboidal to columnar ciliated lining, certain areas of the third ventricle are lined by cells which differ markedly from this pattern. A direct endocrine function has been attributed to some of these areas which have retained few ependymal characters and have been described as organs (*e. g.* sub-commissural organ, sub-fornical organ).

It is interesting to note however that a number of atypical ependymal cells which do not constitute clearly defined organs occupy regions which correspond closely to the "hypophysiotropic area" defined by *Halasz* and coll. (1962). The form and position of these cells has therefore given rise to the speculation that they may be concerned in regulation of the pituitary gland (*Leveque*, 1963; *Vigh* and coll., 1963; *Knowles* and *Vollrath*, 1966; *Knowles*, 1967). Recent discoveries, which will be considered in this review, lend support to this concept that areas of modified ependyma may be directly concerned in the correlation of cerebral and endocrine function.

II. Typical and Atypical Ependyma

A. Typical ependyma

Most ependymal cells lining the central canal of the spinal cord and the ventricles of the brain constitute an epithelial lining and may be described as "normal" or classical ependyma. Many of these cells have cilia which are capable of maintaining currents of flow in the cerebrospinal fluid (*Worthington* and *Cathcart*, 1963). Basally, slender processes containing neuroglial fibres project from each ependymal cell but do not penetrate underlying tissues deeply. In embryonic life basal processes of ependymal cells traverse the whole thickness of the neural wall to become attached to the pia by terminal expansions, the end feet. This condition is still seen in the adult in places where the neural wall is relatively thin (*Cajal*, 1909, 1952), as in the floor plate of the spinal cord, but in regions with thicker walls ependymal fibres usually end within a short distance of the cell bodies.

Under the light microscope ependymal cells appear to form a continuous wall, and preliminary studies using the electron microscope showed that adjoining ependymal elements are connected by junctions which were described by *Brightman* and *Palay* (1963) as *zonulae adhaerentes* and *zonulae occludentes*. *Brightman* has however suggested that these intercellular fusions are labile and can facilitate intercellular passage, since it can be shown that ferritin injected into the cerebrospinal fluid will cross the ependymal barrier and pass into the neuroglial parenchyma (*Brightman*, 1965; see also *Feldberg* and *Fleischhauer*, 1960).

Much of the enzymatic pattern of ependymal cells may be related to the motility of their cilia. Certain studies do however indicate the presence of enzymes which may be concerned with absorption from or secretion into the cerebrospinal fluid (*Nandy* and *Bourne*, 1965). *Smoller* (1965) has remarked that the ultrastructure of ependyma in *Hyla* can indicate either a secretory or a receptive function, or both.

Certainly both absorptive and secretory functions have been claimed for "normal" ependyma. *Klatzo* and coll. (1964) have shown that fluorescine-labelled protein can traverse ependyma and that there are indications of an active transport mechanism in ependymal cells, capable of a differential absorption of proteins. Ependymal cells will also absorb trypan blue injected into the ventricle, but, as *Lups* and *De Haan* (1954) have remarked, in tests such as these one is not working under physiological but under pathological conditions. Nevertheless there are a number of indications that ependymal cells, especially those modified ependymal cells of the choroid plexus, may absorb some cerebrospinal fluid and possibly also contained proteins (*Pollay* and *Davson*, 1963).

The production of cerebrospinal fluid by ependymal cells has also been suggested, and there is reason to believe that the choroid plexus plays a major part in this process (*Ganong*, 1965). The production of cerebrospinal fluid by "normal" ependyma under normal conditions is however less certain, though ependymal cells, in common with the lining of the arachnoid spaces, seem capable of the production of fluid under pathological conditions (see *Lups* and *De Haan,* 1954).

Studies of "normal" ependyma using the light microscope suggest a relatively unmodified ciliated epithelial tissue capable of some synthesis, secretion and absorption, though not apparently very actively engaged in these functions. Electron microscope studies of the cytomembranes and cytoplasmic organelles accord with this view. Many mitochondria lie between the nucleus and the ventricular border of the cell, close to the basal bodies of the cilia, and a few lysosomes lie nearby. Some smooth-surfaced canaliculi and vesicles are found in the cytoplasm, some resembling a Golgi complex, but no organised endoplasmic reticulum is present and ribosomes, though frequent, are rarely associated with membranes, but occur in clusters. Bodian's stain does not reveal any neurofilaments, though under the electron microscope many fine filaments are seen, in a perinuclear position (*Brightman* and *Palay,* 1963).

Little is known about the electrophysiological properties of ependymal cells, beyond the fact that ependymal cells from mammalian brains, in tissue culture, display a resting potential of 40—60 mv., and that this may be reduced by stimulation, the reduction being proportional to the applied stimulus (*Hild* and coll., 1965).

B. Atypical ependyma

Within the ventricles of the brain lie certain circumscribed areas of ependyma which differ markedly from "normal" or classical ependyma. Some of these have already been described as organs, notably the following: a) The sub-commissural organ (*Dendy* and *Nicols,* 1910), b) The sub-fornical organ or intercolumnar tubercule (*Putnam,* 1922), c) The para-ventricular organ (*Roussy* and *Mosinger,* 1938). Possible endocrine functions have been attributed to these regions. Elsewhere also there are less clearly defined regions which contain ependyma which differs markedly from classical ependyma, notably in the floor of the third ventricle, e. g. the median eminence and the infundibular recess; one of these areas has recently been termed the prechiasmatic gland (*Leveque* and coll., 1967).

It is not possible to generalise about all the different forms of atypical ependyma mentioned above for these show many variants of structure and stainability; some of these are considered in detail elsewhere in this volume (see papers by *Akert* and *Oksche*). One may however select for consideration some of the ways in which modified ependymal cells have been shown to differ from the "normal" or classical ependyma (Figs. 1 and 2), choosing for illustration those areas of ependyma which lie in, or adjacent to the hypophysiotropic area, and are therefore of particular interest to neuroendocrinologists.

An absence of cilia distinguishes many areas of modified ependyma including the prechiasmatic gland and other regions lining the floor of the third ventricle. *Schachenmayr* (1967) has drawn attention to differences between the enzymatic patterns of ciliated and non-ciliated ependyma which he considers to be related to the oxidative energy metabolism serving the needs of ciliary movement. There is, for instance, no demonstrable activity of acid phosphatase, succinate dehydrogenase, cytochrome oxidase and hydroxy-butyric acid dehydrogenase in non-ciliated ependyma in contrast to normal ciliated ependyma. *Nandy* and *Bourne* (1965) have suggested that the enzymatic pattern of normal ependyma indicates functions besides those of ciliary activity, possibly concerning absorption

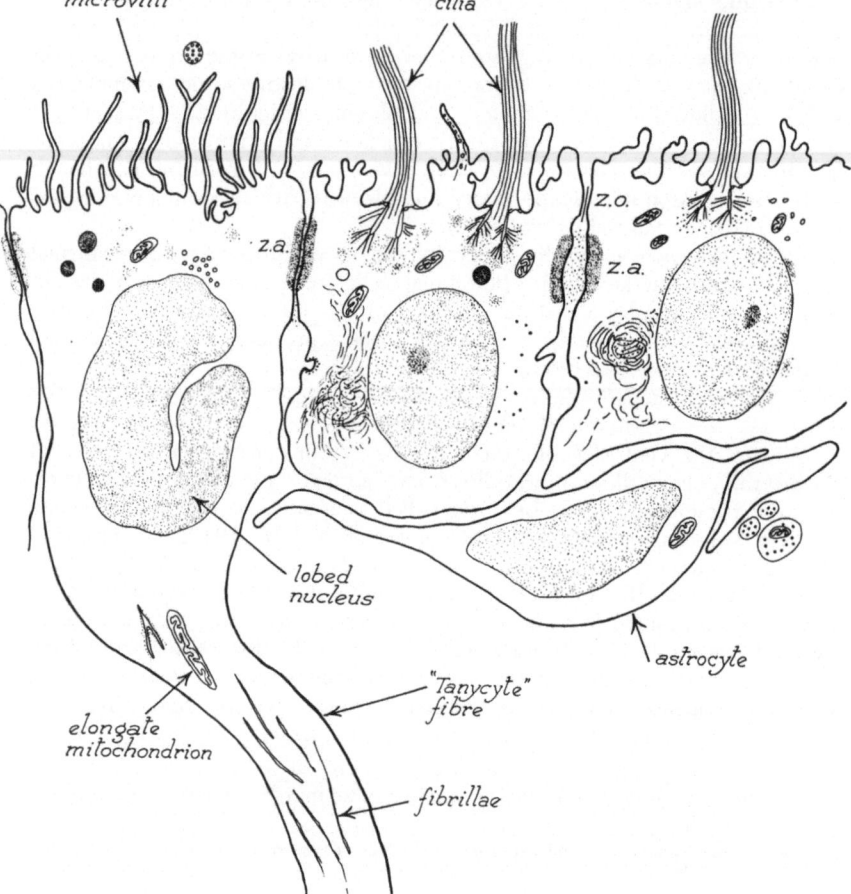

Fig. 1. Some features of "normal" and tanycyte ependyma compared. z.a. zona adhaerens, z.o. zona occludens.

and secretion into the cerebrospinal fluid. In this connection it is interesting to note that *Schachenmayr* (1967) has noted that enzymatic activity in ciliated ependymal cells is predominantly found in the apical part of the cells, whereas in non-ciliated ependyma in the floor of the III. ventricle it is evenly distributed over the cells and in their processes. Such a distribution might be consistent with a view that ciliated ependymal cells are predominantly concerned with events in the cerebrospinal fluid while specialized non-ciliated ependymal cells may serve to link this fluid with other regions of the brain, and notably the blood system.

Many non-ciliated ependymal cells retain their long embryonic basal processes which extend to perivascular basement membranes. *Leonhardt* (1966) has described the ultrastructure of ependymal cells in the wall of the III. ventricle near the entrance to the recessus infundibularis and has shown that their processes make intimate contacts with capillaries (see Fig. 2, sub B). In common with other German workers he uses the term "Tanycyte" (stretch cell) to describe these elongate ependymal cells. *Schachenmayr* (1967) has studied the development of tanycyte ependyma in the ventral region of the ventricle (radix and recessus infundibuli, recessus inframamillaris) and has noted that while classical ciliated ependyma is morphologically and histochemically differentiated between the 5th and 19th day of life in the rat the tanycyte ependyma, though in fact it retains some embryonic characters, does not completely differentiate before the 34th day of life.

Many of the ependymal cells covering the median eminence are characterised by thick basal processes containing microtubules (Fig. 2, sub D), which pass through the substance of the median eminence and terminate perivascularly as ependymal feet (*Rinne*, 1966).

Munroe (1967) has described the median eminence as a "mixture of a large number of axon terminals and ependymal feet converging on the plexus of the portal vessels".

In the ventroposterior part of the infundibular recess ependymal processes extend to pericapillary spaces between the neurohypophysis and adenohypophysis (*Knowles* and *Vollrath*, 1966). The ultrastructure of the ependymal cells lining the medial region of the floor of the III. ventricle suggests that these cells secrete into the cerebrospinal fluid (*Knowles* and *Vollrath*, 1966; *Leveque* and coll., 1967). Membrane-bound vesicles and smooth-surfaced tubules and vesicles resembling a Golgi complex are found in that part of the cytoplasm adjoining the ventricle. This area is stained deeply by the periodic acid-Schiff reagent, and *Knowles* and *Vollrath* have moreover remarked that under certain conditions the cerebrospinal fluid itself in the areas bounded by these cells may become PAS-positive.

Tanycyte ependyma in general stains positively with PAS and with the basic component of the Gomori chrome alum-haematoxylin method, and thus these cells resemble those of the neurosecretory systems of vertebrates and invertebrates. The PAS material does however differ from neurosecretory material in that it is insoluble in an alcohol-chloroform mixture (2 : 1), and the Gomori stain will colour ependymal cells after a very brief oxidation when neurosecretory material remains unstained (*Leveque*, 1963; *Vigh*, 1964). Nevertheless the resemblances between tanycyte ependyma and the neurosecretory neurones of the hypothalamo — hypophysial system are striking. Both constitute glandular "neuronal" elements and have close functional relationships with the cerebrospinal fluid and the blood vascular system. Indeed it has been postulated that Type A neurosecretory neurones may possess a closer cytogenetic relationship to secretory ependyma than to "ordinary" neurones (see *Knowles*, 1967). This suggestion has received some experimental support in the results of extirpation of neurosecretory elements of the urophysis; these are replaced by ependymal elements (*Fridberg* and coll., 1966). The possibility that preoptic neurosecretory neurones are derived from ependymal elements is also suggested by *Smoller* (1965).

Mention should also be made of a striking difference, of evident functional significance, between "normal" and atypical ependyma. No direct innervation of "normal" ependyma has yet been described, but certain glandular ependymal cells of the III. ventricle appear to be innervated by neurosecretory axons. Synaptoid contacts have been seen in the eel (*Knowles* and *Vollrath*, 1966), in the cat (*Hagen* and *Wittkowski*, 1967), and in the Rhesus monkey (*Knowles* and *Anand Kumar*, in preparation).

III. Functional Aspects of Specialized Ependyma

It has been shown that two morphological features of specialised ependyma are especially worthy of note, namely:

a) Many atypical ependymal cells have morphological characteristics very similar to those of embryonic ependymal cells, in that they possess long fibrillar processes which are directed laterally or ventrally away from the ventricle and terminate in close proximity to blood vessels,

b) Evidence for active synthesis and secretion in these cells suggests that this special ependymal elements may secrete either into the cerebrospinal fluid or into the blood stream, or both, or may detect substances in one of these tissues fluids and secrete into the other. Such considerations have led a number of workers to suggest that secretory ependyma rather than neurones might in certain cases serve the correlation of cerebral and endocrine functions (see *Scharrer*, 1965). Currently this possibility is being studied by a group at Birmingham. The work is still in progress and the results have as yet only been presented in abstract form (*Anand Kumar* and *Knowles*, 1967 a, b; *Knowles* and coll., 1967; *Weatherhead*, 1967). Attention has been especially directed towards a group of specialised ependymal cells in the hypothalamus which do not appear to have been previously described specifically or in detail and which seem to bear a special relationship to pituitary function. Some results of these studies will be presented here, as a preliminary survey, in relation to the results of other workers in this field.

Some years ago *Leveque* and *Hofkin* (1960) called attention to two areas in the hypothalamus of the rat, which although known to exhibit morphological differences from normal ependyma (*Cajal*, 1952) were otherwise uncharacterised. *Leveque* and *Hofkin* remarked that these areas, which line the base and lower third of the supraoptic and infundibular recesses, showed a marked affinity for the Schiff reagent in periodic acid Schiff (PAS) preparations. Subsequent light microscopy (*Leveque*, 1963; *Vigh*, 1964) and electron microscopy (see *Leveque* and coll., 1967) have established a secretory activity of these cellular elements.

Experimental procedures were followed by some alteration in the stainability of these cells; cold-stress and cortisone treatment led to an increase in the amount of PAS-positive material (*Leveque*, 1963). These results were suggestive of an involvement of these areas in pituitary function but a correlation with normal function was lacking, and the way in which these special ependymal cells might affect pituitary function was not clear, since the secretory material seemed to be directed mainly towards the ventricle.

A correlation between ependymal secretion and normal physiological activity of the pituitary has recently been demonstrated in the eel (*Knowles* and *Vollrath*, 1966). Conditions of illumination and background which are known to affect MSH release from the pituitary were found likewise to affect secretory activity of that ependyma lining the floor of the infundibular recess. Moreover, synaptoid contacts between neurosecretory fibres and these ependymal cells were demonstrated and shown to alter in relation to MSH release. On the basis of these results *Knowles* and *Vollrath* suggested the possibility that the ependyma in the infundibular recess of the eel might form part of a feedback mechanism between the proximal and distal portions of the hypothalamo-hypophysial system, by means of substances released into the cerebrospinal fluid. This idea of the cerebrospinal fluid as a vehicle for chemical information finds support in the work of *Feldberg* and his fellow workers (see *Feldberg's* paper in this volume).

In 1965 *Hagedoorn* reported that ependyma on the lateral walls of the III. ventricle of the skunk showed seasonal changes in relation to the sexual activity of this animal. The area which she mentioned is close to a region which has been suspected of influencing pituitary gonadotropic function, and it occurred to us that it might be profitable to study ependyma of the anterior ventrolateral walls of the III. ventricle in relation to sexual activity using the electron microscope. These studies in ultrastructure are proceeding on the Rhesus monkey (*Knowles* and *Anand Kumar*, in preparation) and on the ferret (*Knowles* and coll., 1967). Thus far some interesting correlations have been observed between structure and function.

In the ferret an area, slightly higher than those described by *Leveque* and his co-workers has received particular attention. This area lies close to the opening of the infundibular recess and may be seen in optical microscope preparations which have been cut transversely at the level of the ventromedial nucleus. It has, for convenience, been termed the x-area (*Jones*, 1967). A comparable area, at a slightly more anterior position, has been described by *Anand Kumar* and *Knowles* (1967 a, b) in the Rhesus monkey. At the level of ultrastructure the ependymal cells of this region (see Fig. 2, sub B and C) differ markedly from those above and below. Unlike ependymal cells higher on the ventricular wall, they have few or no cilia; instead their ventricular surface is covered with microvilli, some of which in the Rhesus are often swollen to form striking bulbous projections (Fig. 2, sub C). Other striking features are the length of their fibre processes, some of which extend to the primary plexus of the hypophysial portal circulation, and the very numerous branches of these processes containing great quantities of electron-dense granules. The abundance of granules at the distal terminals of these tanycyte fibres indicates that these may secrete into the blood-stream. Granules have been described also in other regions of tanycyte ependyma (see *Kruger* and *Maxwell*, 1966, and

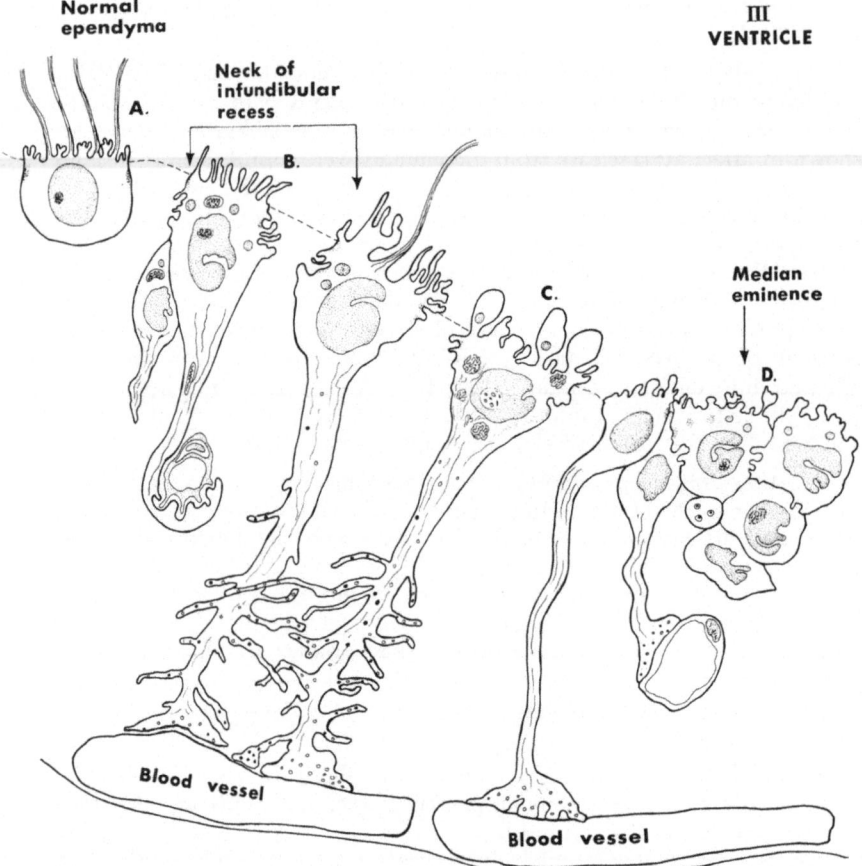

Fig. 2. A semi-diagrammatic representation of some different forms of ependyma found in the anterior region of the infundibular recess of the Rhesus monkey; those of the ferret were substantially similar except that the bulbous projections of cell type C were less evident (for descriptions of cell types A — D see text).

Weatherhead, 1967). In the present studies the finest branches of the ependymal fibres appeared to make contact and form a syncytial meshwork. It is noteworthy that earlier workers using the optical microscope have described similar syncytial arrangements in tanycyte ependyma elsewhere (*Studnička*, 1902; *Cajal*, 1909).

Comparative studies of the brains of female ferrets and monkeys killed at different stages of reproductive activity were made (*Jones,* 1967). Ferrets during the winter months are normally in a state of anoestrus, but they can be brought into oestrus by additional illumination or by injections of oestrogen; in the spring they are in natural oestrus. It has therefore been possible to compare anoestrous and oestrous ferrets simultaneously or seasonally. Interesting differences in the tanycytes at the ventricular border of the x-area were seen. In the anoestrous ferret this area can be recognised under the light microscope by the fact that it stains less densely by the Gomori technique than neighbouring ependyma on either side. In contrast in the oestrous animal, whether natural or artificially induced, the staining of the x-area and nearby areas is similar. Under the electron microscope it may be seen that the cell bodies of x-area ependymal cells of the oestrous ferret contain many more electron-dense inclusions than are found in similar cells of the anoestrous ferret; it seems possible that these may be responsible for the relatively deep staining of these cells in the oestrous animal. At present studies are in progress to find out whether castration of ferrets in the season of normal oestrus is followed by changes characteristic of anoestrus in the x-area ependyma.

Studies are also in progress to compare x-area ependyma at different stages of the menstrual cycle in the Rhesus. Thus far the results suggest that at the time of ovulation, or soon after, the bulbous projections of the x-area tanycytes enlarge greatly and there is a striking increase in the amount of electron-dense material in the perinuclear cytoplasm.

There are some indications that oestrogen injected systemically causes changes in the shape of the ventricular border of the x-area ependyma, and a noticeable increase in the amount of stainable material in the cytoplasm of these cells has been observed in the Rhesus monkey following oestrogen injection. These results would be consistent with a hypothesis that the tanycyte ependymal cells of the x-area are responsive to the amount of oestrogen circulating in the blood and/or the cerebrospinal fluid. The modifications to the ventricular border of the ependyma might seem to favour the cerebrospinal fluid as the more significant vehicle for the transport of oestrogen to the tanycytes and in this connection it is interesting to note that labelled oestrogen injected subcutaneously in the Rhesus has been detected in the cerebrospinal fluid and specifically in and around the x-area (*Anand Kumar* and *Knowles,* 1967 a, b).

In view of the proximity of distal terminals of the x-area ependyma to cells of the pars tuberalis, the ultrastructure of this region received particular attention. The results of these studies will be published in detail later and it will suffice for the present to remark that at moments when the amount of circulating oestrogen might be presumed to be low, in the anoestrous ferret and the preovulatory phase of the monkey-menstrual cycle, cells of the pars tuberalis were packed with electron-dense granules. Conversely in the oestrous ferret and during the post-ovulatory phase of the menstrual cycle of the monkey many cells of the pars tuberalis showed marked vacuolation, and if any granules were present these were few, and small (see *Anand Kumar* and *Knowles*, 1967 a, b).

It would be premature to draw definite conclusions from the studies on the ferret and the monkey until further experiments to determine the specificity of the responses observed can be carried out. The results so far are however sufficiently consistent to point to the tanycyte ependyma of the x-area as a possible receptor site for oestrogen feedback in the relationship between pituitary and gonad.

IV. General Considerations

It is evident that certain ependymal cells in the ventrolateral region of the hypothalamus have glandular characteristics and are so situated that they could absorb from or secrete into the blood and/or the cerebrospinal fluid. In these respects it differs from most other ependyma but resembles some neurosecretory systems (*Knowles*, 1967; see also *Sterba* and *Weiss*, 1965, 1967). The fact that this tanycyte ependyma corresponds in position to the hypophysiotropic area (believed to be concerned in pituitary control), and that some of these tanycytes can be correlated with pituitary function under normal and experimental conditions suggests that it may indeed be implicated in the correlation of cerebral and endocrine function. This aspect of ependymosecretion still requires extensive investigation, but already enough has been discovered to arrive at some interesting theoretical possibilities.

The central nervous system receives and integrates information and stimuli that affect endocrine functions (*Scharrer*, 1965, 1966). In some cases, for example, osmotic control, changes in the blood modify neuroendocrine activity. In other instances, notably in the control of reproduction, information from the external environment is integrated with events in the internal milieu. For example additional illumination of a ferret provokes the release of gonadotropins from the pituitary, but in turn the consequent release of oestrogen from the gonad modifies gonadotropin release.

We must therefore look for neuroendocrine mechanisms capable of responding to synaptic stimulus and/or alterations in the composition of the blood. Within recent years evidence has favoured the hypothesis that some form of neurosecretion provides the final common pathway for pituitary control (*Scharrer*, 1965; *Harris* and *Donovan*, 1966). Comparative studies, especially in the lower vertebrates, have indicated a dual control of some aspects of pituitary function by two distinct types of neurosecretory neurone (A and B), tentatively identified as peptidergic and monoaminergic respectively. These neurosecretory systems have many neuronal characteristics. Their perikarya lie in the brain tissue and are apparently innervated by "normal" neurones (*Knowles*, 1967). It is however debatable whether these alone could carry out the complexity of integration required to maintain the delicate balance of the neuroendocrine system in relation to both the external and internal environments of an animal. In some of his last major works the late *Ernst Scharrer* considered the possibility that the known complexity of neuroendocrine control might be viewed in terms of more than classical neurosecretory systems, and that modified ependymal elements also might play some part in the correlation of cerebral and endocrine function (*Scharrer,* 1965, 1966). Direct proof of this is still lacking, but recent work on the tanycytes of the floor and ventrolateral walls of the III. ventricle does indicate strongly that these may be concerned in pituitary control. It is perhaps noteworthy that, although some evidence of both proximal and distal secretion may be observed in these cells, those tanycytes medially situated appear predominantly to secrete into the cerebrospinal fluid and those more laterally placed show a major evidence of secretion in the direction of the blood vascular system. Thus they could contribute the theoretical essentials for a complete system of feedback control between the pituitary and its target organs, in which both blood system and cerebrospinal fluid may be concerned (see Fig. 3). This at least is a working hypothesis which merits investigation.

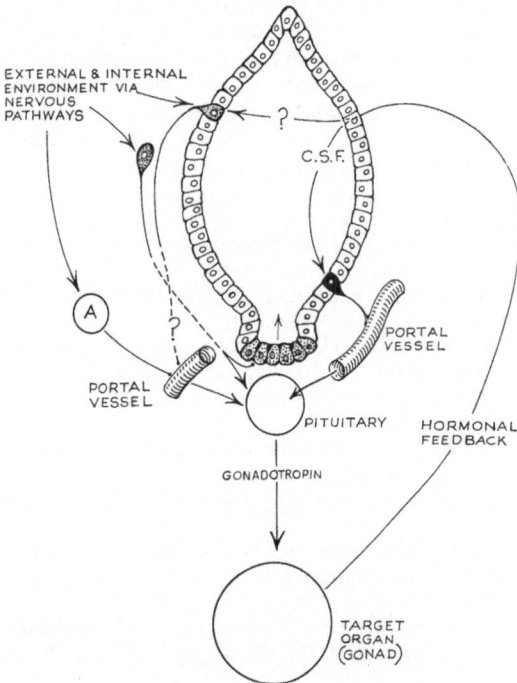

Fig. 3. Some of the postulated neuronal, neurosecretory and hormonal pathways which may be concerned in the regulation of the pituitary-gonad relationship in vertebrates. The left side of the figure depicts neuronal and neurosecretory pathways. It has been suggested that two forms of the latter may be concerned. Type A or peptidergic originating in the supra-optic and paraventricular nuclei, and Type B or aminergic possibly originating in the arcuate and other nuclei (see *Knowles* and *Vollrath,* 1966).
The right side of the figure shows the possible pathways for feedback control, involving specialized ependyma, discussed in the present paper. A the region indicated by *Halasz* et al. (1962), containing the arcuate nucleus.

References

Anand Kumar, T. C. and *Sir F. Knowles:* A system linking the third ventricle with the pars tuberalis of the rhesus monkey. Nature (London) *215*, 54 (1967a).

Anand Kumar, T. C. and *Sir F. Knowles:* Experimental modification of an area of specialised ependyma in the hypothalamus of the rhesus monkey. Abstract. Confer. Europ. Comp. Endocrinol. Carlsbad. Gen. Comp. Endocrinol. 9, 513 1967b).

Brightman, M. W.: The distribution within the brain of ferritin injected into the cerebrospinal fluid compartments. I. Ependymal distribution. J. Cell. Biol. *264*, 99—123 (1965).

Brightman, M. W., and *S. L. Palay:* The fine structure of ependyma in the brain of the rat. J. Cell. Biol. *19*, 415—439 (1963).

Cajal, S. Ramon Y.: Histologie du système nerveux de l'homme et des vertébrés. Paris: Norbert Malcine 1909.

Cajal, S. Ramon Y.: Histologie du systéme nerveux de l'homme et des vertébrés. Madrid: Inst. Ramon y Cajal 1952.

Dendy, A. and *G. E. Nicols:* On the occurrence of a mesocoelic recess in the human brain, and its relation to the sub-commissural organ of lower vertebrates; with special reference to the distribution of Reissner's fibre in the vertebrate series and its possible function. Anat. Anz., Jena *37*, 496—508 (1910).

Feldberg, W., and *K. Fleischhauer:* Penetration of bromophenol blue from the perfused cerebral vesicles into the brain tissue. J. Physiol. *150*, 451—462 (1960).

Fridberg, C., R. S. Nishioka, H. A. Bern and *W. R. Fleming:* Regeneration of the caudal neurosecretory system in the Cichlid teleost *Tilapia mossambica.* J. Exp. Zool. *162*, 311—336 (1966).

Ganong: In: Review of Medical Physiology, 2nd ed. P. 480. Oxford: Blackwell 1965.

Hagedoorn, J.: Seasonal changes in the ependyma of the third ventricle of the skunk, *Mephitis mephitis nigra.* Anat. Rec. *151*, 453 Abstr. (1965).

Hagen, E. and *W. Wittkowski:* Zur Ultrastruktur des Hypothalamus-Hypophysensystems unter besonderer Berücksichtigung neurohämaler und neuroglandulärer Kontaktgebiete der Hypophysis. Abstract. Symposion on Neurohormones and Neurohumors, Amsterdam. 1967.

Halàsz, B., L. Pupp, and *S. Uhlarik:* Hypophysiotropic area in the hypothalamus. J. Endocrinol. *25*, 147—154 (1962).

Harris, G. H. and *B. T. Donovan:* The pituitary gland. London: Butterworths 1966.

Hild, W., T. Takenaka and *F. Walker:* Electrophysiological properties of ependymal cells from the mammalian brain in tissue culture. Exp. Neurol. *11*, 493—501 (1965).

Jones, C. F.: Changes in specialised ependyma of the ferret in relation to the oestrous cycle. Thesis presented for the degree of B. Sc. Birmingham University. 1967.

Klatzo, I., J. Miquel, P. Ferriz, J. D. Prokop and *D. E. Smith:* Observations on the passage of the fluorescein labeled serum proteins (FISP) from the cerebrospinal fluid. J. Neuropath. exp. Neurol. *23*, 18—35 (1964).

Knowles, Sir F.: Neuronal properties of neurosecretory cells. Proceedings IV. Int. Symp. on Neurosecretion. Berlin: Springer 1967.

Knowles, Sir F., and *L. Vollrath:* Neurosecretory innervation of the pituitary of the eels *Anguilla* and *Conger.* Phil. Trans. Roy. Soc. B. *250*, 311—342 (1966).

Knowles, Sir F., T. C. Anand Kumar and *C. F. Jones:* Structure and ultrastructure of an area of specialized ependyma in the hypothalamus in relation to reproductive activity. Abstract. Confer. Europ. Comp. Endocrinol. Carlsbad. Gen. Comp. Endocrinol. *9*, 526 (1967).

Kruger, L. and *D. S. Maxwell:* The fine structure of ependymal processes in the teleost optic tectum. Am. J. Anat. *119*, 479—498 (1966).

Leonhardt, H.: Über ependymale Tanycyten des III. Ventrikels beim Kaninchen in elektronenmikroskopischer Betrachtung. Z. Zellforsch. *74*, 1—11 (1966).

Leveque, T. F.: In Nalbandov's Advances in Neuroendocrinology. Pp. 314—328. Illinois University Press 1963.

Leveque, T. F., and *G. A. Hofkin:* A periventricular PAS-reactive substance in the rat hypothalamus. Anat. Rec. Philadelphia *136*, 232 (1960).

Leveque, T. F., G. A. Hofkin, F. Stutinsky, A. Porte and *M. Stoeckel:* Ultrastructure of the medical prechiasmatic gland in the rat and mouse. Neuroendocrinol. *2*, 56—63 (1967).

Lups, S. and *A. M. F. H. de Haan:* The cerebrospinal fluid. Amsterdam: Elsevier Publishing Company 1954.

Munroe, B. G.: A comparative study of the ultrastructure of the median eminence, infundibular stem and neural lobe of the hypophysis of the rat. Z. Zellforsch. *76*, 405—432 (1967).

Nandy, K. and *G. H. Bourne:* Histochemical studies on the ependyma lining the central canal of the spinal cord in the rat with a note on its functional significance. Acta Anat. (Basel) *60*, 539—550 (1965).

Pollay, M. and *H. Davson:* The passage of certain substance out of the cerebrospinal fluid. Brain *86*, 137—150 (1963).

Putnam, T. J.: The intercolumnar tubercle, an undescribed area in the anterior wall of the third ventricle. Bull. John Hopkins Kosp. *33*, 181—182 (1922).

Rinne, U. K.: Ultrastructure of the median eminence of the rat. Z. Zellforsch. *74*, 98—122 (1966).

Roussy, G. and *M. Mosinger:* Les corrélations épiphyso-hypophysaires. (Le système neuro-docrinien du cerveau). Ann. anat. path. Paris *15*, 847—858 (1938)

Schachenmayr, W.: Über die Entwicklung von Ependym und Plexus chorioideus der Ratte. Z. Zellforsch. *77*, 25—63 (1967).

Scharrer, E. The final common path in neuroendocrine integration. Arch. d'Anat. microsc. *54*, 359—370 (1965).

Scharrer, E.: Principles of neuroendocrine integration. In: Endocrines and the Central Nervous System. Baltimore: Williams and Wilkins Co. 1966.

Smoller, C. G.: Neurosecretory processes extending into third ventricle: secretory or sensory? Science *147*, 882—884 (1965).

Sterba, G. and *J. Weiss:* Beiträge zur Hydrencephalokrinie: I. Hypothalamische Hydrencephalokrinie der Bachforelle (*Salmo trutta fario*). J. f. Hirnforsch. *9*, 4, 359—371 (1967).

Studnička, F. K.: Untersuchungen über den Bau des Ependyms der nervösen Centralorgane. Anat. Hefte, Wiesbaden II, *15*, 301—431 (1902).

Vigh, B.: Ependymosécrétion, sécrétion Gomori-positive de l'épendyme dans l'hypothalamus. Ann. Endocrin. (Paris) *25*, Suppl., 140—141 (1964).

Vigh, B., B. Aros, T. Wenger, S. Koritsanszky, and *G. Cegledi:* Ependymosecretion (Ependymal neurosecretion) IV. The Gomori-positive secretion of the hypothalamic ependyma of various vertebrates and its relation to the anterior pituitary. Acta Biol. Szeged. *13*, 407—419 (1963).

Weatherhead, B.: Ultrastructure of certain ependyma of the lacertilian hypothalamus and pars nervosa. Abstract. Confer. Europ. Comp. Endocrinol. Carlsbad. Gen. Comp. Endocrinol. *9*, 523 (1967).

Worthington, W. C. *Jr.* and R. S. *Cathcart:* Ependymal cilia; distribution and activity in the adult human brain. Science *139,* 221—222 (1963).

Author's address: Sir *Francis Knowles,* Department of Anatomy. King's College, Strand, London, W. C. 2, England.

Discussion

Bargmann: Dr. *Knowles,* can you tell us something concerning the origin of the granules situated in the cytoplasm of the ependymal cells? Are they comparable to elementary granules and do they originate in the Golgi system or not? What about their ultimate fate?

Knowles: I can give but little precize information at this moment. Some spherical electron-dense masses have been seen lying in tubules which resemble a Golgi complex, close to the nucleus, in the basal region of the cell. Some evidence of endoplasmic reticulum has been found in cytoplasm close to the ventricular border. Thus far, the observations indicate synthesis in the perinuclear zone and transport of granules along the distal prolongations of the tanycytes.

Weitzman: Do the whorls of endoplasmic reticulum shown in your micrographs of the pars tuberalis perhaps represent early signs of cell degeneration?

Knowles: It is difficult to be dogmatic on this point, but other indications favour increased synthetic activity of pars tuberalis cells in the oestrous ferret and post-ovulatory Rhesus monkey. Therefore, we believe that the changes in the endoplasmic reticulum are more likely to represent development than degeneration.

Akert: Sir Francis, you mentioned "synaptoid" links which are characterized by apposition of dark material at the presynaptic and not at the postsynaptic site. Could you comment a bit further on this situation?

Knowles: This uneven distribution of electron-dense material has been noted in other "synapses" between neurosecretory fibers and endocrine and other epithelial cells, notably in the dogfish pituitary. It has been suggested that the lack of dark material at the post-synaptic side might represent an absence of a substance or substances inactivating the neurohormone, as might be a prerequisite for the prolonged and continuous release of hormone from neurosecretory neurones.

Journal of Neuro-Visceral Relations, Suppl. IX, 111—139 (1969)

The Subcommissural Organ *

A. Oksche

Department of Anatomy. University of Giessen, Germany

With 15 Figures

Summary

The subcommissural organ (SCO) consists of a secretory ependymal band and secretory hypendymal cells which are derived from the ependymal layer. Because of its intensive secretory activity and its cytochemical properties this general complex of cells assumes, among the ordinary ependymal and glial cells, a position similar to that of the neurosecretory cells among ordinary nerve cells. The secretory materials of the SCO are formed in distended cisternae of the rough endoplasmic reticulum. Using as examples the SCO's of the tree frog (Hyla arborea), the dog, and the human embryo, the anatomical polarity of the SCO, i. e. the orientation to the ventricle as well as to the vascular bed, is discussed. The anatomical situation in some vertebrate species (e. g. the dog) is such as to suggest not only a secretion into the cerebrospinal fluid but also the possibility of direct release of substances into the vascular system. The function of the SCO is not understood with certainty.

Zusammenfassung

Das SCO besteht aus einem sekretorischen Ependymverband und sezernierenden Hypendymzellen, die aus der Ependymschicht ausgewandert sind. Auf Grund seiner exzessiven sekretorischen Aktivität und seiner cytochemischen Eigenschaften nimmt dieser gesamte Zellkomplex unter den gewöhnlichen Ependym- und Gliazellen etwa eine ähnliche Stellung ein wie neurosekretorische Neurone unter gewöhnlichen Nervenzellen. Das Sekret des SCO wird in stark erweiterten Zisternen des rauhen endoplasmatischen Reticulums gebildet. Am Beispiel des SCO von Laubfrosch (Hyla arborea), Hund und Mensch (Embryo) wird die anatomische Polarität des SCO, d. h. Ausrichtung sowohl auf den Ventrikel als auch auf die Blutbahn, diskutiert. Die anatomischen Voraussetzungen sind bei einigen Wirbeltierspezies (z. B. Hund) so, daß eine Abgabe von Stoffen nicht nur in den Liquor cerebrospinalis sondern auch direkt in die Blutbahn denkbar ist. Die Funktion des SCO ist noch nicht mit Sicherheit bekannt.

Introduction

The preceding communication of Professor *Knowles* has demonstrated that with modern methods of ultrastructural research it is possible to demonstrate secretory cells in the ependyma lining the third ventricle of

---

* The investigations reported herein were supported by a research grant from the Deutsche Forschungsgemeinschaft to the author.

the brain. Secretory ependymal cells have also been observed in the
hypothalamic section of the third ventricle and in the infundibular recess
by *Leveque* and coll. (1965, 1967) and *Vigh* (1964) (see also *Löfgren*, 1960 a,
b, 1961). Indications of a regionally restricted secretory activity among
ependymal cells, especially in lower vertebrates, were already known to
Galeotti (1897) and *Studnička* (1900). On the other hand it has now been
recognized (*Takeichi*, 1967; *Röhlich* and *Vigh*, 1967) that in some
ependymal organs, such as the paraventricular organ (*C. U. Ariëns Kappers*,
1920; *Legait*, 1942), formerly assumed to be secretory, the active elements
are actually neurosecretory nerve cells of the subependymal layer. Processes
of such cells extend toward the ventricle, penetrate the ependymal layer,
and project into the lumen of the ventricle in club-shaped, granule-contain-
ing protrusions. Subependymal secretory neurons were described also by
Leonhardt (1967) and reviewed by *Sterba* and *Weiss* (1967).

A specialized ependyma, which is especially active in secretion, occurs
in the subcommissural organ (SCO). This circumventricular organ covers
the posterior commissure in the boundary area between the diencephalon
and mesencephalon; it is notably present in all vertebrates, even in groups
in which the pineal organ is apparently lacking. (For references, see
Oksche, 1965, pp. 18—19).

The general anatomy of the secretory SCO is shown in a cleared total
preparation from a cat (Fig. 1 a). Morphological details can be found in
publications by *Olsson* (1958 a, b), *Talanti* (1958), *Oksche* (1961) and
Lenys (1965 a, b) as well as in the monograph of *Palkovits* (1965). Of
importance is the fact that *Reissner's* fiber (Fig. 1 a), which can extend
to the terminal ventricle of the central canal (*Hofer*, 1963), is now known
to be formed by secretion from the SCO into the ventricular system
(*Sterba* and coll., 1967). *Krabbe* (1925) discovered that certain cells between
the ependyma and the posterior commissure, diffusely distributed or in
clusters, are actually a part of the SCO. This formation of cells was
designated by him as the hypendyma. Further it should be emphasized
that the processes of the ependymal and hypendymal cells can penetrate
the posterior commissure and terminate in club-like endfeet at the external
limiting membrane (Fig. 1 b). *Stutinsky* (1950) first demonstrated that the
secretion of the SCO, like that of the supraoptic and paraventricular nuclei,
can be selectively stained with chromalum-hematoxylin *(Gomori-Bargmann)*.
This method and similar staining techniques have facilitated and intensified
research on the secretory activity of the SCO.

What is known about the function of this organ? Since the demonstration
by *Gilbert* (1956) that electrolytic destruction of SCO in rats disturbs
water balance, several laboratories have conducted investigations on the
basis of this working hypothesis. The results of the investigations of *Gilbert*
(1956, 1957, 1964) and *Farrell* (1958, 1959, 1960) led to the concept of a
center within the posterior commissure — pineal area which exerts some
influence on water and electrolyte metabolism. *Farrell*, on the basis of
stereotactic lesions and experiments with extracts, has suggested that this
center regulates aldosterone output. *Palkovits* (for references, see 1965),

with a variety of physiologic and cytologic methods, has proven that the aldosteronotropic factor arises solely from the SCO. A critical position with respect to these investigations has been adopted by *Crow* (1964) and *Van der Wal* and coll. (1965). A completely isolated removal of secretory subcommissural ependyma and hypendyma cells is, in fact, not

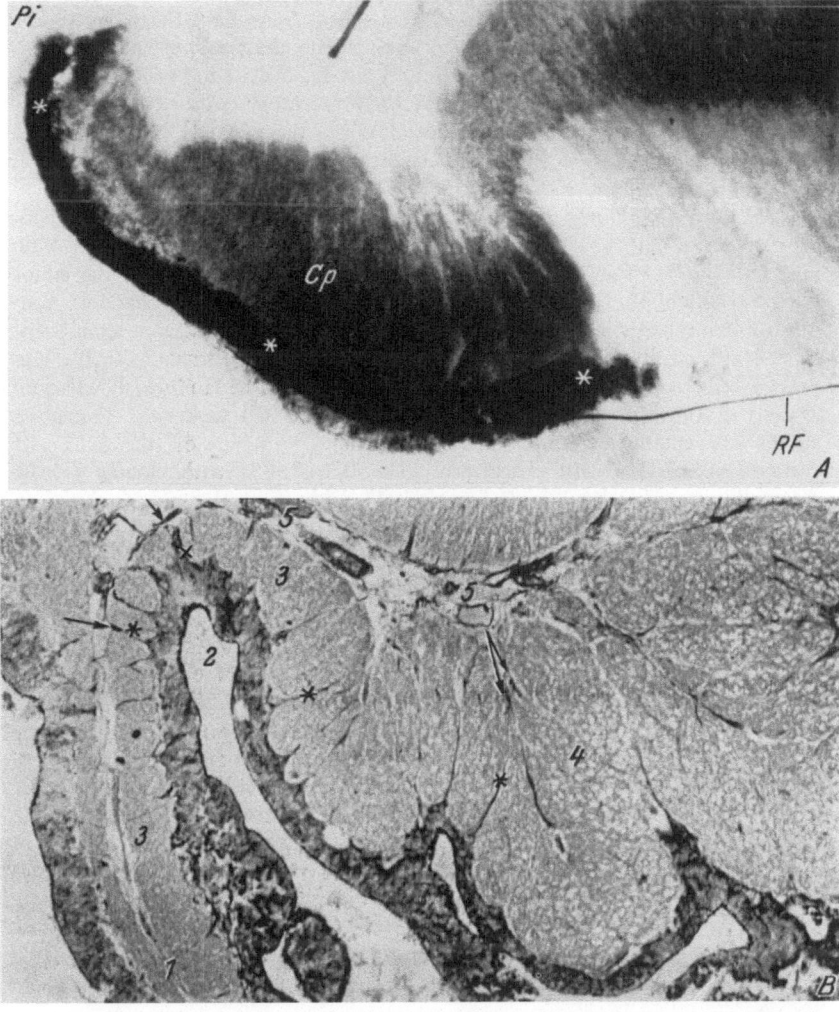

Fig. 1. General view of the subcommissural organ. A: SCO of the cat (°). Cleared total preparation, stained with paraldehyde-fuchsin. RF Reissner's fiber, Cp posterior commissure, Pi pineal area. B: SCO of the dog. Paraffin section, 6 μ, paraldehyde-fuchsin, x 55. Conspicuous regional differentiation. 1, ventral fold; 2, rostral ventricular recess with thin walls (3); 4, thick caudal portion of the posterior commissure. Columns (x) and strands (°) of selectively stained cells can be followed to the external limiting membrane (→); 5, blood vessels.

possible without interfering with the neighbouring pineal organ or disturb-
ing the posterior commissure which contains many different fiber systems.
It seems certain that no ideal neurosurgical, radiological, or pharmacological
approach is available.

Many older hypotheses on the function of the SCO (for reviews see *Olsson*
1958 b, *Palkovits*, 1965), such as that of a sensory organ, can now be regarded as
no longer tenable. Reissner's fiber, which originates from the SCO, is not sensory
(Kolmer), but rather a special form of secretion into the cerebrospinal fluid.

Cytologically it is now undeniably certain that the ependymal cells
of the SCO are secretory and that their secretory products (mucopoly-
saccharide-protein complexes or mucoproteins) pass into the cerebrospinal
fluid (for general review and histochemical references, see *Palkovits*, 1965;
Sterba, 1965; *Sterba* and coll., 1966, 1967). At the ultrastructural level
(Stanka and coll., 1964; *Isomäki* and coll., 1965; *Müller* and *Sterba*, 1965;
Sterba and coll., 1966, 1967; *Barlow* and coll., 1967; *Stanka*, 1967; *Vigh*
and coll., 1967) this is evident from the extensive cisternae of the endoplasmic reticulum. These extensively enlarged intracellular channels, containing a large amount of moderately dense protein material, are reminiscent
of the ultrastructural characteristics of plasma cells *(Fawcett*, 1966). The
precise role of the well-developed Golgi complex and of the intracytoplasmic
system of fine tubules *(Sterba* and coll., 1967) is still unknown. Formative
stages of secretory granules within the Golgi complex of the SCO-cells
were described by *Murakami* and *Tanizaki* (1963) and *Stanka* (1967).
According to *Murakami* and *Tanizaki* (1) dense granules are produced in
the Golgi complex, (2) vacuoles containing a material of low density are
derived from the rough endoplasmic reticulum.

Still open are questions of polarity in the SCO *(Oksche*, 1961), *i. e.*,
the possibility of a second secretory route to the blood vessels in the
commissural area or in the leptomeninges *(meninx primitiva* in lower,
pia-arachnoid in higher vertebrates). The objections to the assumption of a
secretory polarity in the SCO (see *Stanka* and coll., 1964; *Stanka*, 1967;
Müller and *Sterba*, 1965) are, however, primarily based on findings obtained
from species in which the anatomical basis for such a secretory route does
not exist or is at least not apparent. (For positive evidence, see *E. Legait*,
1946, 1949; *Okada*, 1956; *Murakami*, 1959; *Murakami* and *Tanizaki*, 1963;
Mautner, 1965; *Ghiani* and *Uva*, 1965; *Marini*, 1966).

I wish to direct attention to this question now, on the basis of light-
and electron microscopic studies, using material, which, in my opinion,
is of a model nature. Our relevant investigations have been carried out
with *Hyla arborea*, the dog, and the human embryo.

Material and Methods *

I. *Light microscopic studies.*

Bouin's fixation. Staining procedures: (1) *Demonstration of secretory material:*
Chromalum-hematoxylin *(Gomori-Bargmann)*, paraldehyde-fuchsin--orange-G-fast

---

* For details of all procedures see the laboratory manuals of *Romeis* or *Pearse*
and also the papers published by members of our group (see references).

green *(Gomori-Halmi-Dawson)*, alcian blue, PAS-method, pseudoisocyanin-fluorescence method *(Sterba)*. (2) *Demonstration of glycogen:* Lead tetraacetate-Schiff method *(Shimizu* and *Kumamoto)*. The diastase extraction-test was always performed. (3) *Neurohistology: Bodian-Ziesmer* silver impregnation (in some cases with paraldehyde-fuchsin counterstaining); fluorescence microscopy of glial fibers *(Fleischhauer)* after staining with chromalum-hematoxylin. *Fluorescence microscope:* Ortholux-Orthomat *(Leitz)* with BG 12 blue 5 mm and 3 mm + K 530 filters for the pseudoisocyanin procedure and BG 12 blue 5 mm + K 530 filter for demonstration of glial fibers.

II. *Electron microscopic studies*

Fixation: Buffered (pH 7,4) 5% glutaraldehyde (45 min.), 1% OsO₄ (2 hrs). *Embedding:* (1) *Hyla arborea:* Araldite or Vestopal; (2) *dog:* Epon; (3) *human embryo:* Vestopal W. *Ultramicrotome:* LKB, *Porter-Blum MT I* or *MT II. Staining:* Lead hydroxide *(Karnovsky)*. *Electron microscope:* Elmiskop I *Siemens,* 60 kV.

Semi-thin sections of these materials were stained according to *Richardson* or *Rüdeberg* (methylene blue-thionine).

Results

1. *Hyla arborea*

(Fig. 2—4; *Oksche* and *Vaupel-von Harnack,* unpublished results).

Under the light microscope, the SCO of *Hyla arborea* contains numerous fine ependymal processes that penetrate the posterior commissure and pass to the external limiting membrane where they terminate in club- or trumpet-like endfeet *(Oksche,* 1962; *Mautner,* 1965). These fibers, as well as their endfeet, contain a granular substance which is selectively stainable with paraldehyde-fuchsin, chromalum-hematoxylin, and alcian blue. Processes with granules can also be observed in contact with the blood vessels. Some differences in the tinctorial behavior between the apical cell pole (oriented toward the ventricle) and the peripheral cell processes, including the endfeet, raise the question as to whether in these two opposite regions of the SCO the affinities of somewhat similar components for the same selective staining may be misleading with respect to the question of polarity.

The secretory picture in *Hyla arborea* conforms with the scheme presented in the introduction. The secretory material develops in the extensively distended cisternae of the endoplasmic reticulum. On the other hand, there is no indication of the presence of secretory material in the well-developed Golgi complex. Furthermore, nothing has been observed concerning secretion into the cerebrospinal fluid or concerning the formation of *Reissner's* fiber that differs from the observations of *Sterba* and coll. (1967). In the ultrastructure of the long processes of the cells, glycogen granules and glial filaments are prominent. The parallel arrays of distended cisternae, that are characteristic of the perinuclear region and the apical cell pole, are not observed in the long slender process. Instead, large vacuoles filled with low-contrast flocculent material occur in the ependymal endfeet, both at the perivascular and external limiting membranes. These inclusions resemble closely the vacuoles described by *Murakami* (1959) in lizards and by *Murakami* and *Tanizaki* (1963) in toads. After retention for 48 hours

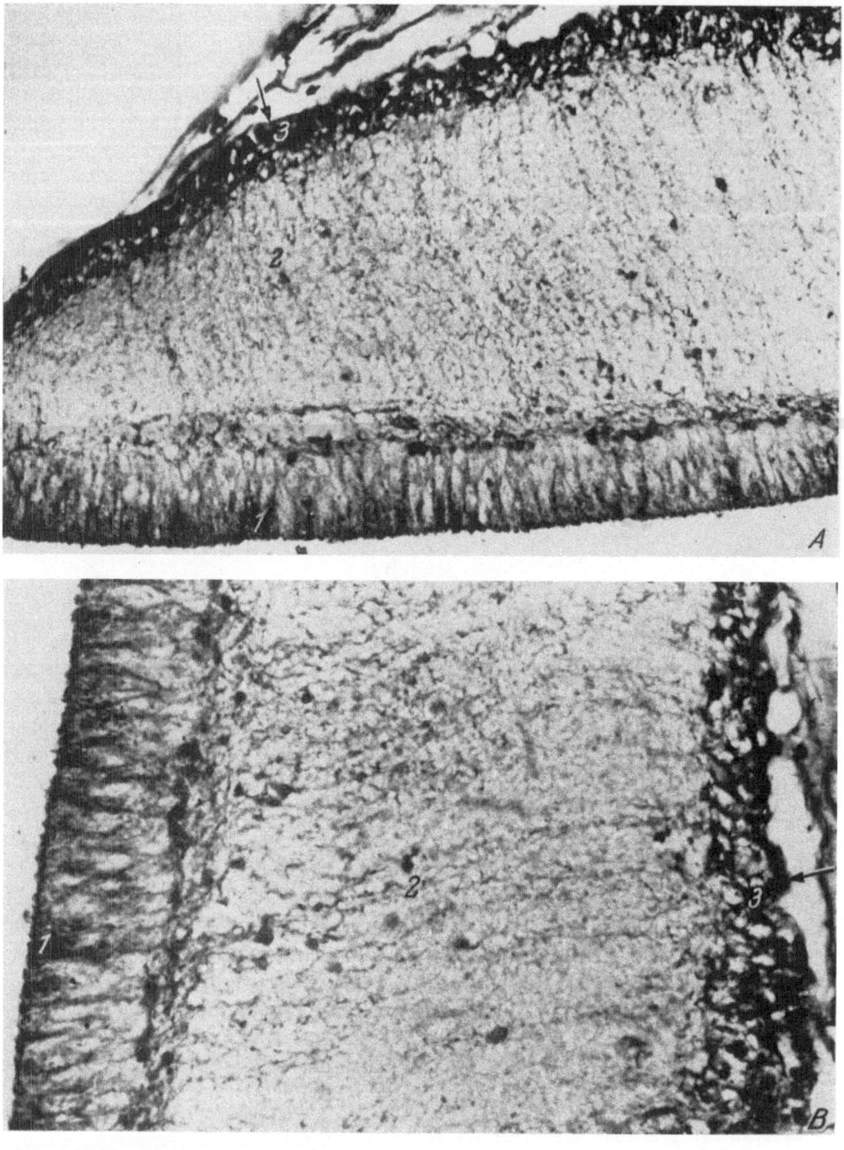

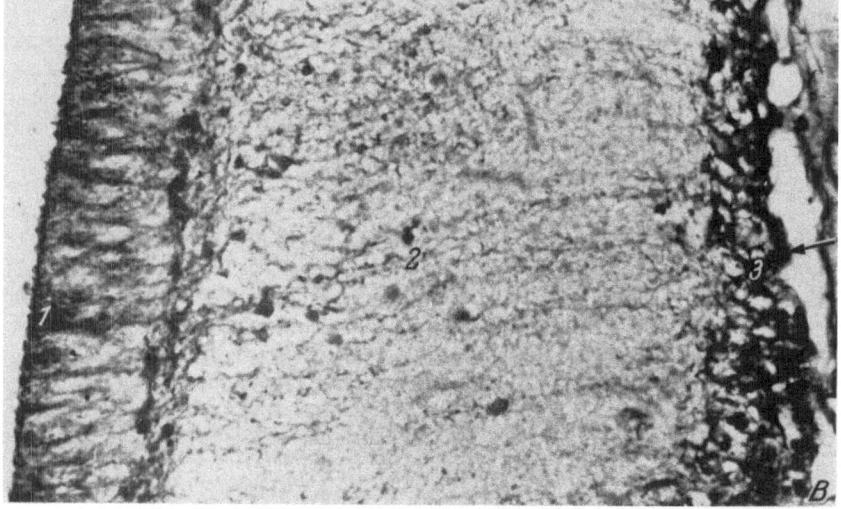

Fig. 2. Subcommissural organ of treefrog, *Hyla arborea*. Selectively stained granules in the apical cytoplasm (1), the ependymal cell processes (2) and the endfeet (3). ↑ External limiting membrane. (Paraldehyde-fuchsin; A: x 350, B: x 560)

Fig. 3. Subcommissural organ of the treefrog, *Hyla arborea*, A: x 17,600, B: x 13,200. Apical cytoplasm in a control animal (A) and after 48 hours in a 1 % NaCl solution (B). Parallel systems of cisternae (1) of the granular endoplasmic reticulum are filled with a flocculent material of low density. 2, Sac-like profiles; 3, multivesicular body; 4, Golgi complex; 5, nucleus. C: Perivascular foot (6) of the subcommissural ependyma with an accumulation of a fine granular material (°); control animal. 7, Distended cisternal profile in a cell process; 8, glycogen particles; 9, myelinated nerve fibers of the posterior commissure. x 13,200.

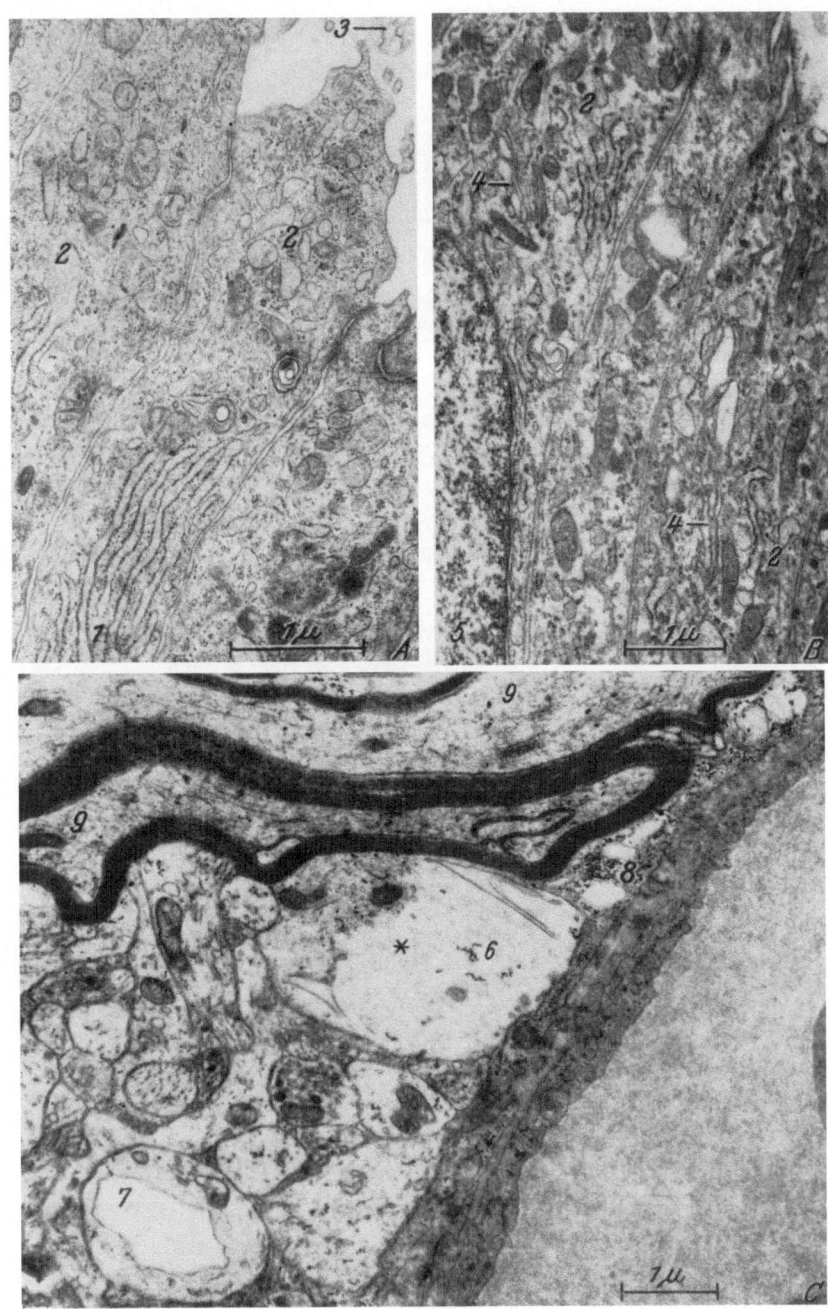

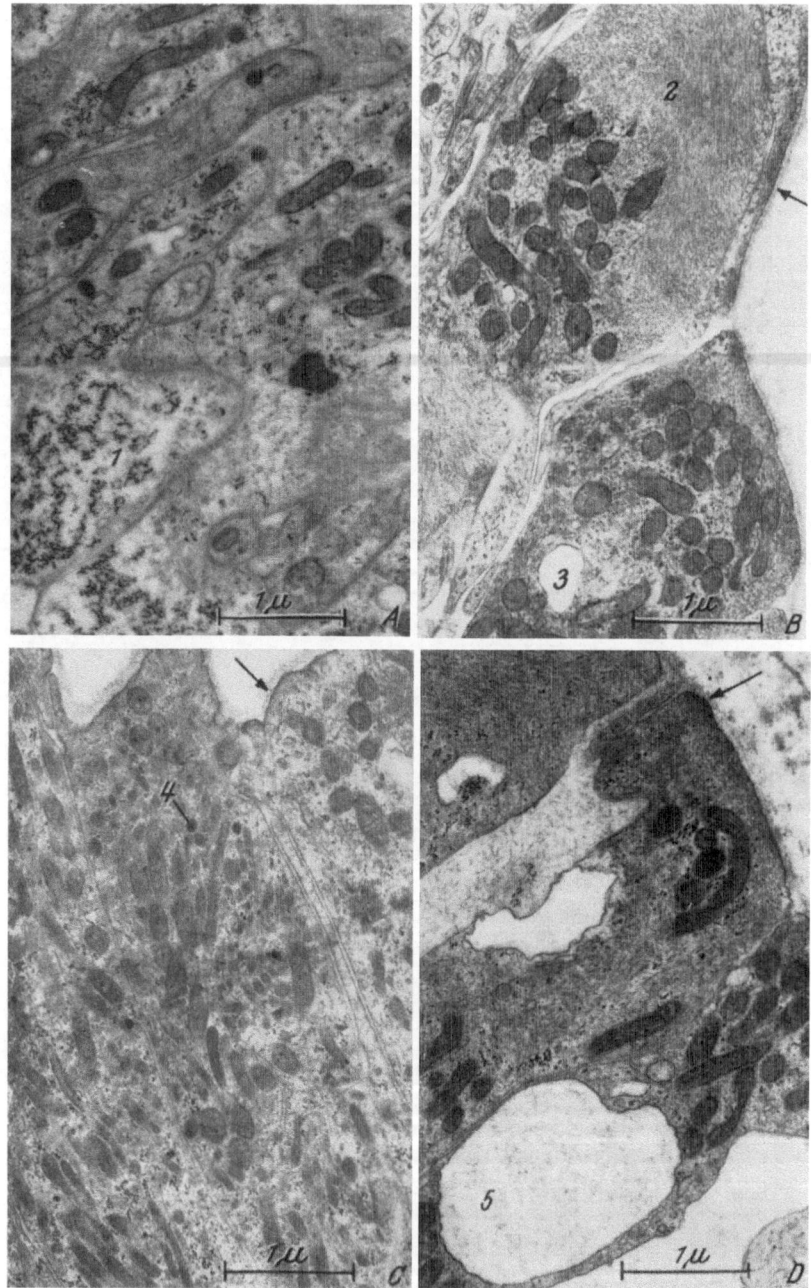

in a 1% NaCl solution, which, in light microscope preparations, leads to an increase in paraldehyde-fuchsin positive substance in the processes and endfeet, these vacuoles, in the electron micrographs, become more conspiciuous than in the control animals. Since the lysosomes in the region of the endfeet are not numerous, it is improbable that an accumulation of these organelles is being mistaken for secretory material.

2. Dog

(Fig. 5—10; *Oksche* and *Kirschstein,* unpublished results).

The canine SCO, in which the activity has been studied by light microscopic and histochemical methods by *Bargmann* and *Schiebler* (1952), *Talanti* (1958), *Oksche* (1961) and *Dellmann* (1965), shows an interesting polar arrangement of secretory elements.

This SCO has a conspicuous regional differentiation in organization as a result of the course and thickness of the posterior commissure (Fig. 1 b). In the thin-walled rostral section there are columns of cells which penetrate the bundles of fibers of the commissure, and which can be followed to the external limiting membrane. These bands of cells can maintain a septal character. The paraldehyde-fuchsin positive granules not only form a dense apical layer in the SCO of the dog but can also be observed in the elements of the cell columns. This is even true of the perikarya lying directly beneath the basal lamina of the pia mater. Delicate, selectively stained strands of cells from the SCO penetrate the thickest parts of the posterior commissure.

Since the paraldehyde-fuchsin procedure is not a very specific staining, the pseudoisocyanin method of *Sterba* (1964) was also used. With this method, which is highly sensitive for SH-groups, only a part of the paraldehyde-fuchsin positive material in the SCO of the dog was stained. Some of the fluorescent substance, however, occurs in perivascular sites and in the most peripheral parts of the cell elements in the vicinity of the basal lamina underlying the pia mater. The distribution of amylase-(diastase-)resistant, PAS-positive material is again different from that of the granules selectively stained by pseudoisocyanin or chromalum-hematoxylin. These observations with the light microscope indicate the desirability of ultrastructural investigations.

In the apical pole of the cell, the electron microscope reveals lighter and darker vesicles filled with dense material; these are derived from the distended cisternal profiles of the granular endoplasmic reticulum. Both types of secretory substance are also identifiable in the perinuclear region

---

Fig. 4. Subcommissural organ of the tree-frog, *Hyla arborea.* Ultrastructure of the endfeet at the external limiting membrane (↑ basal lamina). A: x 17,600, animal. B: x 18,400, after 24 hours in a 1% NaCl solution. C and D: x 17,600, after 48 hours in a 1 % NaCl solution. Note in A aggregates of glycogen (1), in B glial filaments (2) and a vacuole (3), in C small vesicular profiles containing a dense material (4), in D a greatly distended cisternal profile filled with a flocculent material (5).

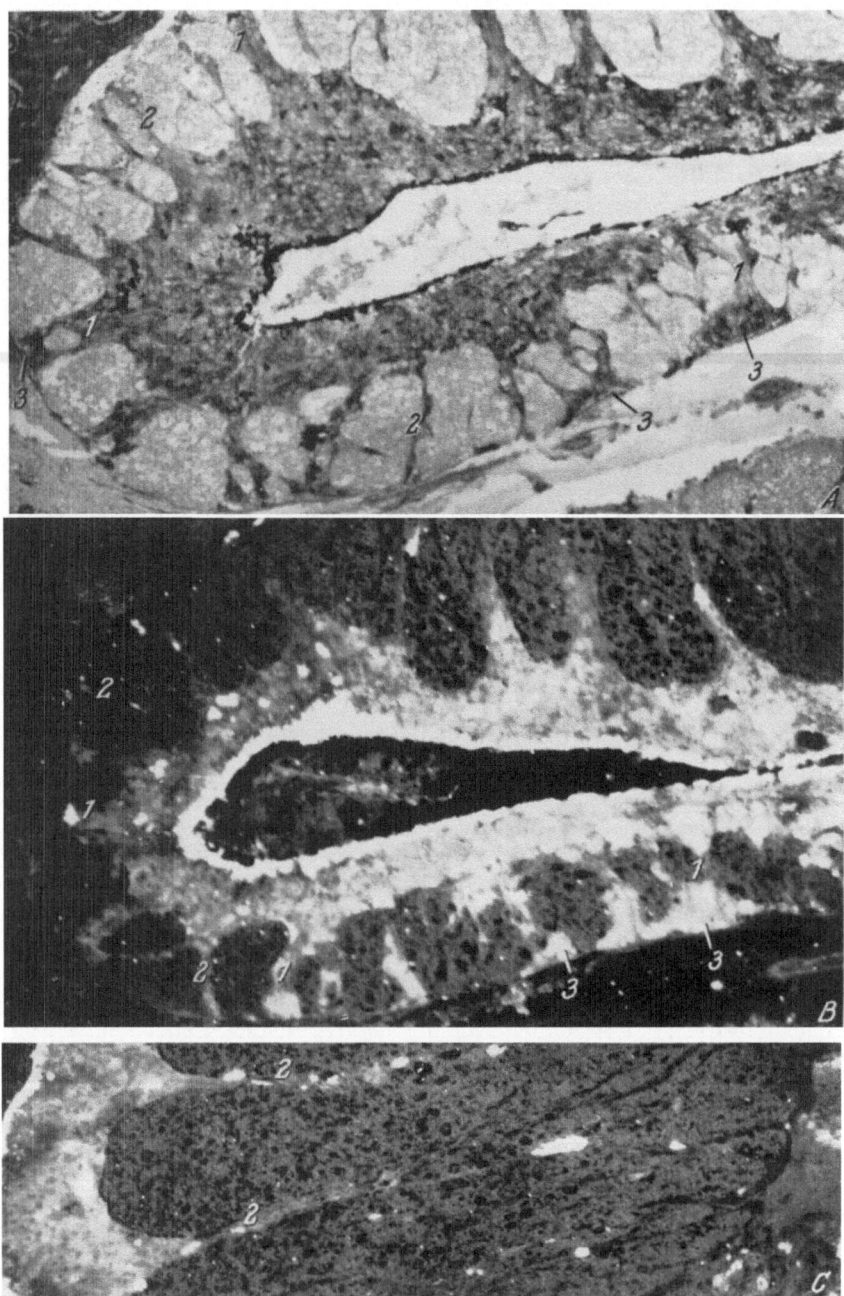

Fig. 5

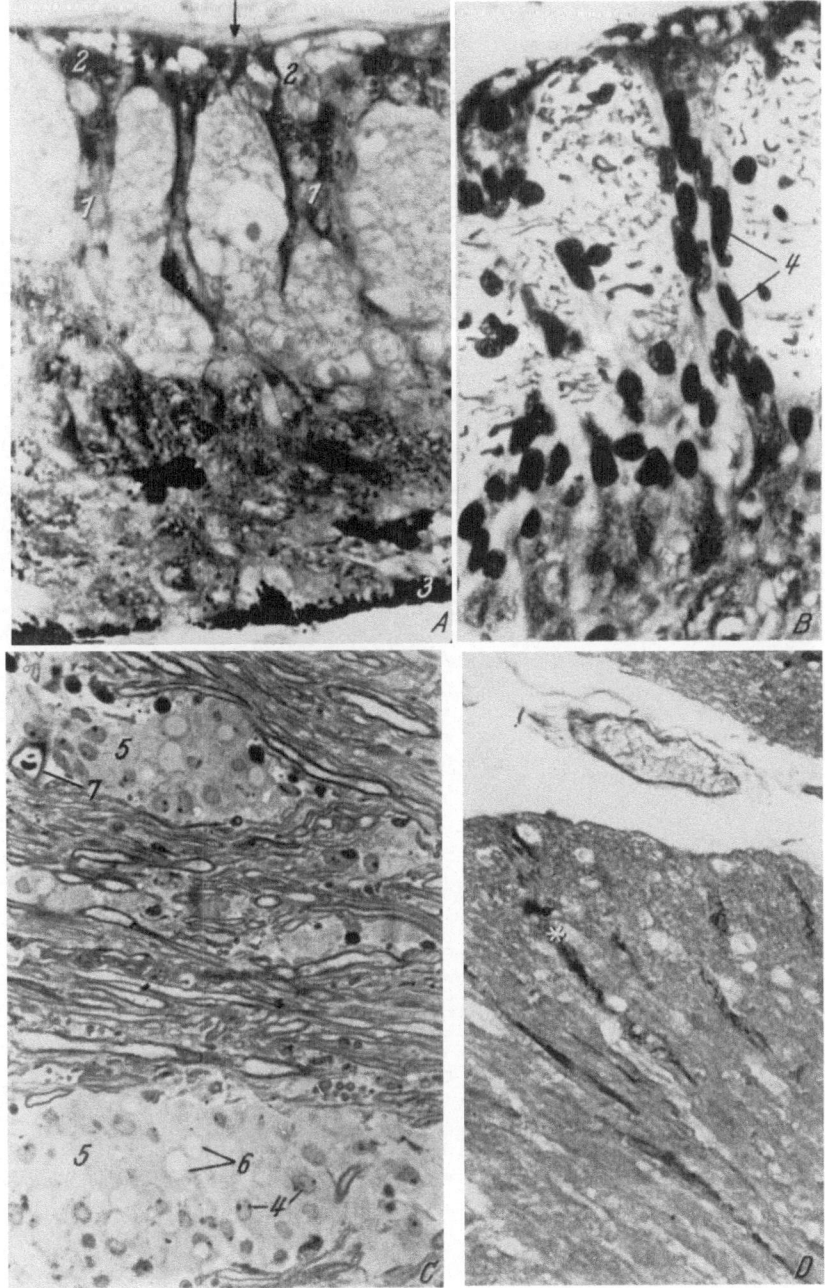

Fig. 6

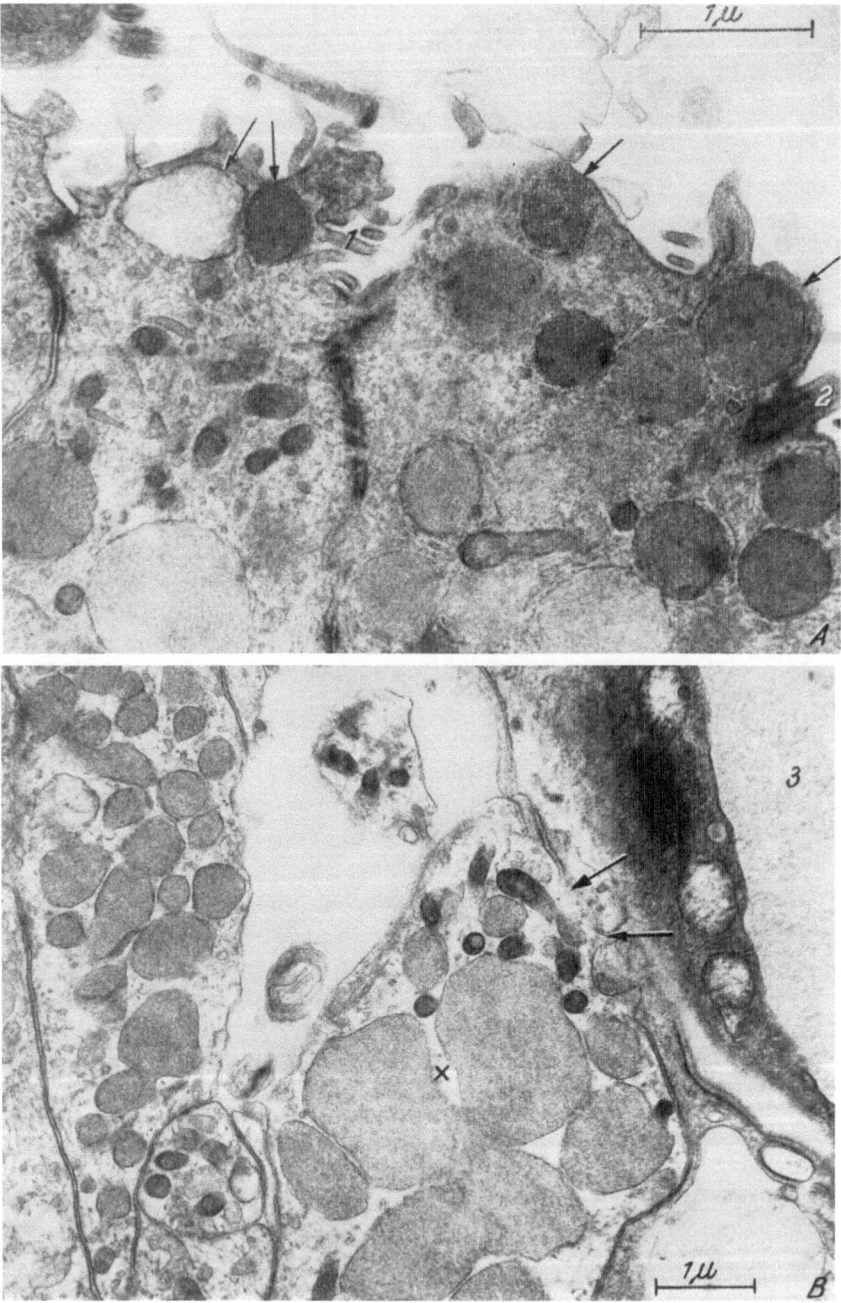

Fig. 7

where they fill tubules and cisternae of various sizes. Whether this indicates separate productions of two different protein materials or of mucopoly-saccharide ° and protein components, cannot be decided at this time. Compared with other species studied thus far, the manner of secretion into the ventricle shows no special peculiarities.

Of special importance is the demonstration that the columns and strands of cells that extend to the external limiting membrane are composed of cells showing a secretory activity very similar to that shown by the ventricular cell elements. The perikarya and processes of the cells, tightly packed with secretory material, are in contact with the vessels of the distinctive subependymal, hypendymal, zone as well as with the vessels accompanying the thicker septa. Accumulated secretory material can be demonstrated even in perikarya and cell processes underlying the connective tissue of the vascularized pia mater.

A part of the trumpet-shaped endfeet do not contain the large spheres of secretory material formed by coalescence, but rather sections of numerous fine tubules filled with a dense material. In the adjacent wide-meshed leptomeningeal tissue there are numerous apparently empty (emptied ?) vesicles, which, from all aspects, suggest this to be a region of exchange of materials.

° According to *Fawcett* (1966, p. 259) the carbohydrate components of muco-polysaccharide secretions may be synthesized in the Golgi complex.

Fig. 5. Subcommissural organ of the dog. A, B: thin-walled rostral recess. C: Caudal bulge of the posterior commissure (see also Fig. 1 B). (A: paraldehyde-fuchsin, x 140; B and C: pseudoisocyanin fluorescence method of *Sterba*, x 140. Preparation and photograph by *P. Zimmermann*). Note (in A) the selectively stained or (in B, C) fluorescent cell columns (1) and strands (2) and their foot-like terminations (3) at the external limiting membrane.

Fig. 6. Subcommissural organ of the dog. A: Columns of selectively stained cells (1) with foot-like terminations (2) at the external limiting membrane (↑); 3, ventric-ular surface. Paraldehyde-fuchsin, x 560. B: Cell nuclei (4) of the columnar cells. *Bodian-Ziesmer*—paraldehyde-fuchsin. x 560. C: Cross section through the columns (5) with cell nuclei (4), vacuolar inclusions (6) and a capillary (7). Some of these columns have a septal character. (Epon, semi-thin section, methylene blue-thionine after *Rüdeberg*, x 350. Light microscopic preparation from a tissue block embed-ded for electron microscopy (Figs. 7—10). D: Longitudinal section through the cell complexes (°) shown in Fig. 6 C. Adjacent area, embedded in paraffin. Paralde-hyde-fuchsin, x 280.

Fig. 7. Subcommissural organ of the dog. A: Free surface with droplets of different size and density in the apical cytoplasm. Different phases of the process of dis-charging dark and light secretory material (↑); 1, microvilli; 2, cilium; x 24,000. B: Perivascular foot (x) with droplets. Fine-textured secretory material of low density. ↑ Zone of fusion (or dehiscence?) of membranes (release of secretion?); 3, blood vessel; x 14,400.

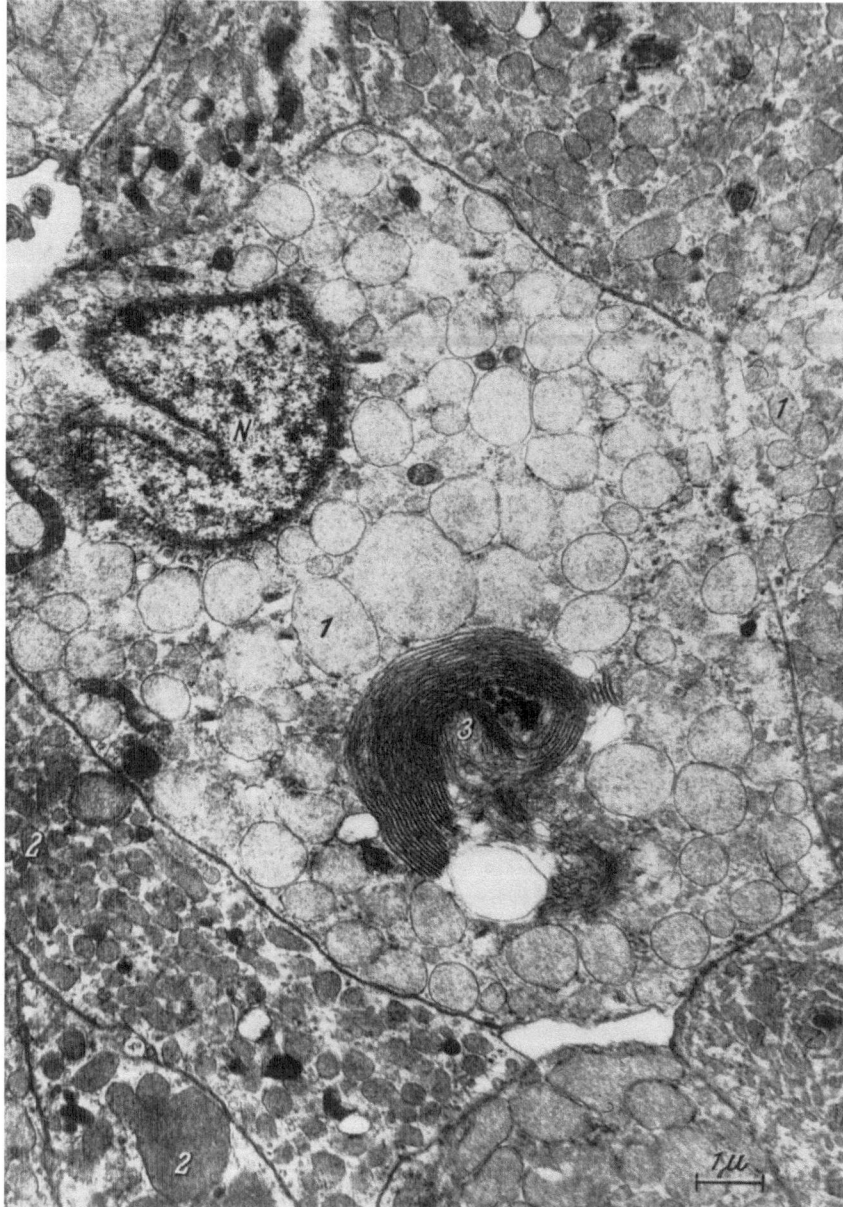

Fig. 8. Subcommissural organ of the dog. Glandular appearance of the cells in a columnar formation (for light microscopic picture see Fig. 6 C). Tubular elements (1) and distended cisternae (2) of the endoplasmic reticulum are filled with a material of moderate density. 3, sacs with a light substance; 4, dark granules of different size; 5, Golgi complex; 6, microtubules; x 18,000.

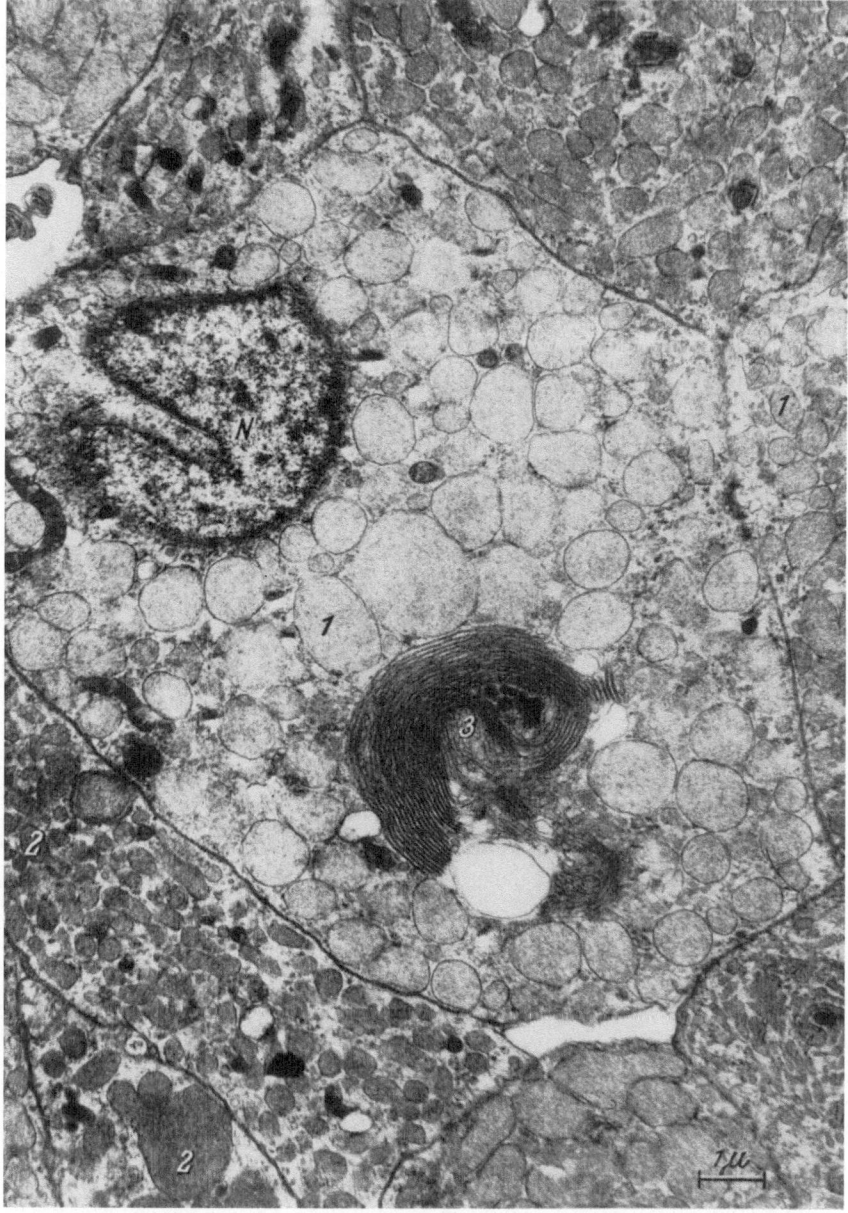

Fig. 9. Subcommissural organ of the dog. Secretory perikarya (N nucleus) and cell processes within a column of cells which penetrates the posterior commissure (for light microscopic picture see Fig. 6 C). Light (1) and dark (2) saccular profiles and droplets of different size. 3, Whorl-like multilamellar body; x 9,600.

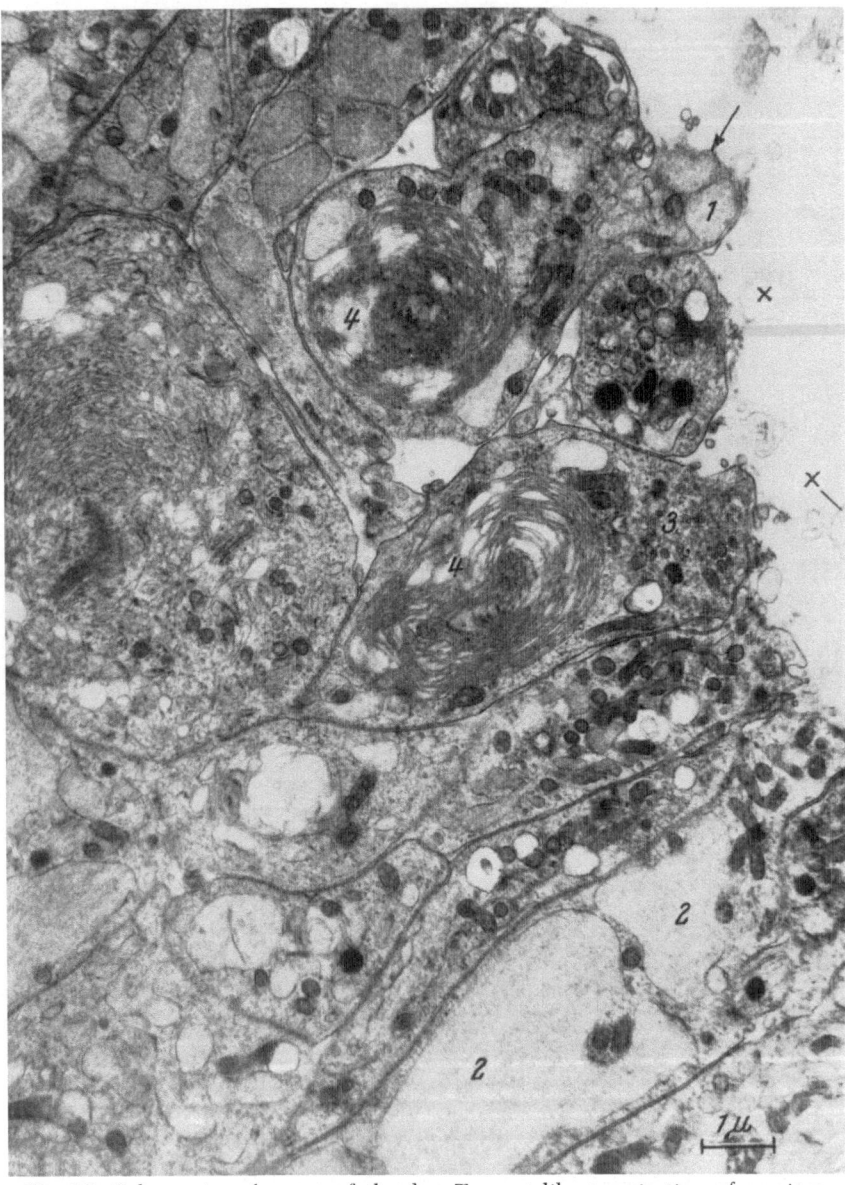

Fig. 10. Subcommissural organ of the dog. Trumpet-like terminations of secretory cells at the external limiting membrane. Some of these endfeet contain droplets (1) or sacs (2) filled with a light secretory material, others show small dark inclusions (3) and whorl-like bodies (4). At some places (↑) secretory material appears to be released. Note the numerous empty vesicles (x), x 10,800.

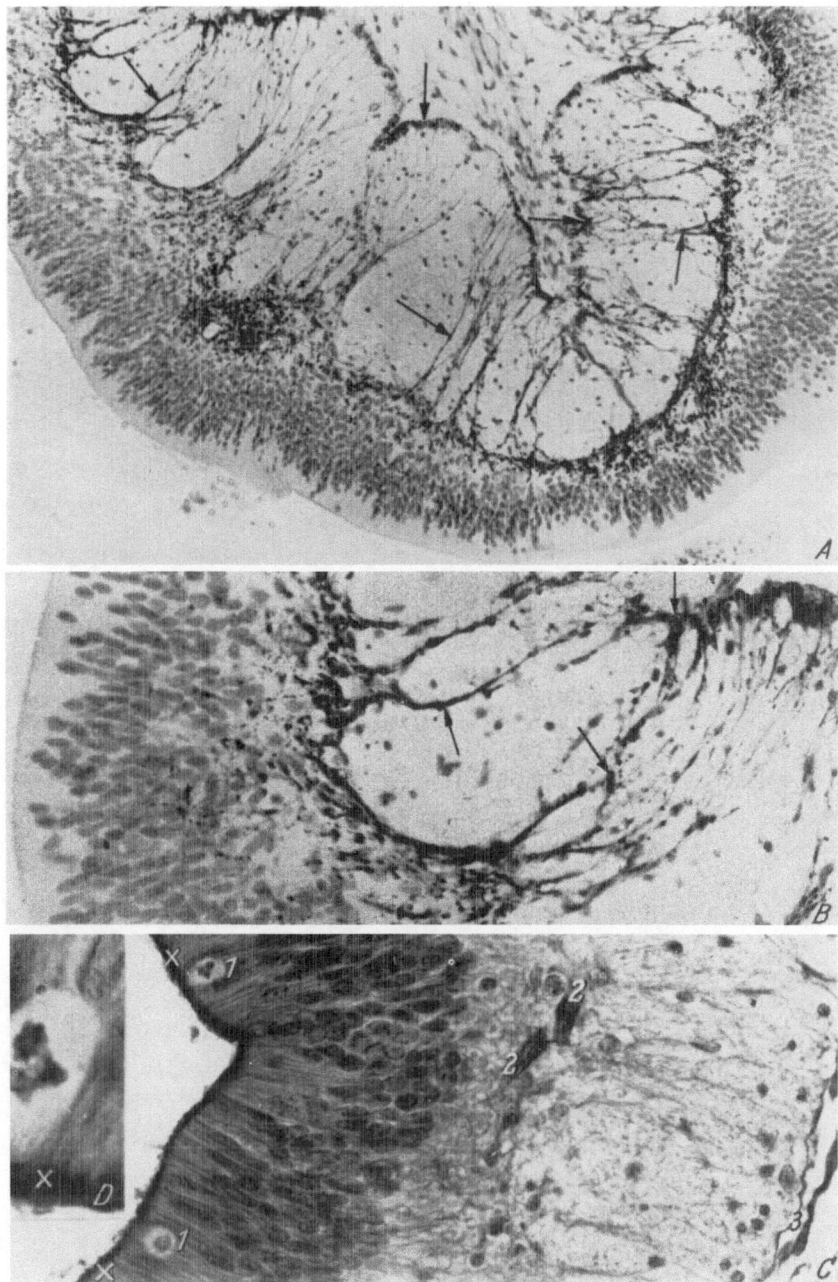

Fig. 11

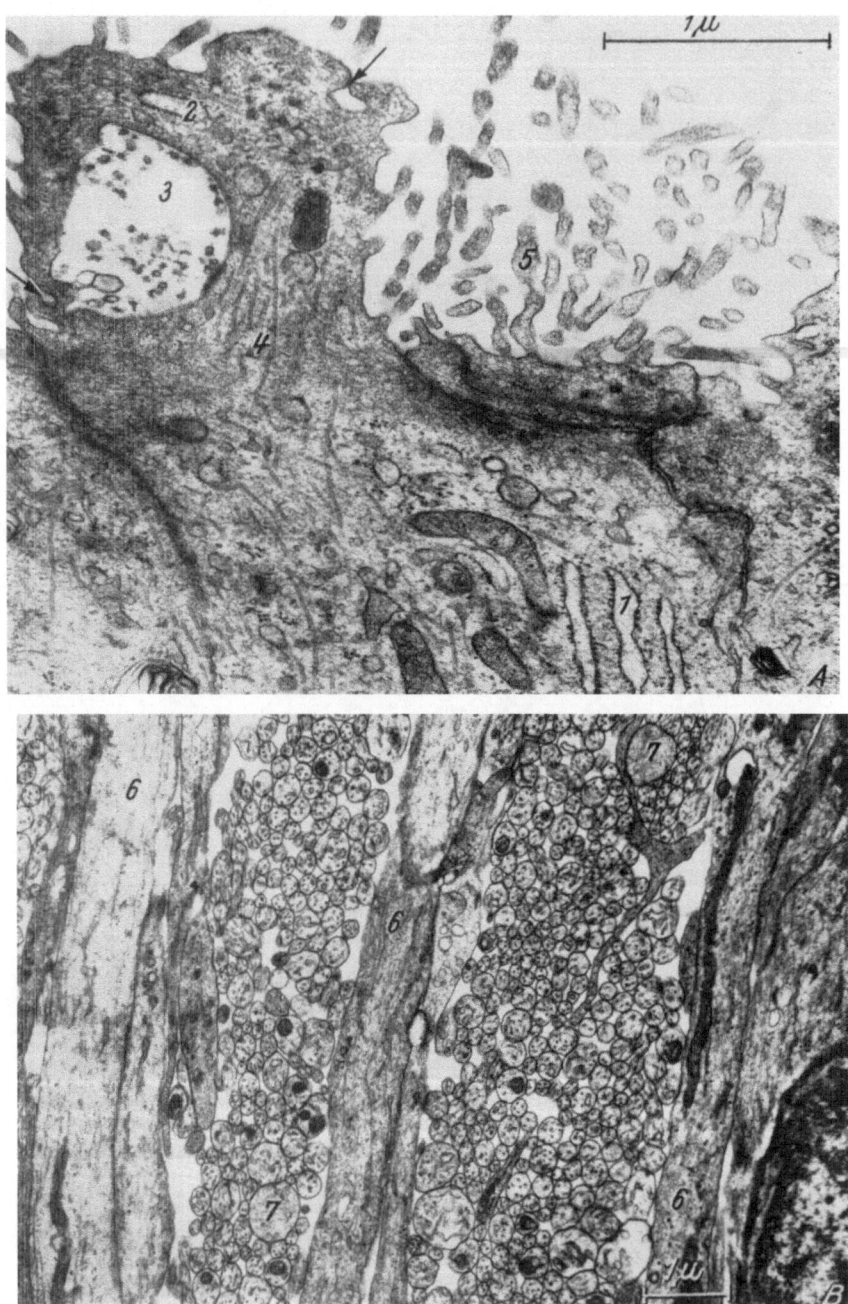

Fig. 12

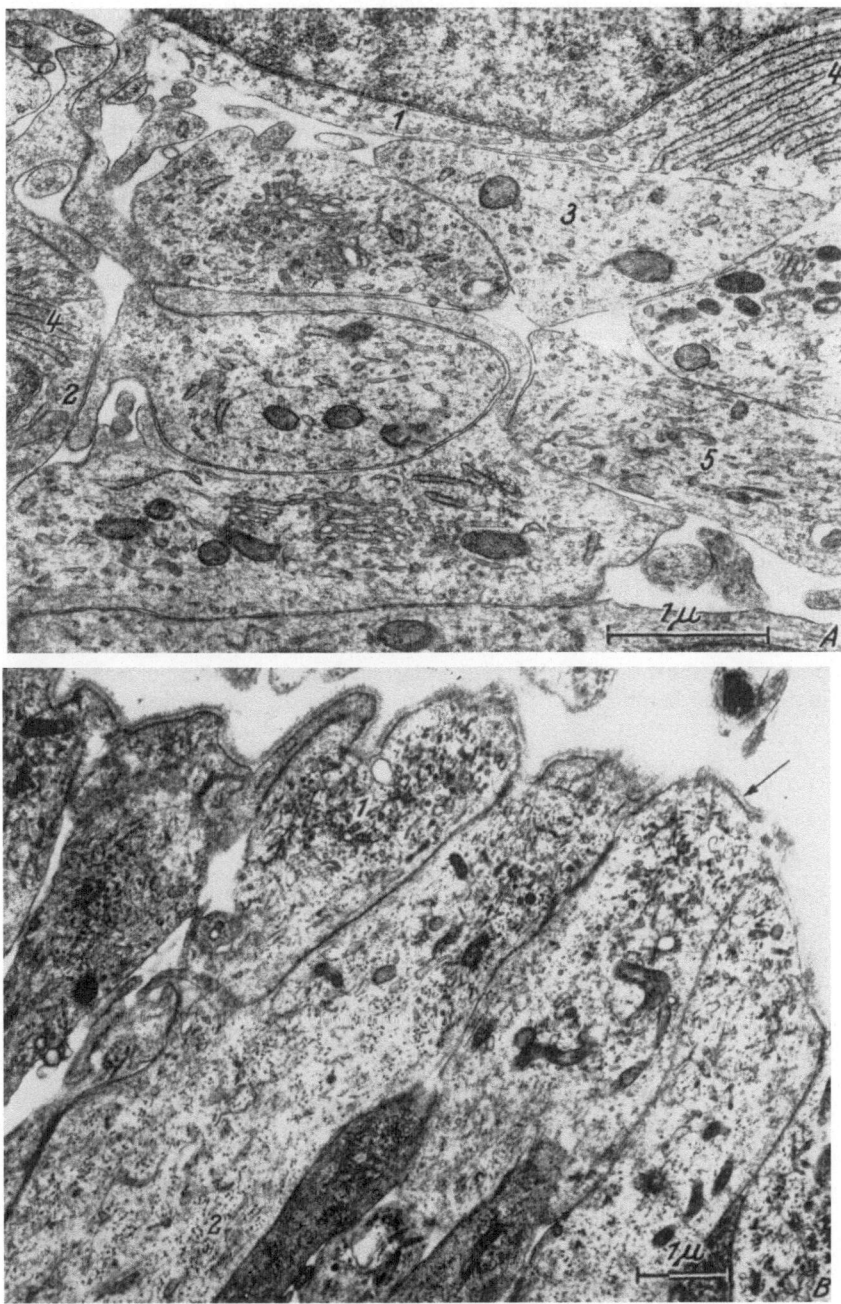

Fig. 13

3. Human embryo, 109 days (18th week of pregnancy)

(Fig. 11—13; *Oksche* and *Kirschstein*, unpublished results.)

Because pregnancy was terminated on recommendation by the attending physician,° it was possible to preserve this material under optimal conditions.

For orientation with respect to the structure and secretory activity of the human SCO during the third and fourth months of pregnancy, the earlier light microscope investigations (*Oksche*, 1956, 1961, 1964) should be reviewed (for literature, see also *Palkovits*, 1965). In this phase of development, selective staining methods demonstrate a conspicuously stained margin on the free ventricular surface of the SCO. The paraldehyde-fuchsin and PAS-positive granules are restricted to the apical cell pole. With the PAS reaction, amylase-resistant, PAS-positive material was demonstrated also in the perinuclear area and in the basal part of the cell. In the extensively ramified system of ependymal fibers there appear to be no materials with the tinctorial properties of the apical secretory material. Glycogen is stored here in conspicuously large amounts.

Our electron microscope studies of the SCO of the 109-day old human embryo have shown the presence of a characteristically arranged and distended endoplasmic reticulum. These cisternae contain a flocculent cell product of moderate density. This substance is similar to that of the SCO of other species. However, the structures of the endoplasmic reticulum are not as numerous and dilated as are the corresponding membrane systems in the tree frog and dog. They are more or less localized in the apical portion of the cell. On the ventricular surface, indications of release of material can be observed. The long ependymal processes and their endfeet contain no cisternae; however, in these structures, distinctive, small (diam-

---

° I am glad to acknowledge the kindness and support of Dr. H. *Langer*, Universitäts-Frauenklinik Giessen (Professor R. K. *Kepp*, Director).

Fig. 11. Subcommissural organ of two human embryos (Mens III). A, B: 90 mm; C, D: 75 mm total length. Numerous glycogen granules (↑) in the slender branched processes and the endfeet of the subcommissural cells; lead tetraacetate-Schiff. A: x 140; B: x 250. In C, selectively stained apical material (x) (paraldehyde-fuchsin, x 350). (For higher magnification see inset D, x 1,400). 1, mitosis; 2, capillaries; 3, external limiting membrane.

Fig. 12. Subcommissural organ of a human embryo (estimated age: 109 days). A: Free surface and apical cytoplasm. 1, Distended cisternae of the endoplasmic reticulum, some of them (2) near the plasmalemma; ↑ pinocytotic invaginations; 3, vacuole (invagination?) containing vesicular inclusions; 4, microtubules; 5, microvilli; x 31,200. B: Long slender processes (6) of the subcommissural cells penetrate the posterior commissure; 7, cross sections of unmyelinated nerve fibers; x 12,000.

Fig. 13. Subcommissural organ of human embryo (estimated age: 109 days). A: Perikarya (1, 2) and cell processes (3) with parallel arrays of cisternae (4). In these profiles of the endoplasmic reticulum a material of low density is stored. 5, Tubules filled with a substance of moderate density; x 21,600. B: Endfeet at the external limiting membrane (↑ basal lamina). 1, Dense vesicular and tubular inclusions; 2, glycogen particles; x 12,000.

etcr about 500 Å) electron-dense bodies (profiles of microtubules?) occur in association with numerous glycogen granules. In spite of the anatomically polar organization of the embryonic human SCO, with one surface to the ventricle and the other in contact with the vascular system and the external limiting membrane, in this case, the processes of the SCO and their endfeet do not contain the usual secretory material.

In our experience the human SCO shows already a regressive development during the second half of pregnancy. Between the third and sixth years of life there remains only an extensively flattened ependymal band. In the adult, only isolated remnants of this parenchyma can be detected (*Oksche*, 1964; for exceptions, see *Palkovits*, 1965).

In order to understand this regressive development of the human SCO, the investigation of material from different primates is most useful (see also *Hofer*, 1958). In our laboratory the research by Dr. *Merker* (unpublished results) on some Platyrrhina, Catarrhina and Pongidae (*Cebus, Ateles, Macaca* and *Pan*) has shown that only in the SCO of Chimpanzees does such a reduced and regionally irregular ependymal band occur, thus resembling the situation in man, especially in childhood. *Cebus, Ateles*, and *Macaca*, on the other hand, have a SCO with a high layer of secretory ependyma. The investigations of *Merker* raise still another problem — that of the resorptive ability of the SCO. That the SCO cells can take up dyes injected into the ventricle has been convincingly demonstrated by *Löfgren* (1965). The surface of the SCO ependyma, as seen in electron microscopic preparations, has numerous microvilli. Even though a strongly developed inner glial fiber layer (Fig. 14) underlies the SCO of *Ateles* and other monkeys, we know from the investigations by *Feldberg* and *Fleischhauer* (1963) that such a layer does not stop the penetration of dye into the deeper layers of brain tissue. In the electron microscopic preparations, which *Merker* has made from the glial fiber layer that covers the SCO of *Macaca*, it is clear that even the astrocyte processes, crowded with great numbers of glial filaments, contain sufficient hyaloplasm to permit active transport.

Discussion

From the more recent literature and from new findings communicated in the present paper, it is obvious that many details of the morphology of the SCO are known. However, on the basis of our present information for the vertebrates as a whole, little is definitely known about its function. According to new results there is no positive anatomical evidence for the existence of a neural or vascular epithalamic complex formed by the SCO and the pineal organ (for references, see *Oksche*, 1965; *Mautner*, 1965). With respect to the significance of contacts between nerve fibers and the SCO the ideas are very contradictory and confusing. For details the papers of *Murakami* and *Tanizaki* (1963), *Stanka* (1964, 1967), J. *Ariëns Kappers* (1965, 1967) and *Oksche* and *Vaupel-von Harnack* (1965) may be consulted. The problem of the vascular supply and connections of the SCO has been discussed by *Duvernoy* and *Koritké* (1964).

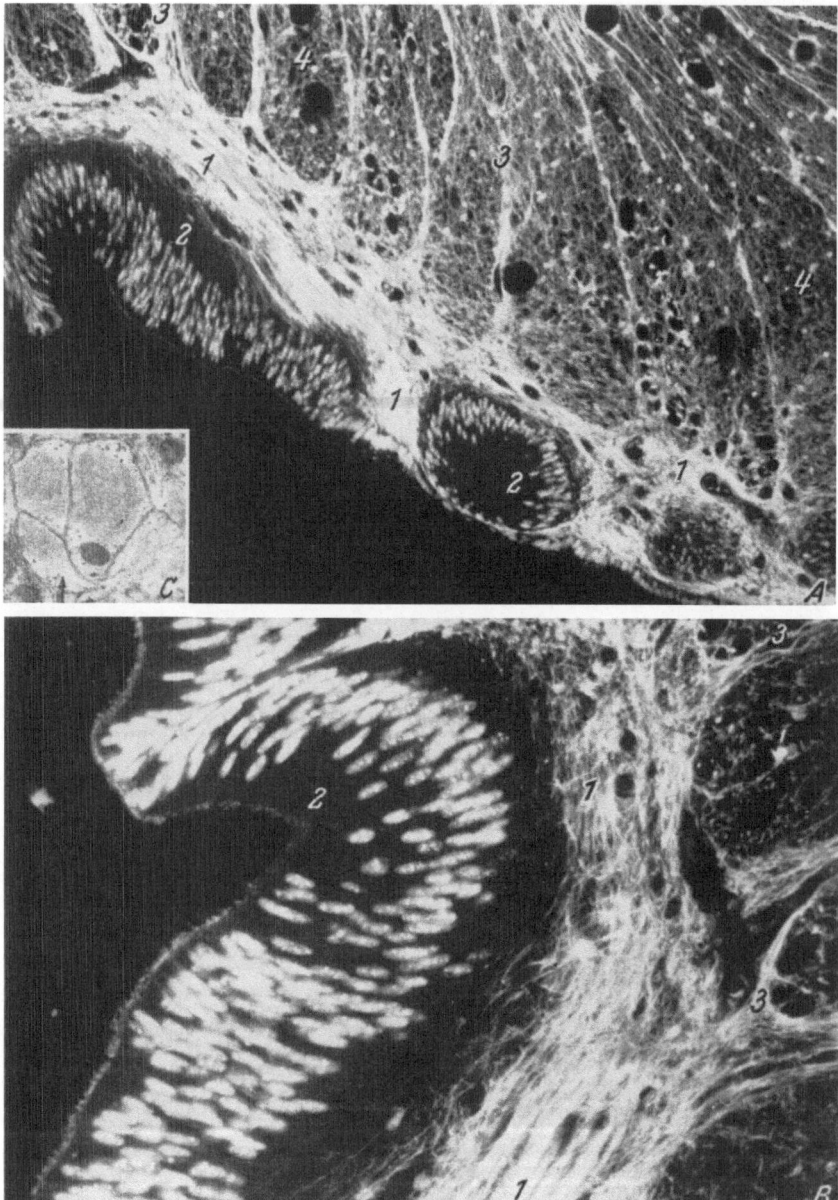

Fig. 14. Subcommissural organ of *Ateles* (A, B). Fluorescence microscopy of glial fibers. A dense meshwork of glial fibers (1) underlies the subcommissural organ (2) and forms septa (3) between the fiber bundles of the posterior commissure (4). Chromalum-hematoxylin, A: x 140; B: x 350. C: Arrangement of glial filaments within the processes of astrocytes located between the subcommissural ependyma and the posterior commissure of *Macaca*. A part of the cytoplasm (↑) is free of glial filaments; x 15,000. (From unpublished material of Dr. G. *Merker*.)

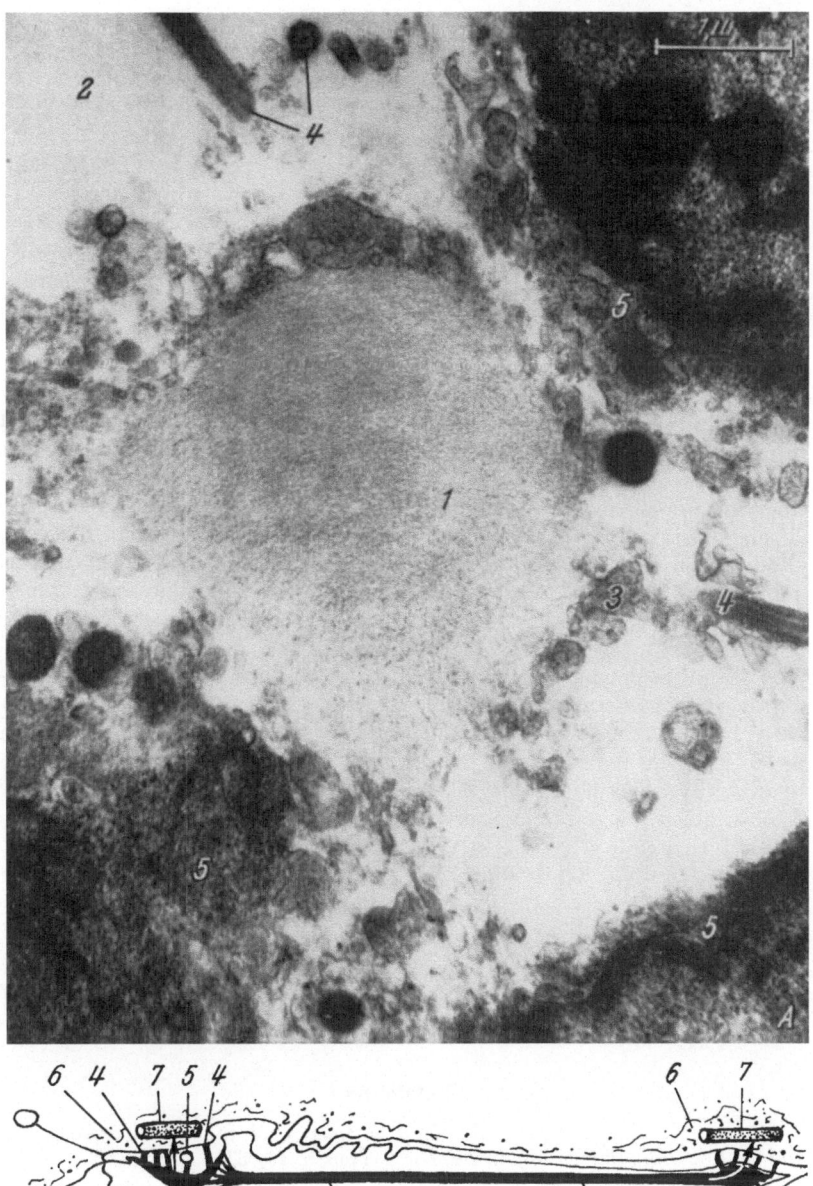

Fig. 15

In my communication I have attempted, above all, to show that, in a few animals (*e. g.* the tree frog, *Hyla arborea,* and the dog) the secretory activity of the SCO is polarized. The close relationship between the secretory perikarya or cell processes and the blood vessels of the wide-meshed pia mater (pia-arachnoid) can be demonstrated nicely in the dog. If the SCO does actually produce a hormon-like material, this could (*e. g.* in the dog) easily reach the blood stream. However, the question as to why this polarity does not appear uniformly in all vertebrates remains unresolved. Also, much is still unknown concerning the property of secretion into the ventricle by all SCO's in general. Moreover, the nature of the exchange of material between the cerebrospinal fluid and *Reissner's* fiber is not really understood. According to our observations in *Rana temporaria* (*Oksche* and *Kirschstein,* unpublished), *Reissner's* fiber is a membrane-free secretory structure (Fig. 15 a) as was first indicated by *Afzelius* and *Olsson* (1957) and later by *Sterba* and coll. (1966, 1967). However, it should be pointed out that the ependymal band in the terminal ventricle of the central canal contains wide intercellular gaps; there is some evidence that material from *Reissner's* fiber may pass to the vascular system of the leptomeninges (see also *Wislocki* and coll., 1956; *Hofer,* 1963; *Mautner,* 1965; *Sterba* and *Naumann,* 1966). If, in fact, such a route exists, then the secretory active cell columns and strands in the area of the SCO would signify nothing more than a shorter pathway to the vascular bed independent of the cerebrospinal fluid (Fig. 15 b). Further speculation concerning such a possibility does not appear useful at present. The hypothesis proposed here is amenable to examination with modern morphological methods, *e. g.* with the use of radioactively labelled materials. The elucidation of the problems of function can be anticipated, however, only with the introduction of refined and sophisticated physiological methods. The investigation of the structure of the SCO is now at such a stage that physiological and biochemical studies can be firmly based on it.

Acknowledgements

The author is greatly indebted to Professor *D. S. Farner,* Seattle, for his generous help in preparing this manuscript and to Miss *H. Kirschstein,* Mrs. *T. Möller,* Miss *M. Langbein,* Miss *I. Lyncker,* Dr. *G. Merker,* Dr. *T. Peters* and cand. med. *P. Zimmermann* for their excellent technical assistance.

References

Afzelius, B. A., and *R. Olsson:* The fine structure of the subcommissural cells and of Reissner's fibre in *Myxine.* Zschr. Zellforsch. *46,* 672—685 (1957).

Bargmann, W., und *Th. H. Schiebler:* Histologische und cytochemische Untersuchungen am Subcommissuralorgan von Säugern. Zschr. Zellforsch. 37, 583—596 (1952).

Barlow, R. M., A. N. D'Agostino, and *P. A. Cancilla:* A morphological and histochemical study of the subcommissural organ of young and old sheep. Zschr. Zellforsch. 77, 299—315 (1967).

Crow, L. T.: Subcommissural organ, lateral hypothalamus and dorsal longitudinal fasciculus in water and salt metabolism. In: Thirst. Proc. 1st Symposium on thirst regulations of body water. Oxford-London-New York-Paris: Pergamon Press 1964.

Dellmann, H. D.: Age variations in the structure of the subcommissural organ of the dog. Anat. Rec., Philadelphia, *151* 449 (1965).

Duvernoy, H. et J. G. Koritké: Contribution à l'étude de l'angioarchitectonie des organes circumventriculaires. Arch. Biol. 75, 693—748 (1964).

Farrell, G.: Regulation of aldosterone secretion. Physiol. Rev., Baltimore, *38,* 709—728 (1958).

Farrell, G.: The physiological factors which influence the secretion of aldosterone. Rec. Progr. Hormone Res. *15,* 275—310 (1959).

Farrell, G.: Adrenoglomerulotropin. Circulation *21,* 1009—1015 (1960).

Fawcett, D. W.: The cell. An atlas of fine structure. Philadelphia-London: Saunders Co., 1966.

Feldberg, W., and K. Fleischhauer: Site of tubocurarine reaching the brain via the cerebral ventricles. In: Progress in Brain Research. 3. The Rhinencephalon and Related Structures. *W. Bargmann* and *J. P. Schadé,* eds. pp. 1—19. Amsterdam-London-New York: Elsevier, 1963.

Galeotti, G.: Studio morfologico e citologico della volta del diencefalo in alcuni vertebrati. Riv. pat. nerv. *2,* 481—517 (1897).

Ghiani, P., and B. Uva: On the comparative physiomorphology of the subcommissural organ. Boll. Mus. Ist. Biol. Univ. Genova *33,* 75—83 (1965).

Gilbert, G. J.: The subcommissural organ. Anat. Rec. *126,* 253—265 (1956).

Gilbert, G. J.: The subcommissural organ: a regulator of thirst. Amer. J. Physiol. *191,* 243—247 (1957).

Gilbert, G. J.: The subcommissural organ and water-electrolyte metabolism. In: Thirst. Proc. 1st Symposium on thirst regulation of body water. Oxford-London-New York-Paris: Pergamon Press, 1964.

Hofer, H.: Zur Morphologie der circumventrikulären Organe des Zwischenhirnes der Säugetiere. Verhandl. Deutsch. Zoolog. Gesellsch. 1958. S. 202—251. Leipzig: Geest und Portig, 1958.

Hofer, H.: Neuere Ergebnisse zur Kenntnis des Subkommissuralorganes, des Reissnerschen Fadens und der Massa caudalis. Verhandlungen der Zoologischen Gesellschaft 1963, S. 431-440.

Isomäki, A. M., E. Kivalo, and S. Talanti: Electron-microscopic structure of the subcommissural organ in the calf *(Bos taurus)* with special references to secretory phenomena. Ann. Acad. Scient. Fenn., Ser. A. V. Medica, 111, p. 3—64 (1965).

Kappers, Ariëns, C. U.: Die vergleichende Anatomie des Nervensystems der Wirbeltiere und des Menschen. I. Haarlem: Bohn, 1920.

Kappers, Ariëns, J.: Survey of the innervation of the epiphysis cerebri and the accessory pineal organs of vertebrates. In: Progress in Brain Research. *10.* Structure

Fig. 15. A: Ultrastructure of Reissner's fiber (1) in *Rana temporaria.* Terminal ventricle of the central canal (2). Microvilli (3) and cilia (4) of ependyma cells (5); x 18,900. B: Diagram showing the location of secretory material in the subcommissural organ of some Anura. 1, Subcommissural organ; 2, Reissner's fiber with its massa caudalis (3) in the terminal ventricle of the spinal cord; 4, selectively stained ependymal processes penetrating the posterior commissure and forming endfeet at the external limiting membrane; 5, vascular foot; 6, wide-meshed leptomeningeal tissue with blood vessels (7). The arrows (↑) indicate the anatomical pathways to the vascular system (sites of release?).

and Function of the Epiphysis cerebri. *J. Ariëns Kappers* and *J. P. Schadé*, eds. pp. 87—153. Amsterdam: Elsevier, 1965.

Kappers, Ariëns, J.: The sensory innervation of the pineal organ in the lizard, *Lacerta viridis*, with remarks on its position in the trend of pineal phylogenetic structural and functional evolution. Zschr. Zellforsch. *81*, 581—618 (1967).

Krabbe, K. H.: L'organe sous-commissural du cerveau chez les Mammifères. Kung. Danske Vidensk. Selesk. Biol. Med. 5, Nr. 4, 1—83 (1925).

Legait, E.: Les organes épendymaires du troisième ventricule. Thèse Méd., Nancy: G. Thomas, 1942.

Legait, E.: L'organe sous-commissural chez la grenouille normale et hypophysoprivé. C. R. Soc. Biol. *140*, 543—544 (1946).

Legait, E.: Le rôle de l'épendyme dans les phénomènes endocrines du diencéphale. Bull. Soc. Sci. Nancy *1*, 1—12 (1949).

Lenys, R.: Contribution à l'étude de la structure et du rôle de l'organe sous-commissural. Thèse. Nancy, 1965 a.

Lenys, R.: Données morphologiques et histochimiques récentes sur l'organe sous-commissural. Ann. Sci. Univ. Besançon, Méd. *1*, 21—29 (1965 b).

Leonhardt, H.: Zur Frage einer intraventrikulären Neurosekretion. Eine bisher unbekannte nervöse Struktur im IV. Ventrikel des Kaninchens. Zschr. Zellforsch. *79*, 172—184 (1967).

Leveque, T. F., F. Stutinsky, M.-E. Stoeckel, et *A. Porte:* Sur les éléments utrastructuraux d'une formation glandulaire périventriculaire dans l'éminence médiane du rat. C. R. Acad. Sci. Paris, *206*, 4621—4623 (1965).

Leveque, T. F., F. Stutinsky, A. Porte, and *M.-E. Stoeckel:* Ultrastructure of the medial prechiasmatic gland in the rat and mouse. Neuroendocrinol. *2*, 56—63 (1967).

Löfgren, F.: On the transport-mechanism between the hypothalamus and the anterior pituitary. Kungl. Fysiogr. Sällsk. i Lund Förhandl. *30*, 115—120 (1960 a).

Löfgren F.: The infundibular recess, a component in the hypothalamo-adenohypophysial system. Acta morph. Neerl.-Scand. *3*, 55—78 (1960 b).

Löfgren, F.: The glial-vascular apparatus in the floor of the infundibular cavity. Further studies on the transport mechanism between the hypothalamus and anterior pituitary. Lunds Univ. Årsskr., N. F., Avd. 2, 57, 1—18 (1961).

Löfgren, F.: On an absorptive ability of the subcommissural organ. A model test. Acta Univ. Lund, Sect. II, No. 29, pp. 3—12 (1965).

Marini, M.: L'organo sottocommissurale degli anfibi. Riv. neurobiol. *12*, 468 to 509 (1966).

Mautner, W.: Studien an der Epiphysis cerebri und am Subcommissuralorgan der Frösche. (Mit Lebendbeobachtung des Epiphysenkreislaufs, Totalfärbung des Subkommissuralorgans und Durchtrennung des Reissnerschen Fadens). Zschr. Zellforsch. *67*, 243—270 (1965).

Müller, H., und *G. Sterba:* Elektronenmikroskopische Untersuchungen des Subkommissuralorganes von *Lampetra planeri* (Bloch). Verhandl. Dtsch. Zoolog. Ges. Jena 1965. S. 441—453.

Murakami, M.: Über die Feinstruktur des Subkommissuralorganes von *Gecko japonicus*. Arch. Histol. Japon. *17*, 411—427 (1959).

Murakami, M., and *T. Tanizaki:* An electron microscopic study on the toad subcommissural organ. Arch. Histol. Japon. *23*, 337—358 (1963).

Okada, M.: On the secretory pathway of the subcommissural organ. Arch. Histol. Japon. *9*, 199—204 (1956).

Oksche, A.: Funktionelle histologische Untersuchungen über die Organe des Zwischenhirndaches der Chordaten. Anat. Anz., Jena, *102*, 404—419 (1956).

Oksche, A.: Vergleichende Untersuchungen über die sekretorische Aktivität des Subkommissuralorgans und den Gliacharakter seiner Zellen. Zschr. Zellforsch. *54,* 549—612 (1961).

Oksche, A.: Histologische, histochemische und experimentelle Studien am Subkommissuralorgan von Anuren (mit Hinweisen auf den Epiphysenkomplex). Zschr. Zellforsch. 57, 240—326 (1962).

Oksche, A.: Das Subkommissuralorgan des Menschen. Verh. Anat. Ges., Jena, *58,* (Genua 1962), 373—383 (1964).

Oksche, A.: Survey of the development and comparative morphology of the pineal organ. In: Progress in Brain Research. *10.* Structure and Function of the Epiphysis cerebri. *J. Ariëns Kappers* and *J. P. Schadé,* eds. pp. 3—29. Amsterdam-London-New York: Elsevier, 1965.

Oksche, A., und *M. Vaupel-von Harnack:* Elektronenmikroskopische Untersuchungen an den Nervenbahnen des Pinealkomplexes von *Rana esculenta* L. Zschr. Zellforsch. *68,* 389—426 (1965).

Olsson, R.: Studies on the subcommissural organ. Acta Zoologica, *39,* 71—102 (1958 a).

Olsson, R.: The subcommissural organ. Thesis. Stockholm 1958 b, pp. 3—15.

Palkovits, M.: Morphology and function of the subcommissural organ. Studia Biologica Hungarica 4., *J. Szentágothai,* ed., Budapest: Akadémiai Kiado, 1965, pp. 1—105.

Röhlich, P., and *B. Vigh:* Electron microscopy of the paraventricular organ in the sparrow *(Passer domesticus).* Zschr. Zellforsch. *80,* 229—245 (1967).

Stanka, P.: Untersuchungen über eine Innervation des Subkommissuralorgans der Ratte. Zschr. mikrosk.-anat. Forsch., Leipzig, *71,* 1—9 (1964).

Stanka, P.: Über den Sekretionsvorgang im Subkommissuralorgan eines Knochenfisches *(Pristella riddlei* Meek). Zschr. Zellforsch. *77,* 404—415 (1967).

Stanka, P., A. Schwink, und *R. Wetzstein:* Elektronenmikroskopische Untersuchung des Subkommissuralorgans der Ratte. Zschr. Zellforsch. *63,* 277—301 (1964).

Sterba, G.: Grundlagen des histochemischen und biochemischen Nachweises von Neurosekret (= Trägerprotein der Oxytozine) mit Pseudoisozyaninen. Acta histochem. *17,* 268—292 (1964).

Sterba, G.: Zur cerebrospinalen Neurokrinie der Wirbeltiere. Verh. Dtsch. Zool. Ges. 1965. pp. 393—440.

Sterba, G., und *W. Naumann:* Elektronenmikroskopische Untersuchungen über den Reissnerschen Faden und die Ependymzellen im Rückenmark von *Lampetra planeri* (Bloch). Zschr. Zellforsch. *72,* 516—524 (1966).

Sterba, G, und *J. Weiss:* Beiträge zur Hydrencephalokrinie: I. Hypothalamische Hydrencephalokrinie der Bachforelle *(Salmo trutta fario).* J. Hirnforsch. *9,* 359—371 (1967).

Sterba, G., H. Müller, und *W. Naumann:* Fluoreszenz- und elektronenmikroskopische Untersuchungen über die Bildung des Reissnerschen Fadens bei *Lampetra planeri* (Bloch). Zschr. Zellforsch. *76,* 355—376 (1967).

Studnička, F. K.: Untersuchungen über den Bau des Ependyms der nervösen Centralorgane. Anat. Hefte, Wiesbaden, *15,* 303—430 (1900).

Stutinsky, F.: Colloïde, corps de Herring et substance Gomori positive de la neurohypophyse. C. R. Soc. Biol., Paris, *144,* 1357—1360 (1950).

Takeichi, M.: The fine structure of ependymal cells. Part II: An electron microscopic study of the soft-shelled turtle paraventricular organ, with special reference to the fine structure of ependymal cells and so-called albuminous substance. Zschr. Zellforsch. *76,* 471—485 (1967).

138   A. Oksche:

Talanti, S.: Studies on the subcommissural organ in some domestic animals. Ann. Med. Exper. et Biol. Fenn. *36*, Suppl. No. 9, pp. 1—97 (1958).

Van der Wal, B., J. Moll, and *D. de Wied:* The effect of pinealectomy and of lesions in the subcommissural body on the rate of aldosterone secretions by rat adrenal glands in *vitro.* In: Progress in Brain Research *10.* Structure and Function of the Epiphysis cerebri. *J. Ariëns Kappers* and *J. P. Schadé,* Editors. pp. 635—645. Amsterdam-London-New York: Elsevier 1965.

Vigh, B.: Ependymosécrétion, sécrétion Gomori-positive de l'épendyme dans l'hypothalamus. Ann. endocr., Paris, *25,* 140—144 (1964).

Vigh, B., P. Röhlich, I. Teichmann, and *B. Aros:* Ependymosecretion (Ependymal Neurosecretion). VI. Light and electron microscopic examination of the subcommissural organ of the guinea pig. Acta Biol. Hung. *18,* (1) 53—66 (1967).

Wislocki, G. B., E. H. Leduc, and *A. J. Mitchell:* On the ending of Reissner's fiber in the filum terminale of the spinal cord. J. Comp. Neurol., Philadelphia, *104* 493—517 (1956).

Author's address: Prof. Dr. *A. Oksche,* Anatomisches Institut der Universität Giessen, Friedrichstraße 24, 63 Giessen, Federal Republic of Germany.

Discussion

Bargmann: Dr. *Oksche,* can you explain briefly your opinion on the functional significance of polarization observed in the subcommissural organ?

Oksche: As far as a close anatomical relationship between the secretory perikarya or cell processes of the SCO and the blood vessels can be demonstrated, some kind of biologically active substance could be released from the accumulations of the secretory material into to vascular system. It is, however, puzzling that this type of vascular contacts does not appear uniformly in all vertebrates. This is the only suggestion I can make about this particular problem on the basis of my findings. I wish to express my gratitude to you for supporting me with some SCO-material from your former experiments performed on the neurosecretory system of the dog. These dogs were treated with water deprivation for 8 to 14 days and in some of them the hypophyseal stalk was transsected. This experimental procedure was followed by a recovery period (water *ad libitum*). In these animals the cell columns and strands of the SCO seem to be more prominent and more heavily stained than in the controls. As the experiments include several steps I would not like to speculate on possible relationships of these findings to the function. I should like to examine the SCO of these specimens more thoroughly with quantitative methods. This work is in progress now. I feel that we should be very careful with our interpretations concerning the influence of environmental factors on the secretory activity of the SCO (primary or secondary effects?). For details of some of these procedures (experimental frogs) the papers of *Legait* (1942, 1946, 1949), *Oksche* (1962), and *Mautner* (1965) may be consulted.

Knowles: I have been greatly impressed by the quality of your electron micrographs. I wonder if you would be kind enough to give us some details of your technique.

Oksche: The material has been fixed with glutaraldehyde in the usual manner. After treatment with 1% OsO_4 the specimens were embedded either in Epon or Vestopal. For better contrast the lead-hydroxyde methode of *Karnovsky* was used (for details see recent papers of my laboratory, for example *A. Oksche* and *H. Kirschstein,* Z. Zellforsch. *78,* 151—166, 1967). Lead-hydroxide gives a good picture of the glycogen. I think in these regions we have to be very careful in respect to any PAS-positive substances. Some of the material appears to be glycogen and a saliva test is very important. For electron microscopy there are even purified diastase preparations which can be used on very small blocks of tissue. In this way all the glycogen can be removed from the tissue.

Knowles: As you know some authors have made much of different forms of PAS-positive material in these cells. Do you have any evidence that there are other substances besides glycogen staining with PAS?

Oksche: In the subcommissural organ one finds several PAS-positive substances and the thing bothering me now is whether this is or is not the same material as showing up in my electron micrographs.

Knowles: Well, that is, I think, a very important general point, because so much is been made about these PAS-substances.

Csillik: Pineal gland, subfornical organ and subcommissural organ are characterized by the absence of pseudocholinesterase activity from their capillaries, whereas other capillaries in the rat brain exert a strong activity of this enzyme. Capillaries in the pineal gland and in the subfornical organ belong to the "fenestrated" type. How about the capillaries in the subcommissural organ? Is there any correlation between the histochemical and ultrastructural features of these special capillaries?

Oksche: As far as our present material is concerned no special features of the capillaries of the subcommissural organ could be observed. After intravenous administration of trypan-blue the dye is not accumulated in the subcommissural organ. On the other side, an accumulation occurs in the pineal body, area postrema, organum vasculosum laminae terminalis and the subfornical organ of these animals. *Weindl* has shown that there is no wide perivascular space containing adventitial cells in the subcommissural organ of the rat and that, in the basal lamina of the subcommissural capillaries, a periodically structured particular type of collagen occurs.

Watermann: Into the third ventricle of the brain of guinea-pigs I injected tuberculosis bacteria. In several parts of the ventricular wall I was able to observe phagocytosis of the bacteria and histopathological alterations, also using the *Ziehl-Neelsen* staining method on paraffin sections. Relatively acid-resistant granules were seen in different parts of the hypendyma, for instance in the region of the subcommissural organ, in the trigonum supra-commissurale, in the tuber cinereum, but not in the interventricular organ. Could these intraplasmatic granules be either lipofuscins, neurosecretory granules or something else?

Oksche: The ependyma of the subcommissural organ is rich in microvilli. By this ependyma substances and even particles can be absorbed from the cerebrospinal fluid. Along this way tuberculosis bacteria, injected into the ventricle, could possibly reach the hypendymal zone. I do not know, however, whether lysosomes stain with the *Ziehl-Neelsen* technique. It is well possible that you demonstrated these organelles by your technique. This supposition includes your demonstration of pigments because the origin of lipofuscin is connected with the lysosomes. Because this material has been observed by you in so many different sites I would not think of secretory granules.

Note. For further discussion, see *P. Stanka:* Z. Zellforsch. *85*, 67—77 (1968); *W. Naumann:* Z. Zellforsch. *87*, 571—591 (1968); *J. F. Leatherland,* and *J. M. Dodd:* Z. Zellforsch. *89*, 533—549 (1968); *N. X. Papacharalampous, A. Schwink,* and *R. Wetzstein:* Z. Zellforsch. *90*, 202—229 (1968); *K. Kohno:* Z. Zellforsch. *94*, 565—573 (1969).

Journal of Neuro-Visceral Relations, Suppl. IX, 140—184 (1969)

The Mammalian Pineal Organ

J. Ariëns Kappers

The Netherlands Central Institute for Brain Research, Amsterdam

With 2 Figures

Summary

This survey is based on selected data from the large literature as well as on own observations and discusses aspects of pineal phylogenetic and ontogenetic development, the fine structure of the mammalian pineal, its biochemistry and physiology. The author has been trying to integrate data, to stress the structural, biochemical and functional specificity of the organ, and to point to problems which should be further investigated. He has not been shy of one or two, perhaps fruitful, speculations.

The phylogenetic development of the organ supports the opinion that the mammalian pinealocyte is a non-nervous element derived from the neurosensory photoreceptor cell present in the submammalian pineal organ. In Sauropsida the photosensory function of the epiphysis is gradually lost while its secretory function develops. This functional change, already foreshadowed in anuran amphibians, is accompanied by a change in pineal innervation. The sensory pineal tract disappears and the organ becomes innervated by peripheral autonomic fibers. Primarily, these fibers reach the pineal exclusively along the pineal vessels, running in the perivascular spaces. Secondarily, an additional contribution of sympathetic fibers is realized by way of the nervi conarii. The latter fibers penetrate directly into the pineal parenchyma their endings making simple appositional as well as specific synaptic contacts with the pinealocytes. In this paper, such synaptic contacts are demonstrated in the rat pineal.

By way of a working hypothesis the mode of innervation of the pineal organ of mammals is compared with that of smooth musculature. Stimulation of pinealocytes is thought to occur by (1) diffusion of neurotransmitter released from the varicosities and endings of sympathetic fibers present in the perivascular spaces, (2) simple appositional contact of sympathetic nerve terminals with pinealocytes in the parenchyma, (3) specialized synaptic contact between sympathetic nerve terminals and pinealocytes in the parenchyma, and, speculatively, (4) by transmission of impulses from one pinealocyte to another which can either be (a) of a neurohumoral or (b) of an electrotonic nature. Probably, (1) is the phylogenetic earliest mode of stimulation.

The topographical relationship between the terminal buds of the processes of pinealocytes and the perivascular space varies in mammals. Although the secretory function of the pinealocyte can scarcely be doubted, the problem of the production, storage and extrusion of secretory products is not quite solved at the ultrastructural level. The compound(s) produced by the pinealocytes most probably reach the blood via the perivascular spaces, a condition which is compared with the release of compounds produced by neurons and neurosecretory cells into the blood via pericapillary spaces in the region of the median eminence and in the neural part of the hypophysis.

In the chapter on pineal biochemistry the presence of catecholamines, indole-amines, lipids, proteins and some enzymes is discussed as is the influence of pineal denervation and of light and darkness on the pineal content of some of them. The neural pathway for the transmission of photic stimuli from the retina of the lateral eyes to the epiphysis is also mentioned. In the pineal some compounds are synthetized which are exclusively produced in this organ.

In the chapter on pineal physiology reference is made to experimental research pointing to the effects of light and darkness, of pinealectomy and of administration of pineal extracts on the size, weight and function of the reproductive organs. It appears that the pineal exerts an inhibiting influence on these organs. A tentative explanation is given of the functional significance of the intact pineal innervation by sympathetic fibers. It is assumed that light and darkness, via this innervation, condition the basically internal regulated function of the organ. External stimuli, such as photic, act in a regulation mechanism which is superimposed on an intrinsic pineal regulation mechanism which is probably of a hormonal nature. Experimentally produced changes in pineal function are, after some time, no longer of any effect on the normal function of the reproductive system. It would seem that the pineal organ is not indispensable for the normal function of this system as appears, on the long run, after pinealectomy.

Regarding the nature of the pineal compound(s) inhibiting the reproductive system different opinions exist. They would be either indole derivatives such as melatonin and 5-methoxytryptophol, or peptides. The finding that by one peptide fraction of a pineal extract hypophyseal FSH secretion *in vitro* is increased whereas by another peptide fraction this secretion is decreased has now proven beyond doubt that (1) the pineal produces an antigonadotropic as well as a gonadotropic principle, (2) these principles act on the gonadotropic cells of the distal part of the hypophysis rather than at the level of the organs of reproduction. Other arguments are also in favor of the opinion that the direct target organ of the pineal compound(s) is the hypophysis. That this pinealo-hypophyseal action may be additionally regulated by hypothalamic centers is suggested by changes in activity of these centers after administration of pineal extracts. It also appears that, most probably, the distal part of the hypophysis rather than the gonads regulates the production of the pineal antigonadotropic principle.

In conclusion it can be stated that the mammalian pineal is a photo-neuro-endocrine organ *(E. Scharrer)* on the understanding that the structure is a neuro-epithelial derivative, that the endocrine function is exerted by non-nervous but specific pineal cells, the pinealocytes, which are phylogenetically derived from neurosensory photoreceptor cells, and that the mammalian pineal is an indirect photosensory organ, photic stimuli reaching it via a neural pathway starting in the retina and ending in the organ by way of its sympathetic innervation. The mammalian pineal functions as a "regulator of regulators" and is not absolutely indispensable for the normal function of the organism.

The facts now known about this intriguing organ should stimulate further research to solve the many problems so far still unsolved.

Introduction

This symposion deals with neurohormones and neurohumors, that is with chemical compounds which, in principle, are present in and eventually produced by neurons. However, cells of neuro-epithelial derivation other than neurons are either known or supposed to produce also substances of physiological consequence. These elements my constitute special organs such as, for instance, the subcommissural organ and the pineal body. Although the structure of most of these organs is known, the function of many is still disputed. This holds also for the mammalian pineal body which, for a long time, has been held to be a mere phylogenetic relic of no physiological significance.

Especially during the past 10 years or so, the results of numerous descriptive and experimental investigations on the pineal have been published. Certainly, we now know much more about the mammalian epiphysis than we knew, say in 1957, on the ground of data and theories published during the two millennia passed before that year. However, notwithstanding our rapidly expanding knowledge it cannot be denied that, especially as the function of the mammalian pineal organ is concerned, we are still more or less at the speculative stage.

The following questions, for instance, cannot be answered quite satisfactorily: (1) Is there only one or are there perhaps more substances produced by the pineal which exert physiological effects, and, if so, which is their exact nature; (2) which is the exact way in which these substances are produced and stored within pineal parenchymal cells at the ultramicroscopical level; (3) which are the ways by which their production and release are influenced; (4) which exactly are the physiological effects exerted by this substance or substances on other organs or systems; (5) which is the direct target organ or organs of the physiologically active pineal compound(s)? Although recent research has done much to unravel at least part of these problems it would be preposterous to assume that we know, at the present time, all about the function of this probably endocrine organ of neuro-epithelial derivation. The story of the discovery of the morphological features and of the functional significance of the mammalian pineal organ recalls, indeed, the similar story concerning the hypophysis which, not so very long ago, was also thought to be a rudimentary organ of no functional consequence whatsoever.

The present survey is meant to give some general information on the phylogenetic and ontogenetic development of the mammalian pineal body, its fine structure, biochemistry and physiology. Owing to the abundance of data recently obtained a selection had to be made so that this account is far from complete. It has been our aim to integrate a number of data, to stress those aspects which point to the structural and functional specifity of the mammalian pineal and to point to some problems which should be the subject of further investigations.

I. Aspects of Pineal Phylogenetic and Ontogenetic Development

In all Vertebrates the embryonic development of the pineal starts with the formation of a neuro-epithelial midplane evagination protruding from the roof of the diencephalon between the anlagen of the habenular and the posterior commissure.

In Anamniota, the matrix layer constituting this diencephalic dorsal evagination produces, in principle, three types of cells: (1) neurosensory elements, (2) sensory nerve cells, (3) cells of an ependymal type or so-called supporting cells. By many authors it has been shown that the ultra-structure of the neurosensory cells is similar to that of the photoreceptor elements in the retina of the lateral eyes. Photic stimuli reaching directly the pineal and accessory pineal organs, such as the parapineal organ of *Petromyzon*, the frontal organ of anuran amphibians and the parietal eye of lizards, are received and transduced by the photosensitive neurosensory cells. The axonic process of these elements is in synaptic contact with the dendritic plexus of the intraepithelial pineal sensory nerve cells. Along the axons of these neurons the transduced photic stimuli travel to the epithalamus. These axons, then, constitute the sensory nerve fiber bundles which are afferent in respect to the brain, but efferent in respect to the pineal organ. There is some indication, at least in anuran amphibians, that some nerve fibers may reach the pineal organ thus being afferent in respect to this structure. Probably, these fibers are of autonomic origin (see below). This is interesting in view of the exclusively afferent autonomic pineal innervation present in mammals.

The course and site or sites of termination of the sensory pineal fibers within the brain are still very imperfectly known (see *Kappers*, 1965, 1967), and so is their functional significance. This is one of the problems to be solved by future research.

In the anamniote adult the epiphysis may show variable degrees of development and differentiation. If well differentiated, its structure and main innervation pattern are remarkably similar to those of the lateral photosensitive organs of diencephalic origin, the lateral eyes, although some characteristic structural and physiological differences exist (*Kappers*, 1965). The photosensitivity of the pineal has also been proven by electrophysiological research, at least in some Anamniota.

Blood vessels vascularizing the organ never penetrate the basement membrane of the pineal epithelium which is separated from the basement membrane lining the capillary endothelium by perivascular spaces of variable width.

The pineal and accessory pineal organs of anamniote Vertebrates are vesicular, saccular or tubular structures which are, in principle, directly photosensitive and show a sensory innervation pattern. It may be that the anamniote epiphysis, at least in some cases, has an additional function, *i. e.* a secretory one. It is, for instance, known that the anuran pineal contains and probably produces melatonin controlling the pigmentation of the animal. For a more detailed discussion of the structure and function of the anamniote pineal organ the reader is referred to *Kappers* (1965, 1969).

Coming now to the Amniota, we will deal first with the pineal organ of reptiles. Rather remarkably, the structure is lacking in *crocodiles*, not only in adults but even at embryonic stages in which no trace of a pineal anlage has been observed. In *lizards* the epiphysis is a saccular or tubular epithelial organ its walls often showing many foldings. Again, the pineal epithelium is constituted of three types of cells: (1) more or less rudimentary neurosensory cells, (2) some scarce sensory nerve cells, (3) supporting cells of an ependymal type. Most of the neurosensory cells do not show all the characteristics of well differentiated photoreceptor elements. This holds especially for the outer segments which are rudimentary. These cells, moreover, contain secretory granules measuring between 1000 and 2000 Å in diameter. Histochemical and fluorescence histochemical investigations point to the probability that these secretory droplets contain serotonin. The number of intraepithelial cell bodies of sensory neurons is very scarce and so is, consequently, the number of pineal sensory fibers reaching the epithalamus. The change in structure of most of the photoreceptor elements, which have been rightly termed secretory rudimentary photosensory cells, as well as the small number of sensory neurons point to a reduction of the direct sensory apparatus in the lacertilian pineal. Ultrastructural research has shown that two types of nerve fibers course in the pericapillary spaces adjacent to the pineal epithelium, myelinated and non-myelinated nerve fibers. Endings of the latter contain clear and dense-core vesicles. The relatively scarce myelinated fibers are the pineal sensory ones being the axons of the intraepithelial pineal neuronal pericarya. Most probably, the rather numerous non-myelinated nerve fibers are postganglionic axons of autonomic origin. They terminate in the perivascular spaces. Very few endings only have been observed between the secretory rudimentary photoreceptor cells within the pineal epithelium. They may lie in close apposition to these cells but, so far, specialized synaptic contacts have not been observed. The autonomic fibers may be assumed to be of functional significance for the secretory and/or excretory processes in the secretory rudimentary photoreceptor cells.

Evidently, the lacertilian pineal shows two functions: (1) the "traditional" direct photosensory function which is, however, very much reduced, and (2) a secretory function which needs further investigation. In consequence, the epiphysis of lacertilians shows also two patterns of innervation: (1) a pinealo-fugal sensory innervation, and (2) a pinealo-petal innervation by autonomic nerve fibers the endings of which contain the characteristic dense-core vesicles which are known to contain biogenic amines. For details, literature references and illustrations the reader is referred to Quay and coll. (1967), Kappers (1967), Collin and Kappers (1968) and Wartenberg and Baumgarten (1968).

The tubular pineal organ of turtles appears to show essentially the same features as that in lizards at least as the reduction of the photosensory apparatus is concerned which is apparent from the relatively small number of pineal sensory nerve fibers and the presence of secretory rudimentary photoreceptor cells (Vivien and Röels, 1967). Further special investigations are, however, needed.

In snakes the epiphysis is a well-vascularized compact parenchymal organ consisting of cell nests. The parenchymal cells have lost the characteristics of neurosensory photoreceptor elements. Probably these cells have become entirely secretory producing a compound which, on the ground of its yellow fluorescence, can be assumed to be serotonin. Autonomic nerve fibers have been observed to enter the snake pineal. So far, nerve cells have not been demonstrated. It is, therefore, questionable whether any sensory pineal fibers reach the epithalamus but this possibility cannot be wholly excluded. Species differences may exist in this respect. For details and literature see *Quay* and coll. (1968).

In *birds* the structure of the pineal organ varies widely. By some light microscopists sensory cells have been described in the organ. Ultramicroscopically, however, these elements do not generally show the characteristics of photosensory cells, especially as their outer segment is concerned. On the other hand, in some birds at least the presence of sensory nerve cells, although much reduced in number, and of a small bundle of pineal sensory fibers reaching the epithalamic region have been clearly demonstrated. Evidently, a similar reduction of the direct pineal photosensory apparatus as is now known to occur in lizards and turtles is also realized in birds, the degree of this reduction varying among the avian families. Besides this reduced sensory function a secretory function of the avian pineal is evident. Pineal cells are known to store serotonin and to be able to produce melatonin. Moreover, several authors observed the presence of an autonomic innervation pattern in the pineal of some avian species. Summarizing shortly all data available it appears that, in the avian pineal, the direct photosensory function is reduced to a variable degree whereas its secretory function is developing just as is the case in the lacertilian and chelonian pineal. For a survey of literature data on the avian pineal see *Kappers* (1967).

In *mammals* the primary pineal anlage is also formed by a neuro-epithelial evagination of the diencephalic roof which is situated between the anlagen of the habenular and posterior commissure. This matrix layer proliferates giving rise to lobules and follicles. Soon, the surrounding embryonic mesenchyma contributes to the formation of the organ producing blood vessels which grow in between the cell cords of neuro-epithelial derivation. Along with these vessels other mesenchymal elements such as fibroblasts, lemmoblasts, and, sometimes, plasma cells and mast cells move into the anlage of the organ. Finally this is shut off from the surrounding tissue by the differentiation of its leptomeningeal covering (*Kappers*, 1960).

We agree with those authors holding that the characteristic cell type of the mammalian pineal organ, the pinealocyte, is a non-nervous element derived from the neuro-epithelial matrix layer of the diencephalic evagination forming the primary anlage of the structure. In our opinion, the arguments brought forward by some authors (*Frauchiger*, 1963; *Meyburg*, 1965) for an at least partly mesenchymal origin of pinealocytes are, as yet, not convincing. The pinealocyte is the specific parenchymal cell of the mammalian pineal organ and belongs to the neuroglial series if one agrees that this series comprises all non-nervous cell types of neuro-epithelial origin.

To this series, then, belong next to the well known astro- and oligodendro-glial elements of ectodermal origin such varying kinds of elements as epen-dymal cells, the epithelial cells of the choroid plexuses, the specialized epen-dymal cells constituting the subcommissural organ, the neurohypophyseal pituicytes, the pineal parenchymal cells and some more cell types all show-ing a specific structural and functional differentiation.

The ultrastructure of the mammalian pinealocyte is different from that of both, neurons and gliocytes. As was suggested by *Bargmann* in 1943, most probably this secretory cell has phylogenetically developed from the neurosensory photoreceptor element of the anamniote pineal organ via the secretory rudimentary photoreceptor cell present in the epiphysis of amniote submammals.

The adult mammalian pineal is a well vascularized parenchymal organ not containing either neurosensory cells or sensory nerve cells. It is primarily constituted of lobules and cords of pinealocytes which are irregularly and often incompletely separated by stromal septa containing blood vessels. Besides pinealocytes and cells of mesenchymal origin mentioned before, another cell type of undoubted neuro-epithelial derivation is present in varying quantities. This mostly resembles a fibrous astrocyte.

In the next chapter the innervation of the mammalian epiphysis will be dealt with. It may, however, be mentioned here that the nerve fibers inner-vating the epiphysis are exclusively of peripheral autonomic origin. Rather remarkably for an organ which is part of the brain it is not functionally connected with the brain either by afferent or by efferent fibers. The mam-malian pineal, indeed, can be considered an endorgan of the autonomic nervous system. The functional significance of the autonomic innervation of the mammalian pineal will be dealt with later.

Summarizing this short survey of the phylogenetic development of the pineal organ in Vertebrates it appears that this shows an interesting struc-tural and functional evolution which is also reflected in its innervation. In anamniote Vertebrates it is predominantly a directly photosensitive epi-thelial structure showing a sensory innervation pattern. In mammals, on the other hand, the epiphysis is a parenchymal secretory organ which is indi-rectly photosensitive by way of its exclusively autonomic innervation as will be shown later. The turning point in this evolutionary trend is most clearly realized in amniote submammalian Vertebrates, that is in Sauropsida (reptiles and birds), in which the photoreceptor function of the organ as well as its sensory innervation pattern are much reduced or even practically absent while secretory phenomena and, in functional association therewith, the autonomic innervation pattern tend to prevail. Recent investigations indicating a secretory function and a possible autonomic innervation, al-though not extensive, of the anuran pineal organ (reviewed by *Kappers*, 1969) may point to the presence of both, a pineal photosensory and secre-tory function associated with their respective innervation patterns in a least some Amphibia.

II. Aspects of the Fine Structure of the Mammalian Pineal

For an extensive and most valuable review of the structure of the mammalian pineal organ based on light microscopical investigations the reader is referred to *Bargmann* (1943). In the present survey some ultramicroscopical aspects of pineal structure will be dealt with, especially in relation to the problems concerning the function of the epiphysis.

a) The pinealocytes

The author agrees with those workers who consider the pinealocyte to be the characteristic cell element of the mammalian epiphysis which is not directly comparable to either a nerve cell or a gliocyte. This parenchymal cell may show one or several processes of different length often ending in terminal clubs or polar terminals. The nuclear shape varies from spherical or oval to lobulated and indented forms. In the latter case, the nuclear membrane shows irregular foldings, the cytoplasm seemingly invaginating into the nucleus. The endoplasmic reticulum is rather similar to that of other cell types but clear tubular profiles are often lacking and there are differences in compactness and branching. Such a close and frequent association of ribosomes with the membranes constituting this reticulum as is characteristic of the typical granular reticulum does not occur in the cytoplasm of pinealocytes. Therefore, this endoplasmic reticulum has been termed a mixed or intergrade reticulum (*Wolfe*, 1965). The Golgi complex is well developed. The pinealocytes are characterized by their large number of large, elongated and rather clear mitochondria.

In the cell bodies and the processes of the pinealocytes structures known as "synaptic ribbons" are often observed. They have also been termed "vesicle-crowned rodlets" (*Wolfe*, 1965) and "vesicle-crowned lamellae" (*Arstila*, 1967) although most of them do not show a lamellar structure (*Wartenberg*, 1968). There organelles are often situated close to the cellular membrane. By *Arstila* and *Hopsu* (1964) and *Hopsu* and *Arstila* (1964) they have been interpreted into functional neurohumoral synaptic contacts between pinealocytes which, by these authors, are considered modified nerve cells (see subchapter on innervation). Similar "synaptic ribbons" were, however, first demonstrated in photoreceptor cells of the retina and have also been observed in the photoreceptor elements of both, the frontal and the pineal organ of *Rana*. In general, they do occur in a number of other receptor cells as well (see *Wartenberg*, 1968, for references). This fact may corroborate the hypothesis that the mammalian pinealocyte has phylogenetically developed from a neurosensory cell. About the functional significance of the "synaptic ribbons" nothing is exactly known.

A main point of interest is whether, in the mammalian pinealocyte, structures are present pointing to a secretory activity of this cell. In this respect earlier light optical findings have been equivocal but some recent results of electron microscopical investigations provide cytological evidence that indeed, the mammalian pineal is a secretory organ. Some literature data pertaining to this question will be briefly summarized here. By *Rodin* and

Turner (1965, 1966) clear vesicles exclusively were observed in rat pinealocytes while, according to *Wolfe* (1965), in the pineal cells of this same
species, acanthosomes or spiny bodies are present, associated with the Golgi
complex. These are probably homologous to the "vesicles with spiked ex-
teriors" demonstrated by *Anderson* (1965) in bovine and ovine pinealocytes
which are likewise in close association with the Golgi complex. *Anderson,*
however, distinguished, besides a multitude of smooth-surfaced clear vesicles, also some dense-core ones which appeared to take origin in the Golgi
apparatus and were also found scattered throughout the cytoplasm of the
cell body and its processes. *Arstila* (1967) described 4 subtypes of membrane-bound inclusion bodies in the terminals of rat pinealocytic processes
the content of which varied from clear to finely granular. By *Sano* and
Mashimo (1966) clear as well as dense-core vesicles measuring 350—1400 Å
in diameter were observed in the pineal parenchymal cells of the dog. Apparently originating in the Golgi complex, their content would be released
at the terminals of the pinealocytic processes. A somewhat similar observation was made by *Leonhardt* (1967) demonstrating clear as well as granular
vesicles in the cells of the rabbit pineal. This author describes and illustrates
the process of extrusion of the secretory product probably contained within
these vesicles. Release of the secretory product would take place either into
intercellular spaces which are in open connection with the perivascular
spaces or directly into the latter. Probably, these intercellular spaces, filled
by the secretory product, are identical with the "osmiophilic enclaves" found
by *Milcou* and *Petrea* (1964 a, b; 1967) to react on experimental procedures.
Halaris and coll. (1967) distinguished clear vesicles, small dense-core and
large dense-core vesicles in the processes of rat pinealocytes, the small
granular vesicles measuring 450 Å and making up almost the total vesicular
population. According to these authors they would contain serotonin or
5-hydroxytryptamine. In the terminals of cat and monkey pinealocytic processes, however, no granulated vesicles could be demonstrated by *Wartenberg* (1968). On the other hand, this author described the presence, in
monkey pinealocytes, of clear vesicles similar to "synaptic vesicles" and of
"dense bodies". The latter contain either amorphous dense aggregates or
lamellated bodies, or they may show an electron dense homogeneous content.
All are limited by a single membrane. Following *Wartenberg* these dense
bodies have to be considered a secretory product. The question as to the
identity of some of these bodies with lysosomes can, however, be raised.

Summarizing these data it seems rather difficult to establish homologies
between the different cytoplasmatic inclusions described by different authors. It appears, however, that the origin of some of them is clearly in the
Golgi complex while their extrusion happens into either the intercellular or
directly into the perivascular spaces. The chemical composition of the content of the agranular and the granular vesicles and of the dense bodies
described is, so far, merely a matter of speculation. Further electron microscopic research, preferably under different experimental conditions, is certainly needed as are finer histochemical investigations. Some of the results

mentioned appear promising and, in the opinion of the present author, the secretory function of the mammalian pinealocyte can now be scarcely doubted. Very probably, differences in cytological observation can, at least in part, be explained by differences in species and/or by differences in physiological condition of the animals examined.

By some authors two types of pinealocytes have been distinguished, "light" and "dark" ones (*Wartenberg* and *Gusek*, 1965, *Gusek* and coll., 1965). Possibly, these types represent functionally different stages of the same kind of cell, but some are of the opinion that they would be glial elements. This question cannot he further dealt with here.

b) The glial elements

Besides the pinealocytes a varying number of elements do occur in the mammalian pineal which are obviously glial cells (*Scharenberg* and *Liss*, 1965; *Wartenberg* and *Gusek*, 1965; *Anderson*, 1965; *Wartenberg*, 1968). By some authors (*Wolfe*, 1965; *Arstila*, 1967) they have been termed "interstitial cells" which is a somewhat confusing name. Most commonly, fibrous astrocytes are present which are ultrastructurally characterized by many and often long processes containing large bundles of fine filaments of indeterminate length. Moreover, these elements differ from pinealocytes in the following respects: (1) their nuclear chromatin is denser and their nuclear shape is more constantly regular than is the nuclear shape of the pinealocytes, (2) the cytoplasma contains more rough ergastoplasm, (3) the mitochondria are smaller, not so elongated as is true for the mitochondria in the pinealocytes of some mammals, and their matrix is more dense, (4) the glial cytoplasm may contain many dense bodies or lysosomes (*Anderson*, 1965; *Wartenberg*, 1968). Besides fibrous or filamentous astrocytes cytoplasmatic astrocytes may also occur. The long processes of these glial elements are intertwined with those of the pinealocytes. The amount of astroglia and, more especially, the number of gliocytic processes contributing to the structure of the mammalian pineal may vary considerably according to the species investigated. This also holds for the topographical relationship between the glial and the pinealocytic processes at the base of the pineal epithelium, that is at the outer border of the perivascular spaces.

c) The perivascular spaces

The perivascular spaces are lined by an internal basement lamina covering the capillary endothelium and an external basement lamina which separates the space from the pineal parenchyma. The perivascular spaces contain, in general, collagenous fibrils, fibrocytes and non-myelinated autonomic nerve fibers, often present in bundles. Locally, these spaces may penetrate the pineal parenchyma. The intraparenchymal spaces, thus formed, are characterized by a basement lamina separating the space from the pineal parenchyma. Other intraparenchymal spaces, not lined by a basement lamina, are extensions of intercellular spaces (*Wolfe*, 1965, rat; *Wartenberg*, 1968, squirrel monkey).

The topographical relationship between the terminals of pinealocytic processes and the perivascular spaces varies. In the squirrel monkey, for instance, a glial barrier lying against the parenchymal side of the external basement membrane of the perivascular space shuts off the terminals of these processes from any contact with this membrane (*Wartenberg*, 1968). In the pineal of rabbit (*Wartenberg* and *Gusek*, 1965) and of sheep and cattle (*Anderson*, 1965) similar relationships exist although, in cattle, terminal endings of pinealocytic processes may contact the external basement membrane. In cat, these terminal buds sometimes protrude into the perivascular space into which they are, so to say, "herniated" still being covered by a thin glial layer and by the external basement membrane of the perivascular space so that these processes do not really enter the space freely (*Duncan* and *Micheletti*, 1966; *Wartenberg*, 1968). In the pineal organ of the Rhesus monkey the enlarged perivascular spaces contain terminals of pinealocytic processes which are only partly covered by glial sheets. Invariably, however, they are covered by the external basement membrane of the perivascular space so that, also in this case, they do not lie freely in this space (*Wartenberg*, 1968). In the rat epiphysis, however, the endings of these processes may be found lying quite freely in the perivascular spaces not covered either by glial processes or by the external basement membrane of the perivascular space which is discontinuous at the site where the pinealocytic process enters the space from the side of the parenchyma. The presence of such free terminals of pinealocytic processes in rat pineal perivascular spaces can be inferred from papers by *Gusek* and *Santoro* (1960), *De Robertis* and *Pellegrino De Iraldi* (1961), *Arstila* and *Hopsu* (1964), *Wolfe* (1965), *Pellegrino De Iraldi* and coll. (1965), and *Arstila* (1967). In this case, therefore, an optimal possibility for an exchange of substances between the cytoplasm of the pinealocytes and the perivascular spaces is realized, the only membrane separating these two compartments being the cellular membrane of the pinealocyte. From the above it appears that the topographical relationship between the terminals of the pinealocytic processes and the perivascular space varies gradually in mammals.

d) The innervation

The mammalian pineal organ is exclusively innervated by autonomic postganglionic nerve fibers which, for the most part, are non-myelinated. In non-Primates, these fibers probably originate exclusively from nerve cell pericarya in the superior cervical ganglia, thus being orthosympathetic. The postganglionic pineal fibers reach the organ either along pineal blood vessels, or by way of two bilateral symmetrically nervi conarii which may or may not fuse before entering the tip of the organ, or along both, the pineal vessels and the nervi conarii (see *Kappers*, 1960, 1965, also for references). It should be stressed that the autonomic nerve fibers entering the organ along the vessels are situated within the pineal perivascular spaces whereas the fibers reaching the pineal in the nervi conarii distribute freely within the pineal parenchyma proper that is between the cell cords and follicles constituted by the pinealocytes, the glial elements and their processes (see f. i. Figs. 25 and 29 in *Kappers*, 1960, and Fig. 27 in *Kappers*, 1965). It is generally supposed that pineal nerve fibers running in the perivascular spaces may enter secondarily the pineal parenchyma traversing the outer limiting membrane of the perivascular space. So far, however, nerve fibers leaving the perivascular spaces have not been clearly demonstrated electron microscopically. Apparently, pineal innervation by way of autonomic fibers coursing within the perivascular spaces is phylogenetically older than that via the nervi conarii and is related with the development of the pineal from an epithelial into a parenchymatous organ.

According to *Kenny* (1961), in the macaque monkey parasympathetic fibers coursing in the greater superior petrosal nerves reach the epiphysis. They would be preganglionic synapsing with intrapineal autonomic nerve cells. Such cells have been demonstrated in Primates only (*Bargmann*, 1943; *Kenny*, 1961; *Kappers*, unpublished). Although it seems that, in sub-Primate mammals, pineal innervation happens entirely by postganglionic ortho-sympathetic fibers it would be interesting to look for some possible para-sympathetic contribution. The innervation of the Primate pineal organ should also be further investigated.

Nerve fibers of cerebral origin do not have endings in the pineal body which are of functional consequence although fibers of habenular and posterior commissural derivation may enter the proximal part of the structure. As has been argued (*Kappers*, 1960, 1965) most of these fibers are aberrant commissural fibers.

A few thinly myelinated pineal nerve fibers, especially present in nerve bundles entering the organ from its periphery, have been observed electron microscopically (*Milofsky*, 1957; *Pellegrino De Iraldi* and coll., 1965; *Anderson*, 1965). By the present author they have also been observed lying scattered in the rat pineal parenchyma. These fibers could either represent aberrant fibers derived from the habenulo-posterior commissural system which penetrate into the pineal body then possibly joining non-myelinated autonomic fibers, or autonomic myelinated postganglionic axons loosing their myelin sheath at deeper pineal levels. The latter possibility is supported by observations of *Pellegrino De Iraldi* and coll. (1965) and *Anderson* (1965), as well as by the fact that occasionally myelinated fibers have been demonstrated in bundles of otherwise non-myelinated postganglionic autonomic fibers in rat and chick iris (*Richardson*, 1964; *Zenker* and *Krammer*, 1967).

Bundles of autonomic fibers embedded in the cytoplasm of lemmocytes are present in the pineal perivascular spaces, but also in the pineal parenchyma proper. Differences in relative numbers of perivascular and intra-parenchymal fibers occur depending on the species investigated. In the squirrel monkey the amount of perivascular autonomic nerve fibers is rather small as compared with other species such as the Rhesus monkey, cat and rat (*Wartenberg*, 1968). By this author this is explained by assuming that, in the squirrel monkey, the perivascular fibers soon leave the perivascular spaces in order to enter the pineal parenchyma. It may, however, be that the number of autonomic fibers entering the organ along the pineal vessels is relatively smaller than that entering with the nervi conarii and distributing directly in the pineal parenchyma (see above).

In the perivascular spaces structures have been described which are generally regarded as autonomic nerve terminals. They are no longer embedded in the cytoplasm of lemmocytes and even their basement lamina may be defective or absent. These terminals are characterized by their "plurivesicular content" (*De Robertis* and *Pellegrino De Iraldi*, 1961; *De Robertis*, 1964; *Pellegrino De Iraldi* and coll., 1965) which means that they contain clear and granular vesicles, both measuring about 400—425 Å in diameter. A similar observation was made by *Rodin* and *Turner* (1965, 1966). On the other hand, three categories of vesicles in rat pineal nerve endings were observed by *Bondareff* (1965 a, b), *Bondareff* and *Gordon* (1966), *Halaris* and coll. (1967) and by *Hassler* and *Bak* (1966), *i. e.* agranular and granular vesicles of approximately 450 Å in diameter as well as larger granular vesicles measuring 840, 750 and 1000 Å, respectively, according to the authors cited. Constantly, the number of large granular vesicles is considerably less than that of the other two categories of vesicles the ratio of which varies between 50/50 and 30/70. In nerve endings present in the pineal perivascular spaces of cat and monkey the three categories of vesicles mentioned have also been observed (*Wartenberg*, 1968), the larger granular vesicles measuring even 1000—12000 Å. In cat this author demonstrated, moreover, "dense bodies" in the nerve endings showing a homogeneous or lamellated structure. According to the illustrations in Wartenberg's paper these bodies are considerably larger than the largest granulated vesicles.

Experimental investigations including superior cervical ganglionectomy and autoradiography following uptake of isotopes have shown that the granular vesicles in the pineal nerve fiber endings contain biogenic amines (see *Collin* and *Kappers*, 1968, for references). Most authors agree that the biogenic amine concerned is noradrenaline but some are of the opinion that these granular vesicles may also contain serotonin. Storage of noradrenaline as well as production of serotonin from 5-hydroxytryptophan by the granular vesicles in pineal nerve endings has been demonstrated by *Taxi* and *Droz* (1966, 1967). As observed by the fluorescence histochemical technique, in some species the autonomic pineal nerve fibers contain almost exclusively noradrenaline whereas, in other species, these fibers contain more serotonin (see later). These differences among the species are worth noting.

It is now generally recognized (see also the discussions in this volume) that the content of the clear vesicles present in autonomic nerve terminals does not necessarily consist of acetylcholine and that, therefore, these vesicles in this respect are not comparable with the "synaptic vesicles" present in presynaptic endings of preganglionic autonomic fibers and in those of cerebrospinal axons. Probably, the clear vesicles which are of the same size as the small granular vesicles represent granular ones which did release their content of biogenic amines. So far, nothing is exactly known about the content of the relatively small number of large granular vesicles which are a fairly constant feature of the pineal autonomic nerve endings. It is supposed that they may likewise contain biogenic amines (see discussion after the paper by *De Robertis* in this volume).

As is now generally accepted, the pinealocytes produce at least one specific compound while it has, moreover, been shown that the autonomic innervation of the organ exerts an influence on the function of these cells (see later). This brings us to the problem how neural impulses coursing along the pineal autonomic nerve fibers are transmitted to the pinealocytes. Evidently, this happens by a transmitter substance, most probably being noradrenaline, which is released from the autonomic nerve endings by the neural impulse. The question now is how the transmitter substance reaches the pinealocytes. Following the opinion of some authors this would only be possible by means of direct synaptic contact between the autonomic nerve terminals and the pinealocytes, not by means of release of transmitter substance from nerve terminals present within the perivascular spaces. This is probably the reason why, for instance, *Wartenberg* (1968) supposes that the club-shaped structures of these nerve fibers in the perivascular spaces do not represent terminals but varicosities which are likewise known to contain clear and granulated vesicles. This author, evidently, does not consider that the varicosities may have a similar function in releasing transmitter substance as have the nerve endings. It is, indeed, known that they have this function in other organs innervated by autonomic fibers. According to *Wartenberg* all autonomic nerve fibers, after running along a shorter or longer distance within the perivascular spaces, will enter the pineal parenchyma, their endings making here specialized synaptic contacts with pinealocytes. If this were true, the number of such synaptic contacts between autonomic nerve terminals and pinealocytes would be rather numerous in consequence of the large number of pineal autonomic fibers present.

As the presence of synapses between autonomic nerve terminals and pinealocytes in the mammalian pineal parenchyma is concerned the following may be remarked. Personally, *Wartenberg* (1968) was not able to detect any synaptic contacts in his material nor was *Gusek* (1968) although this author refers to nerve endings present within intercellular spaces between pinealocytes. *Wolfe* (1965), in the rat epiphysis, observed a close apposition between autonomic nerve endings and pinealocytes. He distinguished a "presynaptic complex" which is characterized by a closer packing of vesicles in the axoplasm immediately subjacent to an area of increased density in the axolemma. Following *Wolfe*, these morphological features and the observed close apposition of the axon ending and the pinealocyte justifies to term this apposition a "synapse". However, such "presynaptic complexes" were also seen by this author in nerve terminals present in the perivascular space lying against a basement membrane. Wolfe's interpretation has not been accepted by other authors such as *Arstila* (1967) who observed similar dense packings of vesicles in nerve endings but no thickenings of plasma membranes. He is of the opinion that it is unlikely that these sites act as specific synapses between the nerve endings and the pinealocytes. *Rodin* and *Turner*

(1966), although observing a close apposition of the axolemma of nerve endings to the cell membrane of pinealocytic processes lying freely within the
perivascular space in the rat pineal, hold that the occasional segmental
thickening of apposed membranes is not sufficiently distinct and structured
to be interpreted into a true synapse. In the opinion of these authors, however, the frequency of appositions between perivascular nerve endings and
pinealocytic processes (rat!) suggests that any transfer of neurohumors from
nerves to pineal cells may occur at these junctures rather than by diffusion
across the perivascular space.

Following *Pellegrino De Iraldi* and coll. (1965), in the rat pineal free
nerve endings are mainly found in the large perivascular spaces "as if the
transmitter could be directly released there". Some endings, however, were
observed in a rather intimate contact with pinealocytes adjacent to the
space that is within the parenchyma. However, pictures that could be interpreted as definite functional contacts between nerve terminals and pineal
cells were not readily observed. Similarly, *Anderson* (1965) was not able to
identify, in ovine and bovine pineals, areas that could unequivocally be
termed synapses. In the rat pineal parenchyma an autonomic nerve ending
terminating in close appositional contact with a pinealocyte has been described and figured by *Wurtman* and *Axelrod* (1965 b). However, no other
special morphological characteristics speaking for the presence of a real
synaptic connection in the sense of this term generally accepted were
present. Following a personal communication by *Wurtman* this was the
only case of such a contact observed by him in his preparations. Frequent
contacts between nerve endings and pineal cells in the rat pineal have also
been mentioned by *Kurosomi* and *Kawabata* (1966) but their exact nature
is not described by these authors.

In electron micrographs of rat pineal parenchyma the present author
observed a great many autonomic nerve fibers and a smaller number of
nerve terminals. Rather infrequently but more often than was expected on
the ground of earlier research, specific synapses were seen between nerve
endings and the cell bodies and processes of pinealocytes. These were
characterized by a close packing of vesicles in the nerve terminal, a slight

Fig. 1. Pineal organ of male rat, 135 g. Intraperitoneal injection of hydrazine, injection of noradrenaline-^{3}H in right cerebral ventricle. Perfusion fixation with a
mixture of paraformaldehyde 1%, glutaraldehyde 1% and $CaCl_2$ 0.002%. Millonig's
phosphate sucrose buffer, pH 7.4. Preparation, not rinsed, in OsO_4 2% and $CaCl_2$,
Millonig's buffer, 60 min. Dehydration in alcohol series and embedding in araldite.
Staining of the sections on the grits with uranylacetate $2^1/_2\%$ and Reynold's lead-
acetate. Philips EM 200. Magnification: Fig. 1 A 30,000×, Fig. 1 B 55,200×, reduced to $^1/_3$. (Division of Electron Microscopy, Netherlands Central Institute for
Brain Research, head: *H. J. Romijn*.) — In both, Figs. 1 A and 1 B, synaptic connections (S) are illustrated between autonomic nerve terminals (NT) and processes of
pinealocytes (P) showing accumulations of vesicles with different contents in the
presynaptic endings and an accumulation of dense material under the postsynaptic
membrane. A slight thickening of the presynaptic membrane and the presence of
particulate matter in the synaptic cleft are seen in Fig. 1 B. M mitochondria,
FA fascia adhaerens.

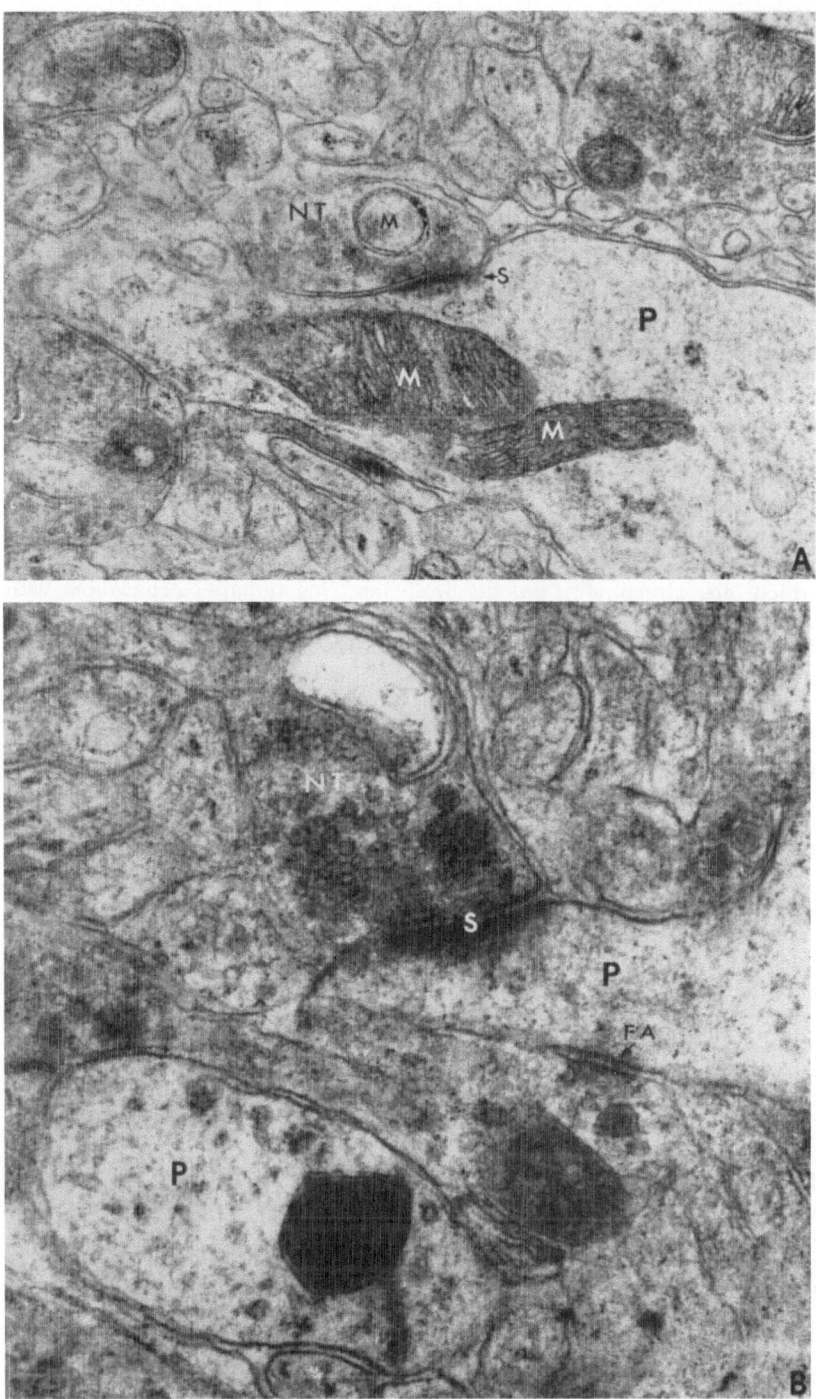

thickening of the presynaptic axolemma, a far more obvious postsynaptic thickening of the cell membrane of the pinealocyte, and the presence of particulate matter within the synaptic cleft. The term "postsynaptic thickening" is, in fact, not quite right as it is clear that electron dense matter has accumulated under the postsynaptic membrane which, itself, is not distinguishable. These characteristics are illustrated in Fig. 1. It is hoped that in the near future better photographs can be obtained.

From all data available it, however, appears that close contacts between nerve terminals and pinealocytes showing all the typical features of specific synaptic connections are rather rare. Specialized synapses are certainly not as frequent as could be expected if the only way by which autonomic fibers can influence the function of the pineal cells would be by synaptic contacts of this kind.

It should, however, be pointed out that autonomic nerve fibers can influence effector cells in other ways than just by means of the highly specialized synapses mentioned. Neurohumors released from either autonomic nerve endings or from the varicosities present along these fibers, which are structurally, biochemically and functionally comparable with terminals, can diffuse into the tissue surrounding the nerve fiber and stimulate effector cells which may lie at a distance of up to about 1000 Å from the nerve fiber. This is known to occur in the innervation of smooth musculature. Some of the effector cells, however, show one or more close (200 Å or less) neuromuscular junctions which are not characterized by other morphological features than just a close apposition of the axolemma to the membrane of the muscle cell. These cells have been termed "active cells" and are more directly affected by transmitter substance released from either the varicosities or the endings of the fibers. Moreover, a third mode of propagation of stimuli is realized is some smooth musculature, i. e. by "low resistance pathways" or "tight junctions" present between the effector cells. "Passive cells" which are not directly affected by transmitter or neurohumor released from the nerve fibers exhibit junction potentials by electrotonic coupling with "active cells". "Passive cells" can also be affected, but somewhat slower, by neurotransmitter substance released from the varicosities and terminals of the autonomic nerve fibers (see *Burnstock*, 1968).

We suggest that this modern conception might be equally applicable to the case of the innervation of the mammalian pinealocytes by autonomic nerve fibers as has, indeed, been partly hinted at by *Wolfe* (1965) and by *Rodin* and *Turner* (1966). As the "tight junctions" known to be present between smooth muscle cells are concerned, a similar kind of junction, i. e. the fasciae adhaerentes, have also been described to occur frequently between the cell bodies and the processes of adjacent pinealocytes. So far, however, these structures have been interpreted in a different way. Possibly these zones are not only of structural but also of functional significance in the sense of the "tight junctions" between smooth muscle cells, propagating impulses electrotonically from one pinealocyte to another.

Moreover, *Hopsu* and *Arstila* (1964) observed a rather frequent close regional apposition of pinealocytes showing, either at one or at both sides of this site of close membrane contact, "synaptic ribbons" ("vesicle-crowned rodlets") or accumulations of vesicles. Membrane thickenings are lacking. The authors considered these morphological findings proof of the presence of interpinealocytic synapses suggesting that synaptic transmission may occur from one pinealocyte to another either in one or in both directions if the accumulations of the vesicles are present at both sides of the close appositional contact between the cell membranes. The presence of vesicles would speak rather for a neurohumoral transmission than for an electrotonic one in this case. Earlier in this survey it has already been mentioned that the present author cannot agree with *Hopsu* and *Arstila* that the pinealocytes would be of nervous origin, an opinion partly based by these authors on the presence of these synaptoid structures.

Accepting, as a working hypothesis, a parallel between the innervation of the smooth musculature and of the pineal organ by autonomic nerve fibers, the following way of propagation of impulses from nerve fibers to pinealocytes can be tentatively suggested. The neurotransmitter(s), once released from the terminals and the varicosities of the autonomic nerve fibers running in the perivascular spaces (1) may reach their effector cells, *i. e.* the pinealocytes, by diffusing through the external basement membrane of the spaces. The neurotransmitter(s) will influence the function of the pinealocytes more readily if the terminal buds of their processes penetrate into the perivascular space either still covered by the external basement membrane mentioned (cat), or, even more directly, if these buds are not covered by this membrane lying freely in the space (rat). If the terminal buds of the pinealocytic processes are not only separated from the perivascular space by the external basement membrane but also by a glial barrier, the neurotransmitter(s) released from the nerve fibers in the space will reach their effector cells only very slowly by diffusion.

Autonomic nerve fibers present in the pineal parenchyma (2) reaching this, possibly, by leaving the perivascular spaces or, certainly, by entering the parenchyma directly by way of the nervi conarii, will affect the pinealocytes much quicker by making close contacts with these elements. A simple close apposition of the axolemma of the nerve ending to the cell membrane of the pinealocyte may suffice for the propagation of neural impulses to this effector cell. However, such a propagation will be still more efficient if specific synaptic contacts between nerve terminals and pinealocytes develop as they have now been shown to occur. Moreover, propagation of impulses from one pinealocyte to another is suggested to be realized either neurohumorally by way of the synaptoid interpinealocytic contacts described by *Hopsu* and *Arstila* or electrotonically by way of the "tight junctions" present between these cells. This is, however, merely a theory.

The main point is that, following this working hypothesis, the pineal autonomic nerve terminals need not necessarily be in synaptic contact with the pinealocytes in order to exert their function. In the most simple way stimulation of these elements may happen by diffusion of the transmitter substance released from the nerve endings in the perivascular spaces through the external basement membrane of these spaces. This is probably the principal way in which stimulation of secretory pineal cells is realized in submammalian vertebrates such as in *Lacerta* in which these elements are still arranged in a rather thin epithelial layer. In mammals, in which a more compact pineal parenchyma has developed, stimulation of pinealocytes by diffusion of transmitter substance released from nerve endings present in the perivascular spaces is also effective, especially when the terminal buds of these elements penetrate into the perivascular space either covered or not by the external basement membrane of this space. In those cases in which the external basement membrane as well as the glial barrier intervene between the terminals of the pinealocytic processes and the perivascular space, the propagation of neural impulses by diffusion of the transmitter released by perivascular fibers will be less efficient.

A next step, also phylogenetically speaking, in the development and effectiveness of pineal autonomic innervation is the penetration of nerve fibers between the effector cells. Within the pineal epithelium of *Lacerta* endings of autonomic fibers making close contact with the cell membrane of the processes of secretory pineal cells have been observed but only very rarely (*Collin* and *Kappers*, 1968). This condition is much more developed in the mammalian pineal parenchyma in which rather simple close appositional contacts of nerve terminals with pinealocytes are much more frequently observed. The development of specialized synaptic contacts showing all of the morphological characteristics of these structures can be considered the phylogenetically last and most efficient step in the development of a direct transmission apparatus for neural impulses to the mammalian pinealocyte. As has been argued the possibility has to be considered that stimulation of pinealocytes in the mammalian epiphysis happens not only by neural impulses but also by neurohumoral and electrotonic impulses generated in other pinealocytes with which they are in close contact.

In turn, as an effect of the stimulation, the secretory products of the pinealocytes will be extruded either directly into the perivascular spaces or into the intercellular spaces mentioned earlier, which are in open communication with the perivascular spaces. From here they can enter the general circulation via the internal basement membrane of the space and the endothelial lining of the capillaries, which, at least in rat (*Milofsky*, 1957), but not in cattle and sheep (*Anderson*, 1965), is fenestrated showing pores which facilitate the uptake of the secretory products into the blood stream.

It should be stressed that the presence of often rather wide perivascular spaces in the mammalian pineal organ and the iuxtapposition of processes of parenchymal cells, *i. c.* of pinealocytes, to the external basement membrane of these spaces is a general characteristic of the structure of endocrine organs (*Lever*, 1962). Similar relationships have, for instance, been observed in the pars distalis of the rat hypophysis (*Rinehart* and *Farquhar*, 1955), the eel hypophysis (*Vollrath, 1966; Knowles* and *Vollrath, 1966*), and in the pars neuralis of the rat hypophysis (*Barer, 1965*, and others). Wide pericapillary spaces surrounding the capillaries of the primary portal hypophyseal plexus within the median eminence and at the border of this eminence and the pars tuberalis of the hypophysis have been lately demonstrated by many authors. Here, nerve terminals of the hypothalamic magnocellular and parvocellular nuclear systems end in close apposition to the external basement membrane of these spaces and it is generally understood that the content of these nerve terminals, once released, reaches the blood stream via this basement membrane, the pericapillary space, the internal basement membrane, and the capillary endothelium which shows pores (a. o. *Szentágothai* and coll., 1962, and *Halász*, this volume, in rat; *Barry, 1965*, and *Wittkowski, 1967* a, b, in guinea pig, and other authors). Thus, the effector substance, released either from peptidergic or aminergic nerve fibers, has to pass the perivascular space to enter the blood. There is question of neurohaemal contact areas and *Wittkowski* (1967 b) speaks of neurocapillary contacts in this case. The part of the external basement membrane of the pericapillary space at which the axons terminate does not show any morphological features characteristic of "true" synapses.

Morphologically as well as functionally, there is a striking similarity between the hypothalamic nerve fibers just mentioned, the axons of neurosecretory cells in the neurohypophysis, and the processes of the pinealocytes all ending, in principle, at the outer surface of the external basement membrane of the pericapillary spaces while the compounds produced, in the one case by neurons, in the other by neurosecretory cells or pinealocytes, have to pass these spaces before entering the blood. There is, however, also a difference: in the pineal organ the pericapillary spaces contain autonomic fibers and their terminals which, apparently, regulate the production and possibly the release of the hormonal effector substances produced in the pinealocytes while, in the pars neuralis of the hypophysis and in the median eminence such fibers are lacking in these spaces. This latter condition is explained when keeping in mind that the production of neurohormones and neurotransmitters, c. q. releasing factors, the release of these compounds from the cells by which they are produced, and, therefore, their passing via the pericapillary spaces into the blood *are regulated a. o. by neuronal pathways originating within the central nervous system* (see f. i. the contributions by *Picard* and *Bargmann* to this volume). In the case of the pineal organ, however, the function of the pinealocytes is *not regulated by central neuronal pathways but by the peripheral autonomic system* which primarily distributes in the pineal along the pericapillary spaces and only secondarily directly in the pineal parenchyma by way of the nervi conarii. In both cases, moreover, the production of compounds by the effector cells — on the one hand the neurosecretory cells and the neurons of, respectively, the magno- and parvocellular hypothalamic nuclear systems, and, on the other, the pinealocytes — and the depletion of these effector elements may be additionally regulated by hormonal feed back via the blood circulation.

III. Aspects of Pineal Biochemistry

A few selected data on pineal biochemistry will be mentioned, more specially those which are important for understanding the function of the organ.

In the mammalian pineal a number of *biogenic amines* have been demonstrated. From the catecholamines, histamine, noradrenaline and dopamine are present.

Pineal *histamine* content varies widely among mammals (*Giarman* and *Day*, 1958). Probably, it is closely related to the pineal mast cell population, the compound being localized in these elements (*Machado* and coll., 1965). Recently, however, it has been demonstrated that this biogenic amine can also be present in small nerve endings (see discussion after *De Robertis'* paper, this volume). The hypotensive effect of pineal extracts has been attributed to their histamine content.

The pineal contains a considerable amount of *noradrenaline*. Deafferentiation of the superior cervical ganglia by cutting the preganglionic fibers produces reduction of pineal noradrenaline content while pineal denervation by superior cervical ganglionectomy is followed by its almost complete disappearance (*Pellegrino De Iraldi* and *Zieher*, 1966). After treatment with nialamide, a monoamine oxydase inhibitor, and the catecholamine precursor 1-dihydroxy-phenyl-alanine (1-DOPA), the pineal noradrenaline content shows a significant increase in normal rats but not after decentralization of the superior cervical ganglia or pineal denervation (*Zieher* and *Pellegrino De Iraldi*, 1966). The transformation of dopamine into noradrenaline by dopamine-β-hydroxylase appears to be directly related to the integrity of both, the pre- and postganglionic neurons.

By fluorescence histochemistry, by autoradiography following uptake of isotopes (see literature cited earlier, and *Wolfe* and coll., 1962), as well as by electron microscopy it has, indeed, been demonstrated that pineal noradrenaline is localized in the autonomic nerve fibers innervating the organ, more especially in the dense-core vesicles accumulated in their endings. The green fluorescence, characteristic of the presence of primary catecholamines, disappears shortly after denervation of the organ while, after treatment with reserpine, both, the granular vesicles and the green fluorescence are no longer apparent in the terminal parts of the autonomic nerve fibers.

In the rat pineal the content of *dopamine* is also relatively high. This was neither affected by decentralization of the superior cervical ganglia nor by pineal denervation after bilateral superior cervical ganglionectomy proving that there is a rich extraneuronal pool of this compound in the rat pineal (*Pellegrino De Iraldi* and *Zieher*, 1966). Synthesis of dopamine after injections of the catecholamine precursor 1-dihydroxy-phenyl-alanine (1-DOPA) has been shown to occur in denervated rat pineals suggesting that the formation of dopamine from 1-DOPA is a more general process not only occurring in pineal nerves but also in extraneuronal pineal elements (*Zieher* and *Pellegrino De Iraldi*, 1966). Probably, pineal dopamine, in contrast to noradrenaline, is not primarily localized in the autonomic nerve fibers but also in pinealocytes. The significance of this finding is not yet clear.

From indoleamines, 5-hydroxy indoles and 5 methoxy indoles have been demonstrated in the mammalian pineal. For an extensive survey on pineal indole derivatives the reader is referred to *Quay* (1965).

In some mammals, the pineal contains a higher amount of *serotonin* (5-hydroxytryptamine) than any other part of the brain (*Giarman* and *Day*, 1958; *Miline* and coll., 1959). The serotonin levels found in human and simian glands are even the highest ever reported for any neural structure of any species investigated while the greatest variability of serotonin content was found in the human pineal (*Giarman* and coll., 1960). As has been demonstrated by fluorescence histochemistry, this yellow fluorescent compound is stored in the pineal in two compartments, *i. e.* in the pinealocytes as well as in the intrapineal part of the pineal autonomic nerve fibers (f. i. *Bertler* and coll., 1963; *Bertler* and coll., 1964; *Owman*, 1964 a, b, 1965; *Falck* and coll., 1966). This, at least, holds in mouse, rat, and guinea-pig. In the hamster and pig pineal parenchyma a yellow fluorescence is observed whereas the pineal nerves are green pointing to the exclusive presence of catecholamines, more especially of noradrenaline, in them. In the rabbit, cat, cow and sheep pineal parenchyma no fluorescence could be demonstrated at all while the pineal nerves in these species show a green fluorescence (*Owman*, 1965). It has been argued that in those cases in which the pineal nerve fibers show an all-over yellow fluorescence their serotonin content is probably larger than their noradrenaline content, the yellow fluorescence masking the green one.

In the rat pineal no significant increase in serotonin levels occurs after injections of the monoamine oxydase inhibitor nialamide (*Quay* and *Halevy*, 1962; see *Falck* and coll., 1966, for possible explanations). After injections of 5-hydroxytryptophan, the intensity of the yellow parenchymal fluorescence was increased. Injections of reserpine cause a complete loss of yellow fluorescence in the pineal nerve fibers. However, parenchymal fluorescence does not change much. All pineal nerve fibers show an intense green fluorescence 45 min after administration of dl-noradrenaline. Concomitantly, a reduction to about 50% of the total pineal serotonin content was found. It appears that noradrenaline replaces serotonin in the nerve fibers because it had been previously demonstrated that about 50% of pineal serotonin is lost after reserpine treatment which is known to cause loss of serotonin from the nerve fibers but not from the pinealocytes. This adds additional support to the opinion that, under normal circumstances, rat pineal nerve fibers are able to store both, noradrenaline and serotonin, the green fluorescence of adrenaline being, however, masked by the yellow fluorescent serotonin.

According to *Owman* (1964 a, b; 1965) and *Falck* and coll. (1966) the content of serotonin in the intrapineal parts of the pineal nerve fibers would be entirely due to the uptake of this compound by the nerves from the pinealocytes. On the other hand, it has been recently demonstrated by *Taxi* and *Droz* (1966, 1967), using injections of tritiated 5-hydroxytryptophan in combination with autoradiography, that serotonin can be synthetized in rat pineal autonomic nerve fibers. By these latter authors it has been presumed that serotonin, just like noradrenaline, may be stored in the dense-core vesicles in their terminal endings. This is, however, no argument against the opinion that at least part of the serotonin in the nerve fibers is of exogenous, pinealocytic origin.

As seen in fluorescent microscopical preparations the cytoplasm of normal rat pinealocytes is crowded with serotonin. So far, the exact localization of this compound in the cytoplasm is not clear. The number of dense-core vesicles described by some authors to be present in pinealocytes seems by far too small to account for the large amount of serotonin observed in these cells. The turn-over rate for serotonin in pinealocytes is higher than in the pineal nerves while it has also been established that an intact pineal autonomic innervation is not essential for the serotonin turn-over in the pinealocytes (*Falck* and coll., 1966).

The conversion of 5-hydroxytryptophan to serotonin is catalyzed by 5-*hydroxytryptophan decarboxylase* which has been demonstrated in bovine and rat pineals. The activity of this enzyme is, indeed, rather considerable in the rat pineal (*Snyder* and *Axelrod*, 1964 a, b; *Snyder* and coll., 1964). This pineal decarboxylase activity increases in rats reared in permanent light and decreases in rats reared in constant darkness in comparison to the activity measured in rats subjected to normal diurnal illumination. Superior cervical ganglionectomy not only causes an increase of the activity of this enzyme (*Pellegrino De Iraldi* and *De Lores Arnaiz*, 1964; *Snyder* and *Axelrod*, 1964 b) but also blocks the increase of this enzyme activity following continuous light (*Snyder* and coll., 1964, 1965). The obvious conclusion has been drawn that the pineal activity of 5-hydroxytryptophan decarboxylase is controlled by pineal innervation (*Snyder* and coll., 1965) although it cannot be simply stated that the innervation inhibits the activity of the enzyme, as is rightly remarked by *Quay* (1965).

The rat pineal also shows a high activity of *monoamine oxidase* which metabolizes serotonin, and of DOPA-*decarboxylase* (*Håkanson* and *Owman*, 1965). After denervation of the gland no change in the activities of these enzymes occurs which led the authors to the conclusion that the neuronal store of these enzymes is small in comparison to the parenchymal store. Another conclusion which could possibly be drawn is that the pineal innervation has no clear effect on the activities of these enzymes in the pineal parenchyma.

Of much interest in mammalian pineal research was the discovery that serotonin is a precursor of the second pineal indole derivate to be mentioned here: N-acetyl-5-methoxytryptamine or *melatonin* (for references see *Wurtman* and *Axelrod*, 1965 a, and *Quay*, 1965). In the pineal, melatonin is synthetized in two steps. First, N-acetylserotonin is produced from serotonin by an acetylating enzyme. During the second step a methylgroup of S-adenosylmethionine is transferred by an enzyme to the hydroxy group of N-acetylserotonin to produce melatonin. This latter enzyme has been termed hydroxyindole-O-methyl transferase or HIOMT for short (for the dependence of its activity on light conditions and pineal innervation see later). In mammals this enzyme is exclusively present in the pineal organ which means that the

pineal is the only organ capable of synthetizing melatonin. It has been calculated that per gram of tissue the rat pineal contains 0.4×10^{-6} grams of
melatonin, the bovine pineal 0.2×10^{-6} grams (*Prop* and *Kappers*, 1961).
Much importance has been attributed to this compound regarding the endocrine function of the mammalian pineal (see below).

Other indole derivatives isolated and identified in the mammalian pineal
are *5-methoxyindole-3-acetic acid* and *5-hydroxyindole-3-acetic acid*. Probably, both are products of serotonin metabolized by monoamine oxydase
(see *Quay*, 1965, Fig. 1, for indole derivatives known and postulated to occur in the pineal and their metabolic pathways). Moreover, *5-hydroxytrypto-
phol* and *5-methoxytryptophol* have been shown to occur in pineal tissue
(*Delvigs* and coll., 1964). The latter compound has a similar and even
stronger inhibitory effect on the pituitary-gonadal axis than has melatonin
(*McIsaac* and coll., 1964).

Of *fats*, phospholipids, triglycerides and fatty acids have shown to be
present in the pineal of several mammalian species. Paper chromatography
and histochemical methods failed, however, to demonstrate acetal lipids
and cerebrosides in rat pineal (*Kappers* and coll., 1964). The either total or
practically total absence of cerebrosides in the pineal of animals and man has
been confirmed by thin-layer chromatography (*Frauchiger* and *Sellei*, 1966).
The rat pineal contains a large amount of lipids, more especially phospholipids, the metabolism of which may be related to the function of the
organ (see below). Of the lipids in the pig pineal 32% are phospholipids.
They have been analysed by *Jouan* and coll. (1964). In the pinealocytes of
rat a high activity of *lipase* and of non-specific *esterase* was demonstrated
by *Prop* (1965).

As has been shown by gel micro-electrophoresis, the concentration of
proteins in rat pineal is especially high in comparison with other parts of the
brain (*Pun* and *Lombrozo*, 1964). Pineal-specific but as yet unidentified
proteins have been demonstrated by these authors. Histochemically, a striking activity of *aminopeptidase* in the rat pineal parenchyma (*Niemi* and
Ikonen, 1960) and in the parenchyma of the human gland (*Bayerova* and
Bayer, 1967) has also been shown. Biochemically, *Jouan* and *Rocaboy* (1966)
demonstrated that the aminopeptidase activity of the pineal is very similar
to that of the hypophysis and the cortex cerebri. On the ground of these
facts it has been postulated that the organ is concerned with an active synthesis of proteins which is of importance since it is assumed by some authors
that the pineal hormone(s) would be of a protein nature (see below). The
intense uptake of labelled amino acids has also been considered proof of a
glandular activity (*Ford*, 1965).

The uptake and turn-over of radioactive *phosphorus* is many times higher
in the pineal than in any other part of the brain (*Borell* and *Örström*, 1945).
This would point to a considerable phosphate-bound metabolism which is
probably involved in a high carbohydrate —, phospholipid —, and nucleic
acid metabolism. An active contribution of the hexosemonophosphate shunt
to total glucose metabolism is, moreover, present in bovine pineals as is also

the case in endocrine tissues. This has been suggested to be correlated with hormone secretion and/or storage (*Krass* and *Labella,* 1966, 1967).

Summarizing the few biochemical data mentioned (see also *Bostelmann,* 1963, and *Arvy,* 1966, for more data and literature) it appears that the mammalian pineal organ, far from being a rudimentary structure or a mere phylogenetic relic of no functional consequence, is, in fact, a veritable factory working at a high metabolic rate. Not only as its structural but also as its biochemical characteristics are concerned the organ shows a high degree of differentiation and specificity.

As has been mentioned earlier, the mammalian pineal is also unique in being a part of the brain which is exclusively innervated by autonomic fibers of peripheral origin. Some examples of the regulation of pineal biochemical metabolism by this innervation will now be mentioned:

1. Bilateral superior cervical ganglionectomy, that is pineal denervation, is not only followed by disappearance of serotonin from the pineal nerves but also by a decrease of the serotonin content of the pinealocytes (*Bertler* and coll., 1963; *Pellegrino de Iraldi* and coll., 1963, 1965).

2. In rats, permanent illumination causes increased activity of tryptophan decarboxylase. This effect is lost after both, ganglionectomy and bilateral enucleation of the eyes (*Pelegrino De Iraldi* and *De Lores Arnaiz,* 1964; *Snyder* and *Axelrod,* 1964 b; *Snyder* and coll., 1964).

3. Pineal serotonin content increases after administration of tryptophan. This increase is greatly diminished after pineal denervation (*Pellegrino De Iraldi* and coll., 1963, 1965).

4. Pineal serotonin content shows a circadian rhythm being highest at noon and lowest just before midnight (*Quay,* 1963 b). This rhythm is neither affected by blinding of the rats nor by rearing them in constant darkness. However, when reared in continuous light, the pineal shows a constant and high serotonin level, the circadian changes being abolished. After pineal denervation pineal serotonin content is reduced (*Bertler* and coll., 1963) loosing its diurnal rhythm and staying at its low, nocturnal level, rising only slowly after some days (*Fiske,* 1964; *Wurtman* and *Axelrod,* 1965 b).

5. The activity of the melatonin producing pineal enzyme HIOMT also shows a circadian rhythm. In contrast to the pineal serotonin content, however, pineal HIOMT-activity is highest at midnight and lowest at noon. Continuous darkness or blinding both abolish the circadian changes in enzyme activity, its level remaining constantly at high nocturnal levels. This same effect is produced by pineal denervation, HIOMT-activity staying at a high level which slightly decreases after some days (*Wurtman* and coll., 1964; *Axelrod* and coll., 1965; *Wurtman* and *Axelrod,* 1965 b). Because the production of melatonin depends on HIOMT-activity which is evidently influenced by pineal innervation, it can be concluded that pineal melatonin production is also influenced by the innervation of the structure.

From the above it follows that (1) the pineal sympathetic innervation regulates the activity and the production of a number of pineal chemical compounds. It is probable, but as yet not quite proven, that the autonomic innervation of the organ exerts also an influence on the depletion of pineal secretory substances. (2) In several instances, enucleation of the eyes or cutting of the optic nerves and pineal denervation show similar effects.

It is quite evident that the innervation of the mammalian pineal by auto-nomic fibers is of considerable biochemical significance and that, apparently, photic stimuli are conveyed to the pineal by the lateral eyes and the pineal orthosympathetic innervation. In the rat, the conduction pathway of photic stimuli to the pineal is now nearly completely known on the ground of experimental investigations since *Moore* and coll. (1968) found the still missing link in the brain stem, *i. e.* the inferior accessory optic tract. This chain of neural connections consists of the following components (see Fig. 2): the retina of the lateral eyes — the optic nerve — the inferior accessory optic tract which joins the median forebrain bundle in the lateral hypothalamus and ends in the rostral part of the midbrain tegmentum — a tract from this tegmental center to the rostral end of the orthosympathetic nuclear column (nucleus intermedio-lateralis) at upper thoracic cord levels — preganglionic fibers terminating in the superior cervical ganglia — postganglionic ortho-sympathetic fibers originating in these ganglia and ending in the pineal organ. The probably rather dispersed fiber tract coursing from the rostral midbrain tegmentum to the cord has, as yet, not been exactly located.

It can be concluded that the mammalian pineal produces at least one but probably more compounds which are organ-specific which means that they are exclusively synthetized in this structure. Apparently, the synthesis, metabolism and possibly the release of these compounds is at least partly regulated by photic stimuli reaching the organ indirectly by way of its autonomic innervation. The importance of photic stimuli for the physiological function of the organ and their possible way of action will be dealt with in more detail in the next chapter.

IV. Pineal Physiology

The last chapter of this survey is concerned with the regulatory function exerted by the pineal organ of mammals on the reproductive organs and with the substances and pathways possibly involved in this regulation which, historically, is the earliest recognized and the best established (*Kitay* and *Altschule*, 1954). Regulatory functions of the pineal exerted on other organs or systems, either proven or tentatively supposed to be present, will not be mentioned in this paper which has to be restricted.

a) Light, pineal and reproductive organs

The effect of excessive quantities of light on the primary and secondary sex organs of mammals is well known. Prolongation of daily photoperiods and continuous light cause, in rats, an increase in weight of the ovaries and precocious estrus (*Fiske*, 1941; *Wurtman* and coll., 1961; *Fiske* and coll., 1962). Permanent light may even cause permanent estrus (*Jöchle*, 1956). Under these conditions pineal serotonin content, pineal lipid content and pineal weight decrease (*Fiske* and coll., 1960; *Wurtman* and coll., 1961; *Quay*, 1961, 1962, 1963 a). The pineal cells become smaller and a decrease of nucleolar seize, RNA-content and HIOMT-activity has been demonstrated (*Quay*, 1961, 1962, 1963 a; *Roth*, 1965) as well as an increase of the activity of 5-hydroxytryptophan decarboxylase (see the foregoing chapter). In rats exposed to excessive quantities of light, pineal weight also decreases

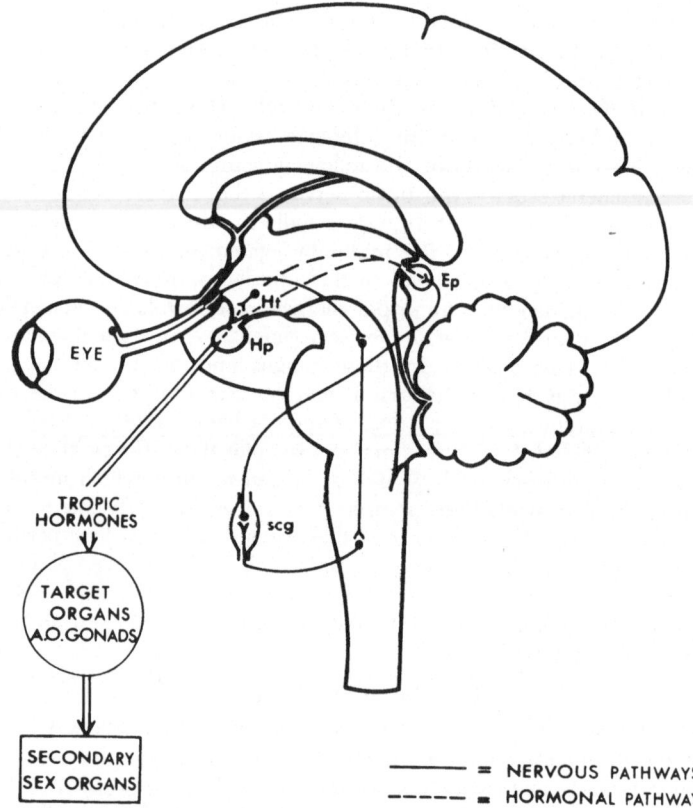

Fig. 2. Schematic diagram of the neural pathway leading from the retina of the lateral eye to the pineal organ (see text) and possible pathways by which the pineal organ may exert its functional influence on the pars distalis of the hypophysis and, via this structure, on the organs of reproduction. Ep epiphysis, Hp hypophysis, Ht hypothalamus, scg superior cervical ganglion.

after removal of the thyroid, hypophysis, adrenal glands and gonads (*Fiske* and coll., 1962). This proves that the decrease in pineal weight is not due to any indirect effect caused by these endocrine organs. It was also shown that the effect of constant light on the pineal happens only if the optic nerves are intact (*Quay*, 1961, 1963 a; *Quay* and *Halevy*, 1962). From these data it can be concluded that pineal function is inhibited in rats exposed to excessive quantities of light while, on the other hand, the function of the organs of reproduction is enhanced.

In contrast, prolongation of daily dark periods or exposure of the animals to constant darkness show opposite effects. In young female rats the beginning of the estrous cycle is postponed (*Fiske*, 1941) while in male hamsters testicular weight decreases significantly (*Hoffman* and *Reiter*, 1965 a; *Hoffman* and coll., 1965). The latter fact also occurs after ablation of the eyes. In female hamsters ovarian weight is about normal but, histologically, these organs show a minimal follicular development while the weight of the uterus is decreased (*Reiter* and coll., 1966 a). Female rats in permanent darkness show a more or less permanent anestrus. Their pineal organ increases in weight, the pineal cells are larger, they contain more lipids and RNA while the activity of the melatonin producing enzyme HIOMT is increased. Blinding shows the same effect as does excessive darkness (*Reiter* and *Hester*, 1966 a). Thus it appears that pineal function is enhanced in animals exposed to excessive darkness while the organs of reproduction are inhibited.

By pinealectomy as well as by superior cervical ganglionectomy atrophy of the sex organs following exposure to reduced quantities of light or ablation of the eyes can be prevented (*Hoffman* and *Reiter*, 1965 a, 1966; *Reiter* and *Hester*, 1966 a). Removal of the pineal organ combined with superior cervical ganglionectomy in non-blinded hamsters had no effect on the size of the reproductive or endocrine organs (*Reiter* and *Hester*, 1966 a). Pinealectomy performed on hamsters during winter time is followed by a more soon resumption of spermatogenetic activity (*Czyba* and coll., 1964, 1965). In prepuberal female rats exposed to normal light conditions, pinealectomy has been stated to result in weight increase of the ovaries and uterus and in precocious opening of the vagina (*Simonnet* and coll., 1951; *Kitay*, 1954; *Wurtman* and coll., 1959). According to *Roth* (1965), pinealectomy in puberal male rats is followed by a significant increase in size and weight of the seminal vesicles and of the ventral prostate, while testicular weight does not change much. However, *Thiéblot* and *Blaise* (1963) and *Thiéblot* (1965) state that the weight of the testes increases also under this condition. Pineal grafts in the anterior chamber of the eye compensate for the structural differences of the ovaries observed after pinealectomy.

The opinion of *Kitay* and *Altschule* (1954) that pinealectomy in rats, but not in old ones, would consistently stimulate the genital system causing gonadal hypertrophy is generally accepted but has been recently attacked by *Wragg* (1967). In his paper in which he also cites literature to which the present author does not refer, *Wragg* severally critizises the scientific value of much earlier work. In female rats pinealectomized at 3 days and sacrificed at 60 days of age, this author was not able to find any difference from normals in the weight of the ovary and of the uterus, nor was there any difference in time of vaginal opening and in incidence of estrous cycle. *Wragg* concludes: "only within a restricted age range is there reasonable evidence that pinealectomy induces ovarian hypertrophy (*i. e.* operated at 26 to 30 days and sacrificed at 50 to 54 days)". Histological control of pinealectomies and evaluation by sound statistical methods of sufficient experimental and control material is certainly needed in all research pertaining to the function of the mammalian pineal.

In contrast to pinealectomy, administration of pineal extract to normal maturing rats causes a significant decrease in ovarian weight and delayed opening of the vagina while, in middle-aged rats, anestrus follows after injections of pineal extract (*Wurtman* and coll., 1959, 1961) which also prevent permanent estrus and pineal weight decrease in rats exposed to permanent light (*Wurtman* and coll., 1963).

Pineal extracts do not show any effect in hypophysectomized rats in which normal ovarian size is maintained by injections of FSH (*Wurtman* and coll., 1961). This would point to the possibility of an influence exerted by the pars distalis of the hypophysis on the epiphysis.

Pineal phospholipid content fluctuates with the periods of the estrous cycle. During pro-estrus it is lowest, rising after ovulation to reach a high level during di-estrus (*Zweens*, 1963, 1964, 1965). Apparently, this finding is in accordance with the fact that pineal lipid content is high in anestrous rats reared in permanent darkness.

The general conclusion can be drawn that the mammalian pineal inhibits the development and normal function of the reproductive system. The effect of pinealectomy is compensated for by either pineal grafts or the administration of pineal extract. Excessive quantities of light depress pineal activity causing increased activity of the reproductive system. In contrast, excessive darkness stimulates pineal function and inhibits the activity of the sex organs.

Interestingly, neither fertility nor pregnancy are impaired in either intact or pinealectomized rats exposed to low quantities of light (*Reiter* and coll., 1966 b) while administration of pineal extract has no influence on pregnancy and lactation. Moreover, alterations in the sex organs produced by changes in the normal circadian light cycle tend to disappear after some time. It seems that a new hormonal equilibrium is established which guarantees normal conditions.

As has been mentioned, pineal metabolism largely depends on the intactness of the neural pathway conveying photic stimuli from the retina to the organ. Because pineal activity, followed by inhibition of the reproductive system, has shown to be caused by excessive darkness, while, on the other hand, excessive light inhibits pineal activity producing hyperactivity of the reproductive system, it would seem that it is the perception of darkness by the retina which, conveyed to the organ by its innervation, is the adequate stimulus for pineal activity. That this theory does not hold is obvious from the fact that interruption of the photic pathway to the pineal, either by enucleation of the eyes or by pineal denervation, shows the same physiological effects as does exposure to darkness. Therefore, the conclusion could be drawn that any absence of photic stimulation of the pineal either caused by darkness in the milieu externe or by interruption at any place of the neural photic pathway from the retina to the organ would result in a hyperactivity of the epiphysis. Photic stimuli, then, would have a positive but inhibiting effect on the activity of the gland.

I would, however, appear that the following tentative explanation of the functional significance of normal intact pineal innervation covers most of the experimental results obtained: The pineal gland shows a high basic activity which is cyclic under normal circadian conditions of lighting and is, probably, primarily dependent on feedback mechanisms in the milieu interne. However, as soon as either the quantity of light reaching the organ from the milieu externe indirectly by way of its innervation is changed, or the neural pathway conducting photic stimuli from the retina to the structure is experimentally impaired, the finely and internally regulated equilibrium of pineal activity is disturbed and this disturbance is followed by dysregulation of the reproductive system. Excessive light shows a positive inhibitory effect on pineal activity, but only if the neural pathway conveying photic stimuli from the retina to the gland is intact. On the other hand, both, excessive darkness and interruption of the photic pathway either by blinding or by pineal denervation, do not tend to stimulate pineal activity but to inhibit the normal inhibiting effect of cyclic diurnal photic stimuli. In this way, light and darkness condition the basically internal regulated function of the mammalian pineal organ. These external stimuli, if excessive, represent a regulation mechanism which is superimposed on the intrinsic regulation set up by internal stimuli being probably of a hormonal nature. It appears that this intrinsic regulation mechanism is the fundamental one. Effects of experimentally produced changes in activity of the pineal, and, therefore, in activity of the reproductive system, tend to disappear from which it can be followed that a new equilibrium in the internal regulation of pineal function has been established. Even after pinealectomy the functioning of the reproductive organs becomes normal again after some time. Thus it would seem that the pineal organ is not quite indispensable for a normal function of the reproductive system.

b) Nature of the pineal compound(s) inhibiting the reproductive system

It stands to reason that melatonin which, in mammals, is exclusively produced by the pineal organ and was shown to be a pineal hormone (*Wurtman* and *Axelrod*, 1965 a), has been held responsible for the inhibitory effect exerted by the pineal on the reproductive system. Daily administration of melatonin to maturing female rats was followed by decrease in ovarian weight (*Wurtman* and coll., 1963; *Wurtman* and *Axelrod*, 1965 a) and in incidence of estrus (*Wurtman* and coll., 1963; *Chu* and coll., 1964). It appeared, moreover, possible to block the increase of the incidence of estrus following after pinealectomy by injections of melatonin. It should, however, be mentioned that *Chu* and coll. (1964) report that the effect of melatonin obtained by these authors was merely statistical, not being present in all of the animals, so that the experimental results were not quite consistent. They may also depend on a different sensitivity to treatment with melatonin among various rat colonies as is mentioned by *Ebels* and *Prop* (1965). When female rats exposed to permanent light were given melatonin, the incidence of estrus was found to be halved the day after the injection while it was also shown that constant light reduced the uptake of circulating labelled melatonin by one half in rat ovaries and the pineal (*Wurtman* and coll.,

1963). Injections of melatonin administered to male adult rats resulted in a decrease in size of the seminal vesicles (*Kappers*, 1962). This preliminary investigation should, however, be repeated in a larger material and using statistical analysis.

Not all experiments in which melatonin was injected gave the results stated above. *Ebels* and *Prop* (1965), for instance, report that melatonin administration to immature and adult female rats exposed to normal light conditions or to continuous light did not cause either an inhibition of ovarian growth in the immature rats or a reduction of ovarian weight in the adults. These authors, moreover, did not observe any effect of melatonin injections on the incidence of estrus in rats under normal light conditions while animals exposed to continuous light did not show a tendency towards a decreased incidence of estrus on the first or second day after the first injection. After melatonin treatment the pineal cells did not show morphological alterations while even large doses of melatonin did not change their lipid content which is known to change during the estrous cycle (see above). In male hamsters, the subcutaneous implantation of melatonin was not found to affect the size of the reproductive organs (*Reiter*, 1967) but this can also be due to a rapid break-down of the compound.

In the opinion of the present author it cannot be stated so far with absolute certainty that melatonin consistently inhibits the reproductive system, even in impuberal mammals, and that this would be the pineal antigonadotropic compound "par excellence". It should also be remembered that another pineal 5-methoxyindole, *i. e.* 5-methoxytryptophol, has been demonstrated to show an inhibiting effect on the reproductive system, likewise decreasing the incidence of estrus and reducing ovarian weight in maturing female rats (*McIsaac* and coll., 1964). It is, therefore, questionable (1) whether melatonin is the only pineal indole derivative exerting this influence, and (2) whether pineal indole derivatives are exclusively responsible for the antigonadotropic effect. It should be mentioned that, in female rats, an antigonadotropic effect of serotonin has also been observed (*O'Steen*, 1964, cited by *Moszkowska*, 1967) so that, apparently, this effect is not only dependent on indole derivatives containing a 5-methoxy group.

There are also arguments in favor of the opinion that pineal peptides show an antigonadotropic effect. *Milcou* and coll. (1963), for instance, isolated a polypeptide from bovine pineals showing not only pressor and oxytocic activity but also an antigonadotropic activity in impuberal mice. These authors concluded that the active pineal principle would possibly be identical with arginine vasotocin or a similarly structured pineal hormone. In the pig pineal *Pavel* (1965) identified a peptide which, regarding its biological, enzymatic and chromatographic characteristics, did not differ significantly from synthetic lysine vasotocin. Following *Pavel* and *Petrescu* (1966), the stimulatory action of pregnant mare's serum gonadotropin on mice uteri and ovaries is inhibited by both, highly purified pineal arginine vasotocin and synthetic arginine vasotocin. However, *Ebels* and coll. (1965 b) have not been able to confirm the results obtained by *Milcou* and coll. (1963). In attempting to isolate a peptide with pressor and oxytocic activity from sheep

and bovine acetone desiccated pineals by gel filtration and column chromatography, no oxytocic activity could be demonstrated in any of the fractions. A slight pressor activity was found only. On the other hand, *Moszkowska* and *Ebels* (1968) confirmed that, indeed, arginine vasotocin inhibits the response of impuberal mice to gonadotropic stimulation possibly acting directly on the gonads or on the gonadotropic hormone(s) and not on the excretion of gonadotropic hormones by the adenohypophysis as the peptide pineal antigonadotropic principle was earlier shown to do by these authors (see below).

Concerning the question which is the direct target organ or organs of the pineal antigonadotropic factor(s) at least two possibilities should be considered: (1) the hormone could either exert its inhibitory influence directly on the organs of reproduction, or (2) indirectly via the adenohypophysis inhibiting the production and/or the excretion of the hypophyseal gonadotropic hormones FSH and LH. In the latter case, the hypophysis could be influenced by the antigonadotropic pineal substances (a) either directly or (b) indirectly via the hypothalamic nuclei responsible for the production of hypophyseal releasing factors. The results of much experimental work are, so far, more in favor of the conception that the pineal antigonadotropic factor acts via the adenohypophysis rather than directly on the organs of reproduction and that the hypothalamus is also involved in this mechanism. For literature we may refer to *Thiéblot* (1965, see also *Thiéblot* and *Blaise*, 1963). Some of the work of *Moszkowska* and coworkers will be surveyed here because this has been performed very systematically and is of considerable importance.

Prepuberal female rats having a male rat adenohypophysis grafted on the ovary, show opening of the vagina within 4 days after grafting, increase in weight of the ovaries and the uterus in 70% of the cases and the presence of many ripe follicles in the ovaries. Additional grafting of male pineals on the ovaries next to the hypophyseal grafts prevented the gonadotropic effect of the hypophyseal grafts on the organs of reproduction (see *Moszkowska*, 1965). These experiments prove that the pineal abolishes the gonadotropic influence of the adenohypophysis on the sex organs but give no evidence whether the pineal inhibits the ovaries directly or counteracts the gonadotropic action of the hypophysis on the gonads. As this question is concerned the following experiments of *Moszkowska* are more conclusive. Injections of incubation fluid of anterior pituitaries of mature male rats into prepuberal female rats caused opening of the vagina after a few days and an increase in weight of the ovaries and the uterus while the ovaries showed many ripe follicles. On the other hand, prepuberal rats treated with injections of fluid in which hypophyses were incubated in the presence of acetonic powder of adult sheep pineals did not show this gonadotropic effect. Positive gonadal stimulation, however, occurred in prepuberal female rats treated with separate injections of adenohypophyseal incubation fluid and of pineal incubation fluid. These experiments prove that the pineal factor cannot counteract the gonadotropic action of the hypophysis at the level of the gonads but that this factor inhibits the gonadotropic effect at the level of the hypophysis.

In 1963 *Moszkowska* demonstrated that acetone desiccated powder of sheep pineals diminishes *in vitro* the hypophyseal secretion of follicle stimulating hormone (FSH). By means of filtration on Sephadex gel of sheep pineal extracts, several fractions were obtained of which one, fraction F_2, was shown to increase hypophyseal FSH secretion *in vitro*, whereas, by another fraction, F_3, FSH secretion was decreased. Both fractions are peptides of a relatively low molecular weight (*Moszkowska* and coll., 1965; *Ebels* and coll., 1965 a; *Ebels*, 1967). It was possible to enrich the active factor in fraction F_2 by special methods while chromatography showed that the antigonadotropic fraction F_3 contains two active zones (*Moszkowska* and coll., 1965; *Ebels*, 1967). Earlier, a *stimulating* influence of the pineal on sexual development had already been observed by several authors (*Jöchle*, 1956; *Reiss* and coll., 1963 a, b; see also *Moszkowska* and coll., 1965, and *Thiéblot*, 1965, for literature). It was, however, thought that this was either a question of a factor showing a synergetic action with gonadotropic hormones or of an estrogenic activity of the pineal compound concerned. *Moszkowska* and coll. (1965), however, could neither demonstrate a synergetic action of fraction F_2 with human chorionic gonadotropin nor an estrogenic effect of this fraction. As was mentioned, F_2 was shown to produce an increase of hypophyseal FSH production *in vitro* by which its direct effect on this organ is demonstrated.

Furthermore, *Moszkowska* (1967), corroborating the results of similar experiments of other authors, found a. o. after administration of hydrosoluble pineal extracts to female guinea pigs a retardation and even a suppression of vaginal opening and an arrest of ovarian follicular growth, a delay of the first estrus in young female rats and mice, and a decrease in weight of the seminal vesicles and the prostate gland in male rats and mice. Moreover, she observed an inhibitory effect of pineal extracts on the permanent estrus and the decrease in pineal weight occurring after exposure to permanent light of the females. She states that, evidently, hydrosoluble pineal extracts are capable of decreasing hypophyseal FSH and LH secretion *in vivo* while, *in vitro*, sheep pineal powder appears to inhibit hypophyseal FSH secretion exclusively. Realizing that by several authors an antigonadotropic effect of serotonin and more especially of melatonin has been observed, *Moszkowska* incubated anterior pituitaries without (h) or with hypothalamus (h + H). In case (h + H) a larger quantity of FSH was produced by the hypophysis than in case (h), while, moreover, in experiment (h + H) she found additional secretion of LH. When she added *serotonin* to the culture (h + H + S) stimulation of FSH and LH secretion did no longer occur. *Moszkowska* assumes that, probably, the inhibition of hypophyseal LH secretion by serotonin is due to the inhibition, by serotonin, of the production of the gonadotropic releasing factor by the hypothalamus.

Moszkowska (1967), injecting very high doses of *melatonin* in female rats, observed ovarian atrophy and retardation of vaginal opening. The fact that the ovaries of these animals did not show any corpora lutea although the many follicles present were of a large size, while also the interstitial tissue was histologically inactive, points to an anti-LH effect of melatonin (see also *Kappers*, 1962). In the anterior pituitaries of these rats, especially the prolactin producing cells were well-developed and FSH producing cells were numerous recalling histological pictures of hypophyses after extensive lesions in the preoptic hypothalamic area. On the contrary, this author got negative results after incubating anterior pituitaries in the presence of melatonin as well as in the presence of both, hypothalamus and melatonin.

Summarizing the results obtained by *Moszkowska* it can at least be concluded (a) that the antigonadotropic as well as the gonadotropic pineal substances, being either peptides or, possibly, a combination of peptides and indole derivatives so far as the antigonadotropic substance is concerned, influence the production and probably also the release of gonadotropic hormones by the gonadotropic adenohypophyseal cells, and (b) that the hypothalamic gonadotropic centers may play an additional role in this mechanism (Fig. 2). In connection with (b) it is of interest that decreased neurosecretory activity was observed in the hypothalamic paraventricular nuclei in the pinealectomized rat (*Miline*, 1963) while hypersecretion and retention of the neurosecretory product in these nuclei followed after administration of pineal extracts (*Aron* and coll., 1961; *Miline*, 1963). From these findings it appears that pineal substance(s) may inhibit the release of neurosecretory substance. In this connection it is interesting to note that, according to *Martini* and coll. (1958), oxytocin, which is produced in the paraventricular nuclei, would represent the hypothalamic releasing factor for hypophyseal gonadotropic hormones. *Milcou* and *Pavel* (1961) observed an increase of the amount of oxytocin in the dog paraventricular nuclei after administration of bovine pineal extract. They hold that this increased concentration is due to blocking of the release of oxytocin from these nuclei by the pineal extract and that the pineal antigonadotropic effect is caused by blocking of the hypothalamic gonadotropic releasing factor normally activating the hypophyseal gonadotropic cells. It would, however, seem that more work is necessary concerning the influence exerted by pineal compounds on the hypothalamus, especially on its parvocellular nuclei which are probably primarily engaged in the regulation of adenohypophyseal gonadotropic activity.

Basing themselves on a number of experiments in which FSH and LH was administered under various experimental conditions, *Reiter* and coll. (1966 a) and *Reiter* (1967) also concluded that in rat as well as in hamster the antigonadotropic pineal factor acts on the hypothalamo-hypophyseal axis rather than at the ovarian or uterine level.

Clementi and coll. (1966), studying changes in ultrastructure and gonadotropin content in the hypophysis occurring after pinealectomy and following the implantation of pineal tissues and of melatonin into the median eminence and the midbrain reticular formation, obtained results suggesting that the pineal exerts its inhibiting influence on the gonadotropic cells of the adenohypophysis via hypothalamic and even mesencephalic centers. The changes in the fine structure of gonadotropic cells induced by implants of melatonin or of pineal tissue were identical to those observed when the synthesis and release of the hypothalamic gonadotropic releasing factors are inhibited.

As has been mentioned before, the atrophic changes of the gonads of the hamster reared in short daily periods of illumination or following ablation of the eyes can be prevented by removal of the pineal organ (*Hoffman* and *Reiter*, 1965 a, b). In male hamsters, living during the dark winter days in a period of sexual inactivity, the hypophyseal gonadotropic cells demonstrate characteristic changes after pinealectomy (*Girod* and coll., 1964). Vacuolated FSH cells, normally present during the non-active period of the reproductive cycle, are absent after the operation while gonadotropic LH (or ICSH) cells are numerous and well developed in contrast to those present in the hypophyses of control animals. In accordance, testicular weight was increased and the testes showed complete spermatogenesis and development of interstitial tissue normal in hamsters during the period of sexual activity. These data support the concept of the functional interrelationship between the pineal and the adenohypophysis and, in this case, would point to an inhibitory influence of the pineal on both, the FSH and the LH cells of the adenohypophysis.

As a possibly feedback mechanism between the adenohypophysis and the pineal is concerned it should be mentioned that, in 10 weeks old rats, hypophysectomy was followed by an increase of nuclear density, indicative of a proliferative phase characteristic of pineals of much younger rats. Pineal weight did not decrease after hypophysectomy as does the weight of the thyroid, the gonads and the adrenals. Moreover, pineal phosphorus turn-over was considerably increased (*Reiss* and coll., 1963 b). These characteristics would rather point to an activation of the pineal after hypophysectomy in young rats in relation to the gonadotropic factor produced by the organ. On the ground of his investigations in which pineal extract (Glanepin) was administered to female rats under various experimental conditions it was already assumed by *Jöchle* (1956) that this extract first stimulates sexual maturation to inhibit afterwards the sexual periodicity established in consequence of this maturation. This would point to the possibility that in very young animals the influence of the pineal gonadotropic factor dominates while, at a more mature age, the antigonadotropic pineal factor is the dominant one.

It has been previously mentioned (see chapter on biochemistry) that the rat pineal contains a large amount of phospholipids and that the pineal lipid content varies according to the phases of the estrous cycle. It has been shown (*Zweens*, 1964, 1965) that in rats hypophysectomy is followed by a decrease of pineal lipid content which is restored to normal by administration of pregnant mare's serum (PMS) or human chorionic gonadotropin (HCG). It also appeared that ovarian hormones do not influence the effect exerted by PMS treatment on the pineal content in hypophysectomized animals. From all the evidence available *Zweens* concludes that the pineal lipid content is influenced by the serum gonadotropin level which is also responsible for the

changes in lipid content during the estrous cycle. Although, following *Zweens*, a possible influence of FSH on this content could not be excluded, the effect of LH was convincingly demonstrated. Long-term changes in serum gonadotropin level cause corresponding changes in pineal lipid content implicating that a raised or lowered antigonadotropic pineal activity corresponds with a raised or a lowered pineal lipid content, respectively. It is of interest that, evidently, the pineal lipid content can be used as a parameter for the antigonadotropic activity of the organ, at least in rat, the pineal of which contains much phospholipid.

Electron microscopical investigations of the pineal organ in female rats after administration of chorion gonadotropin, ovariectomy and ovariectomy in combination with chorion gonadotropin administration showed differences in the number and qualitative composition of the cell organelles in the pinealocytes. Especially, an increase of the pineal lipid content wos observed (Gusek and *Buss*, 1966) which is in agreement with the observation by *Zweens*.

That the mammalian pineal organ shows, in general, a strong cytological reactions to stress (*Miline*, 1957) can, at least partly, be explained by the fact that the structure is an endorgan of the peripheral autonomic nervous system. It is also evident that stress, the fact that a number of pineal compounds show circadian changes in quantity and activity, differences in the preparation of pineal extracts, differences in doses of pineal extracts administered, differences in the period of administration of pineal substances, differences in age of the experimental animals, as well as species differences are all factors which may influence the results of investigations on pineal morphology and function. It is also noteworthy that, evidently, the pineal produces an gonadotropic factor next to an antigonadotropic factor which may complicate the interpretation of some results.

It appears that, on the long run, removal of the pineal, pineal denervation, and exposure to either excessive light or excessive darkness is of but little consequence for the normal function of the reproductive system. Apparently, a new and effective hormonal equilibrium is established after some time. It has, therefore, been suggested that the pineal organ functions as a regulator of regulators (*Reiter* and *Hester*, 1966 b) not being quite indispensable for normal life.

Conclusions

In concluding it appears that the mammalian pineal organ, far from being a phylogenetic relic of no functional consequence, shows characteristic structural, biochemical and functional features. It produces some specific compounds and exerts an influence, besides on other systems and organs not mentioned in this survey, on the organs of reproduction. The pineal produces an antigonadotropic as well as a gonadotropic substance which have been shown to affect the adenohypophyseal gonadotropic cells, probably directly as well as via hypothalamic centers (Fig. 2). The antigonadotropic effect of the pineal organ has been most studied. The nature of the pineal compounds influencing the hypothalamo-hypophyseo-gonadal axis is still somewhat conjectural. Although indole derivatives, some of which are specifically produced by the organ, have been shown to inhibit the reproductive system it has also been established that some active compounds are of a protein nature. Possibly, a combination of indole derivatives and polypeptides acts under normal physiological circumstances.

It is suggested that pineal function is primarily regulated by several internal regulation mechanisms which are, as yet, not very well known. Superimposed on these mechanisms is the regulation by external stimuli such as photic stimuli which are conducted via a nervous pathway leading from the retina of the eyes to the organ. The last part of this chain consists of the sympathetic pathways innervating the pineal organ. It is surmised that the pinealocytes are stimulated by the sympathetic nerve terminals in two ways: (1) a phylogenetically old and less efficient way realized by diffusion of neurotransmitter released from the perivascular nerve fibers, and (2) a phylogenetically younger and more efficient way also realized neurohumorally but by the terminals of nerve fibers penetrating directly into the pineal parenchyma via the nervi conarii. These terminals make simple close appositional contacts with pinealocytes as well as specific synaptic contacts which are described and illustrated for the first time in this paper. "Tight junctions" between pinealocytes may possibly conduct stimuli electrotonically.

At the ultrastructural level the production and the excretion of the specific compounds produced by the pinealocytes is not yet quite clear. These products are probably released into the blood stream via the pericapillary spaces.

In view of all of its characteristics it is justified to call the mammalian pineal a photo-neuro-endocrine organ, a term first coined by *Ernst Scharrer* (1964). It should, however, be understood that, although the structure is of neuro-epithelial derivation, the endocrine function is not exerted by neurons but by pinealocytes, specific pineal cells which derive phylogenetically from neurosensory photoreceptor elements. Moreover, the mammalian pineal is not a direct but an indirect photosensory organ, photic stimuli reaching it via the neural pathway mentioned above.

Much future multidisciplinary research in the fields of electron microscopy, biochemistry and endocrinology is needed to solve the many problems still existing as this intriguing and fascinating organ is concerned.

References

Anderson, E.: The anatomy of bovine and ovine pineals. Light and electron microscopic studies. J. Ultrastruc. Res., Suppl. 8, 1—80 (1965).

Aron, E., C. Combescot, J. Demaret, L. Guyon, et *R. Y. Mauvernay:* Modifications de la neurosécrétion observées dans l'encéphale du rat après injection d'un extrait épiphysaire. C. R. Soc. Biol., Paris, *155,* 593—595 (1961).

Arstila, A. U.: Electron microscopic studies on the structure and histochemistry of the pineal gland of the rat. Suppl. Neuroendocrinol. 2, 1—101 (1967).

Arstila, A. U., and *V. K. Hopsu:* Studies on the rat pineal gland. I. Ultrastructure. Ann. Acad. Sc. Fenn., Ser. A, V. *113,* 1—21 (1964).

Arvy, L.: La glande pinéale, centre régulateur des amines biogènes. Ann. Biol., Paris, *5,* 565—594 (1966).

Axelrod, J., R. J. Wurtman, and *S. H. Snyder:* Control of hydroxyindole O-methyltransferase activity in the rat pineal gland by environmental lighting. J. Biol. Chem. *240,* 949—954 (1965).

Barer, R.: The ultrastructure of small blood vessels of the posterior pituitary gland in relation to neurosecretion. 3rd Europ. Conf. Microcirc., Jerusalem 1964. In: Bibl. anat. 7, 304—309. Basel-New York: Karger, 1965.

Bargmann, W.: Die Epiphysis cerebri. In: Handbuch der mikroskopischen Anatomie des Menschen. W. von Möllendorff, ed. Bd. VI/4, 309—502. Berlin: Springer, 1943.

Barry, J.: Rapports et structure des capillaires du plexus porte primaire de l'hypophyse. Pathologie-Biologie *13,* 866—886 (1965).

Bayerova, G., und *A. Bayer:* Beitrag zur Fermenthistochemie der menschlichen Epiphyse. Acta Histochem. *28,* 169—173 (1967).

Bertler, Å., B. Falck, and *Chr. Owman:* Cellular localization of 5-hydroxytryptamine in the rat pineal gland. K. fysiogr. Sällsk. Lund Förh. *33,* 13—16 (1963).

Bertler, Å., B. Falck, and *Chr. Owman:* Studies on 5-hydroxytryptamine stores in pineal gland of rat. Acta Physiol. Scand. *63,* Suppl. 239, 1—18 (1964).

Bondareff, W.: Electron microscopic study of the pineal body in aged rats. J. Geront. *20,* 321—327 (1965 a).

Bondareff, W.: Submicroscopic morphology of granular vesicles in sympathetic nerves of rat pineal body. Z. Zellforsch. *67,* 211—218 (1965 b).

Bondareff, W., and *B Gordon:* Submicroscopic localization of norepinephrine in sympathetic nerves of the rat pineal. J. Pharmacol. exp. Therap. *153,* 42—47 (1966).

Borell, U., and *Å. Örström:* Metabolism in different parts of the brain, especially in the epiphysis, measured with radioactive phosphorus. Acta Physiol. Scand. *10,* 231—242 (1945).

Bostelmann, W.: Beitrag zur Kenntnis der Epiphysis cerebri unter Berücksichtigung ihrer Zytochemie. Wiss. Z. Univ. Rostock *12,* 437—453 (1963).

Burnstock, G.: The autonomic muscular junction. In: Proc. of the XXIV International Congress of Physiological Sciences, Washington, *3,* 7—8 (1968).

Chu, E. W., R. J. Wurtman, and *J. Axelrod:* An inhibitory effect of melatonin on the estrous phase of the estrous cycle in the rodent. Endocrinology 75, 238—242 (1964).

Clementi, F., G. de Virgiliis, F. Fraschini, and *B. Mess:* Modifications of pituitary morphology following pinealectomy and the implantation of the pineal body in different areas of the brain. In: Proceedings of the Sixth International Congress for Electron Microscopy, Kyoto, 1966, 539—540. Tokyo: Maruzen Co., 1966.

Collin, J. P., and *J. Ariëns Kappers:* Electron microscopic study of pineal innervation in lacertilians. Brain Research *11,* 85—106 (1968).

Czyba, J. C., C. Girod et *N. Durand:* Sur l'antagonisme épiphyso-hypophysaire et les variations saisonnières de la spermatogenèse chez le Hamster doré (*Mesocricetus auratus*). C. R. Soc. Biol., Paris, *158,* 742 (1964).

Czyba, J. C., C. Girod, M. Curé et *N. Durand:* Sur les corrélations épiphyso-testiculaires chez le Hamster doré (*Mesocricetus auratus* Waterh.). Bull. Assoc. Anatom., 50e Réun., Lausanne, 324—333 (1965).

Delvigs, P., W. M. McIsaac, and *R. G. Taborsky:* The metabolism of 5-methoxytryptophol. J. Biol. Chem. *240,* 348—350 (1965).

De Robertis, E.: Histophysiology of Synapses and Neurosecretion. Oxford: Pergamon, 1964.

De Robertis, E., and *A. Pellegrino de Iraldi:* Plurivesicular secretory processes and nerve endings in the pineal gland of the rat. J. Biophys. Biochem. Cytol. *10,* 361—372 (1961).

Duncan, D., and *G. Micheletti:* Notes on the fine structure of the pineal organ of cats. Tex. Rep. Biol. Med. *24,* 576—587 (1966).

Ebels, I.: Étude chimique des extraits épiphysaires fractionnés. Biol. Méd. *56,* 395—402 (1967).

Ebels, I., and *N. Prop:* A study of the effect of melatonin on the gonads, the oestrous cycle and the pineal organ of the rat. Acta Endocrinol. *49,* 567—577 (1965).

Ebels, I., A. Moszkowska et *A. Scemama:* Étude *in vitro* des extraits épiphysaires fractionnés. Résultats préliminaires. C. R. Acad. Sci., Paris, *260,* 5126—5129 (1965 a).

Ebels, I., D. V. G. Versteeg, and *J. F. G. Vliegenthart:* An attempt to isolate arginine vasotocin from sheep and bovine pineal body. Proc. Kon. Ned. Akad. Wet., Amsterdam, Ser. B, *68,* 1—4 (1965 b).

Falck, B., Chr. Owman, and *E. Rosengren:* Changes in rat pineal stores of 5-hydroxytryptamine after inhibition of its synthesis or break-down. Acta Physiol. Scand. *67,* 300—305 (1966).

Fiske, V. M.: Effect of light on sexual maturation, estrous cycles and anterior pituitary of the rat. Endocrinology *29,* 187—196 (1941).

Fiske, V. M.: Serotonin rhythm in the pineal organ: control by the sympathetic nervous system. Science *146,* 253—254 (1964).

Fiske, V. M., G. K. Bryant, and *J. Putnam:* Effect of light on the weight of the pineal in the rat. Endocrinology *66,* 489—491 (1960).

Fiske, V. M., J. Pound, and *J. Putnam:* Effect of light on the pineal organ in hypophysectomized, gonadectomized, adrenalectomized or thiouracil-fed rat. Endocrinology *71,* 130—133 (1962).

Ford, D. H.: Uptake of [131]-I-labelled triiodothyronine in the pineal body as compared with the cerebral grey and other tissues of the rat. Progr. Brain Res. *10,* 530—539 (1965).

Frauchiger, E.: Altes und Neueres über die Zirbeldrüse (Epiphysis cerebri). Schweiz. Arch. Tierheilk. *105,* 183—194 (1963).

Frauchiger, E., und *K. Sellei:* Dünnschichtchromatographie der Lipoide in der Glandula pinealis. Schweiz. Arch. Neurol. Neurochir. Psych. *98,* 240—243 (1966).

Giarman, N. J., and *M. Day:* Presence of biogenic amines in the bovine pineal body. Biochem. Pharmacol. *1,* 235 (1958).

Giarman, N. J., D. X. Freedman, and *L. Picard-Ami:* Serotonin content of the pineal glands of man and monkey. Nature, Lond., *186,* 480—481 (1960).

Girod, C., J. C. Czyba et *N. Durand:* Influence de l'épiphysectomie sur les cellules gonadotropes antéhypophysaires du Hamster doré *(Mesocricetus auratus* Waterh.). C. R. Soc. Biol. Paris *158,* 1636 (1964).

Gusek, W.: Neue Befunde zur Morphologie und Funktion der Epiphysis cerebri. Ergebn. allg. Path. path. Anat. *50,* 104—148 (1968).

Gusek, W., und *H. Buss:* Morphologische und histochemische Veränderungen der Zirbeldrüse unter dem Einfluß von Prolan und nach Ovarektomie. Frankf. Z. Pathol. *75,* 172—186 (1966).

Gusek, W., und *A. Santoro:* Elektronoptische Beobachtungen zur Ultramorphologie der Pinealzellen bei der Ratte. Arch. int. Biol. norm. e pat. *13,* 451—464 (1960).

Gusek, W., H. Buss und *H. Wartenberg:* Weitere Untersuchungen zur Feinstruktur der Epiphysis cerebri normaler und vorbehandelter Ratten. Progr. Brain Res. *10*, 317—330 (1965).

Håkanson, R., and *Chr. Owman:* Effect of denervation and enzyme inhibition on DOPA decarboxylase and monoamine oxidase activities of rat pineal gland. J. Neurochem. *12*, 417—429 (1965).

Halaris, A., E. Rüther, and *N. Matussek:* Effect of benzoquinolizine (RO 4—1284) on granulated vesicles of the rat brain. Z. Zellforsch. *76*, 100—107 (1967).

Hassler, R., and *I. J. Bak:* Effects of amine-depleting and amine-storing substances on the axon terminals of the pineal gland. In: 5th int. Congr. Electron Microscopy, Kyoto, 1966, 521—522. Tokyo: Maruzen Co., 1966.

Hoffman, R. A., and *R. J. Reiter:* Pineal gland: influence on gonads of male hamsters. Science *148*, 1609—1611 (1965 a).

Hoffman, R. A., and *R. J. Reiter:* Influence of compensatory mechanisms and the pineal gland on dark-induced gonadal atrophy in male hamsters. Nature, Lond., *207*, 685—659 (1965 b).

Hoffman, R. A., and *R. J. Reiter:* Responses of some endocrine organs of female hamsters to pinealectomy and light. Life Sci. *5*, 1147—1151 (1966).

Hoffman, R. A., R. J. Hester, and *C. Towns:* Effect of light and temperature on the endocrine system of the golden hamster *(Mesocricetus auratus* Waterh.). Comp. Biochem. Physiol. *15*, 525—533 (1965).

Hopsu, V. K., and *A. U. Arstila:* An apparent somato-somatic synaptic structure in the pineal gland of the rat. Exp. Cell Res. *37*, 484—487 (1964).

Jöchle, W.: Über die Wirkung eines Epiphysenextractes (Glanepin) auf Sexualentwicklung und Sexualcyclus junger weiblicher Ratten unter normalen Haltungsbedingungen und bei Dauerbeleuchtung. Endokrinologie, Leipzig, *33*, 287—295 (1956).

Jouan, P., et *J.-C. Rocaboy:* Étude de l'activité peptidasique de la glande pinéale du porc. C. R. Soc. Biol., Paris, *160*, 859 (1966).

Jouan, P., T. Viem Dai et *M. Cormier:* Sur la nature des phospholipides de la glande pinéale (épiphyse) du porc. Bull. Soc. Chim. biol., Paris, *46*, 1121—1129 (1964).

Kappers, J. Ariëns: The development, topographical relations and innervation of the epiphysis cerebri in the albino rat. Z. Zellforsch. *52*, 163—215 (1960).

Kappers, J. Ariëns: Melatonin, a pineal compound. Preliminary investigations on its function in the rat. Gen. Comp. Endocrinol. *2*, 610—611 (1962).

Kappers, J. Ariëns: Survey of the innervation of the epiphysis cerebri and the accessory pineal organs of vertebrates. Progr. Brain Res. *10*, 87—153 (1965).

Kappers, J. Ariëns: The sensory innervation of the pineal organ in the lizard, *Lacerta viridis,* with remarks on its position in the trend of pineal phylogenetic structural and functional evolution. Z. Zellforsch. *81*, 581—618 (1967).

Kappers, J. Ariëns: Morphological and functional evolution of the pineal organ. In: Proc. of the Third Internat. Congr. of Endocrinology, Mexico, 1969. In press.

Kappers, J. Ariëns, N. Prop, and *J. Zweens:* Qualitative evaluation of pineal fats in the albino rat by histochemical methods and paper chromatography and the changes in pineal fat contents under physiological and experimental conditions. Progr. Brain Res. *5*, 191—199 (1964).

Kenny, G. C. T.: The "nervus conarii" of the monkey. (An experimental study.) J. Neuropath. exp. Neurol. *20*, 563—570 (1961).

Kitay, J. I.: Effects of pinealectomy on ovary weight in immature rats. Endocrinology *54*, 114—116 (1954).

Kitay, J. I., and *M. D. Altschule:* The pineal gland. Commonwealth Fund Book, Cambridge, Mass., Harvard Univ. Press, 1954.

Knowles, F., and *L. Vollrath:* Neurosecretory innervation of the pituitary of the eels *Anguilla* and *Conger*. I. The structure and ultrastructure of the neurointermediate lobe under normal and experimental conditions. II. The structure and innervation of the pars distalis at different stages of the life-cycle. Phil. Trans. Roy. Soc., Lond., Ser. B, *250*, 311—342 (1966).

Krass, M. E., and *F. S. Labella:* Biochemical evidence for a secretory role of the pineal body. Oxidation of 1-^{14}C- and 6-^{14}C-glucose *in vitro* by pineal body, pituitary, and brain from young and adult animals. J. Neurochem. *13*, 1157—1162 (1966).

Krass, M. E., and *F. S. Labella:* Hexosemonophosphate shunt in endocrine tissues. Quantitative estimation of the pathway in bovine pineal body, anterior pituitary, posterior pituitary, and brain. Biochim. Biophys. Acta *148*, 384—391 (1967).

Kurosomi, K., and *I. Kawabata:* Electron microscopic studies of the fine structure of pineal glands in normal and experimental rats. In: Proc. 6th International Congress for Electron Microscopy, Kyoto, 1966. 519—520. Tokyo: Maruzen Co., 1966.

Leonhardt, H.: Über axonähnliche Fortsätze, Sekretbildung und Extrusion der hellen Pinealozyten des Kaninchens. Z. Zellforsch. *82*, 307—320 (1967).

Lever, J. D.: Fine structural organization in endocrine tissues. Brit. Med. Bull. *18*, 229—232 (1962).

Machado, A. B. M.: Ultrastructure of the pineal body of the newborn rat. Anat. Rec. *154*, 381 (1966).

Machado, A. B. M., L. C. M. Faleiro, and *W. Dias da Silva:* Study of mast cell and histamine contents of the pineal body. Z. Zellforsch. *65*, 521—529 (1965).

Martini, L., L. Mira, A. Pecile, and *S. Saito:* Neurohypophyseal hormones and gonadotrophins release. Acta Endocrinol., Suppl. 38, 81—82 (1958).

McIsaac, W. M., R. G. Taborsky, and *G. Farrell:* 5-Methoxytryptophol: effect on estrus and ovarian weight. Science *145*, 63—64 (1964).

Meyburg, P.: Zur Frage der Herkunft der Parenchymzellen des Pinealorganes. Schweiz. Arch. Neurol. Neurochir. Psych. *95*, 245—270 (1965).

Milcou, S. M., and *S. Pavel:* Antigonadotropic function of the pineal gland and the oxytocin of the neurosecretory hypothalamic system. Nature, Lond., *187*, 950—951 (1961).

Milcou, S. M., and *I. Petrea:* Electron microscopic studies on the secretory cytodynamics of the pineal gland. Epiphysis-hypophysis-gonadic correlations. Rev. Roum. Endocrinol. *1*, 109—114 (1964 a).

Milcou, S. M., and *I. Petrea:* Some aspects of the endocrinosecretory cytodynamic in the pineal gland at the electron microscope. In: Proc. 3rd Europ. Reg. Conf. on Electron Microscopy. Czechoslovak Acad. Sci., Prague, 1964 b, 481—482.

Milcou, S. M., and *I. Petrea:* Cytology of the pineal gland. In: Proc. Congr. Nat. Endocrinologie, Bucharest, 1967, 177—185.

Milcou, S. M., S. Pavel, and *C. Neacsu:* Biological and chromatographic characterization of a polypeptide with pressor and oxytocic activities isolated from bovine pineal gland. Endocrinology *72*, 563—566 (1963).

Miline, R., P. Stern et *S. Huković:* Sur la présence de la sérotonine dans la glande pinéale. Bull. Sci. *4*, 75 (1959).

Miline, R.: La part de l'épiphyse dans le syndrome d'adaptation. In: Congrès National des Sciences Médicales, Bucarest, 5—11 Mai 1957, 421—444. Édition de l'Académie de la République populaire roumaine, 1957.

Miline, R.: La part du noyau paraventriculaire dans l'histophysiologie corrélative de la glande thyroïde et de la glande pinéale. Ann. Endocr., Paris, *24*, 255—269 (1963).

Milofsky, A. H.: The fine structure of the pineal in the rat, with special reference to parenchyma. Anat. Rec. *127*, 435—436 (1957).

Moore, R. Y., A. Heller, R. K. Bhatnager, R. J. Wurtman, and *J. Axelrod:* Central control of the pineal gland: visual pathways. Arch. Neurol. *18*, 208—218 (1968).

Moszkowska, A.: L'antagonisme épiphyso-hypophysaire. Ann. Endocrinol., Paris, *24*, 215—226.

Moszkowska, A.: Contribution à l'étude du mécanisme de l'antagonisme épiphyso-hypophysaire. Progr. Brain Res. *10*, 564—575 (1965).

Moszkowska, A.: Étude des extraits épiphysaires fractionnés — Physiologie. Biol. Méd. *56*, 403—412 (1967).

Moszkowska, A., and *I. Ebels:* A study of the antigonadotrophic action of synthetic arginine vasotocin. Experientia *24*, 610—611 (1968).

Moszkowska, A., I. Ebels et *A. Scemama:* Étude *in vitro* des extraits fractionnés d'épiphyses d'agneau. C. R. Soc. Biol., Paris, *159*, 2298 (1965).

Niemi, M., and *M. Ikonen:* Histochemical evidence of amino peptidase activity in rat pineal gland. Nature, Lond., *185*, 928 (1960).

Owman, Chr.: Sympathetic nerves probably storing two types of monoamines in the rat pineal gland. Int. J. Neuropharmacol. *2*, 105—112 (1964 a).

Owman, Chr.: New aspects of the mammalian pineal gland. Acta Physiol. Scand. *63*, Suppl. 240, 1—40 (1964 b).

Owman, Chr.: Localization of neuronal and parenchymal monoamines under normal and experimental conditions in the mammalian pineal gland. Progr. Brain Res. *10*, 423—453 (1965).

Pavel, S.: Evidence for the presence of lysine vasotocin in the pig pineal gland. Endocrinology *77*, 812—817 (1965).

Pavel, S., and *S. Petrescu:* Inhibition of gonadotrophin by a highly purified pineal peptide and by synthetic arginine vasotocin. Nature, Lond., *212*, 1054 (1966).

Pellegrino De Iraldi, A., and *R. De Lores Arnaiz:* 5-Hydroxytryptophan decarboxylase activity in normal and denervated pineal glands of rats. Life Sci. *3*, 589—593 (1964).

Pellegrino De Iraldi, A., and *L. M. Zieher:* Noradrenaline and dopamine content of normal, decentralized and denervated pineal gland of the rat. Life Sci. *5*, 149—154 (1966).

Pellegrino De Iraldi, A., L. M. Zieher, and *E. De Robertis:* The 5-hydroxytryptamine content and synthesis of normal and denervated pineal gland. Life Sci. *9*, 691—696 (1963).

Pellegrino De Iraldi, A., L. M. Zieher, and *E. De Robertis:* Ultrastructure and pharmacological studies of nerve endings in the pineal organ. Progr. Brain Res. *10*, 389—421 (1965).

Prop, N.: Lipids in the pineal body of the rat. Progr. Brain Res. *10*, 454—463 (1965).

Prop, N., and *J. Ariëns Kappers:* Demonstration of some compounds present in the pineal organ of the albino rat by histochemical methods and paper chromatography. Acta Anat., Basel, *45*, 90—109 (1961).

Pun, J. Y., and *L. Lombrozo:* Microelectrophoresis of brain and pineal proteins in polyacrylamide gel. Analyt. Biochem. *9*, 9—20 (1964).

Quay, W. B.: Reduction of mammalian pineal weight and lipid during continuous light. Gen. comp. Endocrinol. *1*, 211—217 (1961).

Quay, W. B.: Metabolic and cytologic evidence of pineal inhibition by continuous light. Amer. Zool. *2,* 550 (1962).

Quay, W. B.: Cytologic and metabolic parameters of pineal inhibition by continuous light in the rat, *Rattus norvegicus.* Z. Zellforsch. *60,* 479—490 (1963 a).

Quay, W. B.: Circadian rhythm in rat pineal serotonin and its modification by estrous cycle and photoperiod. Gen. comp. Endocrinol. *3,* 473—479 (1963 b).

Quay, W. B.: Indole derivatives of pineal and related neural and retinal tissues. Pharmacol. Rev. *17,* 321—345 (1965).

Quay, W. B., and *A. Halevy:* Experimental modification of the rat's pineal content of serotonin and related indole amines. Physiol. Zool. *1,* 1—7 (1962).

Quay, W. B., J. F. Jongkind, and *J. Ariëns Kappers:* Localizations and experimental changes in monoamines of the reptilian pineal complex studied by fluorescence histochemistry. Anat. Rec. *157,* 304—305 (1967).

Quay, W. B., J. Ariëns Kappers, and *J. F. Jongkind:* Innervation and fluorescence histochemistry of monoamines in the pineal organ of a snake *(Natrix natrix).* J. Neuro-Viscer. Relat. *31,* 11—25 (1968).

Reiss, M., R. H. Davis, M. B. Sideman, I. Mauer, and *E. S. Plichta:* Action of pineal extracts on the gonads and their function. J. Endocrin. *27,* 107—118 (1963 a).

Reiss, M., I. Mauer, M. B. Sideman, R. H. Davis, and *E. S. Plichta:* Pituitary-pineal-brain interrelationships. J. Neurochem. *10,* 851—857 (1963 b).

Reiter, R. J.: The pineal gland: a report of some recent physiological studies. Edgewood Arsenal Technical Report, 4110, 1—108. Res. Laborat. US Army Edgewood Arsenal, Maryland (1967).

Reiter, R. J., and *R. J. Hester:* Interrelationships of the pineal gland, the superior cervical ganglia and the photoperiod in the regulation of the endocrine systems of hamsters. Endocrinology *79,* 1168—1169 (1966 a).

Reiter, R. J., and *R. J. Hester:* Neuroendocrinological interrelationships. In: Metabolic Regulation of Physiological Activity, ed. by *B. Sacktor, R. J. Reiter, J. E. Wilson, H. J. Smith, C. G. Tiekert,* and *R. J. Hester,* 13—18. Medical Research Laboratory, Research Laboratories US Army, Edgewood Arsenal, Maryland, USA, 1966 b.

Reiter, R. J., R. A. Hoffman, and *R. J. Hester:* The role of the pineal gland and of environmental lighting in the regulation of the endocrine and reproductive system of rodents. Edgewood Arsenal Technical Report, 4032, 1—43. Res. Laborat. US Army Edgewood Arsenal, Maryland (1966 a).

Reiter, R. J., R. A. Hoffman, and *R. J. Hester:* The effects of thiourea, photoperiod and the pineal gland on the thyroid, adrenal and reproductive organs of female hamsters. J. exp. Zool. *162,* 263—268 (1966 b).

Richardson, K. C.: The fine structure of the albino rat iris with special reference to the identification of adrenergic and cholinergic nerves and nerve endings in its intrinsic muscles. Amer. J. Anat. *114,* 173—205 (1964).

Rinehart, J. F., and *M. G. Farquhar:* The fine vascular organization of the anterior pituitary gland. An electron microscopic study with histochemical correlations. Anat. Rec. *121,* 207—239 (1955).

Rodin, A. E., and *R. A. Turner:* The relationship of intravesicular granules to the innervation of the pineal gland. Lab. Invest. *14,* 1644—1651 (1965).

Rodin, A. E., and *R. A. Turner:* The perivascular space of the pineal gland. Tex. Rep. Biol. Med. *24,* 153—163 (1966).

Roth, W. D.: Metabolic and morphologic studies on the rat pineal organ during puberty. Progr. Brain Res. *10,* 552—562 (1965).

Sano, Y., und *T. Mashimo:* Elektronenmikroskopische Untersuchungen an der Epiphysis cerebri beim Hund. Z. Zellforsch. *69,* 129—139 (1966).

Scharenberg, K., and *L. Liss:* The histologic structure of the human pineal body. Progr. Brain Res. *10,* 193—217 (1965).

Scharrer, E.: Photo-neuro-endocrine systems: general concepts. New York Acad. Sci. *117,* 13—22 (1964).

Simonnet, H., L. Thiéblot, et *T. Melik:* Influence de l'épiphyse sur l'ovaire de la jeune rate. Ann. Endocrinol., Paris, *12,* 202—205 (1951).

Snyder, S. H., and *J. Axelrod:* A sensitive assay for 5-hydroxytryptophan decarboxylase. Biochem. Pharmacol. *13,* 805—806 (1964 a).

Snyder, S. H., and *J. Axelrod:* Influence of light and the sympathetic nervous system on 5-hydroxytryptophan decarboxylase (5-HTPD) activity in the pineal gland. Fed. Proc. *23,* 206 (1964 b).

Snyder, S. H., J. Axelrod, J. E. Fischer, and *R. J. Wurtman:* Neural and photic regulation of 5-hydroxytryptophan decarboxylase in the rat pineal gland. Nature, Lond., *203,* 981 (1964).

Snyder, S. H., J. Axelrod, R. J. Wurtman, and *J. E. Fischer:* Control of 5-hydroxytryptophan decarboxylase activity in the rat pineal gland by sympathetic nerves. J. Pharmacol. exp. Ther. *147,* 371—375 (1965).

Szentágothai, J., B. Flerko, B. Mess, and *B. Halász:* Hypothalamic control of the anterior pituitary. P. 97. Budapest: Akadémiai Kiadó, 1962.

Taxi, J., et *B. Droz:* Étude de l'incorporation de noradrénaline-^{3}H (NA-^{3}H) et de 5-hydroxytryptophane-^{3}H (5-HTP-^{3}H) dans l'épiphyse et le ganglion cervical supérieur. C. R. Acad. Sci., Paris, *263,* 1326—1329 (1966).

Taxi, J., et *B. Droz:* Localisation d'amines biogènes dans le système neurovégétatif périphérique. (Étude radioautographique en microscopie électronique après injection de noradrénaline-^{3}H et de 5-hydroxytryptophane-^{3}H.) In: Neurosecretion, Proc. 4th int. Symp. Neurosecretion, Strasbourg, 25—27 Juillet, 1966, *F. Stutinsky* ed., 191—202. Berlin: Springer, 1967.

Thiéblot, L.: Physiology of the pineal body. Progr. Brain Res. *10,* 479—488 (1965).

Thiéblot, L., et *S. Blaise:* Influence de la glande pinéale sur les gonades. Ann. Endocrinol., Paris, *24,* 270—286 (1963).

Vivien, J. H., et *B. Roëls:* Ultrastructure de l'épiphyse des chéloniens. Présence d'un «paraboloïde» et de structures de type photorécepteur dans l'épithélium sécrétoire de *Pseudemys scripta elegans.* C. R. Acad. Sci., Paris, *264,* 1743—1746 (1967).

Vollrath, L.: The ultrastructure of the eel pituitary at the elver stage with special reference to its neurosecretory innervation. Z. Zellforsch. *73,* 107—131 (1966).

Wartenberg, H.: The mammalian pineal organ: electron microscopic studies on the fine structure of pinealocytes, glial cells and on the perivascular compartment. Z. Zellforsch. *86,* 74—97 (1968).

Wartenberg, H., und *H. G. Baumgarten:* Elektronenmikroskopische Untersuchungen zur Frage der photosensorischen und sekretorischen Funktion des Pinealorgans von *Lacerta viridis* und *L. muralis.* Z. Zellforsch. *127,* 99—120 (1968).

Wartenberg, H., und *W. Gusek:* Licht- und elektronenmikroskopische Beobachtungen über die Struktur der Epiphysis cerebri des Kaninchens. Progr. Brain Res. *10,* 296—315 (1965).

Wittkowski, W.: Kapillaren und perikapilläre Räume im Hypothalamus-Hypophysen-System und ihre Beziehungen zum Nervengewebe. Eine elektronenmikroskopische Studie am Meerschweinchen. Z. Zellforsch. *81,* 344—360 (1967 a).

Wittkowski, W.: Synaptische Strukturen und Elementargranula in der Neurohypophyse des Meerschweinchens. Z. Zellforsch. *82,* 434—458 (1967 b).

Wolfe, D. E.: The epiphyseal cell: an electron-microscopic study of its intercellular relationships and intracellular morphology in the pineal body of the albino rat. Progr. Brain Res. *10*, 332—376 (1965).

Wolfe, D. E., L. T. Potter, K. C. Richardson, and *J. Axelrod:* Localizing tritiated norepinephrine in sympathetic axons by electron microscopic autoradiography. Science *138*, 440—442 (1962).

Wragg, L. E.: Effects of pinealectomy in the newborn female rat. Amer. J. Anat. *120*, 391—402 (1967).

Wurtman, R. J., and *J. Axelrod:* The formation, metabolism and physiologic effects of melatonin in mammals. Progr. Brain Res. *10*, 520—528 (1965 a).

Wurtman, R. C., and *J. Axelrod:* The pineal gland. Sci. Amer. *213*, 50—60 (1965 b).

Wurtman, R. J., M. D. Altschule, and *U. Holmgren:* Effects of pinealectomy and of a bovine pineal extract in rats. Amer. J. Physiol. *197*, 108—110 (1959).

Wurtman, R. J., J. Axelrod, and *E. W. Chu:* Melatonin, a pineal substance: effect on the rat ovary. Science *141*, 277—278 (1963).

Wurtman, R. J., J. Axelrod, and *J. E. Fischer:* Melatonin synthesis in the pineal gland: effect of light mediated by the sympathetic nervous system. Science *143*, 1328 (1964).

Wurtman, R. J., W. Roth, M. D. Altschule, and *J. J. Wurtman:* Interactions of the pineal organ and exposure to continuous light on organ weights of female rats. Acta endocrinol. *36*, 617—624 (1961).

Zenker, W., und *E. Krammer:* Untersuchungen über Feinstruktur und Innervation der inneren Augenmuskulatur des Huhnes. Z. Zellforsch. *83*, 147—168 (1967).

Zieher, L. M., and *A. Pellegrino De Iraldi:* Central control of noradrenaline content in rat pineal and submaxillary glands. Life Sci. *5*, 155—161 (1966).

Zweens, J.: Influence of the oestrous cycle and ovariectomy on the phospholipid content of the pineal gland in the rat. Nature, Lond., *197*, 1114—1115 (1963).

Zweens, J.: The pineal lipid content in rat and the involvement of the epiphysis cerebri in the hypophyseo-gonadal interrelation. Med. Thesis, 1—107. Groningen: Van Denderen, 1964.

Zweens, J.: Alterations of the pineal lipid content in the rat under hormonal influences. Progr. Brain Res. *10*, 540—551 (1965).

Address of the author: Prof. Dr. *J. Ariëns Kappers,* The Netherlands Central Institute for Brain Research, Ydijk 28, Amsterdam-O., The Netherlands.

Discussion

Smelik: If I did not misunderstand you, you said that denervation of the pineal gland depletes noradrenaline, but not dopamine. Would this mean that dopamine is not situated in the nerve endings, but in the chief cells? And if this is so, could dopamine be considered to be produced by the pineal, and is anything known about the effects of pineal dopamine?

Kappers: I cited a paper by *Pellegrino De Iraldi* and *Zieher.* Their conclusion is indeed that dopamine is present in pinealocytes. So far, nothing is known about physiological effects of pineal dopamine.

Neurohumors (Neurotransmitters)

Journal of Neuro-Visceral Relations, Suppl. IX, 187—211 (1969)

Acetylcholine

B. Csillik

Department of Anatomy, University Medical School, Szeged, Hungary

With 21 Figures

Summary

Structural correlates of acetylcholine effects are discussed in the terms of molecular anatomy. Polarization microscopy proves that acetylcholine induces a structural desorganisation of synaptic membranes, probably responsible for the increased ion flux during synaptic transmission. — Synaptic vesicles within pre-synaptic terminals are not identical with the transmitter substance; rather they can be regarded as "containers" that bind acetylcholin on their surface membranes. — The enzyme responsible for acetylcholine synthesis (acetylcholine acyltransferase = choline acetylase) cannot be demonstrated at present by histological techniques. On the other hand, the enzyme hydrolyzing the transmitter (acetylcholinesterase) can readily be stained by means of the *Koelle*-technique, both at the level of light and electron microscopy. Presence of acetylcholinesterase in some excitatory and inhibitory synapses of the central nervous system suggests the "pretransmitter" role of acetylcholine. — The fine structural localizations of specific and non-specific esterases in the synaptolemmal layers of the motor end plate, as studied by means of electron histochemistry, proves that the "middle dense layer" contains arylesterase, whereas pre- and post-synaptic membranes are loaded with acetycholinesterase. Therefore, the theory of a cascade-connected enzyme system, responsible for junctional transmission, is forwarded.

Since the early observations of *Hunt* and *Taveau* (1906) and *Dixon* (1907), and since the great discoveries of *Dale* (1914) and *Loewi* (1921), the substance *acetylcholine* is continuously raising interest of thousands of biologists. In spite of its well-known stereochemical structure (Fig. 1),

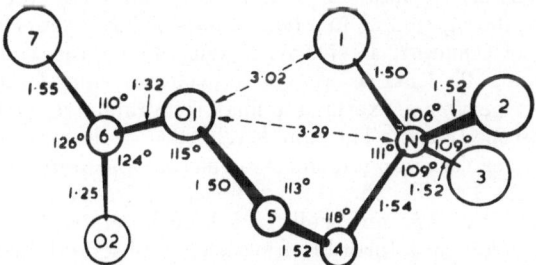

Fig. 1. Stereochemical picture of the acetylcholine molecule. (From *Canepa* and coll., 1966.) Inter-atomic distances in Ångstrom units.

its function is still enigmatic. However, the biological structures whereupon acetylcholine exerts its action are more responsible for the secrets inherent in acetylcholine action than the structure of the acetylcholine molecule itself. In other words: the action of acetylcholine cannot be treated separately from the structure of biological membranes.

Electrogenic Effect

More than a quarter of a century ago *Feldberg* and coll. (1940) published their important experiment on the electrogenic effect of acetylcholine injected intra-arterially in *Torpedo marmorata*. Since the electric organs are homologous with motor end plates both, phylogenetically and ontogenetically (*Ranvier*, 1878), the idea appears more than plausible that similar electrochemical effects may take place in humorally mediated junctional tissues of higher vertebrates.

Membrane Effects

The exact nature of the interaction between acetylcholine and cytological structures is unknown. Acetylcholine increases the permeability of synaptic membranes, first of all for Na ions. In the motor end plate, however, penetration of other ions is also greatly increased. Increased permeability of the post-synaptic membrane induces an ionic flux that results in the EPSP (excitatory post-synaptic potential). Acetylcholine is effective only when applied externally to the post-synaptic membrane; acetylcholine injected directly into the post-synaptic cell evokes only a small catelectrotonic potential (*Del Castillo* and *Katz*, 1955).

Whether the changes in membrane function are due to the effect of acetylcholine upon proteins or rather upon lipids, is also only partially understood. *Beutner* and *Barnes* (1941) described a negative electrical potential evoked by acetylcholine on water-lipid interphases. More recently, *Hyono* and *Kuriyama* (1966) observed acetylcholine-induced changes in the surface tension of lipid monolayers, similar to those evoked by K-ions.

Most probably, however, acetylcholine influences more vigorously the molecular structure (configuration or, more properly, conformation) of membrane proteins. Polarization microscopic studies on the fine structure of the post-synaptic membrane in the motor end plate clearly point at such a possibility (*Csillik*, 1963). In frog muscles, the post-synaptic membrane of the myoneural junction consists of large semicircular lamellae ("organites" of *Couteaux*, 1947). When using an appropriate fixation (fixatives containing Pb ions, as proposed by *Sávay* and *Csillik*, 1958), the post-synaptic membrane exerts a brilliant birefringence under the polarization microscope (Fig. 2). The double refraction of the synaptic membranes is due to the regular arrangement of submicroscopic units, the "organites", consisting of 2—3 parallel sheaths (junctional folds) as demonstrated electron microscopically by *Birks* and coll. (1960). Such relatively large submicroscopic units result in a form-birefringence, as observed in the imbibition experiment. Since, however, the imbibition curve does not decline to zero even at the point of minimum (Fig. 3) the system possesses also an "intrinsic birefringence", due to the regular organization of junctional folds at the molecular level. Apparently, lipid molecules are oriented transversely to the non-lipid particles since the imbibition curve after lipid extraction (B) runs considerably higher than before (A). Accordingly, the molecular organi-

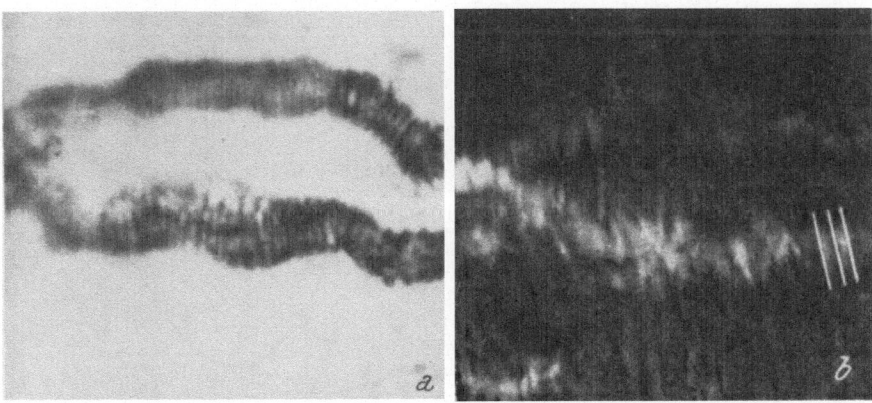

Fig. 2. Synaptic membranes of the myoneural junction, frog; a, stained for acetylcholinesterase; b, birefringence after lead fixation. ×2,000 (Csillik, 1963).

zation of the junctional membranes is essentially similar to that of other biological "unit membranes" (Robertson, 1960).

Application of acetylcholine, combined with the AChE-inhibitor physostigmine or Neostigmine, or a long-lasting supramaximal stimulation of the junction by an electric current results in a striking molecular re-arrangement. Though the optically visible polarization pattern does not undergo any spectacular changes, imbibition experiments prove that the intrinsic birefringence is considerabely altered. After lipid extraction (Fig. 4, curve B) the imbibition curve of the excited post-synaptic membrane declines to zero, proving that the molecular organization of non-lipid particles became random in lieu of the original, geometrically regular organization. Thus acetylcholine appears to change the conformation of membrane proteins, i. e. that of the acetylcholine receptor, as suggested already by Nachmansohn (1963). It stands to reason to assume that a membrane, consisting of regularly arranged molecules, lets pass ions of smaller dimensions only (e. g. K^+), whereas a random system might be also permeable for larger ions (e. g. the strongly hydrated Na^+) or even molecules.

It would, of course, be preposterous to go into a mechanistically detailed interpretation of these polarization optical studies. They might reveal the rough outlines of structural alterations at the molecular level. Yet it should be kept in mind that biological membranes are far from being as simple constructions as the circle-and-bar schemes of the textbooks. Our factual knowledge with regards to membrane structure is derived from indirect information that can be used only with a cautious extrapolation (Robertson,

190   B. Csillik:

1960). It is mainly the watery structure of living systems that makes any far-reaching conclusions irrealistic. *Szent-Györgyi* was perhaps the first who envisaged the immense role of "ordered water" in the maintenance of biological processes and *Fernández-Morán* (1962) applied the concept of "liquid ice" to the architecture of lipoprotein membranes. It is evident

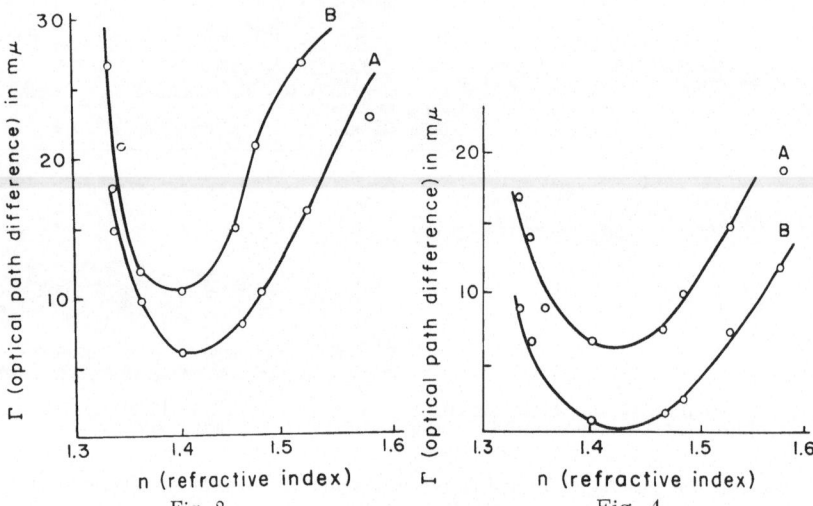

Fig. 3   Fig. 4

Fig. 3. Imbibition curves of the synaptic membranes in the myoneural junction, frog. A, before lipid extraction; B, after mild lipid extraction with 56⁰ C acetone. Note that the point of minimum of curve B is considerably above the abscissa. *(Csillik, 1963)*

Fig. 4. Imbibition curves of the synaptic membranes in the myoneural junction after supramaximal stimulation with tetanizing stimuli (90 min., 60 c/s). A, before lipid extraction; B, after acetone treatment. Note that the minimum point of curve B declines to zero. Virtually the same curve characterizes muscles treated with acetylcholine + eserine. *(Csillik, 1963)*

that any perturbation of the lipid or protein components of a membrane will inevitably result in drastic changes of structurally bound water dipoles that, in turn, will also change the ephemeric and continuously changing crystalline and fluid compartements of "flickering water clusters" (Fig. 5). And, since the solubility of ions is, first of all, dependent of the structural organization of "bound" water, it stands to reason to assume that any alterations in lipid or protein structures will change the solubilities of sodium and potassium ions within the membrane and, accordingly, will change also the possibilities of transport movements of these same ions through the membrane.

Synaptic Vesicles

It would be one of the greatest tasks in histology to develop a technique for the visualisation of acetylcholine itself. Trials to locate choline Reineckeate *(Coujard*, 1948) failed to have success. The hydroxylamineferrichloride technique (*Hestrin,* 1949) has not been employed until now in histochemistry. Thus the electron microscopic observation of synaptic vesicles in nerve terminals is, at present, the only morphological guide for the demonstration of acetylcholine.

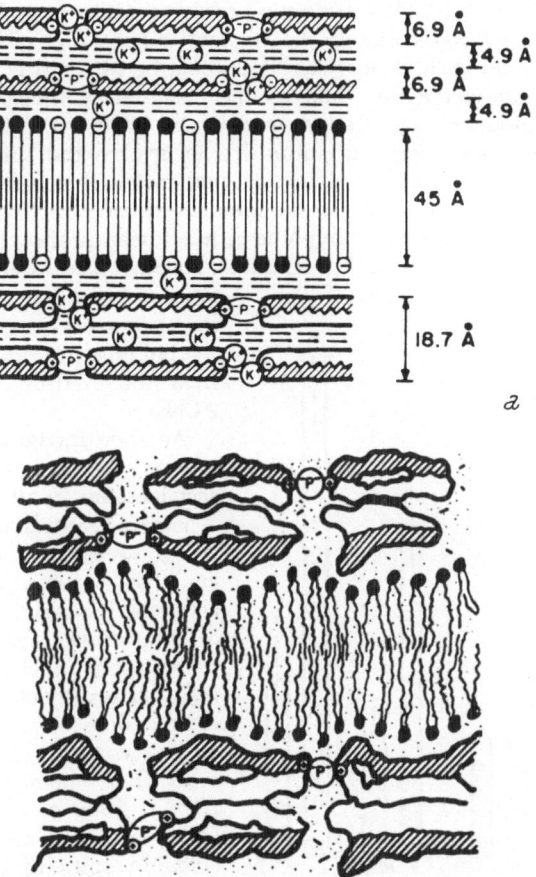

Fig. 5. Possible effect of stimulation on the molecular structure of a lipoprotein "unit" membrane. A, under normal conditions, not only protein and lipid molecules but also water dipoles are in an extremely oriented pattern (liquid ice). B, stimulation results in molecular disarrangements of protein and lipid units as well as a random or less ordered water structure. (*Hechter,* 1965)

Synaptic vesicles measuring 350—450 Å in diameter are well known since the very beginnings of electron microscopy. Vesicles were tentatively identified as small stores of neurochemical transmitter substances, first of all as of acetylcholine (*De Robertis* and *Bennett,* 1955; *Palade* and *Palay,* 1954; *Palay* and *Palade,* 1955; *Fernández-Morán,* 1957). Studies performed by the aid of differential centrifugation proved that acetylcholine was actually bound to vesicle-containing fractions (*De Robertis* and coll., 1963; *Whittaker* and *Gray*, 1962). Since it had been supposed already by *Fatt* and *Katz* (1952) that the release of acetylcholine from nerve endings was not continuous but "quantal", it was an obvious conclusion drawn by *Del Castillo* and *Katz* (1956) that synaptic vesicles contained quanta of acetylcholine, *i. e.* 400 molecules of the transmitter substance. Thus, the intravesicular concentration of acetylcholine would be about 0.11 M. It is more difficult to interpret recent data on the quantum. *Krnjevic* and *Mitchell* (1963), for instance, suggested a quantum of 8000—40.000 molecules of acetylcholine. Such a large quantum has to involve the simultaneous discharge of large amounts of vesicles, since otherwise, if it were concentrated in a single 400 Å body, it would require the extreme concentration of up to 2—10 Mols. As an alternative hypothesis, it has been proposed by us (*Csillik* and *Joó*, 1967) that acetylcholine might be bound to the external surfaces of

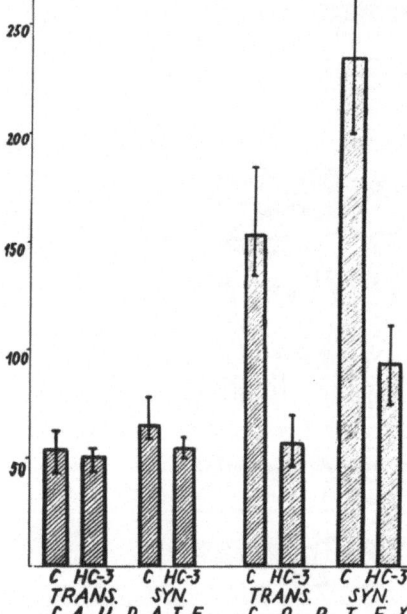

vesicles (measuring 500.000 Å² or more) possibly in such a way that the binding protects acetylcholine from the hydrolyzing action of AChE.

Another intriguing fact is the relation of synaptic vesicles to acetylcholine synthesis. The pioneering studies of *McIntosh* (1963) revealed that hemicholinium (HC-3) compounds inhibit acetylcholine synthesis by means of interfering with the choline transporting mechanism. Accordingly, after injection of HC-3 a drastic decrease of the acetylcholine content of various brain areas has been observed by *Hebb's* group as well as by *Metz* (1962) and by *Quastel* and *Curtis* (1965).In spite of this, the number of synaptic vesicles showed only a slight, non-significant decrease in the caudate nucleus after injection of HC-3 (Fig. 6) whereas the acetylcholine content of this same nucleus decreased to 10% of the original (*Hebb* and coll., 1964). Thus, though synaptic vesicles are undoubtedly

Fig. 6. The effect of hemicholinium (HC-3) upon the numbers of synaptic vesicles in the rat caudate nucleus and parietal cortex. There is no significant change in the caudate nucleus, whereas, in the cortex, the number of synaptic vesicles decreased considerably both in transient and in synaptic axon profiles. (*Csillik* and *Joó*, 1967)

related to the storage and transport of acetylcholine, this relation is just somewhat similar to that of a cable-car to its content (*Csillik* and *Joó*, 1967).

Finally it should be noted that, though most authors regard synaptic vesicles as structural units transporting acetylcholine or other substances to their site of action, the direction of this movement has not been demonstrated unequivocally. *Andres* (1964), for instance, is of the opinion that synaptic vesicles, being pinocytotic units, are moving in a retrograde direction.

Cholin Acetylase

The enzyme system synthesizing acetylcholine has not been demonstrated by means of morphological techniques. An immunohistochemical procedure, utilizing fluorescein isothiocyanate as a marker for light microscopy and ferritine for electron microscopy, is presently being elaborated in our department. The main difficulty in this task is to obtain chemically a cholin acetylase (acetylcholine acyltransferase) enzyme preparation as pure as possible.

Acetylcholinesterase (AChE)

The enzyme responsible for the hydrolytic breakdown of acetylcholine, on the other hand, can readily be demonstrated histochemically. The *Koelle* acetylthiocholine technique (*Koelle* and *Friedenwald*, 1950; *Koelle*, 1951, 1954) can be regarded as a standard light microscopic staining procedure. The main advantage of this ingenious technique is that the chromogeneous substrate is virtually identical with the physiological substrate of the enzyme, the only difference being an "S" atom in place of an "O" atom in the ester link of acetylcholine.

The application of this technique for neurohistological studies has been widely discussed by *Koelle* (1963) and by *Gerebtzoff* (1959), *Couteaux* (1958), *Taxi* (1965), *Csillik* (1965) and others. Such studies revealed the concentration of this enzyme in synaptic regions throughout the nervous system.

With regard to the innervation apparatus of the autonomic periphery, the histochemical organization of the rat iris can be regarded as characteristic. As shown by *Csillik* and *Koelle* (1965), AChE is strictly localized in cholinergic nerve fibers. Such fibers are more abundant in the sphincter territory of the iris, but also the dilator area contains quite a dense plexus of cholinergic nerve fibers (Fig. 7). The beads or varicosities of these nerve fibers display the strongest AChE activity (Fig. 8). Schwann cells are devoid of AChE activity containing, however, high amounts of pseudo-(butyryl)-cholinesterase, the physiological substrate of which is presently unknown (Fig. 9).

Electron Histochemistry of Acetylcholinesterase

It is one of the objectives of up-to-date histochemistry to cut down the barriers of the resolving power of the light microscope. Since the first trial of *Lehrer* and *Ornstein* (1959), at least ten different techniques were published aiming at an electron microscopic localization of AChE. Such studies were initiated in our laboratory in 1962. After many unsuccessful trials we realized that, even for electron microscopy, the *Koelle* technique is unsurpassible in specificity, due to the unique structure of the chromogeneous substrate. The only modification introduced by my colleagues and myself was the introduction of two bivalent cations in the original *Koelle*

reaction. Copper was supplemented by a cation of a high atomic number (lead or uranyl, resp.) in order to yield a high-grade electron contrast (*Csillik* and coll., 1966; *Kása* and *Csillik*, 1967; *Knyihár* and *Csillik*, 1968). The introduction of a second cation not only increased the electron density of the reaction but, at the same time, inhibited the formation of needleshaped thiocholine crystals, common in usual histochemical specimens.

The histochemical reaction requires pre-fixation of the sample. Formalin and/or glutaraldehyde was most suitable for such purposes. After the reaction had been completed, we had to use also a post-fixation in osmic acid, indispensable for a general electron microscopic contrast. In order to prevent the decolorizing effect of osmic acid, special mixtures of osmium with

Fig. 7. Acetylcholinesterase activity in the rat iris. Stretch specimen, low power. Note the strong reaction in the sphincter area (sph) and the loose nerve plexus in the dilator area (Dil).

various buffer solutions and glutaraldehyde were introduced for postfixation.

By means of this technique, AChE, in the terminal autonomic innervation apparatus, is observed to be confined to the axolemmal membranes (Fig. 10). Pre-terminal myelinated postganglionic fibers of the ciliary ganglion as well as a fair number of non-myelinated fibers exert an AChE reaction. However, the majority of non-myelinated fibers is blank. This fact disproves the view of those authors who sought to support *Burn's* theory (1966) on the acetylcholine-mediated release of norepinephrine from autonomic terminals by demonstrating an AChE reaction of adrenergic nerve fibers under the light microscope.

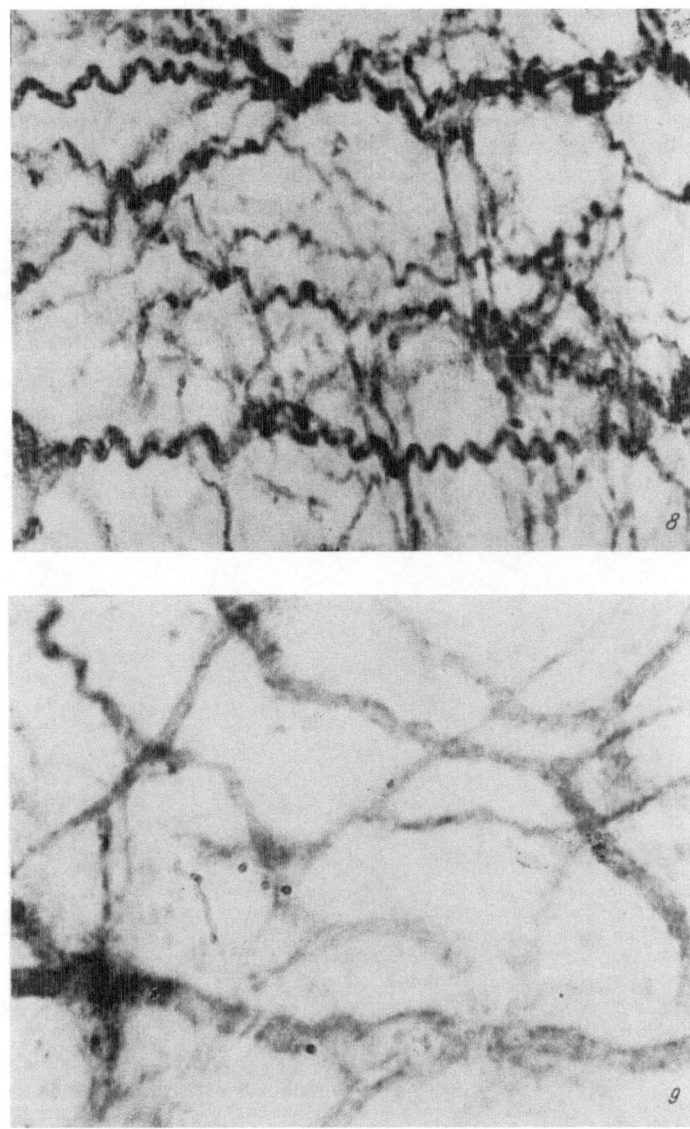

Fig. 8. Acetylcholinesterase in autonomic nerve fibers of the dilator area in the rat iris. Note the heaviest activity in the varicosities (beads) of the fibers. × 2,000. (*Csillik* and *Koelle*, 1965)

Fig. 9. Butyryl-(pseudo-)cholinesterase activity in the interstitial (Schwann) cell cytoplasm in the rat iris. The area is virtually the same as depicted in Fig. 8. × 2,000. (*Csillik* and *Koelle*, 1965)

Pre-transmitter Modulatory Role of Acetylcholine

The general neurotransmitter role of acetylcholine, as proposed some years ago by *Koelle* (1962), cannot be proved at the level of the peripheral autonomic innervation apparatus. Yet it appears that *Koelle's* theory holds for higher nervous centers where some kind of interaction between acetyl-

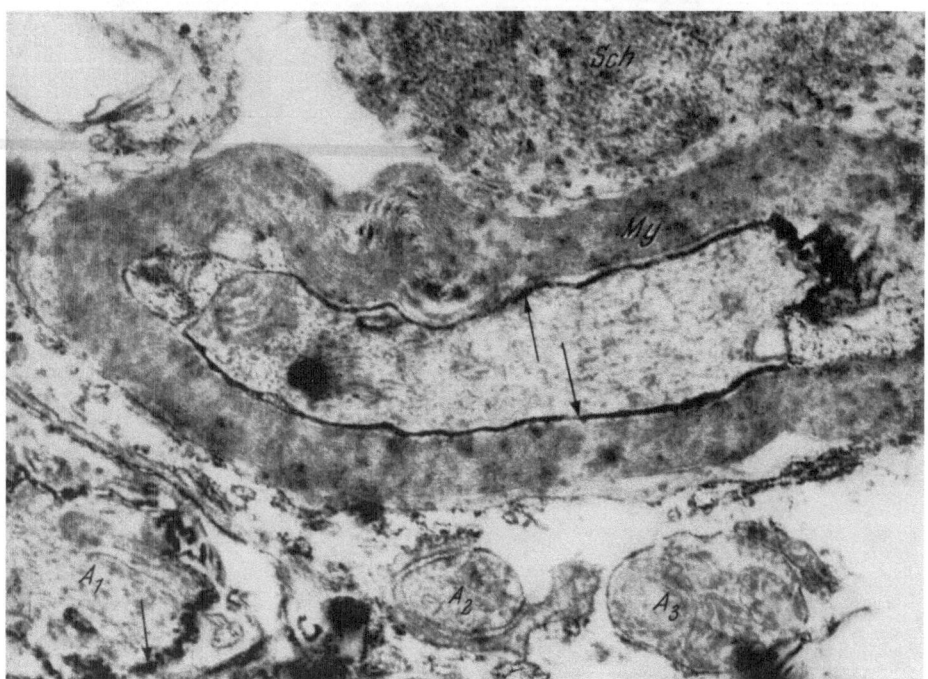

Fig. 10. Acetylcholinesterase activity in nerve fibers of the rat iris. Enzyme activity can be seen in the axolemmal sheath of the myelinated fiber (arrows), whereas myelin lamellae (My) are entirely non-reacting. One of the non-myelinated fibers (A-1) also exhibits an enzyme reaction at the axolemmal membrane, whereas other non-myelinated fibers (A-2, A-3, etc.) are devoid of an enzyme reaction. Sch nucleus of a Schwann (interstitial) cell. × 30,000.

choline and the true excitatory and inhibitory substances may actually take place.

Widespread neurophysiological and pharmacological studies initiated by Sir *John Eccles* (1964) and *D. R. Curtis* (1965) have disclosed the very function of several morphologically identified cerebellar synapses. Thus, the mossy fiber → granule cell synapse, as well as the granule axon (parallel fiber) → Purkinje dendritic spine synapse proved to be excitatory, whereas synapses between basket cell terminals and Purkinje somata proved to belong to the inhibitory type. Electron microscopic studies by *Hámori* and *Szentágothai* (1965) revealed that this latter junction is actually located at the very origin of the Purkinje axon, *i. e.* at the base of the Purkinje cell.

Histochemical studies performed with Dr. *P. Kása* (1965, 1966) revealed that in the cerebellar cortex of the guinea pig a strong AChE activity is present at the parallel axon → Purkinje dendrite synapses (Fig. 11). Electron histochemistry has shown that, here again, the enzyme is located at the surfaces of the axons (Fig. 12). It turned out, however, that not only in these excitatory junctions but also at the inhibitory synapses between basket cell terminals and Purkinje somata, an AChE reaction can be observed (Fig. 13). Thus, if one accepts the premiss that activity of an enzyme points to the presence of its substrate, one inevitably arrives at the conclusion that, in the guinea pig cerebellum, acetylcholine acts as a transmitter substance, both in excitatory and inhibitory synapses.

Such a dual role of acetylcholine has been described by *Tauc* and *Gerschenfeld* (1961, 1962) in *Aplysia* ganglia, where acetylcholine in low concentrations results in a hyperpolarization followed by inhibition of ganglion cells, whereas in a higher concentration the same drug results in depolarization followed by excitation.

In the guinea pig cerebellar cortex, however, a more delicate mechanism is probable. It is a commonplace in neurochemistry that, in spite of its high AChE activity, the content in acetylcholine and cholin acetylase of the cerebellum is extremely low. It is unlikely, therefore, that acetylcholine would act as a transmitter substance in this organ since in that case a continuous outflow and replacement of acetylcholine would require high amounts of acetylcholine and cholin acetylase. On the other hand, acetylcholine applied by a multibarrelled coaxial microelectrode does not evoke either excitation or inhibition of cerebellar synapses, except if extremely high non-physiological concentrations are applied (*Curtis*, 1965).

Therefore, the assumption appears plausible that acetylcholine exerts an intracellular action in cerebellar synapses, an action similar to that proposed by *Nachmansohn* (1955). Accordingly, the participation of acetylcholine in synaptic transmission would be restricted to the increase of the permeability of synaptic membranes for other substances of molecular size such as for amino acids, peptides, and possibly polypeptides in excitatory synapses and for gamma-amino-acid or related compounds in inhibitory synapses. Such a functional organization of central (muscarinic) cholinergic synapses has been forwarded by us (*Csillik* and *Joó*, 1965).

Electron Histochemical Structure of the Motor End Plate

In order to test the applicability of this "pre-transmitter" theory for nicotinic cholinergic junctions we recently undertook a systematic study of the myoneural junction, using different chromogeneous substrates [*] to

---

[*] Acetylthiocholine for AChE, indoxylacetate and indoxyl butyrate for non-specific esterases, and thiol acetic acid for esterases plus Cathepsin C.

locate various esterases (*Knyihár* and *Csillik*, 1968). The light microscopic
pattern obtained by all of these methods is just the same. In mammalian

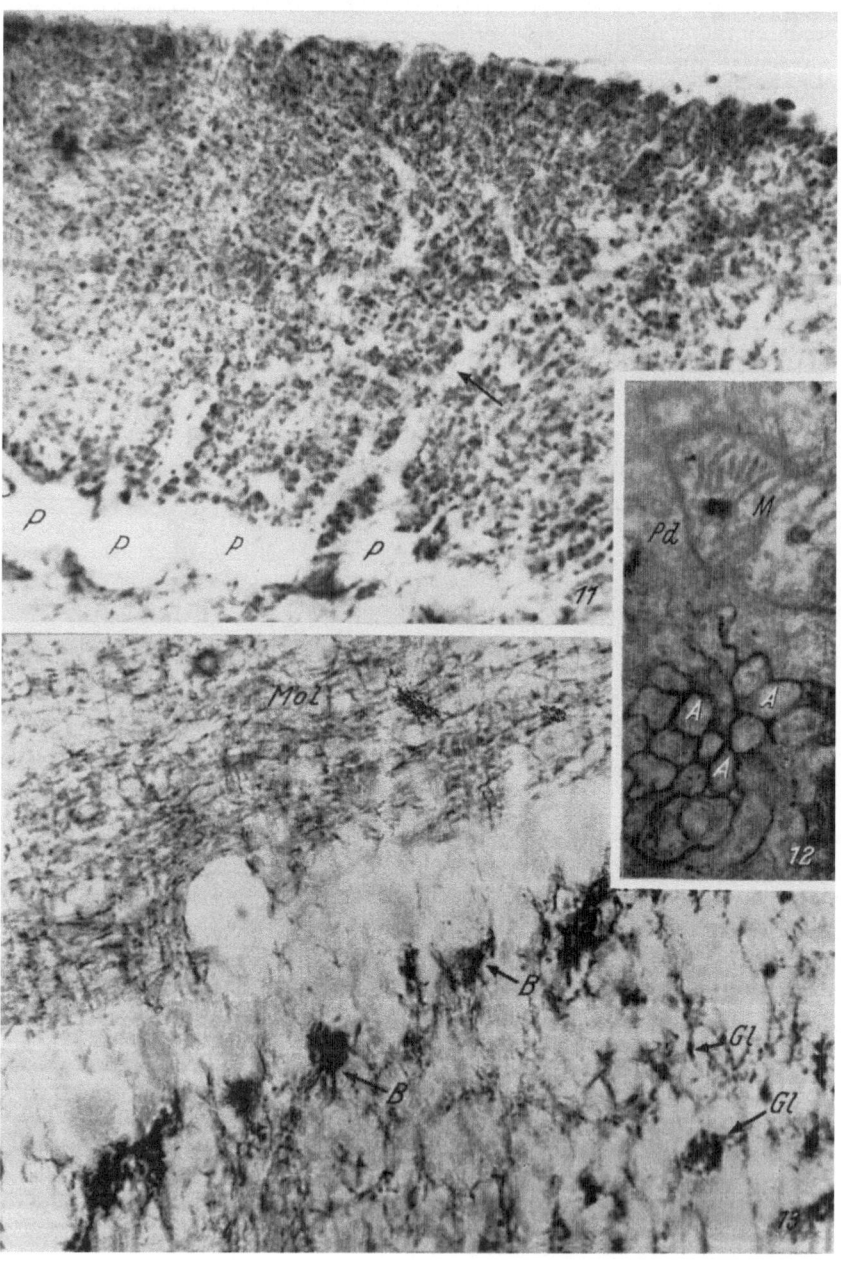

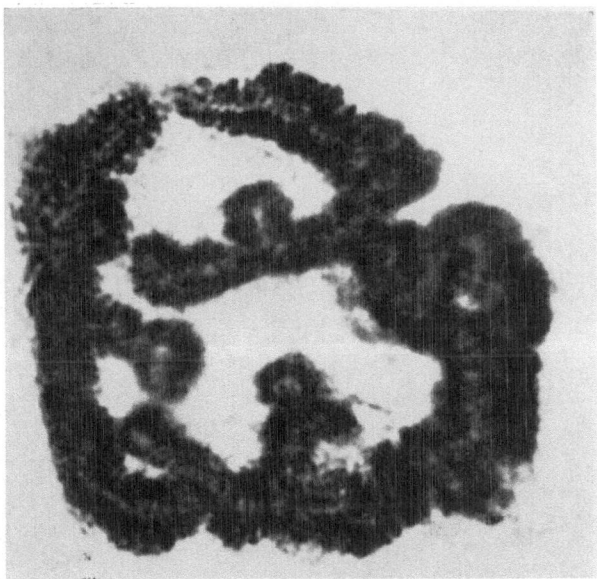

Fig. 14. Acetylcholinesterase activity of the subneural apparatus in the rat's gastrocnemius. The telodendrial nerve fiber proceeds in the gutter formed by the semi-circular organites. × 2,000.

muscles the reactions outline the pretzel-shaped "subneural apparatus" (*Couteaux*, 1947). When inspected under high power, the enzyme reaction appears to be confined to the "organites" that furnish a gutter under the telodendrial nerve fiber (Fig. 14). Especially in low power electron micrographs, the localization of synaptic enzymes in the junctional folds is clearly visible (Fig. 15).

The patterns obtained at higher resolution are, however, far from being identical and they appear to be entirely dependent on the chromogeneous substrate. When using acetylthiocholine, the resultant electron dense precipitate of the AChE reaction is present in both the pre- and the post-

---

Fig. 11. Acetylcholinesterase activity in the cerebellar cortex of the guinea pig. Note the reaction in the molecular layer, outlining the non-reacting Purkinje dendrites (arrow). Purkinje cells (P) are non-reacting. Sagittal section. (*Csillik* and *Kása*, 1966)

Fig. 12. Acetylcholinesterase activity of parallel fibers (A) in the guinea pig cerebellar cortex. Sagittal section. M mitochondrion, Pd Purkinje dendrite. × 30,000. (*Kása* and *Csillik*, unpublished picture)

Fig. 13. Acetylcholinesterase activity of basket terminals (B) at the bases of Purkinje cells. Frontal section. Gl: glomerulus cerebellaris. Note the regular course of parallel fibres in the molecular layer (Mol). (*Csillik* and *Kása*, 1966)

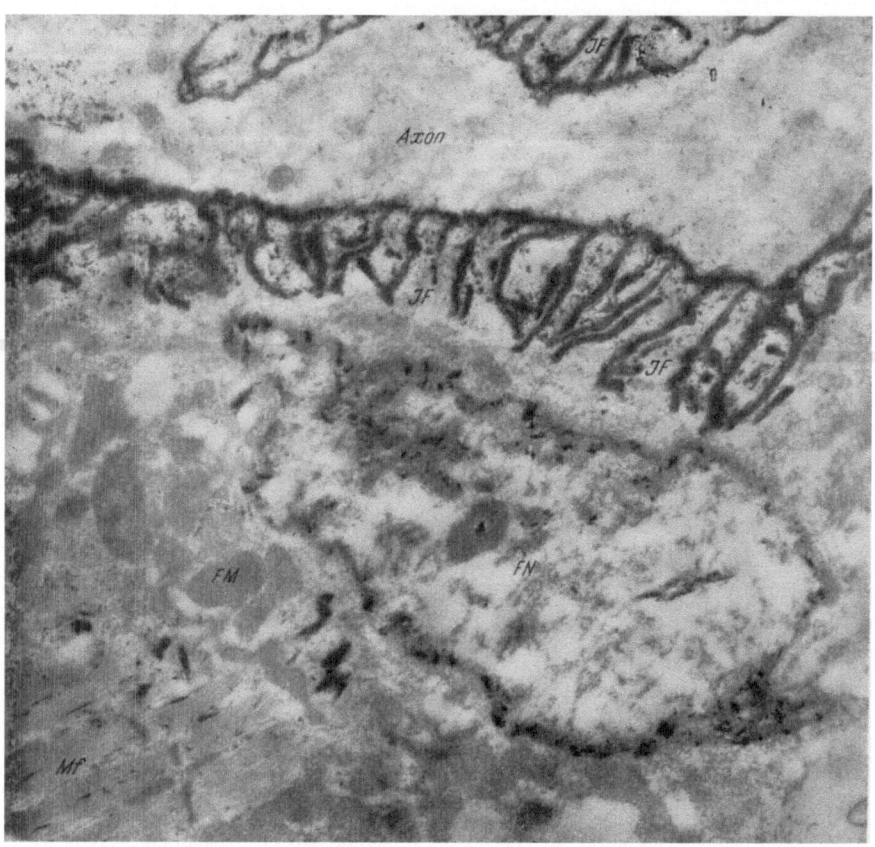

Fig. 15. Acetylcholinesterase activity of the subneural apparatus in the rat's diaphragm. Like in the preceding figure, the axon and the synaptic gutter appear in a longitudinal section. The enzyme activity is concentrated in the junctional folds (JF). FN fundamental nucleus; FM fundamental mitochondria; Mf myofibrils. × 23,000.

synaptic membranes (Fig. 16) *. On the other hand, indoxyl acetate yields a more or less homogeneous reaction product filling the synaptic gap (Fig. 18). Neither pre- nor post-synaptic membranes show any distinct

---

* Thus the textbook opinion that AChE of the myoneural junction is located post-synaptically can be accepted only as a rough approximation. The post-synaptic membrane, due to its numerous folds, is 10—20 times larger in surface area than the pre-synaptic one. Thus, the main bulk of AChE is really present at the post-synaptic side. Yet, the amount per surface area of AChE in the pre-synaptic membrane is just the same; in this context, it should be recalled that *Couteaux* in one of his first papers on this subject held the opinion that the axon itself also exerts an AChE activity — *prima cogitatio est optima*.

reaction. In striking contrast to this, indoxyl butyrate, when used as a substrate for motor end plate enzyme activity, yields a well-defined end product in the "middle dense layer of the synaptolemma", the function of which was entirely unknown hitherto (Fig. 19). Finally, thiol acetic acid,

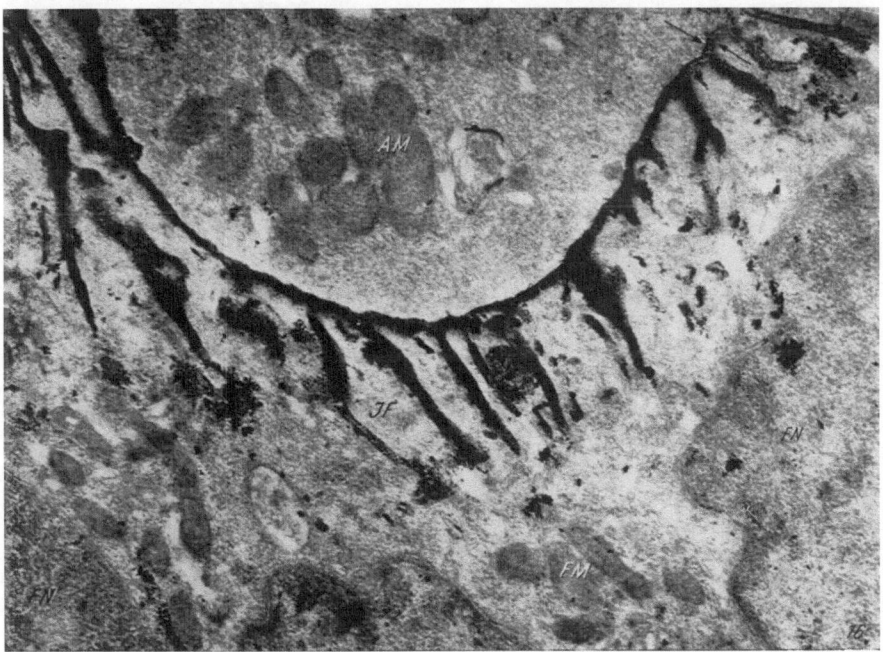

Fig. 16. Virtual cross section of the subneural apparatus. In the semi-circular profile of the axon, several mitochondria (AM) are present. Acetylcholinesterase activity is apparent in both pre- and post-synaptic membranes (arrows) and in the junctional folds (JF). Fundamental nuclei (FN) and fundamental mitochondria (FM) are free of the enzyme reaction product. Pb^{++}-Cu^{++}-thiocholine technique; diaphragm of the rat. $\times$ 40,000.

Especially characteristic is the appearance of AChE on pre- and post-synaptic membranes in electron histochemical specimens stained for a brief period when using the uranyl thiocholine procedure. In such sections, not only the general disposition of the enzyme but also the particular units where the enzyme reaction started can readily be seen and counted. These units, presumably identical with active AChE groups in the membrane, are located at even distances (300—500 Å) from each other. By means of a simple "unit per square" count it can be stated that in a cubic micron of the junctional fold area the number of such primary AChE spots is 8000—10.000. This figure is strikingly similar to that obtained by *Rogers* and coll. (1966), who used autoradiographic, beta-absorption and scintillation techniques.

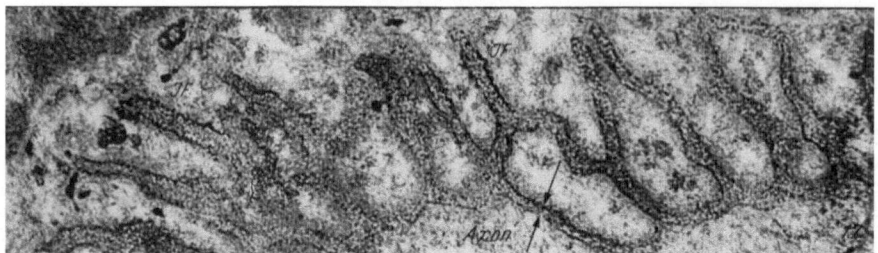

Fig. 17. Detail of the junctional folds (JF); acetylcholinesterase reaction (Uranyl-Cu^{++}-thiocholine technique). Note the regular localization of the electron dense reaction product in pre- and post-synaptic membranes and in the membranous "walls" of the junctional folds. The granular appearance of the reaction product (primary reaction centers) is clearly visible. $\times$ 70,000.

used under the same parameters as the other reactions, yields a rough particulate reaction attached mainly to the post-synaptic membrane, accompanied by a slighter reaction of the pre-synaptic membrane and the pre-synaptic axoplasm (Fig. 20).

When evaluating these various enzyme localizations (Fig. 21) it may be stated that the variability of the results is not due to the substantivity of the end products. In other words, the differences in the electron microscopic localizations of the above four endplate enzymes cannot be ascribed to the differences in the solubilities of the end products of these reactions. For instance, the end products of reactions 1 and 4 are identical (PbS). Yet the localizations are distinctly different. Reactions 2 and 3 result in the very same indoxyl-hexazo dye. Yet the end product occupies quite different localizations. Thus it appears that different layers of the synaptolemma are equipped with different enzyme spectra. At present, it would be hazardous to give, even in the form of a working hypothesis, a detailed interpretation of these cascade-connected enzymes. Yet one cannot escape the suspicion that not only these enzymes but also their, unknown, physiological substrates may participate in a chain reaction during impulse transmission. Recent neurophysiological data obtained by *Riker* (1966) actually suggest that it is not acetylcholine alone that is responsible for neuromuscular impulse transmission. Further studies will decide whether the action of AChE is also connected with other esterases in other cholinergic junctions.

Acknowledgement

The author is deeply indebted to Miss *Elizabeth Knyihár*, M. D., for her invaluable help in the electron histochemical studies and in preparing the text of this paper.

References

Andres, K. H.: Mikropinocytose im Zentralnervensystem. Zschr. Zellforsch. 64, 63—73 (1964).

Beutner, R., and *T. C. Barnes:* Science, 94, 241 (1941). Cit. *A. S. V. Burgen* and *F. C. MacIntosh:* The physiological significance of acetylcholine. In: Neuro-

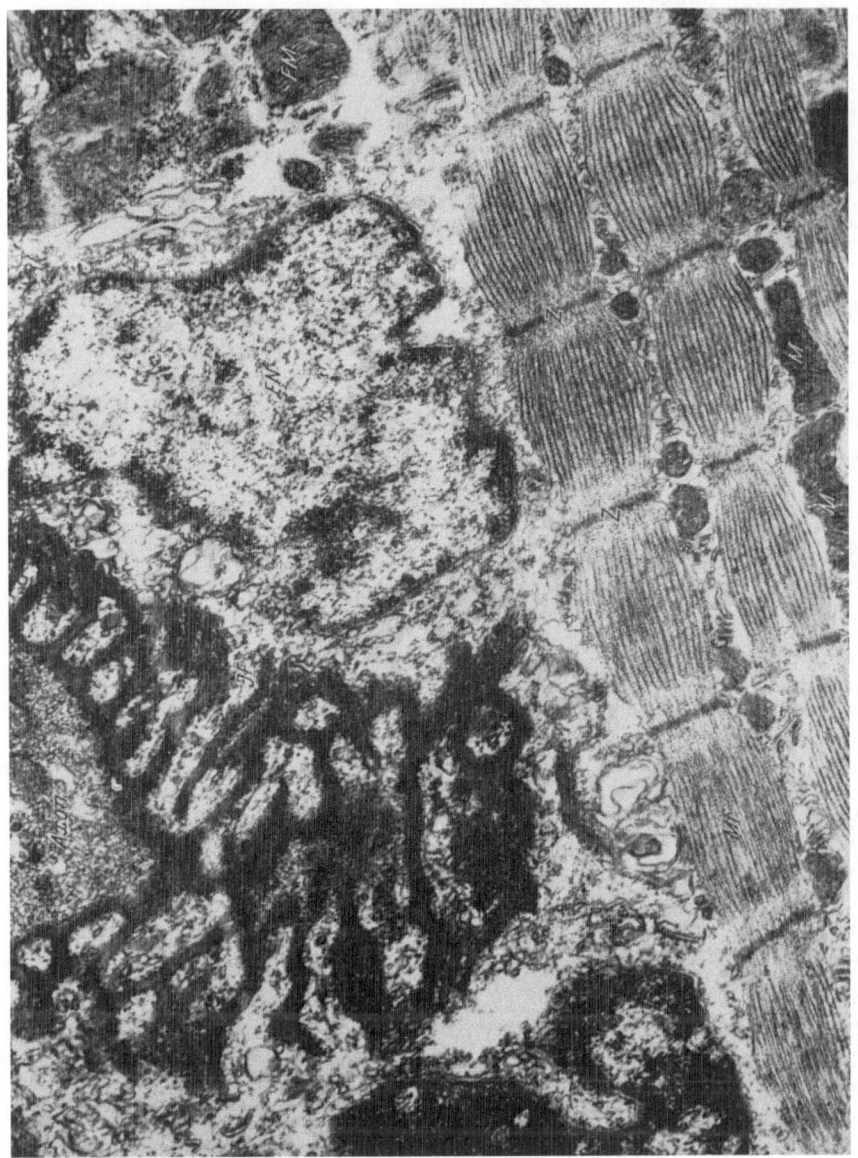

Fig. 18. Indoxylacetate-esterase reaction (hexazo coupling according to *Holt*) in the subneural apparatus. Note the homogeneous reaction product filling the primary and secondary synaptic gaps. JF junctional folds, FN fundamental nucleus, FM fundamental mitochondria, Mf myofibrils with Z lines, M muscle mitochondria. × 40,000

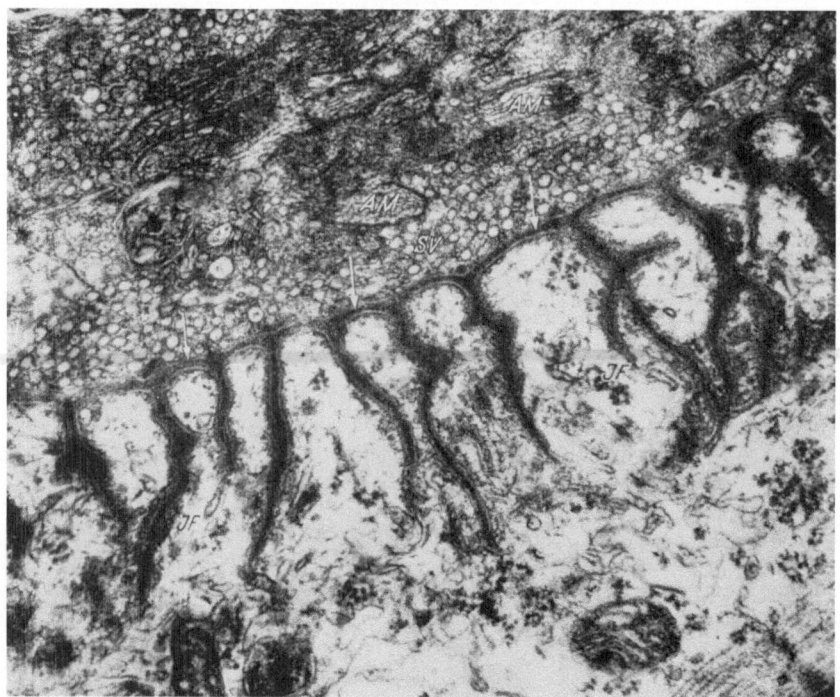

Fig. 19. Indoxylbutyrate-esterase reaction (hexazo coupling according to *Holt*) in the subneural apparatus. In the axon, a faint reaction takes place in between axonal mitochondria (AM) and synaptic vesicles; the heaviest reaction occurs, however, in the "middle dense layer" of the synaptolemma (arrows) both in the primary synaptic gap and in the junctional folds (JF). × 52,000.

chemistry. *K. A. C. Elliot, I. H. Page* and *J. H. Quastel,* Eds. Thomas, Springfield 1955.

Birks, R., II. E. Huxley, and *B. Katz:* The fine structure of the neuromuscular junction of the frog. J. Physiol. (Lond.) *150,* 134—144 (1960).

Burn, J. II.: Adrenergic transmission. Introductory remarks. Pharmacol. Rev., Baltimore, *18,* 459—470 (1966).

Canepa, F. G., P. Pauling, and *H. Sörum:* Structure of acetylcholine and other substrates of cholinergic systems. Nature (Lond.), *210,* 907—909 (1966).

Coujard, R.: Essais sur la signification chimique de quelques méthodes histologiques (Réactions phénoliques, colorations nerveuses et mitochondriales). Bull. histol. appl. *20,* 161—173 (1943).

Coujard, R.: Démonstration histochimique de la présence de choline au niveau de terminaisons nerveuses. C. R. Soc. Biol. Paris, *142,* 16—17 (1948).

Couteaux, R.: Contribution à l'étude de la synapse myoneurale. Rev. canad. biol. *6,* 563—711 (1947).

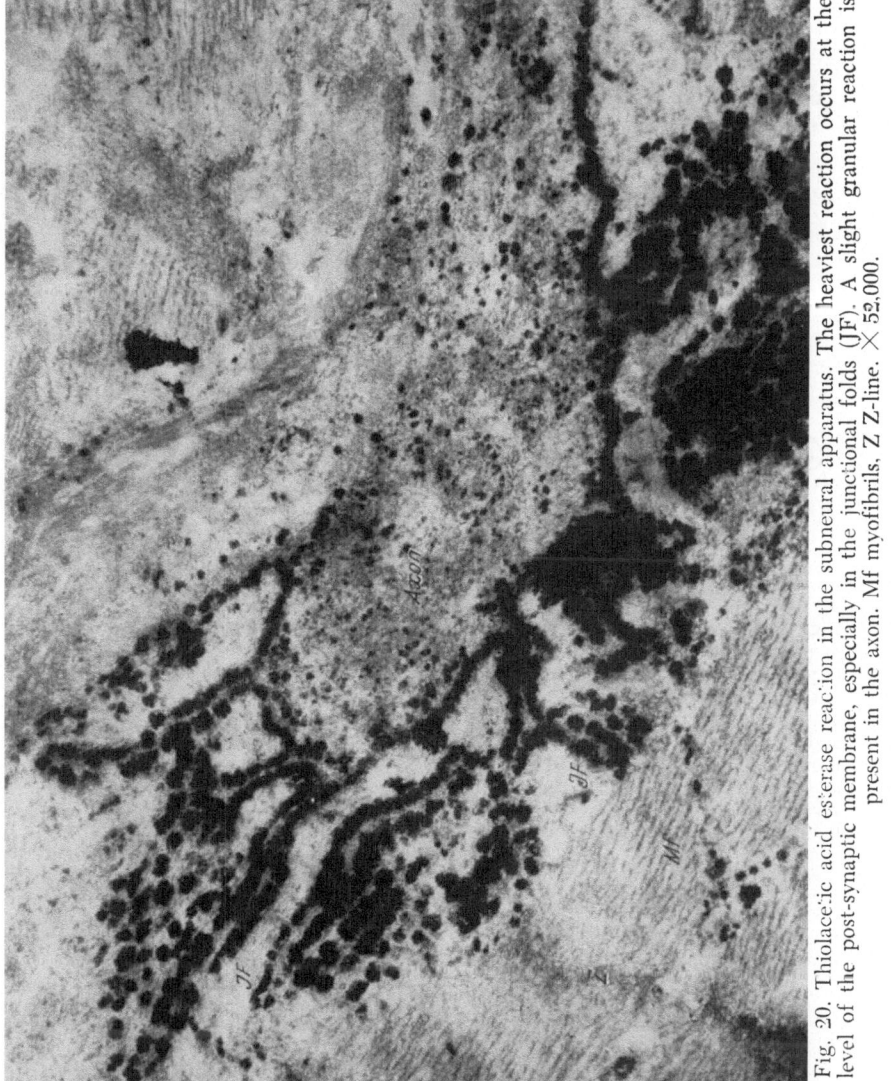

Fig. 20. Thiolacetic acid esterase reaction in the subneural apparatus. The heaviest reaction occurs at the level of the post-synaptic membrane, especially in the junctional folds (JF). A slight granular reaction is present in the axon. Mf myofibrils, Z Z-line. × 52.000.

Couteaux, R.: Morphological and cytochemical observations on the post-synaptic membrane at motor end-plates and ganglionic synapses. Exp. Cell Research. Suppl. 5, 293—322 (1958).

Csillik, B.: Submicroscopic organization of the post-synaptic membrane in the myoneural junction. J. Cell Biol. *17,* 571—586 (1963).

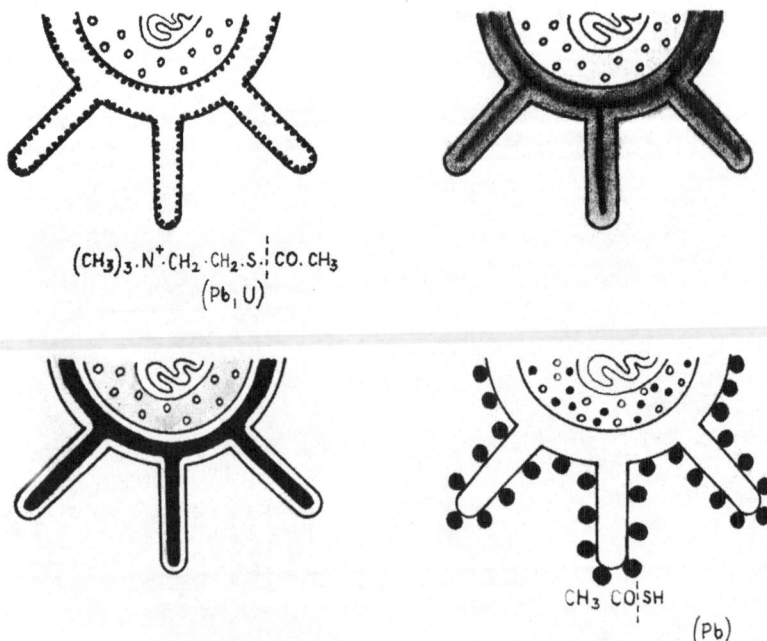

$$(CH_3)_3 \cdot N^+ CH_2 \cdot CH_2 \cdot S \cdot | CO \cdot CH_3$$
$$(Pb, U)$$

$$CH_3 \ CO | SH$$
$$(Pb)$$

Fig. 21. Summary of the electron histochemical enzyme reactions in the synaptic membranes of the motor end plate. Acetylcholinesterase (1) is located in both pre- and post-synaptic membranes; non-specific acetate-esterase occupies the synaptic cleft (2); non-specific butyrate-esterase is located in the middle dense layer of the synaptolemma (3) and thiol acetic esterase comprising various esterases and the proteolytic enzyme Cathepsine C, is bound to the post-synaptic membrane (4). The semi-circular profile represents the axon, with mitochondrium and synaptic vesicles; the finger-like processes represent junctional folds. The substrate used for the reaction and the site of the enzymic hydrolysis (dotted line) is indicated under the corresponding figures; in brackets, the coupling agent is indicated.

Csillik, B.: Functional structure of the post-synaptic membrane in the myoneural junction. Akadémiai Kiadó, Budapest, 1965, 1967.

Csillik, B., and *F. Joó*: Cholinesterase, arylesterase and the structural basis of neurohumoral transmission in the central nervous system. Acta biol. Hung. *16*, 185—205 (1965).

Csillik, B., and *F. Joó*: Effect of hemicholinium on the number of synaptic vesicles. Nature *213*, 508—509 (1967).

Csillik, B., and *P. Kása*: Localization of acatylcholinesterase in the guinea pig cerebellar cortex. Acta neuroveget. Wien, *29*, 289—296 (1966).

Csillik, B., and *G. Koelle*: Histochemistry of the adrenergic and the cholinergic autonomic innervation apparatus as represented by the rat iris. Acta Histochem. *22*, 350—363 (1965).

Csillik, B., F. Joó, P. Kása, and *G. Sávay:* Pb-Thiocholine techniques for the electron histochemical localization of acetylcholinesterase. Acta Histochem. 25, 58—70 (1966).

Curtis, D. R.: The actions of amino acids upon mammalian neurons. Studies in Physiology, Springer, Berlin, pp. 34—42 (1965).

Dale, H. H.: The action of certain esters and ethers of choline, and their relation to muscarine. J. Pharmacol. Exper. Therap., Baltimore, 6, 147—190 (1914).

Del Castillo, J., and *B. Katz:* On the localization of acetylcholine receptors. J. Physiol., Lond. 128, 157—181 (1955).

Del Castillo, J., and *B. Katz:* Biophysical aspects of neuromuscular transmission. In: *J. A. V. Butler,* ed. Progr. in Biophys. Biophysic.-Chem. 6, 122—170 (1956).

De Robertis, E. D. P., and *H. S. Bennett:* Some features of the submicroscopic morphology of synapses in frog and earthworm. J. biophys. biochem. Cytol. 1, 47—58 (1955).

De Robertis, E., C. R. Lores Arnaiz, L. Salganicoff, A. P. De Iraldi, and *L. M. Zieher:* Isolation of synaptic vesicles and structural organization of the acetylcholine system within brain nerve endings. J. Neurochem. 10 225—235 (1963).

Dixon, W. E.: Vagus inhibition. Brit. Med. J. 2, 1807 (1907).

Eccles, J. C.: The physiology of synapses. Springer, Berlin-Göttingen-Heidelberg 1964.

Fatt, P., and *B. Katz:* Spontaneous subthreshold activity at motor nerve endings. J. Physiol., London, 117, 109—128 (1952).

Feldberg, W. A., A. Fessard., and *D. Nachmansohn:* The cholinergic nature of the nervous supply to the electrical organ of the Torpedo *(Torpedo marmorata).* J. Physiol., London, 97, 3—5 (1940).

Fernández-Morán, H.: Electron microscopy of nervous tissue. In: *D. Richter,* Ed., Metabolism of the nervous system. 1—34. Pergamon Press, London 1957.

Fernández-Morán, H.: Cell membrane ultrastructure: low temperature electron microscopy and X-ray diffraction studies of lipoprotein components in lamellar systems. In: Ultrastructure and metabolism of the nervous system. S. R. *Korey, A. Pope, E. Robins,* Eds. pp. 235—267. Williams and Wilkins, Baltimore 1962.

Gerebtzoff, M. A.: Cholinesterases: A histochemical contribution to the solution of some functional problems. Pergamon Press, Oxford 1959.

Hámori, J., and *J. Szentágothai:* The Purkinje cell baskets: ultrastructure of an inhibitory synapse. Acta biol., Szeged, 15, 465—479 (1965).

Hebb, O. C., G. M. Ling, E. G. McGeer, P. L. McGeer, and *D. Perkins:* Effect of locally applied hemicholinium on the acetylcholine content of the caudate nucleus. Nature, London, 204, 1309—1311 (1964).

Hechter, O.: Intracellular water structure and mechanisms of cellular transport. Annal N. Y. Acad. Sci. 125, 625—646 (1965).

Hestrin, S.: The reaction of acetylcholine and other carboxylic acid derivatives with hydroxylamine and its analytical application. J. Biol. Chem., Baltimore, 180, 249—261 (1949).

Hunt, R., and *R. M. Taveau:* Brit. Med. Journal 2, 1788 (1906). Cit.: *J. H. Quastel:* Acetylcholine synthesis in the central nervous system. In: Neurochemistry. *K. A. C. Elliot, I. H. Page* and *J. H. Quastel,* Eds. Thomas, Springfield, pp. 153—172 (1955).

208   B. Csillik:

Hyono, A., and *S. Kuriyama:* Properties and structures of lecithin monolayers containing potassium ions or acetylcholine. Nature, London, *210,* 300—301,
(1966).

Kása, P., and *B. Csillik:* Cholinergic excitation and inhibition in the cerebellar
cortex. Nature, London, *208,* 695 (1965).

Kása, P., and *B. Csillik:* Electron microscopic localization of cholinesterase
by a copper-lead thiocholine technique. J. Neurochemistry *13,* 1345—1349 (1966).

Knyihár, E., and *B. Csillik:* Ultrastructural basis of excitation and inhibition
in mammalian autonomic ganglia. Acta biol. Hung. *19,* 227—238 (1968).

Koelle, G. B.: The elimination of enzymatic diffusion artifacts in the histochemical localization of cholinesterases and their localizations in the tissues of
the cat. J. Pharmacol. Exper. Therap., Baltimore, *103,* 153—171 (1951).

Koelle, G. B.: The histochemical localization of cholinesterases in the central
nervous system of the rat. J. Comp. Neurol., Philadelphia, *100,* 211—228 (1954).

Koelle, G. B.: A new general concept of the neurohumoral functions of acetylcholine and acetylcholinesterase. J. Pharm. Pharmacol. *14,* 65—90 (1962).

Koelle, G. B.: Cytological distribution and physiological functions of cholinesterases. Hdb. der experimentellen Pharmakologie. Ergänzungswerk. Vol. 15,
187—298. Springer, Berlin 1963.

Koelle, G. B., and *J. S. Friedenwald:* The histochemical localization of cholinesterase in ocular tissue. Amer. J. Ophth. *33,* 253—256 (1950).

Krnjevic, K., and *J. F. Mitchell:* Unpublished experiments. Cited by *Hebb, C.,*
in Hdb. d. exp. Pharmakologie, Ergänzungswerk, Vol. 15, p. 80. Springer, Berlin 1963.

Lehrer, G. M., and *L. Ornstein:* A diazo coupling method for the electron
microscopic localization of cholinesterase. J. biophys. biochem. Cytol. *6,* 399 to
406 (1959).

Loewi, O.: Über humorale Übertragbarkeit der Herznervenwirkung. Pflügers
Arch. Physiol., Bonn, *189,* 239—242 (1921).

McIntosh, F. C.: Synthesis and storage of acetylcholine in nervous system.
Canad. J. Biochem. Physiol. *41,* 2555—2571 (1963).

Metz, B.: Correlation between respiratory reflex and acetylcholine content
of pons and medulla. Amer. J. Physiol. *202,* 80—82 (1962).

Nachmansohn, D.: Die Rolle des Acetylcholins in den Elementarvorgängen
der Nervenleitung. Brg. Physiol. *48,* 575—683 (1955).

Nachmansohn, D.: Actions on axons, and evidence for the role of acetylcholine in axonal conduction. In: Cholinesterases and anti-cholinesterase agents.
G. B. Koelle, Subeditor, pp. 701—740. Springer, Berlin-Göttingen-Heidelberg
1963.

Palade, G. E., and *S. L. Palay:* Electron microscope observations of interneuronal and neuromuscular synapses. Anat. Rec. *118,* 335 (1954).

Palay, S. L., and *G. E. Palade:* The fine structure of neurons. J. biophys.
biochem. Cytol. *1,* 69—88 (1955).

Quastel, D. M. J., and *D. R. Curtis:* A central action of hemicholinium.
Nature, London, *208,* 192—194 (1965).

Ranvier, L.: Leçons sur l'Histologie du système nerveux. Librairie F. Savy,
Paris 1878.

Riker, W. F.: Actions of acetylcholine on mammalian motor nerve terminal.
J. Pharmacol. Exp. Therap. *152,* 397—416 (1966).

Robertson, J. D.: The molecular structure and contact relationships of cell
membranes. Progress in Biophysics, *10,* 343—418 (1960).

Rogers, A. W., Z. Darzynkiewicz, E. A. Barnard, and *M. M. Salpeter:* Number and location of acetylcholinesterase molecules at motor end plates of the mouse. Nature, London, *210,* 1003—1006 (1966).

Sávay, G., and *B. Csillik:* Lead-reactive substances in myoneural synapses. Nature, London, *181,* 1137 (1958).

Tauc, L., and *H. M. Gerschenfeld:* Cholinergic transmission mechanisms for both excitation and inhibition in molluscan central synapses. Nature, London, *192,* 366—367 (1961).

Tauc, L., and *H. M. Gerschenfeld:* A cholinergic mechanism of inhibitory synaptic transmission in a molluscan nervous system. J. Neurophysiol., Springfield, *25,* 236—262 (1962).

Taxi, J.: Contribution à l'étude des connexions des neurones moteurs du système nerveux autonome. Ann. Sci. Naturelles, Zool. *7,* 413—674 (1965).

Whittaker, V. P., and *E. G. Gray:* The synapse: biology and morphology. Brit. Med. Bull. *18,* 223—228 (1962).

Author's address: Prof. Dr. *B. Csillik,* Department of Anatomy, University Medical School, Szeged, Hungary.

Discussion

Aalberse: What is your opinion of the influence of the mode of fixation on the appearance of synaptic vesicles?

Csillik: Fixation has a very marked effect on shape and appearance of vesicles. Osmic acid renders "normal" synaptic vesicles (*i. e.* non-granulated vesicles) uniform, whereas, after aldehyde fixation, there is a clear-cut distinction between round (spherical) and flattened (ovoid) vesicles. On the other hand, osmic acid does not reveal the dense core of granulated vesicles in several areas of the autonomic periphery and the same might hold true for several regions of the CNS. In the autonomic periphery, *e. g.,* the rat iris is well known to contain a dense network of adrenergic fibers, exerting a brilliant fluorescence with the *Falck* paraformaldehyde technique. In spite of this, there are scarcely any dense-core vesicles in these fibers. When using osmic acid fixation pre-fixation in aldehyde may improve the pattern, but, as *Richardson* has pointed out recently and as proved by our studies also, $KMnO_4$ fixation reveals the dense core in all the vesicles present in adrenergic axons of the rat iris. Therefore, I regard fixation as the crucial point of vesicle morphology.

Carlsson: Prof. *Csillik* mentioned that his electron micrographs argued in favor of a "backward" movement of the vesicles. How is such a movement visualized in an electron micrograph? I should like to take this opportunity to ask if any of the electron microscopists has observed increased frequency of signs suggesting "negative pinocytosis" induced by nerve stimulation?

Csillik: In 1964, *Andres* published a paper on the micropinocytosis in the central nervous system. Accordingly, it may be assumed, as an alternative hypothesis for the genesis of synaptic vesicles, that they may arise from such pinocytotic vesicles. Pictures similar to those in *Andres'* paper have been seen in our material. Though it is very difficult to judge unequivocally whether a vesicle is moving in one direction or in an other, their structural relations to the membrane may suggest either a forward or a backward motion. It should be noted that, according to the majority of authors, the electron micrographs should be interpreted as equivalents of a forward movement.

With regard to stimulation-induced alterations in number and microtopographical distribution of synaptic vesicles, I refer to the classical studies of *De Robertis* and coll. (1958, 1959).

Graf: After administration of reserpine in a large dose, adrenal medullary granules of mice almost completely disappear. In this period a high frequency of pinocytotic activity, probably in the negative direction, can be seen. On the other hand, tyramine only decreases the osmiophilia of the core of the granules. This may be correlated to different secretory mechanisms under these pharmacological conditions. One can imagine that depletion of catecholamines in adrenergic nerves is accompanied by analogous structural alterations (*Graf, J.:* Zur Strukturbindung und Sekretion der Catecholamine. Arch. int. Pharmacodyn. *159,* 170—184, 1966).

Schümann: You showed us that the postganglionic sympathetic neurons of the iris do not have any acetylcholine esterase activity. I should like to know whether you have investigated also other sympathetic postganglionic neurons in this respect, for instance those of the hypogastic nerve?

Csillik: Not as yet.

Tanabe: On the histochemical study of true AChE activity: 1. have you any observation or consideration upon the afferent, especially sensory pathways, including sensory receptors in the skin, tongue etc.? 2. Peripheral motor nerves in the portion near the endplate do not show true AChE activity in some kinds of adult animals and adult human cases, but positive activity is shown in case of infants or fetus. What is your comment on this?

Csillik: Ad 1. Afferent nerve fibers, according to *G. B. Koelle,* do not contain AChE. Wagner-Meissner corpuscles in the skin were shown to contain an esterolytic enzyme (*Csillik* and *Sávan,* 1954) which, however, turned out to be pseudocholinesterase.

Ad 2. The pre-terminal portion of motor axons is non-myelinated. Accordingly, it is more sensitive to formalin fixation. Thus in usual histochemical sections this part does not show AChE activity which is evident, however, in unfixed cryostat sections. In embryonic and early post-natal material axonal AChE is, in general, stronger; thus every part of the motor axon stains for AChE, even in formalin-fixed specimens.

Graf (with demonstration of slides): Acetylcholinesterase can be demonstrated in adrenergic nerves by electron microscope cytochemistry. I did apply the method of *Karnovsky* on the heart of the mouse and I could localize the enzyme in all the nerve bundles, especially at axon membranes. There is no difference between nerves containing small clear vesicles and such containing dense-core vesicles too. Inhibitor experiments prove the specifity of the enzyme. The presence of clear vesicles and dense-core vesicles in the same axon, the first perhaps storing acetylcholine, the second catecholamines, and the simultaneous localization of acetylcholinesterase suggests a role of acetylcholine in adrenergic transmission, as it has been proposed by *Burn* et al. Details concerning the specifity of the method and the vesicles are published in: *Graf. J.,* Demonstration of acetylcholine-esterase in adrenergic nerves by electron microscope cytochemistry. J. Neurochem. 1967, in press.

Csillik: I am reluctant in accepting the projected slides as a proof in favor of *Burn* and *Rand's* theory. First, the occurrence of normal and granulated vesicles in the very same nerve fibers is a common finding, which, due to the uncertainty as to the chemical content of the two kinds of vesicles cannot be regarded as demonstrating the occurrence of acetylcholine and noradrenaline in the very same fiber. Second, in the electron histochemical section stained by the Karnovsky-Roots technique, the activity of pseudo-ChE has not been inhibited by DFP (the advisable concentration being 3×10^{-8} M). Until the participation of a pseudo-ChE cannot be excluded, the presence of a staining in the axolemmal sheath of nerve fibers containing dense-core vesicles cannot be regarded as a proof for the Burn theory, since the physiological substrate of pseudo-ChE is not known. Finally, I feel the dense-core vesicles in these axonal profiles belong rather to the 800 Å group than to the 450 Å group; and these large granulated vesicles contain substances different from noradrenaline. Therefore, I think at present that no morphological proofs are available to support *Burn's* theory on the acetylcholine-induced catecholamine-release from postganglionic adrenergic nerve fibers.

Feldberg: In connection with this discussion I want to point out that it can be dangerous to make conclusions about transmitter functions of pharmacologically active substances from histological findings. You gave us a beautiful electron microscopic picture of vesicles in the cerebellum, as prototype, I understood, of acetylcholine containing vesicles. But the cerebellum contains very little acetylcholine or choline acetylase. You yourself pointed out that the number of vesicles in the caudate nucleus did not decrease after depletion of their acetylcholine content by metacholine bromide, and that the caudate nucleus, the part of the brain with the highest acetylcholine and choline acetylase content, contains a smaller number of vesicles than the cerebral cortex. I also want to emphasize that the presence of true acetylcholinesterase is no evidence for cholinergic neurons. The staining of the dendrites around the basket cells you showed thus needs not necessarily be evidence of cholinergic nerve endings, and therefore it is not necessary to accept the theory you developed. True acetylcholinesterase is present in all nerve fibers, whether they are cholinergic or not, as well as in skeletal muscle fibers, and its presence is not associated with cholinergic transmission.

Csillik: I completely agree with you — and that was the main argument of my paper — that vesicles cannot be identified with transmitter substances. With regard to the cerebellum, I have pointed out, in spite of its high AChE activity, that the cerebellum contains very little acetylcholine and choline acetylase, though in this respect there are marked species differences. I agree that the activity of AChE is not a proof for a "cholinergic" transmission. Yet I believe an intracellular effect of even extremely small amounts of acetylcholine might promote the transport and action of other metabolic products exerting exitation and inhibition, respectively.

Journal of Neuro-Visceral Relations, Suppl. IX, 212—235 (1969)

Catecholamines and Tryptamines

W. B. Quay

Department of Zoology, University of California, Berkeley, California, USA

With 11 Figures

Summary

Catecholamines are represented in the vertebrate nervous system most notably by noradrenaline (NA) and dopamine (DA), but also by the precursor compounds and related ones. Tryptamines are represented primarily by 5-hydroxytryptamine (5-HT or serotonin), but tryptamine itself has been detected there as well. These two groups of so-called "biogenic" or neurohumoral amines have been shown by fluorescence histochemical methods to occur normally within specific neurons, and by physiological and pharmacological methods to affect specific neuronal systems. These results have led to two major functional generalizations about these compounds in the central nervous system. Dopamine has been implicated in central generalized motor stimulation mechanisms and 5-hydroxytryptamine in central induction of sleep.

These neurohumoral amines are not restricted to neurons, and under certain experimental conditions their neuronal distribution may be modified. Although currently intensively studied as possible neurotransmitters, some of these compounds, such as the catecholamines released by the adrenal medulla, function as modulators of neuronal activity in some parts of the body. The additional role of neurotransmitters seems most probable for noradrenaline, in certain specific sites in both central and peripheral nervous systems. Nevertheless, the proof for transmitter roles of any of the compounds remains incomplete.

A survey of the many biological systems in which noradrenaline and 5-hydroxytryptamine are active, suggests that the molecular basis for their activity, at least in many instances, may stem from interactions with specific nucleotides involved in energy transfer. More speculative is a suggested functional association of 5-hydroxytryptamine with microtubular and related systems of possible significance in intracellular conductile and contractile mechanisms. Study of the molecular interactions of these compounds in simpler biological systems will aid in understanding the basis of their actions in nervous tissue.

I. Introduction

Catecholamines and tryptamines are two of the most important chemical categories of neurohumoral compounds in man and other animals. Their roles, however, cannot yet be clearly delineated and their mechanisms and precise sites of action are still obscure. Satisfaction of all the criteria for neurotransmitters is still imperfect as far as these compounds are concerned (*Florey*, 1967). The one that probably comes closest to congruence as a transmitter is noradrenaline (NA). For a basis of understanding of the synaptic structural and physiological relations of neurotransmitters the reader is referred to the books by *De Robertis* (1964) and *Eccles* (1964). Our attention can be focused here in introduction to neurohumoral-neurotransmitter characteristics on, (A) characteristics of chemical synapses as compared with more classically defined electrical ones; and (B) characteristics required of a chemical neurotransmitter. Both listings are largely derived and modified from the excellent recent review by *Gerschenfeld* (1966). A chemical synapse should possess the following physiological characteristics: 1. between the pre- and post-synaptic structures functional electrical coupling should be lacking; 2. the post-synaptic membrane should be electrically inexcitable; 3. for post-synaptic potentials there should exist an inversion potential. Structural characteristics expected of a chemical synapse are the following: 1. pre-synaptic storage vesicles containing the transmitter; 2. special desmosome-like structures at both the pre- and post-synaptic membranes; 3. a well-developed synaptic cleft between the pre- and post-synaptic membranes; and 4. the absence of so-called "tight junctions" between these membranes.

The characteristics expected of a neurotransmitter are as follows:

1. Presence and storage within synaptic vesicles and the larger structural divisions containing vesicles or their precursors, such as synaptic endings, nerve fibers, ganglia and specific portions of nervous tissue.

2. Presence of the enzyme or enzyme systems for the biosynthesis of the compound within or immediately adjacent to the synaptic vesicles.

3. Presence of an enzyme or of enzyme systems for biological inactivation of the compound in the region of the synapse.

4. The compound should be released from the synaptic structure following stimulation of the nerve leading to it.

5. In the post-synaptic cell there should exist specific "receptors" to the substance. Such "receptors" might be revealed by the following pharmacological techniques:

a) a crossed desensitization between the compound and the natural transmitter;

b) blockade of the transmitter's receptors by competitive antagonists or analogues of the compound;

c) microinjection or microiontophoresis of the compound on the post-synaptic membrane.

6. The physiological actions and biochemical characteristics of the test compound and the natural transmitter should be identical, as might be demonstrated through the following lines of evidence:

a) They should cause similar specific ionic permeability changes in the post-synaptic membrane.

b) They should have similar inversion potentials of their action.

c) Membrane resistance must be changed by the compound.

d) Drugs inhibiting the inactivating enzyme found in the synaptic region prolong the synaptic actions of both the test compound and the natural transmitter.

f) The compound must be active in very low concentrations, approaching those that might·be expected of a natural transmitter *in vivo*.

7. The compound must have the additional properties of:

a) free diffusion,

b) causing desensitization of "receptors",

c) capability of being refixed by storage mechanisms.

These characteristics and events may be visualized in part with the aid of the final illustration (Fig. 11) in this article, where, in conclusion, we will trace the probable localizations of the steps in the life history of noradrenaline as a possible neurotransmitter.

Neurohumoral activities may be construed, according to current usage, to include possibilities other than the direct agency of chemical neurotransmission at particular synapses. These other possibilities may be tentatively collected within a common description as modulation of neural, and indeed perhaps other, cell activities (*Florey,* 1967). One might well expect to find an evolution in the precision and efficiency of chemical transmission of information between cells, and especially between nerve cells. Compounds such as 5-hydroxytryptamine (5-HT), found from the most primitive and simple to the most advanced and complex animals, have as neurohumoral amines such a diversity of actions and physiological relations that each cellular context must be studied in detail before generalizations can be made. Our attention here will be directed primarily to the neural contents and sites of action. Nevertheless, it will be pointed out that the best experimental systems for defining the molecular basis for the biological and neural actions of neurohumoral amines are probably those derived from exceedingly simple humoral systems of lower organisms.

This short report is primarily a review and as such must be highly selective, seeking to ascertain certain common basic features and mechanisms. The truly phenomenal and detailed knowledge that has been amassed concerning neurohumoral amines, primarily of a pharmacological nature and dealing almost exclusively with complex mammalian systems, can be appraised in recent monographs (for catecholamines: Pharmacological Reviews *18* [1] : 1—804, 1965; *Wurtman,* 1965; *Iverson,* 1967; for tryptamines: *Erspamer,* 1961, 1966; *Garattini* and *Valzelli,* 1965).

II. Biosynthesis and Metabolism

A. Biosynthesis of Neurohumoral Catecholamines

Vertebrate central nervous systems contain as known neurohumoral catecholamines, dopamine, noradrenaline and adrenaline. This listed sequence portrays their primary biosynthetic relation to one another (Fig. 1). Phylogenetically, however, it has been suggested that the sequence in neuro-humoral evolution may be the reverse, that is, with adrenaline significant

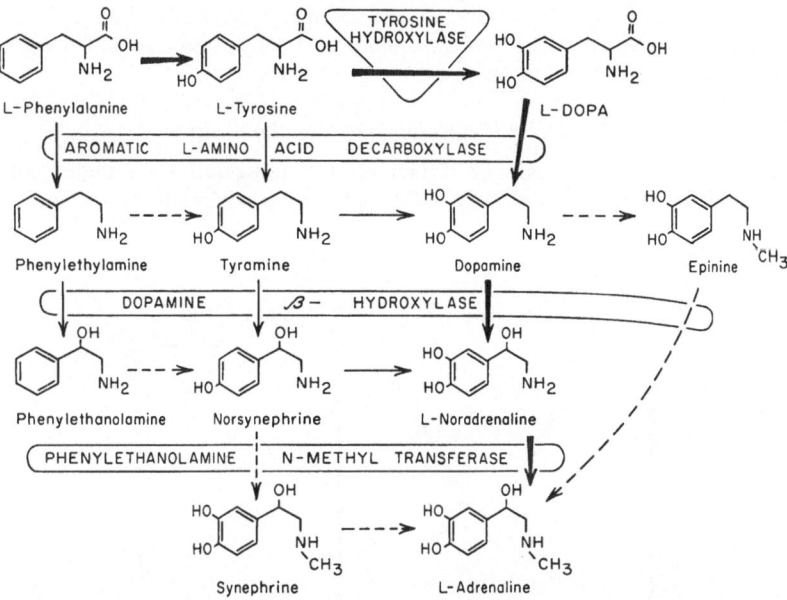

Fig. 1. Biosynthetic pathways to dopamine, noradrenaline and adrenaline in the mammalian body. Heavy arrows show the primary route; solid arrows some of the demonstrated secondary routes, and dashed-lined arrows *in vitro* or theoretically possible routes.

as a central neurohumor only in lower vertebrates and dopamine assuming prominence in the higher groups (*Pscheidt*, 1963).

The history of our knowledge of catecholamine biosynthesis may be said to start with the discovery of dopadecarboxylase by *Holtz* (1939) along with the proposition of the true biosynthetic pathway from tyrosine by *Blaschko* (1939). In animals, although tyrosine can be derived from phenyl-alanine in the diet by means of liver phenylalanine hydroxylase, bio-synthesis of dopamine and its derivatives can be considered to start with tyrosine. The specific enzyme, tyrosine hydroxylase, of the sympathetic nervous system for converting tyrosine to dopa was isolated by *Nagatsu* and coll. (1964). Its properties have been reviewed recently by *Udenfriend* (1966) and it appears likely that it is the limiting factor in the rate of synthesis of NA occurring at any one time. The tissue concentrations of tyrosine even during starvation apparently are always sufficient to saturate the hydroxylase and therefore do not affect the rate of NA synthesis. But inhibitors of tyrosine hydroxylase decrease the synthesis of NA in direct proportion to their hydroxylase inhibition. A possible natural mechanism for control of this critical step is end product inhibition.

The additional steps involved in the synthesis of dopamine, nor-adrenaline and adrenaline are portrayed in Fig. 1. It can be appreciated that the local synthesis of these neurohumors within particular cells of the brain or of the peripheral sympathetic system and adrenal medulla will depend upon the presence of one or more of the enzymes in the sequence: aromatic L-amino acid decarboxylase, dopamine β-hydroxylase and phenyl-ethanolamine N-methyl transferase.

B. Metabolic Degradation of Catecholamines

In Fig. 1 the metabolism of dopamine to noradrenaline is singled out. The possible alternate route of NA to adrenaline via epinine involves a

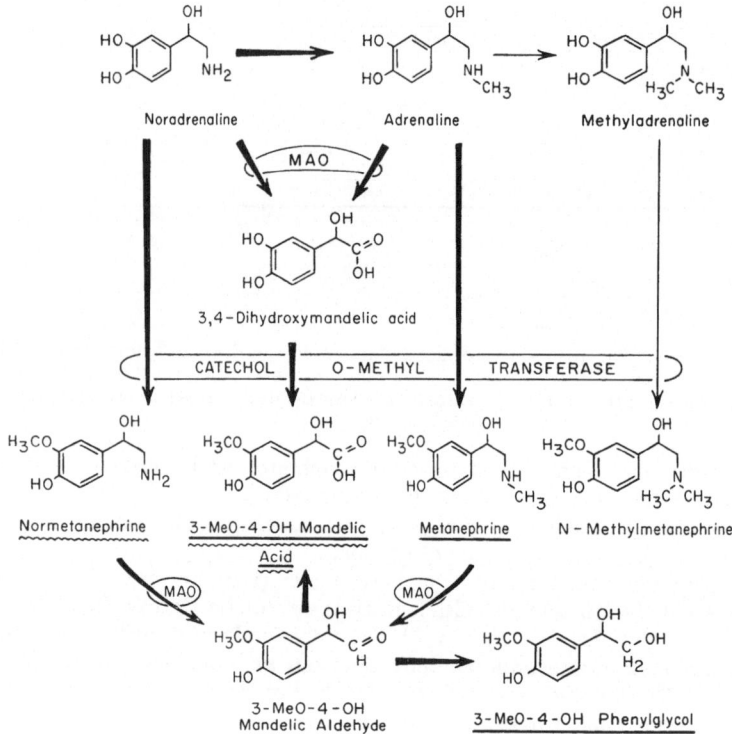

Fig. 2. Primary pathways of metabolic degradation of noradrenaline and adren-aline in the mammalian body. Additional explanation occurs in the text.

methylating enzyme known at present only in rabbit lung. It should be emphasized that the main metabolic breakdown product of dopamine in the brain is homovanillic acid (4-hydroxy-3-methoxyphenylacetic acid). Both routes to homovanillic acid from dopamine involve the actions of monamine oxidase (MAO) and catechol-O-methyl transferase (COMT). The metabolism and actions of dopamine in the brain have been reviewed recently by *Hornykiewicz* (1966).

In Fig. 2 primary routes of degradation of noradrenaline and adrenaline are portrayed, and as in the case of the degradation of dopamine, MAO and COMT figure prominently. Indicated by wavy underlinings are those compounds that are recovered as the primary products in free or conjugated form in the urine after intravenous administration of DL-H³-noradrenaline to rats (*Kopin* and coll., 1961; *Kopin* and *Gordon*, 1963). Indicated by solid underlinings are those compounds that are recovered as the primary products after similar administrations of DL-H³-adrenaline (*La Brosse* and coll., 1961; *Kopin* and coll., 1961). Metabolism of NA and other catecholamines in nervous tissue has been reviewed recently by *Axelrod* (1966), *Glowinski* and *Baldessarini* (1966), *Iverson* (1967) and others.

C. Biosynthesis and Metabolism of Tryptamines

The enzymatic steps in the synthesis and breakdown of neurohumoral tryptamines, and especially of 5-hydroxytryptamine (5-HT), the most significant one of these, show parallels with those just described for the catecholamines (Fig. 3). Similarly, the initial hydroxylation step is the rate-limiting one in the synthesis of 5-HT. The decarboxylation of 5-HTP to form 5-HT occurs through the agency of the same enzyme as is responsible for formation of dopamine from DOPA, at least in the investigated mammals. The formation of tryptamine itself in the brain through the activity of the same enzyme may be demonstrable at least in certain experimental conditions, but 5-HT is the primary if not normally the only brain tryptamine compound. Normally, its major degradation product in brain as well as peripherally is 5-hydroxyindole-acetic acid (5-HIAA). However, the formation of 5-hydroxytryptophol from 5-HT by liver (*Feldstein* and *Wong*, 1965) and by rat and human brain homogenates *in vitro* (*Eccleston* and coll., 1966) has been demonstrated.

The methylation of these products as well as of N-acetyl serotonin is an event that in mammals is restricted to the pineal organ. The enzyme responsible for this step, hydroxyindole O-methyl transferase or acetylserotonin methyl transferase was discovered and studied in detail by *Axelrod* and *Weissbach* (1960, 1961). Recent studies show that within the mammals this enzyme differs between genera or families both in demonstrable level of activity and in substrate specificity (*Quay* and *Smart*, 1967). As one descends the scale through the submammalian vertebrates, a progressively wider tissue distribution of the methylating capacity is observed. From birds and reptiles down through cyclostomes it frequently occurs in the retina of the lateral eyes as well as in the pineal complex (*Quay*, 1965 a). In amphibians, fish and cyclostomes it occurs in brain, and in cyclostomes it is found in many other tissues as well. Just as adrenaline has been considered the relatively most primitive of the catecholamine neurohumors, the 5-hydroxymethylated derivatives of 5-HT may possibly be relatively primitive in their phylogenetic occurrence as neurohumors of major physiological significance.

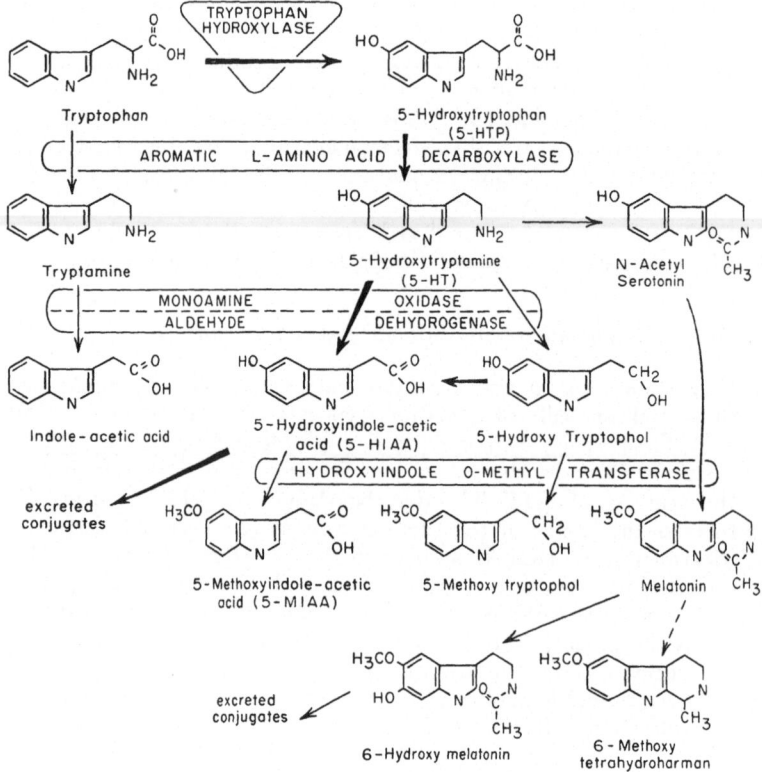

Fig. 3. Primary (heavy lines and arrows) and possible and proposed secondary routes of synthesis and metabolism of neurohumoral tryptamines in mammals.

III. Localizations and Their Physiological Significance

A. Peripheral and Protean Localizations

The distributions of the two categories of neurohumoral compounds, catecholamines and tryptamines may be broadly compared in terms both of occurrences in diverse organisms and of localizations within different kinds of cells and their products. A broad survey of this kind would probably show that on both accounts 5-HT and related indoles have a far greater and more protean distribution in organisms and their cells than do the catecholamine neurohumors. Although 5-HT is found in nerve tissue of various groups of organisms from coelenterates to man, it is also found in a bewildering variety of other cell types and would seem to be associated with multiple cellular activities. Possible molecular common denominators for the seemingly widely disparate actions of 5-HT and some of its derivatives will be suggested in a later section of this review.

The peripheral distribution of catecholamine neurohumors and especially of noradrenaline is largely that of particular links of the visceral nervous system. On this account as well as on the basis of many other lines of evidence, the functional significance of noradrenaline conforms rather well to the pattern expected of a *neuro*-humor and possible neurotransmitter, in contrast with 5-HT and its congenors. Thus, in a later section of this review, consideration of the possible mechanism and control of the neurohumoral action of catecholamines will focus on the events in presumptive noradrenergic nerve terminals.

B. Localizations and Quantitative Changes in the Central Nervous System

The history of the methodological basis for our knowledge of central localizations of neurohumoral amines has within its brief span witnessed many exciting advances and a number of observations of great physiological interest. The recent advance that undoubtedly has contributed the largest amount of information and that is still in its growth phase is the technique perfected by *Hillarp* and *Falck* (*Falck*, 1962) for the histochemical demonstration of catecholamines and 5-hydroxytryptamine by means of their fluorescent condensation products with formaldehyde. This method has been reviewed recently by *Corrodi* and *Jonsson* (1967). DOPA, dopamine, noradrenaline and adrenaline under appropriate conditions give rise to green or yellowish-green fluorescing products by this method, and both 5-HT and its precursor 5-HTP provide purely yellow fluorescing products. To illustrate and extend some of the general results and conclusions from the use of this method on brain, especially by the Swedish workers, I will use some reptilian materials from work that I performed recently at the Netherlands Central Institute for Brain Research in association with Professor *J. Ariëns Kappers* and Dr. *J. F. Jongkind*.

The use of the fluorescence method quickly provided visual proof that catecholamines and 5-HT in the mammalian central nervous system are accumulated in high concentrations in synaptic terminals belonging to special systems of neurons and that their changes after treatment with appropriate drugs also supported the view that they represented special systems of monoaminergic neurons (*Carlsson* and coll., 1962; *Carlsson* and coll., 1964; *Dahlström* and *Fuxe*, 1964 a; *Fuxe*, 1965 a). Experiments with transections or constrictions of the axons of the monoaminergic neurons provided evidence for the idea that the amine granules are formed in the neuron perikarya and are transported down their axons (*Dahlström* and *Fuxe*, 1964 b—d). Most of the cell bodies of the monoamine neurons are found in the lower brainstem, the 5-HT neurons almost entirely in the raphe nuclei and the catecholamine neurons more ventrally and laterally disposed. This same general pattern is observed in at least two groups of

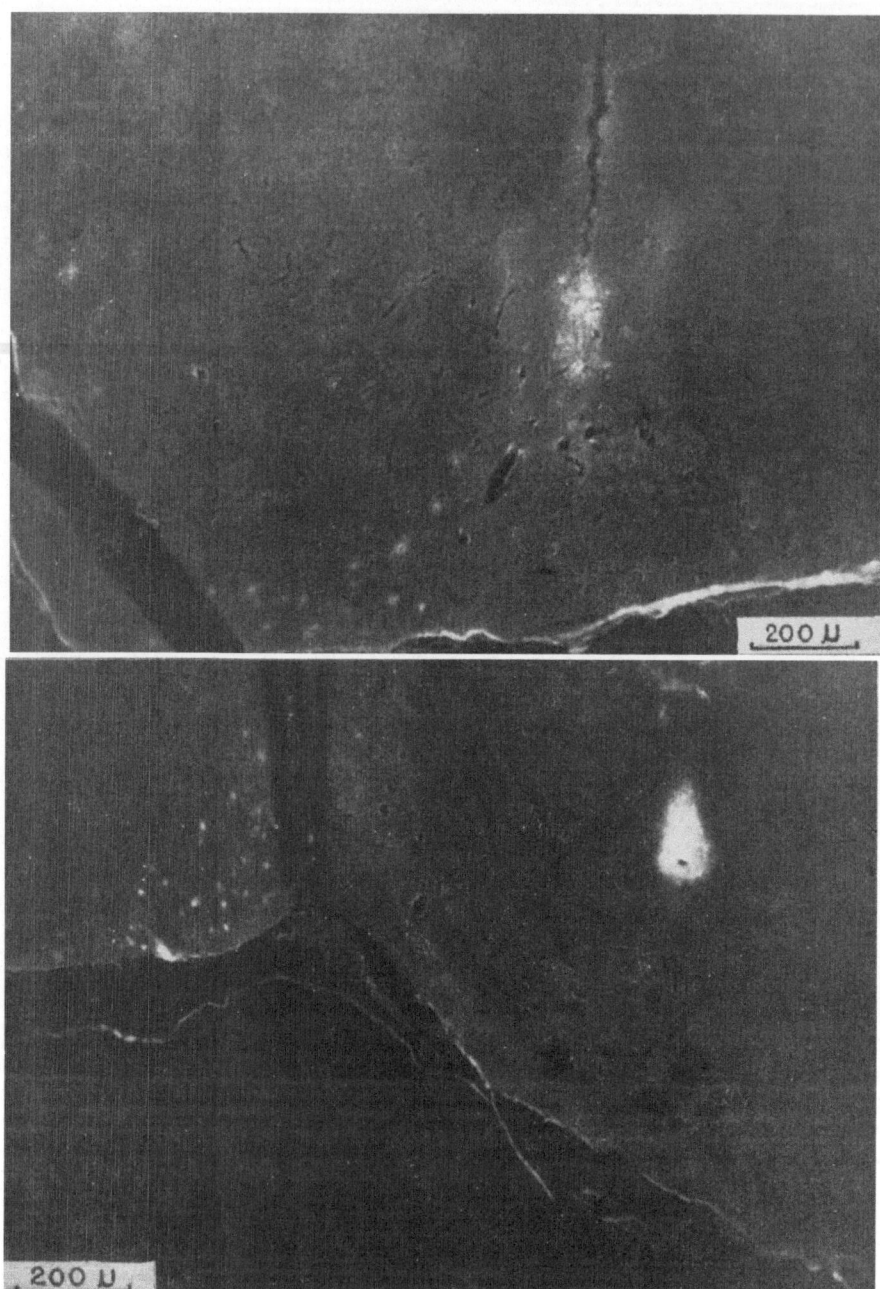

Fig. 4

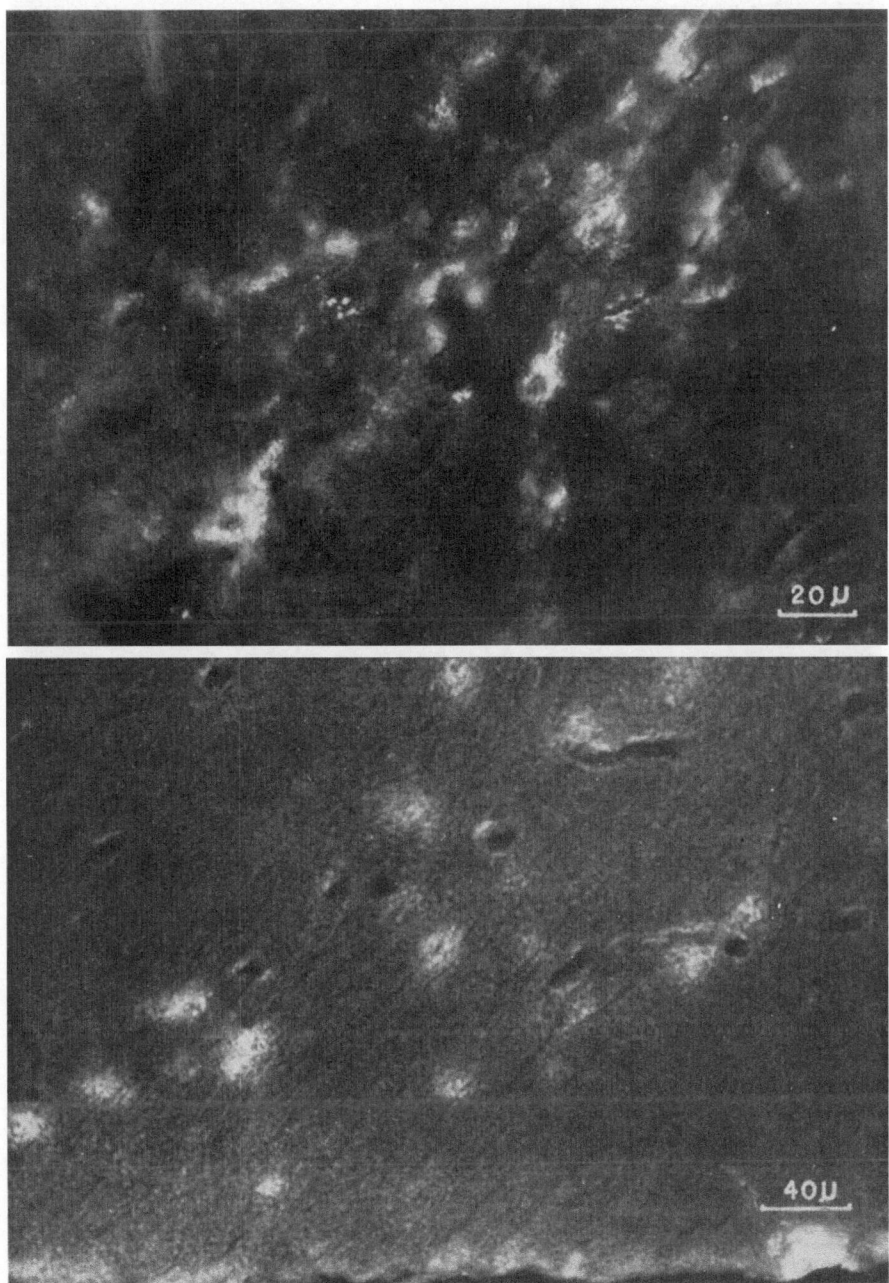

Fig. 5

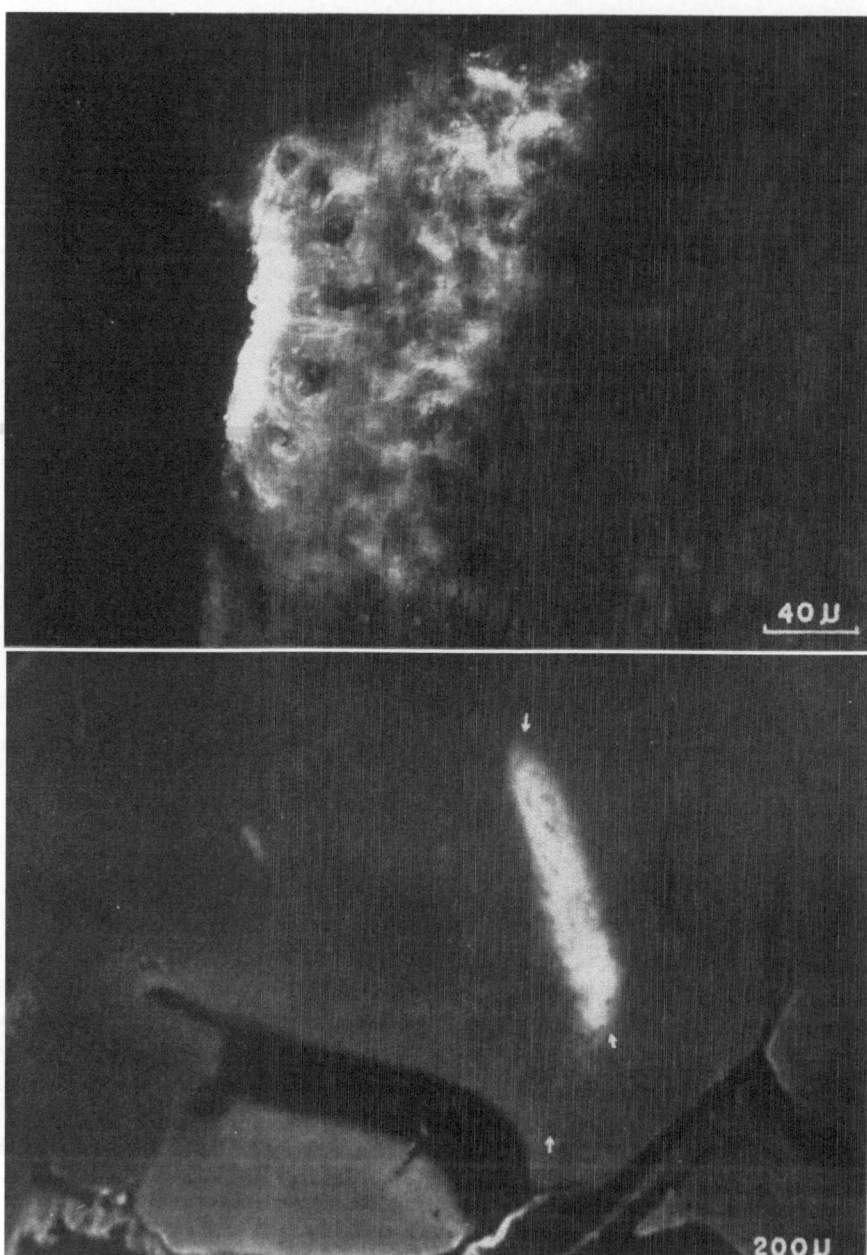

Fig. 9. Top. Paraventricular organ of a control lizard *(Lacerta viridis)* seen in a cross-section of the hypothalamus with the third ventricle at the left and the fluorescent cell processes extending into the neuropil at the right. — Bottom. Parasagittal section of the hypothalamus of a snake *(Natrix natrix)* previously injected intraperitoneally with iproniazid and DOPA and showing the tangentially sectioned and fluorescent paraventricular organ (arrows).

reptiles (Fig. 4, 5 and 6), although in my experience it has so far not usually been possible to sharply distinguish green and yellow fluorescing types on the basis of their color.

The monoamine-containing fibers in the mammalian spinal cord, as usually studied in the rat, include both NA- and 5-HT-containing types and are especially concentrated in the sympathetic lateral column. The NA nerve terminals in the spinal cord have been noted particularly in both the ventral horn and the substantia gelatinosa. Spinal cord 5-HT terminals have been observed in the ventral horn, especially at lumbosacral levels, and to synapse apparently on both cell bodies and processes of the α-motoneurons. For further information the reader is referred to the original work of *Fuxe* (1965 b).

In the mammalian brainstem, 5-HT-containing nerve terminals are seen particularly in the nucleus suprachiasmaticus and the nucleus geniculati lateralis, where at least some of them probably participate in axodendritic contacts. NA-containing nerve terminals in the brainstem are most apparent as dense plexuses in the vegetative hypothalamus (*Sano* and coll., 1967), certain parts of the limbic system and many of the visceral efferent and afferent nuclei of the cranial nerves. A plexus of catecholamine terminals occurs in the pars intermedia of the hypophysis and stems from axons descending the infundibular stalk (*Fuxe*, 1964). These terminals may possibly inhibit the release of MSH (melanocyte-stimulating hormone) (*Iturriza*, 1966). The hypophyseal anterior lobe on the other hand does not contain monoamine nerve terminals, but certain of its PAS-positive parenchymal cells have the capacity to store primary catecholamines and also to form them from L-DOPA (*Dahlström* and *Fuxe*, 1965).

Fig. 4. Top. Cross-section of ventral part of the mesencephalon of a snake *(Natrix natrix)* injected intraperitoneally with iproniazid and DOPA, and showing some of the yellow-fluorescing raphe neurons (right of center of field) as well as more ventral and lateral yellow-green fluorescent neurons. Bottom. Parasagittal section of snake *(Natrix natrix)* tegmentum (left) and hypothalamus (right) after intraperitoneal injection of iproniazid and DOPA. Fluorescent ventral tegmental neurons are seen at left of emerging cranial nerve fibers. In the hypothalamus a part of the paraventricular organ is visible.

Fig. 5. Top. Yellow-fluorescent neurons of anterior-most part of mesencephalic raphe complex (parasagittal section) in an untreated lizard (*Lacerta muralis*, ♂, sexually inactive, Nov. 23). The intensely fluorescent cytoplasmic granules in the perikarya of two neurons in particular can be seen. — Bottom. Fluorescent neurons and blood vessel borders in ventro-lateral mesencephalic tegmentum of a snake *(Natrix natrix)*, injected intraperitoneally with iproniazid and DOPA.

Fig. 6. Top. Transverse section through a portion of the third ventricle and hypothalamus of an untreated snake, showing fluorescent nerves and fibers in the vicinity of the paraventricular nucleus. — Bottom. Ventral tegmentum (upper left) and neurohypophysis (lower right) of a lizard (*Lacerta viridis*) injected intraperitoneally with iproniazid and 5-HTP. The fluorescence of tegmental neurons is increased and the neurohypophysis becomes intensely yellowish-green fluorescent with yellow-fluorescent luminal material and circumferential membrane.

The mammalian brainstem's dopamine (DA)-containing neurons are localized most notably in the substantia nigra, chiefly in the pars compacta. Their axons ascend in the crus cerebri and internal capsule and are completely uncrossed. They terminate primarily in the caudate nucleus and the putamen, forming thus a nigro-neostriatal DA tract, which is responsible for about 70% of the DA content of the rat brain (*Andén* and coll., 1964, 1965). Unilateral electrolytic lesions of the tract in the rat leads within 10 days to a 70 to 80% reduction of the ipsilateral forebrain's DA as compared with the intact controlateral side. Both the localizations and the behavioral effects of lesions and drug treatments with this DA fiber system support the view that it is chiefly concerned in central regulation of somatic motor systems (*Andén*, 1966). However, DA-nerve terminals have been claimed to occur also in the tuberculum olfactorium, the median eminence and probably the nucleus accumbens (*Fuxe*, 1965 a, b).

Catecholamine- and 5-HT-containing nerve fibers, according to *Dahlström*, *Fuxe* and coll. using the fluorescence histochemical technique, ascend from the lower brainstem not only to the hypothalamus and the preoptic area, but also to the limbic forebrain and even the neocortex (*Andén* and coll., 1965). These systems are thought to be mostly uncrossed. A forebrain release and decrease of 5-HT and increase in 5-HIAA after electrical stimulation of the midbrain raphe system has been claimed by *Aghajanian* and coll. (1967). Other investigators have studied the effects of lesions in the medial forebrain bundle and in various tegmental regions on neurohumoral amine levels in diverse brain regions (*Heller* and *Moore*, 1965; *Moore* and coll., 1965). The complexity of their results is not yet readily accommodated to the view of continuity of monoaminergic fiber systems from lower brainstem to cerebral cortex. Rather, transynaptic and more complex cause and effect relations seem probable. These ascending neurohumoral neuron systems, whether direct or indirect, are of very great physiological interest and appear to have regulatory influences on cortical neurons. The most striking and completely studied of these is that having to do with the mechanism of sleep, recently most beautifully illucidated by *Jouvet* (1967, 1968). The midbrain raphe system through its projection of 5-HT on higher centers seems thus responsible for slow-wave sleep and the region of the locus coeruleus through its neurohumor, NA, is thought to be responsible for so-called paradoxical sleep. Recently too, experiments on the neurohumoral amine fibers and contents in the hypothalamus favor the likelihood of physiological control mechanisms of certain hypothalamo-hypophyseal functions including a catecholaminergic link or release (*Donoso* and coll., 1967).

The chief centers of neurohumoral amine concentration are certain brainstem neurons (note Fig. 4, 5 and 6), and some other neuroepithelial cells of perhaps less certain nature. These diverse centers of neurohumor content might be viewed as forming a relatively primitive and axial or periventricular system. In addition to the above described presumptively monoaminergic neuronal systems there are as well the monoamine-containing systems of the subfornical organ (*Lichtensteiger*, 1967), the paraventricular organ (organum vasculosum hypothalami of some authors) (*Fuxe* and *Ljunggren,* 1965), (Fig. 4 bottom, 7 and 9), the pineal organ (epiphysis cerebri) (Fig. 10), the area postrema (*Fuxe* and *Owman,* 1964) and probably others. In pointing out some of the most interesting features of these, I will limit myself to brief remarks concerning those with which I am most familiar.

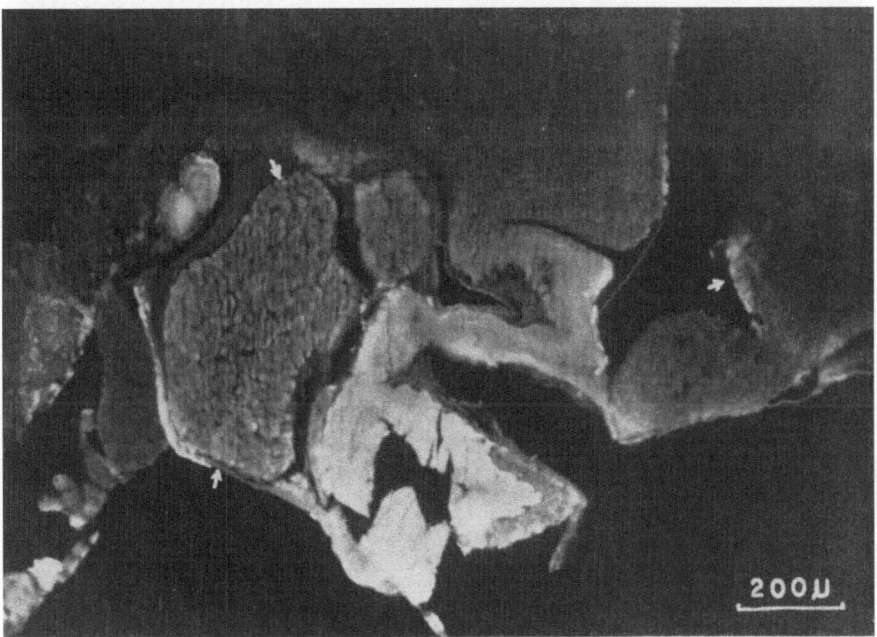

Fig. 7. Oblique-longitudinal section of hypothalamus and hypophysis of a lizard *(Lacerta viridis)* injected intraperitoneally with iproniazid and DOPA. Ventral end of paraventricular organ is shown at right (arrow) and intensely greenish-fluorescent neurohypophysis is shown at lower center. The adenohypophysis (left, arrows) is only weakly fluorescent.

Small fluorescent neurons of the lower vertebrate's paraventricular organ include both yellow and green fluorescent types. But after intraventricular injection of certain but not all catecholamines I have found that in the lacertilian reptiles studied the paraventricular organ's cells are all transformed to green fluorescence. Some other investigators as well have suggested a selective resorptive activity of these cells for certain contents of the cerebrospinal fluid as it circulates by ciliary propulsion ventrally and posteriorly in the anterior and then hypothalamic parts of the third ventricle, and thence dorsally and posteriorly in that part of the third ventricle channelized along the opposing surfaces of the paraventricular organ, especially in some fishes, reptiles and birds. The distal cellular processes of these monoamine-containing cells extend and ramify far laterally with respect to the third ventricle and the most vascular part of the organ. It seems likely that their resorption and transport of particular amines from the cerebrospinal fluid may have some physiological significance as far as the hypothalamus is concerned and particularly have effects on those monoamine-sensitive portions of the hypothalamo-hypophyseal system.

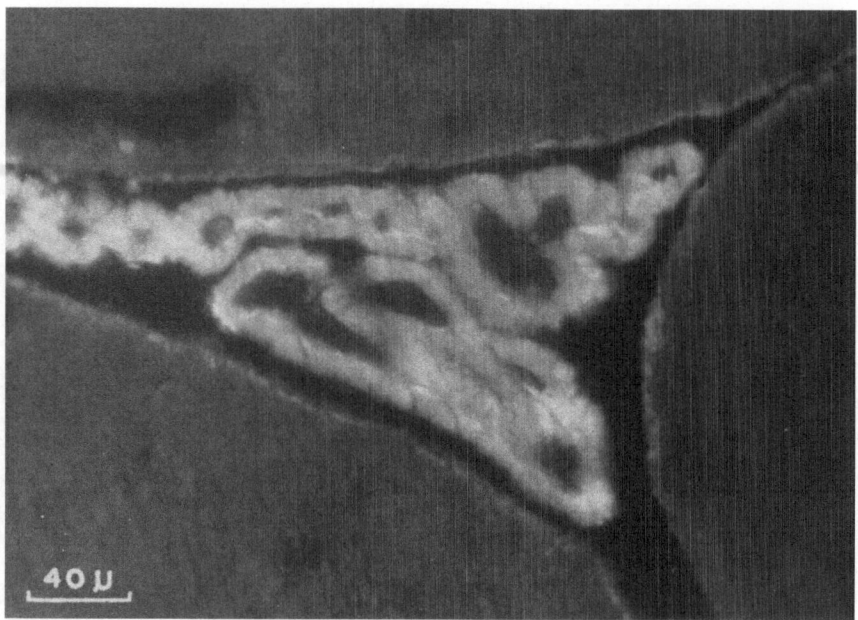

Fig. 8. Intensely yellow-fluorescent ependyma and choroid plexus of the lateral ventricle of a lizard injected intraperitoneally daily for the first two days with iproniazid (2 mg/10 g) and 5-HTP (0.2 mg/10 g) and then two days with 5-HTP alone (0.2 mg/10 g body wgt).

The parenchyma of the pineal organ of vertebrates can be considered a derivative of the brainstem's neuroectoderm, as noted by *Kappers* in this symposium. As such, it shows an especially high concentration of indolealkylamines, especially of 5-HT, within its specialized neuroectodermal and probably endocrine parenchymal cells. The pineal's neurohumoral amine content shows rhythmic changes of greater amplitude than those of brain regions insofar as is known, and these are more closely timed to the onset and offset of environmental illumination (*Quay*, 1965 b). The experimental manipulation and cytochemical localization of these in selected vertebrate types will probably provide us with productive insights regarding the basic intercellular actions of neurohumoral amines. Stimulated by the pineal localization in mammals of acetylserotonin-O-methyl transferase (hydroxyindole-O-methyl transferase) which forms melatonin (Fig. 3), some investigators have urged an endocrine role for melatonin

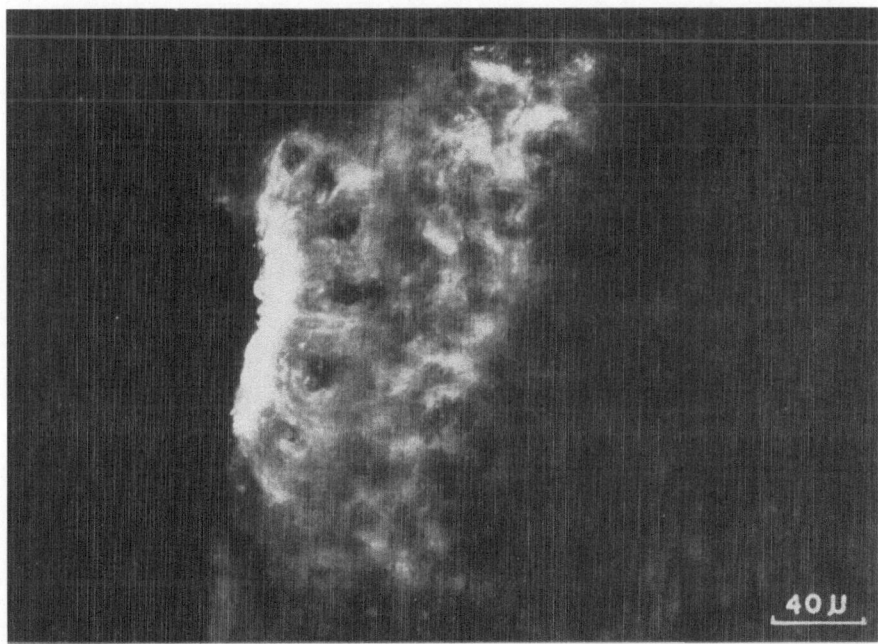

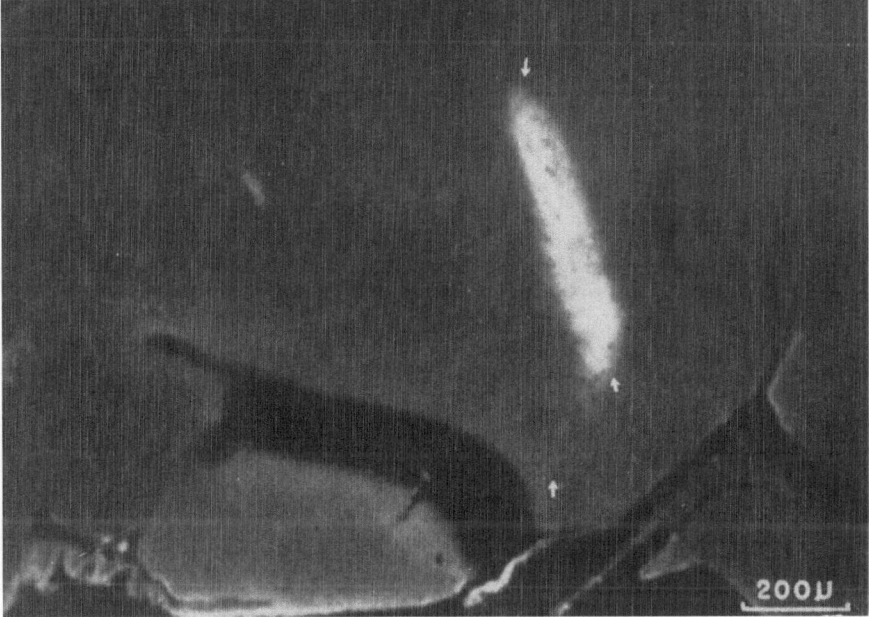

Fig. 9. Top. Paraventricular organ of a control lizard (*Lacerta viridis*) seen in a cross-section of the hypothalamus with the third ventricle at the left and the fluorescent cell processes extending into the neuropil at the right. — Bottom. Parasagittal section of the hypothalamus of a snake (*Natrix natrix*) previously injected intraperitoneally with iproniazid and DOPA and showing the tangentially sectioned and fluorescent paraventricular organ (arrows).

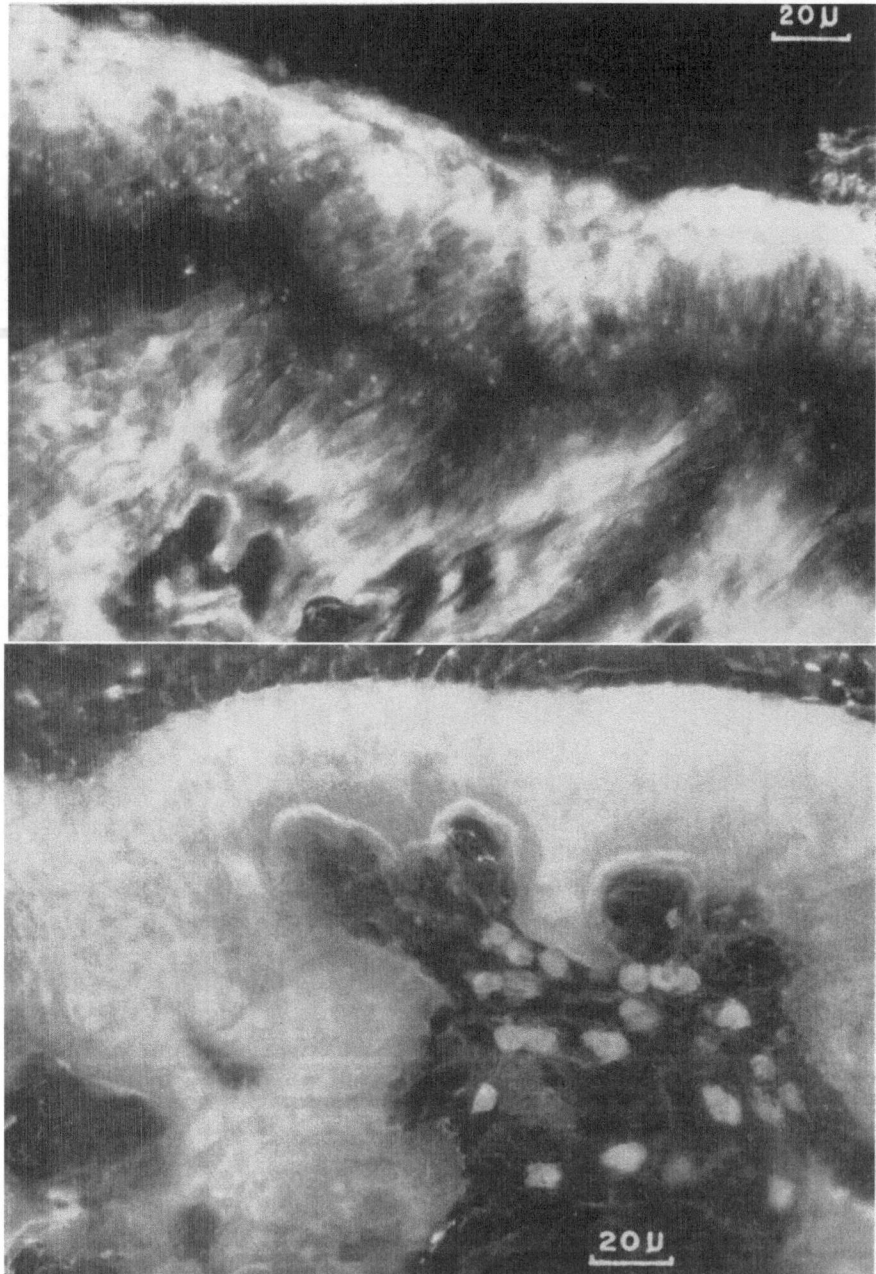

in mammals, and sometimes similarly for related possible indole derivatives. Nevertheless, in no animal has a pineal concentration of melatonin plus its precursors approached the levels of catecholamines found in the adrenal medulla, a proven endocrine secretor of such amines. I consider it more likely that the pineal's neurohumoral amines have to do with intramural or intrapineal intercellular actions (*Quay*, 1965 b). In relation to this hypothesis it may be pertinent to note that, in a number of primarily protein-secreting endocrine organs, special indoleamine or 5-HT-containing cells exist, and these are shown to be physiologically correlated in their activity with the truly endocrine secretory activity of the organ concerned. The 5-HT may possibly have the action in these instances of causing some change in capillary permeability or transport mechanisms for protein hormones (*Pearse*, 1966). The mammalian pineal, at least provisionally, might also be referable to this kind of situation.

The various kinds of cells which normally may contain neurohumoral amines have each been demonstrated to have special capabilities in selective synthesis and uptake of particular neurohumoral amines or their precursors. However, experimentally, uptake or high levels of neurohumoral amines can be induced in other cell types, such as those of the ependyma and choroid plexuses (Fig. 8). Exaggerated levels can be induced in such structures as the neurohypophysis and pineal organ by systemic administration of precursor and a monoamine oxidase inhibitor (Fig. 6 bottom, 7 and 10). There has accumulated in the scientific literature at this time a confusing array of results in regard to localizations of uptake and synthesis of neurohumoral amines by neural tissue. Many of these results must be construed as demonstrations of potentialities *in vitro* or in abnormal circumstances, but possibly not accurately representative of events *in vivo* in the normal or untreated individual.

Fig. 10. Top. Pineal organ of a lizard (*Lacerta muralis*, ♂, Nov. 28) killed during the middle of the daily period of darkness and after one week on a reversed photoperiod schedule. The pineal luminal material (extending left to right through the middle of the figure) is negative. The more superficial of the pineal epithelial cells contain yellow-fluorescent vesicles or ovoid bodies. The basal epithelial cells are intensely yellow-fluorescent throughout but for their nuclei. The outer yellow- and green-fluorescent fibrous layers are not readily distinguished here without the aid of color reproduction. — Bottom. One side of the pineal organ of a lizard (*Lacerta viridis*) injected intraperitoneally with iproniazid and 5-HTP. The lumen (top edge of field) contains yellow-fluorescent material and the pineal epithelium is yellow-fluorescent throughout. The outer yellow fibrous layer is distinct but the normally green layer is faint or absent. Yellow-fluorescent mast cells are seen in the surrounding connective tissue (lower right quadrant)

IV. Neurohumoral Mechanisms and Fates at Terminations

The mechanisms and fates of neurohumoral amines at peripheral nerve terminations, presumably using a particular amine as a transmitter substance, might be provisionally projected for central synapses of a like nature. Likewise, the mechanisms and fates of one such provisional transmitter may serve as a model for studies of others. Inasmuch as the dynamics of noradrenaline's activity and metabolism in nerve terminations is somewhat better understood than that of other neurohumoral amines, we will concentrate our attention on it. In Fig. 11 the pathways are shown believed to be taken by noradrenaline as a neurotransmitter in a termination of a sympathetic postganglionic neuron. It can be seen that the various transport, binding and metabolic steps can be selectively modified or controlled by appropriate pharmacological intervention, enabling stepwise analysis of transmitter mechanisms and fates. Ten sites of action and control can be listed, and are depicted in Fig. 11: 1. Action potentials propagated within the nerve terminal may be blocked by such adrenergic blocking drugs as bretylium and guanethidine. 2. Tyrosine as the precursor of NA is taken up from the circulation by some portions of the adrenergic neuron by some probably carrier-mediated transport mechanism. Structural analogues of tyrosine, such as α-methyl-m-tyrosine, can take the place of tyrosine and thereby be converted by the biosynthetic machinery of the neuron to false adrenergic transmitters (see *Carlsson* in this volume). 3. False transmitters can also be formed from drug analogues of intermediate compounds in the synthesis of NA, as for example α-methyl-DOPA in place of DOPA. 4. Within the nerve terminal's axoplasm free NA is normally destroyed by monoamine oxidase (MAO) within certain of the axoplasm's mitochondria. This enzyme can be inhibited in varying degree and imperfect specificity by a wide selection of so-called MAO-inhibitors, such as pheniprazine, nialamide and iproniazid. 5. Tyrosine hydroxylase, the enzyme responsible for the rate-limiting step in the biosynthesis of NA, can be inhibited by such drugs as α-methyl-p-tyrosine and 3-iodotyrosine. 6. Storage of NA within special particles or vesicles is prevented by reserpine. 7. NA is released from the nerve terminal by nerve impulses. At this stage it can be taken up again by the nerve terminal through a membrane transport mechanism which can be blocked by many drugs, including cocaine, desipramine and various sympathomimetic amines. 8. There are as well drugs that can substitute for NA as substrates for this uptake mechanism, and then within the axoplasm displace NA from the storage granules and act as false neurotransmitters (metaraminol, octopamine, α-methyl-noradrenaline). 9. Catechol-O-methyl transferase (COMT), apparently outside of the terminal and responsible for the most significant step in the extraneuronal metabolism of NA, can be inhibited by such compounds as pyrogallol and the tropolones. 10. NA's interactions with postulated α- and β-adrenergic receptors can be both blocked by appropriate receptor blocking drugs (phenoxybenzamine, phentolamine and pronethalol) and mimicked by directly effective sympathomimetic amines (adrenaline, synephrine, etc.). For more detailed consideration of this subject the reader is referred to *Carlsson* (1966), *Iverson* (1967), *Kopin* (1966), and *Weiner* and *Rutledge* (1966).

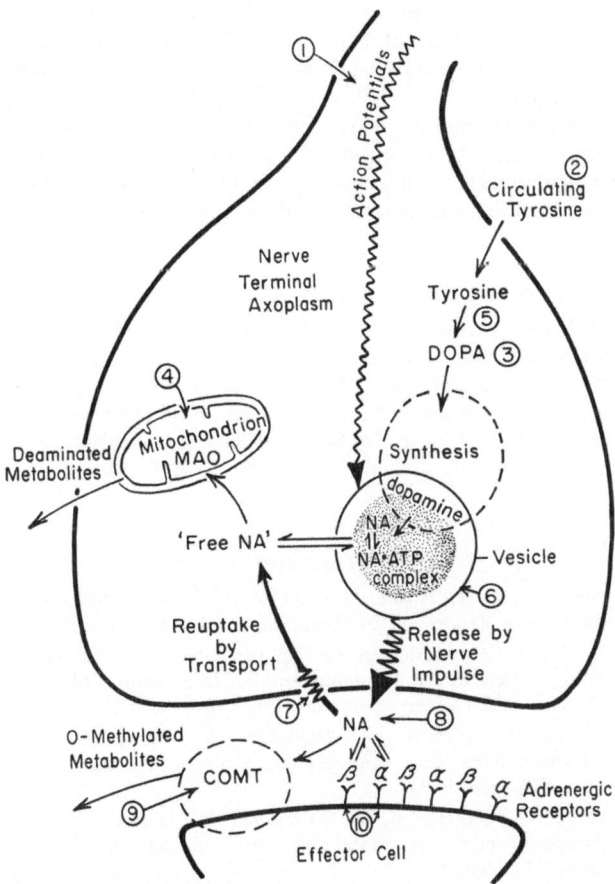

Fig. 11. Diagrammatic representation of a noradrenergic nerve terminal showing possible sites of control mechanisms and drug actions. Further explanation in text. (Modified from *Iverson*, 1967, and others).

V. Molecular Mechanism of Neurohumoral Actions

The complexity in mammals of the membranes and cytoplasmic systems within nervous tissue, and especially at synaptic junctions and nerve terminals, makes it very difficult to determine here the primary molecular mechanisms of neurohumoral actions. Certain structurally and chemically simpler systems offer more hope. Also, neurohumorally regulated and very simple effector systems in lower organisms appear to provide some ideal materials for experimental analysis.

Considering first the possible cytoplasmic sites of action of an almost ubiquitous neurohumor such as 5-HT, one can speculatively note a possible common denominator in these in terms of microtubular and possibly related structures currently tentatively thought to be agents in cytoplasmic movements and some kinds of cellular contractions. Within the many biological systems within which 5-HT has either natural or pharmacological actions, association with microtubules or their possible derivatives in mechanisms of cytoplasmic movements seem possible in the cases of: 1. separation of mitotic centers in dividing cells of certain lower animals; 2. movement of particular regions of cilia in some invertebrate animals; 3. axonal or axoplasmic flow in some neurons; and 4. movement of pigment granules in some of the chromatophores of lower animals. Further discussion and references relating to these possibilities may be found in a recent review of the comparative physiology of 5-HT and melatonin (*Quay*, 1968).

Recent evidence pertaining to more basic molecular mechanisms of the actions of neurohumoral amines increasingly points to interactions with nucleotides or with steps in the metabolism of particular nucleotides. Most of the molecular interactions which have been noted, however, were studied in non-neuronal sites of neurohumor storage and metabolism. A brief review of some of these may be consulted (*Quay*, 1968). See also *Westermann* in this volume.

References

Aghajanian, G. K., J. A. Rosecrans, and *M. H. Sheard:* Serotonin: release in the forebrain by stimulation of midbrain raphe. Science *156,* 402—403 (1967).

Andén, N.-E.: On the function of the nigro-neostriatal dopamine pathway. In: Mechanism of Release of Biogenic Amines. *U. S. von Euler, S. Rosell,* and *B. Uvnäs,* eds., 357—359. Oxford: Pergamon Press, 1966.

Andén, N.-E., A. Carlsson, A. Dahlström, K. Fuxe, N.-A. Hillarp, and *K. Larsson:* Demonstration and mapping out of nigro-neostriatal dopamine neurons. Life Sci. *3,* 523—530 (1964).

Andén, N.-E., A. Dahlström, K. Fuxe, and *K. Larsson:* Further evidence for the presence of nigro-neostriatal dopamine neurons in the rat. Amer. J. Anat. *116,* 329—333 (1965).

Axelrod, J.: Methylation reactions in the formation and metabolism of catecholamines and other biogenic amines. Pharm. Rev. *18,* 95—113 (1966).

Axelrod, J., and *H. Weissbach:* Enzymatic O-methylation of N-acetylserotonin to melatonin. Science *131,* 1312 (1960).

Axelrod, J., and *H. Weissbach:* Purification and properties of hydroxyindole-O-methyl transferase. J. biol. Chem. *236,* 211—213 (1961).

Blaschko, H.: The specific action of l-dopa decarboxylase. J. Physiol., London, *96,* 50 P—51 P (1939).

Carlsson, A.: Physiological and pharmacological release of monoamines in the central nervous system. In: Mechanism of Release of Biogenic Amines. *U. S. von Euler, S. Rosell* and *B. Uvnäs,* eds., 331—345. Oxford: Pergamon Press, 1966.

Carlsson, A., B. Falck, and *N.-Å. Hillarp:* Cellular localization of brain monoamines. Acta physiol. Scand. *56,* Suppl. 196, 1—28 (1962).

Carlsson, A., B. Falck., K. Fuxe, and *N.-Å. Hillarp:* Cellular localization of monoamines in the spinal cord. Acta physiol. Scand. *60,* 112—119 (1964).

Corrodi, H., and *G. Jonsson:* The formaldehyde fluorescence method for the histochemical demonstration of biogenic amines. A review of the methodology. J. Histochem. Cytochem. *15,* 65—78 (1967).

Dahlström, A., and *K. Fuxe:* Localization of monoamines in the lower brain stem. Experientia, Basel, *20,* 398—399 (1964 a).

Dahlström, A., and *K. Fuxe:* Evidence for the existence of monoamine-containing neurons in the central nervous system. 1. Demonstration of monamines in the cell bodies of brain stem neurons. Acta physiol. Scand. *62,* Suppl. 232, 1—55 (1964 b).

Dahlström, A., and *K. Fuxe:* A method for the demonstration of monoamine containing nerve fibers in the central nervous system. Acta physiol. Scand. *60,* 293—295 (1964 c).

Dahlström, A., and *K. Fuxe:* A method for the demonstration of adrenergic nerve fibres in peripheral nerves. Zschr. Zellforsch. *62,* 602—607 (1964 d).

Dahlström, A., and *K. Fuxe:* Monoamines and the pituitary gland. Acta Endocrinol. *51,* 301—314 (1965).

De Robertis, E. D. P.: Histophysiology of Synapses and Neurosecretion. Oxford: Pergamon Press, 1964.

Donoso, A. O., F. J. E. Stefano, A. M. Biscardi, and *J. Cukier:* Effects of castration on hypothalamic catecholamines. Amer. J. Physiol. *212,* 737—739 (1967).

Eccles, J. C.: The Physiology of Synapses. New York: Academic Press; Berlin: Springer-Verlag, 1964.

Eccleston, D., A. T. B. Moir, H. W. Reading, and *I. M. Ritchie:* The formation of 5-hydroxytryptophol in brain in vitro. Brit. J. Pharmacol. Chemotherap. *28,* 367—377 (1966).

Erspamer, V.: Recent research in the field of 5-hydroxytryptamine and related indolealkylamines. Progr. Drug Research *3,* 151—367 (1961).

Erspamer, V. (sub-editor): 5-hydroxytryptamine and related indolealkylamines. Handbuch der experimentellen Pharmakologie *19,* 1—928 (1966).

Falck, B.: Observations on the possibilities of the cellular localization of monoamines by a fluorescence method. Acta physiol. Scand. *56,* Suppl. 197, 1—25 (1962).

Feldstein, A., and *K. K.-K. Wong:* Enzymatic conversion of serotonin to 5-hydroxytryptophol. Life Sci. *4,* 183—191 (1965).

Florey, E.: Neurotransmitters and modulators in the animal kingdom. Fed. Proc. *26,* No. 4, 1164—1178 (1967).

Fuxe, K.: Cellular localization of monoamines in the median eminence and the infundibular stem of some mammals. Zschr. Zellforsch. *61,* 710—724 (1964).

Fuxe, K.: Evidence for the existence of monoamine neurons in the central nervous system: III. The monoamine nerve terminal. Zschr. Zellforsch. *65,* 573—596 (1965 a).

Fuxe, K.: Evidence for the existence of monoamine neurons in the central nervous system. IV. Distribution of monoamine nerve terminals in the central nervous system. Acta physiol. Scand. *64,* Suppl. 247, 37—85 (1965 b).

Fuxe, K., and *L. Ljunggren:* Cellular localization of monoamines in the upper brain stem of the pigeon. J. Comp. Neurol., Philadelphia, *125,* 355—381 (1965).

Fuxe, K., and *C. Owman:* Cellular localization of monoamines in the area postrema of certain mammals. J. Comp. Neurol., Philadelphia, *125,* 337—353 (1965).

Garattini, S., and *L. Valzelli:* Serotonin. Amsterdam: Elsevier Publ. Co., 1965.

Gerschenfeld, H. M.: Chemical transmitters in invertebrate nervous systems. Symposia Soc. Exper. Biol., Cambridge, *20,* 299—323 (1966).

Glowinski, J., and *R. J. Baldessarini:* Metabolism of norepinephrine in the central nervous system. Pharmacol. Rev. *18,* 1201—1238 (1966).

Heller, A., and *R. Y. Moore:* Effect of central nervous system lesions on brain monoamines in the rat. J. Pharmacol. Exp. Therap. *150,* 1—9 (1965).

Holtz, P.: Dopadecarboxylase. Naturwissensch. *27,* 724 (1939).

Hornykiewicz, O.: Dopamine (3-hydroxytyramine) and brain function. Pharmacol. Rev. *18,* 925—964 (1966).

Iturriza, F. C.: Monoamines and control of the pars intermedia of the toad pituitary. Gen. Comp. Endocrinol. *6,* 19—25 (1966).

Iverson, L. L.: The Uptake and Storage of Noradrenaline in Sympathetic Nerves. Cambridge: Cambridge University Press, 1967.

Jouvet, M.: Neurophysiology of the states of sleep. Physiol. Rev. *47,* 117—177 (1967).

Jouvet, M.: Insomnia and decrease of cerebral 5-hydroxytryptamine after destruction of the raphe system in the cat. Adv. Pharmacol. *6 B,* 265—279 (1968).

Kopin, I. J.: Biochemical aspects of storage and release of biogenic amines from sympathetic nerves. In: Mechanism of Release of Biogenic Amines. *U. S. von Euler, S. Rosell* and *B. Uvnäs,* eds., 229—246. Oxford: Pergamon Press, 1966.

Kopin, I. J., and *E. K. Gordon:* Metabolism of administered and drug released norepinephrine-7-H³ in the rat. J. Pharmacol. Exper. Therap., Baltimore, *140,* 207—216 (1963).

Kopin, I. J., J. Axelrod, and *E. K. Gordon:* The metabolic fate of H³-epinephrine and C¹⁴-metanephrine in the rat. J. Biol. Chem., Baltimore, *236,* 2109—2113 (1961).

La Brosse, E. H., J. Axelrod, I. J. Kopin, and *S. S. Kety:* Metabolism of 7-H³-epinephrine-d-bitartrate in normal young men. J. Clin. Invest. *40,* 253 to 260 (1961).

Lichtensteiger, W.: Monoamines in the subfornical organ. Brain Research *4,* 52—59 (1967).

Moore, R. Y., S.-L. R. Wong, and *A. Heller:* Regional effects of hypothalamic lesions on brain serotonin. Arch. Neurol. *13,* 346—354 (1965).

Nagatsu, T., M. Levitt, and *S. Udenfriend:* Tyrosine hydroxylase — the initial step in NE biosynthesis. J. Biol. Chem., Baltimore, *239,* 2910—2917 (1964).

Pearse, A. G. E.: 5-hydroxytryptophan uptake by dog thyroid 'C' cells, and its possible significance in polypeptide hormone production. Nature, London, *211,* 598—600 (1966).

Pscheidt, G. R.: Biochemical correlates with phyletic division of the nervous system. J. Theoret. Biol. *5,* 52—56 (1963).

Quay, W. B.: Retinal and pineal hydroxyindole-O-methyl transferase activity in vertebrates. Life Sci. *4,* 983—991 (1965 a).

Quay, W. B.: Indole derivatives of pineal and related neural and retinal tissues. Pharmacol. Rev. *17,* 321—345 (1965 b).

Quay, W. B.: Comparative physiology of serotonin and melatonin. Adv. Pharmacol. *6 A,* 283—297 (1968).

Quay, W. B., and *L. I. Smart:* Substrate specificity and post-mortem effects in mammalian pineal acetylserotonin methyltransferase activity. Arch. Int. Physiol. *75,* 197—210 (1967).

Sano, Y., G. Odake, and *S. Taketomo:* Fluorescence microscopic and electron microscopic observations on the tuberohypophyseal tract. Neuroendocrinol. *2,* 30—42 (1967).

Udenfriend, S.: Tyrosine hydroxylase. Pharmacol. Rev. *18,* no. 1, part 1, 43—51 (1966).

Weiner, N., and *C. O. Rutledge:* The actions of reserpine on the biosynthesis and storage of catecholamines. In: Mechanism of Release of Biogenic Amines. *U. S. von Euler, S. Rosell,* and *B. Uvnäs,* eds., 307—318. Oxford: Pergamon Press, 1966.

Wurtman, R. J.: Catecholamines. N. England J. Med. *273,* 637—646, 693 to 700, 746—753 (1965).

Author's address: Professor Dr. *W. B. Quay,* Department of Zoology, University of California, Berkeley, Calif. 94720, USA.

Journal of Neuro-Visceral Relations, Suppl. IX, 236—248 (1969)

Substance P and Its Central Effects*

P. Stern

Department of Pharmacology, Medical Faculty, Sarajevo, Yugoslavia

Summary

A survey is given of the present state of research concerning substance P. The chemical properties of this polypeptide are shortly mentioned and, moreover, its most important peripheral activities are dealt with. Furthermore, the experiments pointing to a central activity of SP are discussed. Special significance is ascribed to its central sedative activity and its influence on sensory transmission.

Pharmacological analysis in rats and mice in which, using different methods, a condition of excitation or hypermobility has been produced, demonstrates that, in such conditions, SP shows a tranquillizing effect.

It is stressed that, after purification of SP, testing of the activity of the newly-obtained fractions on isolated ileum only does not suffice. At the same time, a test should be performed concerning the central sedative activity of the fractions. As the best method for this, testing the aggression of isolated male rats is proposed.

In the fields of pharmacology and physiology, the importance which, in recent times, is attributed to polypeptides is becoming ever greater. Let us only mention various polypeptides that are hormones, e. g. bradykinin and angiotensin and also some new synthetic polypeptides. One of the first of these compounds is substance P (SP), which was discovered by *Euler* and *Gaddum* as long ago as in 1931. These authors noticed that the molecules of SP are larger than those of histamine (H) or acetylcholine (ACh). After some time, the attention paid to this polypeptide grew less. The few papers dealing with SP were mainly written by *Euler* and his coll. Later, the findings of *Pernow* (1953), *Lembeck* (1953), and *Zetler* (1956) proved of great importance and aroused much interest in the substance. *Pernow* drew attention to the selective distribution of this polypeptide in the central nervous system (CNS), while *Lembeck* pointed out that a larger amount of SP is present in the dorsal than in the ventral roots of the spinal cord. From this *Lembeck* concluded that SP might play a certain role in the transmission of sensory impulses. Likewise, the results of *Zetler* were rather significant since he showed that SP has several essential effects. The substance is, for instance, antistrychninic and sedative, it prolongs bulbocapnine catalepsy, etc. Besides these central effects,

* Supported by the National Institute of Health, Bethesda, U.S.A., grant MH 11944.

SP has also peripheral effects, especially on the intestines. It is, in fact, known that both, intestine and CNS, contain SP as well as many other transmitters (*Euler*, 1936). The next important step in the study of this polypeptide was the isolation of pure SP by *Franz* and coll. (1961), by *Vogler* and coll. (1962), and by *Zuber* and *Jaques* (1962). In the mean time, symposia on SP were held in Sarajevo (*Stern*, 1961), in New York (*Erdös*, 1963), and in Florence (*Erdös* and coll., 1966).

In the literature, a great many data occur about different characteristics of SP. For surveys of the literature concerning its chemical properties we may refer to *Lembeck* and *Zetler* (1962), and to *Christensen* and *Haley* (1966). Molecular weight of SP is about 1650. It contains 13 aminoacids: lysine, arginine, asparaginic acid, glutamic acid, proline, glycine, alanine, valine, leucine, isoleucine, phenylalanine, treonine, and serine. Quite pure SP is very unstable. In a dilution of 10^{-6} it loses half of its activity within two minutes. This polypeptide is freely soluble in water, methanol, ethanol and acetic acid, but insoluble in ether and chloroform. At a pH of 1—7 it is thermostable, quickly decomposing in an alkaline medium. The substance dialyses through cellophane and collodion membranes.

Unfortunately, there is no specific chemical reaction on SP. Pure substance contains about 120—150.000 U/mg. From the biological tests that on isolated guinea-pig ileum is best. One unit of SP has the activity of 2—4 threshold doses on isolated rabbit jejunum in 30 ml of Tyrode's solution; 20—30 mg of horse small intestine contains on average one unit of SP. SP is usually tested in a 2—5 ml organ-bath containing Tyrode solution (32—37° C) with, as antagonists, ACh, H and serotonin. It is kept in the bath for 30 sec and then washed out. The superfusion method is quite suitable for studying very small amounts of SP (*Gaddum*, 1953). There is also a micromethod for determing 1/100—1/1000 unit of SP described by *Gaddum* (1961) in which goldfish intestine is used.

The methods of SP purification have been described by *Franz* and coll. (1961), *Vogler* and coll. (1962), *Zuber* and *Jaques* (1962), and by *Meinardi* and *Craig* (1966). SP is decomposed by chymotrypsin and slightly less by trypsin and diamino-oxydase. Both, the impure and the purest SP, are decomposed by chymotrypsin, which shows that the substance is really a polypeptide (*Christensen* and *Haley*, 1966; *Lembeck* and *Zetler*, 1962). Likewise, SP is decomposed by the serum of pregnant woman, but not by normal serum (*Stern* and coll., 1961). It is worth mentioning that SP obtained from intestine is identical to that obtained from the CNS. It is possible to differentiate between SP and other polypeptides (*e. g.* bradykinin, etc.) by several methods (*Horton*, 1959; *Christensen* and *Haley*, 1966; *Lembeck* and *Zetler*, 1962).

Attention must be drawn to the way in which the quality of the new fraction obtained by purification, is checked. All authors concerned have done this using isolated guinea-pig ileum. This method, however, often leads to erroneous conclusions, for, similarly, a great number of various other polypeptides cause contraction of guinea-pig ileum. From this rather important fact it appears that the very same contraction may well be caused by impurities of SP owing to the presence of several other poly-peptides. Consequently, if the rate of increase in the contraction of guinea-pig ileum caused by the new fraction is higher than the one caused by the initial fraction, this may not necessarily be due to the process of purification. Purified SP was first studied by *Euler* and *Gaddum* (1931) and, later, also by *Zetler* (1956), *Stern* and *Dobrić* (1957), *Krivoy* and *Kroeger* (1963), and some other authors, who, moreover, obtained certain central effects. *Stern* (1966) found that both, impure and pure SP, act as tranquillizers when administered to aggressive mice. For this reason, the present author is of the opinion that a pure fraction of SP needs to be tested both on isolated guinea-pig ileum and on aggressive mice. In case of obtaining a stronger effect of a new fraction in both tests, the conclu-sion could be drawn that the SP in question is actually purer. This way of establishing the purity of SP is, however, only permissible if the poly-peptide would penetrate into the CNS through the blood-brain barrier. The findings of *Serafimov* and *Stern* (1956) showed that this is really the case. Similarly, the method described by *Krivoy* and *Kroeger* (1963) may be used for the same purpose, as it was shown that both, impure and very pure SP intensify the potentials of action currents of dorsal roots of cat spinal cord. Nevertheless, this method is more complicated than ours and, consequently, less suitable for screening tests.

The effects of SP on intestine are very interesting the pure SP (50.000 U/mg.) in doses of 10^{-9} g./ml. causing a very strong contraction of isolated guinea-pig ileum (*Vogler* and coll., 1962). Converted into molar value, the effect of this SP is 10 times stronger than that of ACh.

Stern and *Huković* (1961) investigated a great number of substances in order to find a possible antagonist of SP. They showed that among several substances only cystin-di-β-naphthylamide and trimethaphan-d-camphorsulfonate were effective. LSD intensifies the effect of SP, both on isolated guinea-pig ileum (*Krivoy*, 1957) and on the dorsal roots of the spinal cord (*Krivoy* and *Kroeger*, 1963).

Euler (1936) held that SP is a hormone stimulating the motility of the digestive tract. The ileum, colon and rectum, especially their muscularis mucosae, contain the greatest quantity of this polypeptide (*Pernow*, 1953; *Douglas* and coll., 1951). SP exerts the strongest effect if it is injected into the lumen of the intestine (*Beleslin* and *Varagić*, 1958). A stronger motility of the intestine also results if SP is administered intravenously (*Gernandt*, 1941/42). In patients with paralytic ileus, the polypeptide causes peristaltic waves only during its infusion (*Liljedahl* and coll., 1958).

It is interesting that those segments of the intestine which show the thickest muscularis mucosae contain the largest amount of SP. Muscularis mucosae manifests the greatest spontaneous activity. Since peristalsis persists even after the administration of atropine, antiserotonic substances and antihistaminics, it is quite possible that peristalsis is caused by SP. The mechanism of its action in these parts is independent of the nervous plexus, for it acts even after the administration of tetramethylamonium and hexamethonium (*Pernow*, 1953). This indicates that SP acts directly on muscular cells. The findings of *Ehrenpreis* and *Pernow* (1953) are also of much interest. These authors, investigating the amount of SP in the wall of the large intestine in patients suffering from Hirschsprung's disease, observed an appreciable concentration of SP in the distal part of the rectosigmoid which, in this disease, does not show spontaneous peristalsis and which contains degenerated ganglion cells. On the other hand, the concentration of SP was far greater in the proximal part which was histologically normal showing also a normal or even higher peristalsis.

These findings as well as the fact that a large amount of SP is concentrated in the submucosa which contains many nerve fibres and cells of *Meissner's* plexus, indicate that there is a correlation between the presence of ganglion cells, the concentration of SP and intestinal peristaltic activity. Irrespective of the problem concerning the origin of SP, *i. e.* whether this polypeptide is either formed within nerve cells or outside of them, it is well possible that the automatism of the digestive tract is due to SP because *Ivančević* and *Stern* (1949) showed that cat intestine is active in spite of the presence of atropine or antihistaminics. However, in rats with a megacolon the content of SP is not diminished (*Potkonjak* and *Stern*, unpublished results).

Unlike serotonin, SP can re-establish a peristaltic reflex blocked by D-tubocurarine (*Radmanović*, 1961). *Beleslin* and *Varagić* (1959) found that SP increases the amount of serotonin in the isolated guinea-pig and rat intestine. On the other hand, *Radmanović* (1964 a) showed that inhibitors of cholinesterase diminish the SP content in the intestine of hare, while this was significantly increased in the small intestine of hare after vagotomy.

Some other peripheral effects of SP are worth mentioning. The polypeptide causes bronchoconstriction and peripheral vasodilatation with a temporary fall of blood pressure (*Christensen* and *Haley*, 1966; *Lembeck* and *Zetler*, 1962). After a intra-arterial injection of SP (1 U/min) the circulation in the hand shows an increase of three times while the saturation of the blood with oxygen in the superficial and deep veins is also increased (*Pernow*, 1963). SP provokes a strong pain when applied on a blister surface caused by cantharidin (*Christensen* and *Haley*, 1966; *Lembeck* and *Zetler*, 1962). Moreover, it highly increases the permeability of capillaries (*Lembeck* and *Starke*, 1963).

With regard to the subject of this symposion, the role of SP in the CNS is of great interest. The CNS of all vertebrates, especially that of amphibians and fish, contains SP (*Christensen* and *Haley*, 1966; *Lembeck* and *Zetler*, 1962). The more developed the brain is, the less is the amount of SP. Interestingly, the difference between the amount of SP in the forebrain and that in the brainstem is smaller in lower than in higher animals. This may be due to the fact that, in the forebrain of fish, amphibians and birds, the part constituted by the basal ganglia is relatively large. In general, phylogenetically older parts of brain contain more SP.

Kopera and *Lazarini* (1953), *Pernow* (1953), *Amin* and coll. (1954), *Zetler* and *Schlosser* (1954), and *Paasonen* and *Vogt* (1956) studied the distribution of SP in the CNS. Grey substance contains much more of it than white substance. The globus pallidus, and the striatum in general, the substantia nigra and the area postrema are rather rich in SP. SP is intracellularly localized. There is no SP in glioma of humans (*Graubner* and *Lembeck*, 1960). Like ACh, the substance is present in synaptic vesicles and probably associated with phosphatides (*Christensen* and *Haley*, 1966; *Lembeck* and *Zetler*, 1962). The retina, the dorsal roots of the spinal cord, the nucleus cuneatus and the nucleus gracilis (sensory) pathways and centers contain also S.P It is worth mentioning that regions which are rich in SP contain no or quite little cholineacetylase (*Zetler* and *Schlosser*, 1954). The peripheral nervous system also contains SP (*Christensen* and *Haley*, 1966).

Among the central effects of SP, the most important is a sedative one. *Zetler* (1956) was first in noticing some central effects of this polypeptide, e. g. antistrychninic, the prevention of picrotoxic convulsions, the inhibition of harmine tremor, the prolongation of hexobarbital narcosis and of bulbo-capnin catalepsy. SP is an antagonist of morphinic analgesia and provokes hyperalgesia. *Stern* and *Huković* (1960) showed that a purer SP containing 100 or 270 U/mg. had no antistrychninic effect. On the contrary, the effect of strychnin was intensified. Moreover, hexobarbital narcosis was not prolonged, but morphinic analgesia was antagonized. These experiments lend support to *Lembeck's* hypothesis (1953), that SP is a transmitter substance in sensory pathways. This was confirmed by experiments of *Krivoy* and *Kroeger* (1963), and of *Lembeck* (1957). The authors first mentioned experimented on dorsal roots of the spinal cord using a very pure fraction, while *Lembeck* showed that SP intensifies sensory impulses of the nervus auricularis in hare. *Holton* (1960), cutting this nerve, found that the amount of SP was diminished in its distal part, but increased in the proximal part of the nerve. *Serafimov* (1958) obtained the same results.

Stern and coll. (1958) showed that SP has an inhibitory effect on poly-synaptic reflexes, but not on monosynaptic ones. These findings are in good agreement with earlier results of the present author showing that SP is a synergist of mephenesine and meprobamate (*Stern* and *Dobrić*, 1957). However, *Kissel* and *Domino* (1959) could not confirm this. This may be due to the fact that these authors used rather small doses of SP so that their findings cannot be compared with those made by *Stern* and coll.

Physiostigmine and other cholinesterase poisons that penetrate the blood-brain barrier diminish the amount of SP in the CNS (*Radmanović*, 1964 b). Amphetamine and morphine increase the amount of this polypeptide in the brain of mice, while chloroform, urethane, and phenobarbitone diminish its amount. LSD did not show any effect (*Zetler* and *Ohnesorge*, 1957). *Stern* and *Kocić-Mitrović* (1960) observed that the content of this poly-peptide was increased after the administration of reserpine, while *Radmano-vić* (1964 b) could not prove this. However, the former authors performed their experiments with rats while the latter used rabbits so that the different results may be due to the difference in species.

We could show that the amount of SP in the nervous system varies with sensory impulses. The SP content in the retina is, for instance, increased when the animals are in the dark, but a decrease is observed in daylight (*Stern* and *Kocić-Mitrović*, 1958); SP prevents audiogenic stress (*Čatović* and *Stern*, 1963); moreover, besides being an antagonist of morphinic analgesia, it also prevents the abstinence syndrome to morphine in rats (*Stern*, 1963). According to *Przić* (1961), SP weakens the effect of morphine in the so-called "writhing effect" provoked by HCl. In connection with this, it is of interest that morphine and its derivates prevent the effect of SP on peristalsis while the reverse is not true (*Medaković* and *Radmanović*, 1959). As has already been mentioned, very pure SP (10.000 U/mg.) intensifies the action potentials in dorsal pathways of the spinal cord after administration of LSD (*Krivoy* and *Kroeger*, 1963).

Besides these central effects of SP in the sensory zone, there is an impressive sedative effect which could even be obtained with the purest SP. Both, impure (*Zetler*, 1956) and pure SP (*Haefely* and *Hürlimann*, 1962) tranquillize spontaneous motility in mice, although the latter authors did not pay much attention to this supposing that SP has no tranquillizing effect.

On the basis of these findings we investigated the sedative effect of SP in different ways. SP was given to aggressive mice in which agressiveness was provoked by isolation without administering any drugs (*Stern*, 1966). Moreover, SP was given to rats in which restlessness was provoked by destroying the septum pellucidum (*Stark* and *Henderson*, 1966), by excessive amounts of alcohol (*Igić* and coll., unpublished results), or by morphinisation (*Stern*, 1963). Finally, spontaneous motility of mice was registered both daily and nightly (*Čatović*, in press). As can be seen from Table 1, SP tranquillized aggressive mice, rats in which restlessness was provoked by morphinisation, and also the strong spontaneous motility normally occurring during the night. SP had no effect on restlessness in septal rats and in those in which restlessness was provoked by alcohol. Meprobamate had also the same effect, while chlorpromazine tranquillized both aggressive mice and rats showing restlessness provoked in several ways. From these results, therefore, the conclusion can be drawn that SP and meprobamate are synergists. This fact had already been earlier shown (*Stern* and *Dobrić*, 1957). On the other hand, the results obtained suggest that one has to be very careful in forming a judgement in studies on the effect of tranquillizers. It is necessary to use many different tests during the investigation of a depressant or antidepressant in order to draw final conclusions.

Furthermore, we have recently shown that SP tranquillizes aggressive and restless mice which were given very large doses of LD-DOPA, these doses provoking aggressiveness and restlessness in mice (*Katz* and coll., 1967). Meprobamate, even in this case, had the same effect. The restlessness reappeared after administration of desmethylimipramine, although the animals were under the influence of SP.

Due to the fact that both, impure and pure SP, have a sedative effect and that this polypeptide penetrates the blood-brain barrier (*Serafimov* and *Stern*, 1956), the conclusion may be drawn that SP acts centrally. Furthermore, the following facts lend support to this conclusion: desmethylimipramine is a centrally acting drug and, as has already been mentioned, the restlessness reappeared after administration of this compound to animals which had been tranquillized with SP.

Table 1

Kind of Excitation	Substance	Effect	Remarks
Fighting mice (isolated)	SP 5000 J/kg. i. p.	depression	Desmethylimipramin brings back the fighting in depressed mice
	Meprobamat 100 mg./kg. i. p.	depression	
Fighting mice (induced by DOPA)	SP 5000 J/kg. i. p.	depression	Desmethylimipramin brings back the fighting in depressed mice
	Meprobamat 50 mg./kg.	depression	
Motility of mice registered during the day or during the night	SP 3250 J/kg. i. p.	depression	Registration during the night
	SP 3250 J/kg. i. p.	without effect	Registration during the day
"Septal" rats	SP 4000 i 8000 J/kg. i. p.	without effect	
	Meprobamat 50 mg./kg. i. p.	without effect	
Rats accustomed to alcohol	SP 4000 i 8000 J/kg. i. p.	without effect	
	Meprobamat 50 mg./kg. i. p.	without effect	
Abstinence syndrome to morphine	SP 4000 J/kg. i. p.	depression	Rat
	Meprobamat 100—150 mg./kg.	depression	Mice (*Maggiolo* and *Huidobro*, 1962)

Mechanism of action of SP. There are a great many studies about the effects of different drugs on the content of SP in the CNS. *Lembeck* and *Zetler* (1962) mention the effects of more than 60 compounds on the action of SP on guinea-pig intestine. However, no one had any specific inhibitory effect. The results were so different that it was impossible to make any conclusion on the action of SP. At this moment, only the compounds mentioned by *Stern* and *Huković* (1961) are known as specific SP antagonists.

Stern and *Huković* (1960) showed that relatively pure SP has not the identical effect as impure SP, the antistrychninic effect and the prolonga-the findings of *Stern* and coll. will be mentioned briefly. These authors tion of hexobarbital narcosis being, for instance, different. For these reasons studied the effect of SP on the content of biogenic amines in rat brain. Two fractions were used, the impure (6.5 U/mg.) and the rather pure one (330 U/mg.). Both fractions significantly increased the cerebral contents of serotonin and dopamine 180 min. after their administration. On the other hand, SP did not show any effect on the cerebral content of ACh, H and noradrenaline. It is worth mentioning that *Matussek* (1965) found that passiveness of animals is induced by an increase of the content of serotonin in the brain. This fact might be a cause of the sedative effect of SP as this polypeptide increases the content of serotonin in the CNS (*Stern* and coll., 1966). On the other hand, SP did not tranquillize the septal rat although *Heller* and coll. (1964) observed that the serotonin content is diminished in the brain of septal rats and it was shown by *Stern* and coll. (1966) that SP increases the content of serotonin in the brain as has been mentioned.

The increase of dopamine in the CNS after the administration of SP is rather interesting because dopamine prevents rigor and akinesia in man (*Hirschmann* and *Mayer,* 1964) which leads to psychic appeasement. We could not explain this important finding in the light of the role of SP as a transmitter substance in sensory pathways. LSD, which intensifies the effect of SP (*Krivoy,* 1957), increases the amount of serotonin in the CNS (*Freedman,* 1961). In this connection, the results of *Pepeu* (1966) and of *Radmanović* (1964 b) should be mentioned. The former author found that SP releases ACh in cat brain while the latter showed that inhibitors of cholinesterase diminish the SP content in rabbit brain, which is also in agreement with his finding concerning the effect of physostigmine on the amount of SP in rabbit intestine. *Ramwell* and *Shaw* (1963) observed that direct excitation of the brain cortex in anaesthetized cats releases a substance in the perfusing solution which is chemically similar to SP. *Baldauf* and coll. (1967) showed that there is a fraction of SP in the phylogenetically youngest regions of the brain that releases ACh.

Since the amounts of serotonin and dopamine in the brain are increased after the administration of SP (*Stern* and coll., 1966), the question arises whether there is any relationship between SP and resting tremor. It is known that the amounts of both of these biogenic amines are decreased in the brain of patients suffering from Parkinson's disease (*Hornykiewicz,* 1962). However, *Stern* (1963) showed that SP did not prevent the occurrence of resting tremor. *Birkmayer* and *Hornykiewicz* (1964) found that the SP content was not diminished in the nucleus caudatus in patients who died of parkinsonism. It had already been shown that tremorine increases the serotonin content (*Friedman* and coll., 1963), but that it has no effect on the dopamine content of the brain in experimental animals (*Everett*, 1964). However, *Efron* (1965) found a diminished amount of dopamine in the brain of rats having pseudo-parkinsonism provoked by chlorpromazine. The question now arises whether a relationship exists between the amount of dopamine and rigor. Dopamine content is always diminished in the brain of patients who died of Parkinson's disease, even in those who did not show any tremor but other symptoms of parkinsonism exclusively (*Hornykiewicz*, private communication).

On purpose we stressed the central effects of SP because these may be rather important for the physiology and the therapy of CNS diseases. There is, however, great need of further investigation in this field. Nevertheless, there are many proofs that SP plays an important role as a transmitter substance in sensory pathways and that this substance exerts some sedative central effects.

Future studies using the purest SP should be directed firstly on the problem whether this polypeptide indeed penetrates the blood-brain barrier and, if so, in which regions of the CNS. Secondly, the question is of importance whether the distribution in the CNS of exogenous SP is identical with that of the endogenous one. Thirdly, it should be investigated whether purified SP, injected into various regions of the brain, exerts identical effects as does purified SP when administered parenterally.

Addendum. After submitting the manuscript for publication, the paper by *Baile* and *Meinardi* (Brit. J. Pharmac. Chemother. *30*, 302, 1967) came to the attention of the author. The observation of a central sedative activity of highly purified SP in goat brain, made by *Baile* and *Meinardi*, supports the opinion of the present author, advanced during several years, that SP indeed has a central sedative effect.

References

Amin, H., Crawford, and *H. Gaddum:* The distribution of Substance P and 5-hydroxytryptamine in the central nervous system of the dog. J. Physiol., London, *126*, 596—618 (1954).

Baldauf, J., P. Harnacke, and *G. Zetler:* Darmkontrahierende Peptide in Substanz P-haltigen Extrakten aus Cortex. Globus pallidus und Substantia nigra des menschlichen Gehirns. Arch. Pharm. exp. Path. *257*, 263—264 (1967).

Beleslin, D., and *V. Varagić:* The effect of Substance P on the peristaltic reflex of the isolated guinea-pig ileum. Brit. J. Pharmac. *13*, 321—325 (1958).

Beleslin, D., and *V. Varagić:* The effect of Substance P on the amount of 5-Hydroxytryptamine in the gut. Experientia, Basel, *15*, 186 (1959).

Birkmayer, W., and *O. Hornykiewicz:* Weitere experimentelle Untersuchung über L-DOPA beim Parkinson-Syndrom und Reserpin-Parkinsonismus. Arch. Psychiatrie u. Z. ges. Neurol. *206*, 367—381 (1964).

Catović, S., and *P. Stern:* Odnos tremora i audiogenog stresa. Acta pharmaceutica Iugoslavica *13*, 51—56 (1963).

Catović, S., and *P. Stern:* Djelovanje Supstance P na ritam pokretljivosti miseva. Neurofiziologija (in press).

Christensen, H. D., and *T. J. Haley:* Distribution and biological effects of substance P. J. Pharmac. Sci. *55*, 757—757 (1966).

Douglas, W. W., W. Feldberg, E. M. Paton, and *M. Schachter:* Distribution of histamine and Substance P in the wall of the dog's digestive tract. J. Physiol., London, *115,* 163—176 (1951).

Efron, D.: Quoted from „Discussion on extrapyramidal side effects" Psychopharmacology Semvice Center Bulletin *3,* 19—23 (1965).

Ehrenpreis, T., and *B. Pernow:* On the occurrence of Substance P in the rectosigmoid in Hirschsprung's disease. Acta physiol. Scand. *27,* 380 (1953).

Erdös, E. G.: In: Structure and function of biologically active peptides: bradykinin, kallidin and congeners. B. B. Brodie, ed. (1963).

Erdös, E. G.: N. Back, and *F. Sicuteri:* Hypotensive peptides. New York Springer Verlag, 1966.

Euler, U. S. von: Untersuchung über Substanz P, die atropinfeste darmerregende und gefäßerweiternde Substanz aus Darm und Hirn. Arch. exper. Path. Pharmak., Leipzig, *181,* 181—197 (1936).

Euler, U. S. von, and *J. H. Gaddum:* An unidentified depressor substance in certain tissue extracts. J. Physiol., London 72, 74—87 (1931).

Everett, G. M.: Pharmacological studies on tremorine. In: Biochemical and neurophysiological correlation of centrally acting drugs. *Trabucchi, E., R. Paoletti,* and *N. Canla,* eds. 69—74, Pergamon Press, London, 1964.

Franz, J. R. A. Boissonnas, and *E. Stürmer:* Isolierung von Substanz P aus Pferdedarm und ihre biologische und chemische Abgrenzung gegenüber Bradykinin. Helv. chim. Acta *44,* 881—883 (1961).

Freedman, D. X.: Effects of LSD-25 on brain serotonin. J. Pharm. *134,* 160—166 (1961).

Friedman, A., R. Aylesworth, and *G. Friedman:* Tremorine, its effect on amines in the central nervous system. Sciences *14,* 1188—1189 (1963).

Gaddum, J. H.: The technique of superfusion. Brit. J. Pharmacol. *8,* 321—339 (1953).

Gaddum, J. H.: The estimation of Substance P in tissue extracts. In: Int. Symposium on Substance P. P. Stern ed., Sarajevo, 7—14 (1961).

Gernandt, B.: Untersuchung über die biologische Wirkung der Substanz P. Acta physiol. Scand. *3,* 270—274 (1941/42).

Graubner, K., and *F. Lembeck:* Untersuchungen über den Gehalt cerebraler Tumoren an Substanz P und Serotonin. Arch. exper. Path. Pharmak., Leipzig, *240,* 157—161 (1960).

Haefely, W., and *A. Hürlimann:* Substance P, a highly active naturally occurring polypeptide. Experientia *18,* 297—303 (1962).

Heller, S., J. A. Harvey, and *R. Y. Moore:* The effect of central nervous system lesions in the rat on brain serotonin. In: Biogenic Amines. *Himwich, M. E.,* and *W. A. Himwich,* eds. Progress in Brain Research, Amsterdam, London, New York Elsevier Publ. Comp., *8,* 53—60 (1964).

Hirschmann, J., and *E. Mayer:* Zur Beeinflussung der Akinese und anderer extrapyramidal-motorischer Störungen mit L-Dopa (L-Di-hydrophenylalanin). Deut. Med. Wschr. 89, 1877—1880 (1964).

Holton, P.: Substance P concentration in degenerating nerve. In: Polypeptides which affect smooth muscles and blood vessels. Pergamon Press, Oxford, London, New York, Paris, 192—196 (1960).

Hornykiewicz, O.: Dopamin (3-Hydroxytyramin) im Zentralnervensystem und seine Beziehung zum Parkinson-Syndrom des Menschen. Dtsch. med. Wschr. *87,* 1807—1810 (1962).

Horton, E. W.: Human urinary kinin excretion. Brit. J. Pharmacol. *14,* 125—133 (1959).

Ivančević, I., and *P. Stern:* Über Einwände zur Histamintheorie der Allergie. Arch. int. Pharmacodyn. *78,* 225—228 (1949).

Katz, M. R., H. C. Yen-Koo, and *S. Krop:* The effects of psychoactive drugs on 3,4-Dihydroxyphenylalanine (DL-DOPA) induced excitation in mice. Federation Proceed. *26,* 289 (1967).

Kissel, J. W., and *E. Domino:* The effect of some possible neurohumoral agents on spinal cord reflexes. J. Pharmacol. Exper. Therap., Baltimore, *125,* 168—177 (1959).

Kopera, H., and *W. Lazarini:* Zur Frage der zentralen Übertragung afferenter Impulse. Arch. exper. Path. Pharmak., Leipzig, *219,* 214—222 (1953).

Krivoy, W. A.: The preservation of Substance P by lysergic acid diethylamide. Brit. J. Pharmacol. *12,* 361—364 (1957).

Krivoy, W. A., and *D. Kroeger:* The neurogenic activity of high potency Substance P. Experientia *19,* 366—368 (1963).

Lembeck, F.: Zur Frage der zentralen Übertragung afferenter Impulse. Arch. exper. Path. Pharmak., Leipzig, *219,* 197—213 (1953).

Lembeck, F.: Untersuchung über die Auslösung afferenter Impulse. Arch. exper. Path. Pharmak., Leipzig, *230,* 1—9 (1957).

Lembeck, F., and *K. Starke:* Substance P content and effect of capillary permeability of extracts of various parts of human brain. Nature *199,* 1295—1296 (1963).

Lembeck, F., and *G. Zetler:* Substance P: a polypeptide of possible physiological significance, especially within the nervous system. In: Int. Rev. of Neurobiology *4,* 159—215 (1962).

Liljedahl, S., O. Matson, and *B. Pernow:* The effect of Substance P on intestinal motility in man. Scand. J. Clin. Laborat. Invest. *10,* 16—25 (1958).

Maggiolo, C., and *F. Huidobro:* The influence of some drugs on the abstinence syndroms to morphine in mice. Arch. int. Pharmacody. *138,* 157—168 (1962).

Matussek, N.: Wirkung von 5-Hydroxytryptophan auf zentrale adrenerge Erregungen. Naturwissensch. *52,* 85—86 (1965).

Medaković, M., and *R. Radmanović:* The antagonism of the morphine-like analgetics and Substance P on the peristaltic reflex of the isolated guinea-pig ileum. Arch. int. Pharmacodyn. *122,* 428—433 (1959).

Meinardi, H., and *L. C. Craig:* Studies of Substance P. In: Hypotensive peptides. *Erdös, E., N. Back,* and *F. Sicuteri,* eds. New York: Springer-Verlag, 595—607 (1966).

Paasonen, M. K., and *M. Vogt:* The effect of drugs on the amounts of Substance P and 5-hydroxytryptamine in mammalian brain. J. Physiol., London, *131,* 617—626 (1956).

Pepeu, G.: Discussion. In: Hypotensive peptides. *Erdös, E. G., N. Back,* and *F. Sicuteri,* eds., New York Springer Verlag, 631 (1966).

Pernow, B.: Studies on Substance P. Acta physiol. Scand. *29,* suppl. 105, pp. 1—90 (1953).

Pernow, B.: Pharmacology of Substance P. Ann. New York Acad. Sci. *104,* 393—403 (1963).

Przić, R.: The influence of Substance P on attack against the "Writhing syndrome" in the mouse. In: Int. Symposium of Substance P, Sarajevo, Scientific Society of Bosnia and Herzegowina — Yugoslavia, 71—74 (1961).

Radmanović, B. Z.: Ispitivanje dejstva nekih relaksantnih sredstava na izolovanom ileumu zamorca. Acta med. Jugosl. *14,* 144—152 (1961).

Radmanović, B. Z.: Effect of vagotomy, vagus stimulation and various drugs on the Substance P content in the small intestine of the rabbit. Acta physiol. Scand. *61,* 272—278 (1964 a).

Radmanović, B. Z.: The effect of drugs on the amount of Substance P in rabbits brain. Nature *204,* 1205—1206 (1964 b).

Ramwell, P. W., and *J. E. Shaw:* The nature of non-cholinergic substances released from the cerebral cortex of cats on direct and indirect stimulation. J. Physiol. *169,* 51P (1963).

Serafimov, N.: Uticaj novocainske blokade na sadrzaj Supstance P u zadnjim korjenima kicmene mozdine. In: The influence of the novocain blockade on the content of the Substance P in dorsal roots of the spinal cord. 3rd Conference of the Yugoslavian Physiologists, Skoplje (1958).

Serafimov, N., and *P. Stern:* Influence des médiateurs chimiques sur la teneur en „Substance P" du cerveau du rat. Arch. intern. de Physiol. et de Biochem. *66,* 653—657 (1956).

Stark, P., and *J. K. Henderson:* Differentiation of classes of neurosedatives using rats with septal lesions. Int. J. Neuropharmacol. *5,* 385—389 (1966).

Stern, P.: Intern. Symposium on Substance P. Sarajevo, Yugoslavia. Sc. Soc. of the Republic of Bosnia and Herzegowina, 1961.

Stern, P.: Substance P as a sensory transmitter and its other central effects. Ann. New York Acad. Sci. *104,* 403—414 (1963).

Stern, P.: Contribution to the sedative action of Substance P. In: Hypotensive peptides. *Erdös, E. G., N. Back,* and *F. Sicuteri,* eds., New York Springer Verlag, 633—640 (1966).

Stern, P., and *V. Dobrić:* Über die Wirkung der Substanz P im Zentralnervensystem. In: Psychotrophic Drugs, Amsterdam, London, New York, Princeton Elsevier Publ. Comp., 448—452 (1957).

Stern, P., and *S. Huković:* Beziehungen zwischen zentraler und peripherer Wirkung der Substanz P. Medicina Experimentalis *2,* 1—7 (1960).

Stern, P., and *S. Huković:* Specific antagonists of Substance P. In: Int. Symposium on Substance P, Sarajevo, 83—88 (1961).

Stern, P., and *D. Kocić-Mitrović:* Über die Wirkung des Lichtes und der Dunkelheit auf die Substanz P in der Retina. Naturwissenschaften *45,* 213 (1958).

Stern, P., and *D. Kocić-Mitrović:* Wirkung des Reserpins auf die Menge der Substanz P im Rattengehirn. Arch. exper. Path. Pharmak., Leipzig, *238,* 57—58 (1960).

Stern, P., I. Gasparović, and *J. Kovac:* A factor inactivating Substance P present in the serum of pregnant women. In: Int. Symposium on Substance P. *P. Stern* ed. Sarajevo, Yugoslavia. Sc. Soc. of the Republic of Bosnia and Herzegowina, 141—143 (1961).

Stern, P., S. Hadzović, and *I. Gasparović:* On the relationship between Substance P and biogenic amines. Iugoslav. Physiol. Pharmacol. Acta *2,* 31—38 (1966).

Stern, P., R. Igić, and *J. Jelicić:* Uticaj Supstancije P na zivotinje u stanju ekscitacije. I, Yugoslaw Psycho-Pharmacological Congress. April 1967 (in press).

Stern, P., A. Misirlija, V. Dobrić, and *D. Kocić-Mitrović:* Beitrag zur Wirkungsweise der Substanz P. Acta med. Jugosl. *12,* 153—157 (1958).

Vogler, K., W. Haefely, A. Hürlimann, R. Studer, W. Lergier, R. Strässle, and *K. Berneis:* A new purification procedure and biological properties of Substance P. Ann. N. Y. Acad. Sc. *104,* 378—390 (1962).

Zetler, G.: Substanz P, ein Polypeptid aus Darm und Gehirn mit depressiven, hyperalgetischen und morphinantagonistischen Wirkungen auf das Zentralnerven-system. Arch. exper. Path. Pharmak., Leipzig, *228,* 438—513 (1956).

Zetler, G., and *G. Ohnesorge:* Die Substanz P Konzentration im Gehirn bei verschiedenen Funktionszuständen des Zentralnervensystems. Arch. exper. Path. Pharmak., Leipzig, *231,* 199—210 (1957).

Zetler, G., and *L. Schlosser:* Substanz P im Gehirn des Menschen. Natur-wissenschaften *41,* 46 (1954).

Zuber, H., and *R. Jaques:* Isolierung von Substanz P aus Rinderhirn. Angew. Chem. *74,* 216 (1962).

Author's address: Prof. Dr. *P. Stern,* Department of Pharmacology, Medical Faculty, Sarajevo, Yugoslavia.

Discussion

Hauswirth: I should like to ask Prof. *Stern* whether his septal rats, he spoke of, were agitated or quiet. Our great German poet Friedrich Hölderlin had after a very spiritual life a nervous break down. Afterwards he remained nearly fourty years in a stuporous state. He did not speak, write or move, and when he died, a section was performed. This revealed a cyst in the septum pellucidum and so I guess his excessive tranquillity was caused by a destruction of this structure.

Stern: It is indeed a well-known pharmacological fact that damage of the septum leads to this kind of tranquillity.

Journal of Neuro-Visceral Relations, Suppl. IX, 249—260 (1969)

Pseudo-Transmitters

Arvid Carlsson

Department of Pharmacology, University of Göteborg, Sweden

With 6 Figures

Summary

The pseudo-transmitter concept is discussed with particular reference to the antihypertensive action of α-methyldopa. It is pointed out that a pseudo-transmitter mechanism in the peripheral sympathetic system cannot account for the antihypertensive action of α-methyldopa. Probably the site of action of this drug has to be sought in the central nervous system. Here the drug may well act by a pseudo-transmitter mechanism *.

The pseudo-transmitters and their precursors have proven important tools for clarifying the various processes of uptake, storage and release in the monoamine carrying neurons. By means of these tools it has, for example, been possible to demonstrate two distinct uptake mechanisms in these neurons, one located at the cell membranes and sensitive to imipraminelike antidepressant drugs, the other located in the amine storage granules and sensitive to reserpine.

A pseudo-transmitter, or false transmitter, may be defined as an unnatural transmitter analogue mimicking the true transmitter to such an extent as to be able to replace it in the store and to be released by the nerve impulse. If the physiological activity of the pseudo-transmitter is equal to that of the true one, the replacement should not necessarily result in loss of nerve function. If the activity is lower, partial blockade of nerve function might occur.

The current discussion on pseudo-transmitters goes back to an observation made six years ago (*Carlsson* and *Lindqvist*, 1962) that α-methyldopa, a drug with antihypertensive and moderately sedative properties, is undergoing conversion in the body to form α-methyl noradrenaline, which displaces the transmitter noradrenaline (NA) from its stores in the peripheral and central nervous system:

---

* Recently *M. Henning* has presented strong evidence that decarboxylation of α-methyldopa within the central nervous system is a prerequisite for its antihypertensive action (Brit. J. Pharmacol. *34*, 233 P [1968]).

COOH
|
CH₂ — C - NH₂
        |
        CH₃   →
   OH
OH
α-methyldopa

CH₂ – CH – NH₂
              |
              CH₃   →
   OH
OH
α-methyldopamine

OH
|
CH - CH—NH₂
        |
        CH₃
   OH
OH
α-methylnoradrenaline

This conversion is analogous to the formation of NA from its physiological precursor dopa and is brought about by the enzymes dopa decarboxylase and dopamine-β-oxidase. The pronounced and sustained depletion of NA caused by α-methyldopa is not due to inhibition of synthesis or a reserpine-like effect as suggested earlier, but to an approximately mole-for-mole displacement by α-methyl NA. After some initial controversy these findings were confirmed in several laboratories. Furthermore, it was demonstrated that after α-methyldopa treatment stimulation of sympathetic nerves resulted in release of α-methyl NA (for review see *Muscholl*, 1966).

Carlsson and *Lindqvist* also reported that α-methyl metatyrosine (α-MMT) behaves like α-methyldopa. The prolonged and sustained depletion of NA induced by this compound is due to formation of metaraminol (*via* α-methyl metatyamine) which displaces NA in the store. Displacement can also be induced directly by administered α-methyl NA, metaraminol and several other analogues, which can then be released by nerve stimulation (for review see *Shore*, 1966).

The Antihypertensive Action of α-Methyldopa and the Pseudo-Transmitter Concept

Since α-methyl NA possesses only one third of the pressor activity of NA in man, it seems logical to suggest, as did *Day* and *Rand* (1964) that the antihypertensive action of α-methyldopa is due to release of this less potent analogue, by the nerve impulse, leading to partial blockade of sympathetic nerve function. Some experimental support for this hypothesis is available (*Day* and *Rand*, 1964; *Muscholl* and *Maître*, 1963; *Haefely* and coll., 1966; *Muscholl* and *Rahn*, 1966). However, certain obstacles to this view will be briefly mentioned.

1. α-Methyldopa — in contrast to agents with well established adrenergic neuron blocking properties — does not cause the syndrome typical of loss of sympathetic function, *e. g.* ptosis, relaxation of the nictitating membrane and pronounced orthostatic hypotension. Even if a moderate loss of peripheral sympathetic function has sometimes been observed, it is doubtful whether it can account for the rather pronounced blood-pressure lowering activity.

2. α-MMT, whose decarboxylation products are much less potent pressor agents than NA, is at least as efficient as α-methyldopa in displacing NA, and yet it is not as potent in reducing blood pressure (*Horwitz* and *Sjoerdsma*, 1964; *Henning*, unpublished data, Fig. 1).

3. α-Methyldopamine is able to deplete NA stores efficiently, which is not unexpected since it is the intermediate product in the conversion of α-methyl NA from α-methyldopa. However, it has not yet been possible to demonstrate a decrease in blood pressure following treatment with this agent (*Henning,* unpublished data). In fact, several NA analogues with

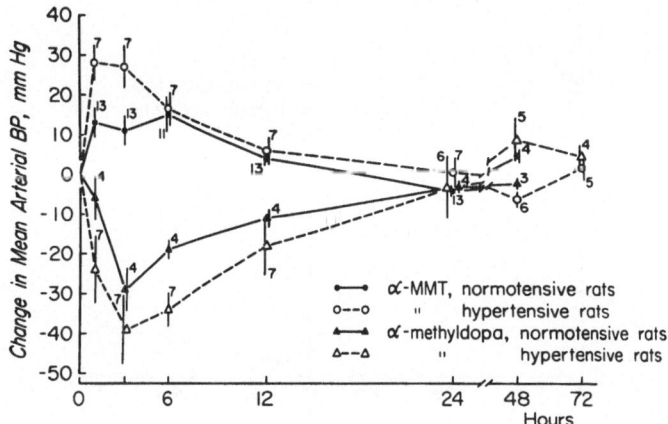

Fig. 1. Effect of L-α-methyldopa (200 mg/kg i. p.) and DL-α-methylmetatyrosine (α-MMT; 400 mg/kg i. p.) on arterial blood pressure in normotensive and renal hypertensive rats *(Henning, M.,* unpublished data).

relatively low sympathomimetic potency have been found to cause severe depletion of NA with but little blockade of sympathetic nerve function (*Andén* and *Magnusson,* 1965). Admittedly, certain sympathomimetic amines possess some antihypertensive activity along with NA depletion (*Crout* and coll., 1965; *Maître* and *Brunner,* 1966; *Brunner* and coll., 1967; *Henning,* unpublished data), but the mechanism of these effects has not yet been elucidated.

4. After α-methyldopa, the time correlation between NA depletion and decrease in blood pressure is poor, the latter effect lasting much shorter than the former (*Henning,* unpublished data, Fig. 2).

These four points require some clarification before the false transmitter concept for the antihypertensive action of α-methyldopa, in the simple formulation given above, can be accepted. On the other hand, there are good reasons for believing that the pseudo-transmitter concept, generally spoken, has come to stay, and also that it may finally help to clarify the mode of action of α-methyldopa. There is, however, another quite important aspect of the pseudo-transmitter problem. The pseudo-transmitters have proved to be very important tools for elucidating the various processes going on in the transmission apparatus. In our laboratory we have also been interested in this aspect. Our studies in this area are not yet completed. So far, most of our work has dealt with α-methyl NA and metaraminol. Tritiated material of these compounds has been kindly synthesized by Dr. *Corrodi* and his coworkers, and this has greatly facilitated our investigations.

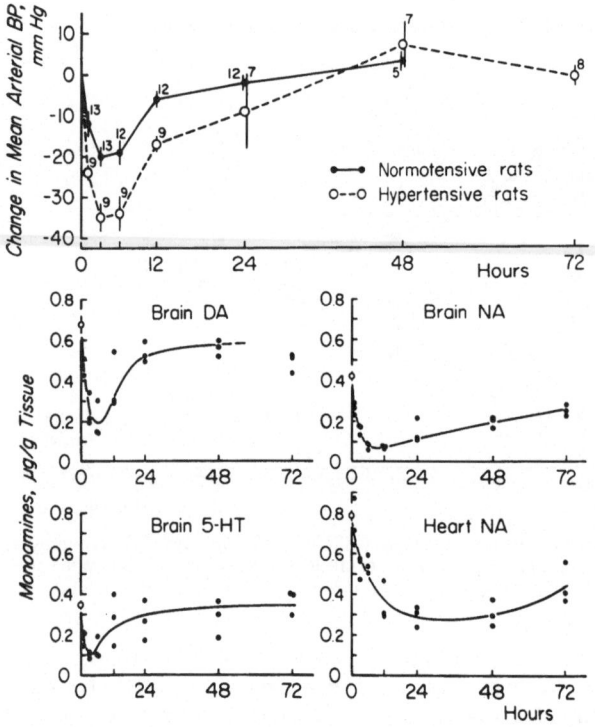

Fig. 2. Effect of L-α-methyldopa (400 mg/kg i. p.) on arterial blood pressure and monoamine levels in tissues (*Henning, M.,* unpublished data).

Storage and Release of Pseudo-Transmitters

Like NA itself and many other NA analogues, α-methyl NA and metaraminol are rapidly taken up and concentrated in adrenergic nerves. An illustration of this is shown in Fig. 3 (*Almgren* and *Waldeck,* 1967). ^{3}H-Metaraminol is rapidly taken up by the rat's submandibular gland. After chronic postganglionic sympathetic denervation, the uptake and retention are very poor, showing that the uptake occurs in the sympathetic nerves. A clearcut difference between the normal and preganglionically sympathectomized gland is also seen. Normally there is a biphasic loss of ^{3}H-metaraminol but this is no longer seen after stopping the nerve impulse flow by decentralization; only a slow apparently monophasic loss occurs. This suggests that metaraminol is released by the nerve impulse. In the course of time, however, it moves to sites which are less available for the nerve impulse. These observations support earlier data of *Crout* and coll. (1964).

By means of drugs, and by subcellular fractionation, it can be shown that different mechanisms take part in the accumulation of amines in the adrenergic nerves.

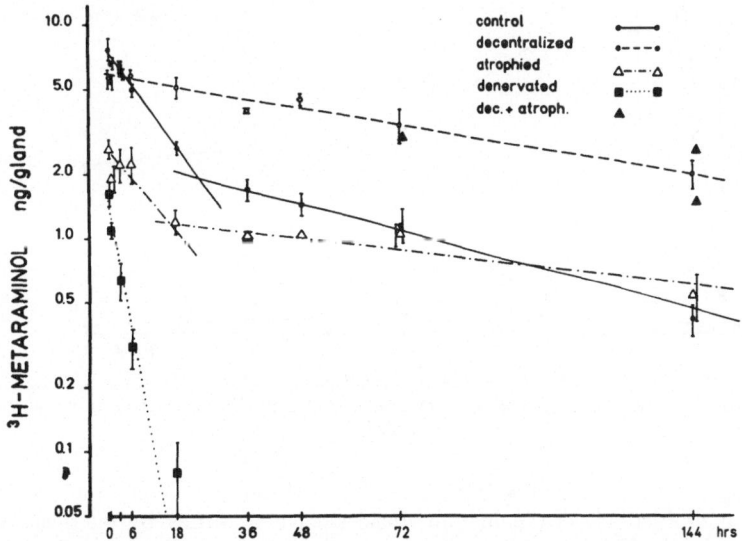

Fig. 3. Disposition of ³H-metaraminol in the rat salivary gland. ³H-metaraminol 0.01 mg/kg was given i. v. to rats whose left submaxillary and sublingual glands had been either denervated, decentralized or brought to atrophy by ligation of the excretory ducts. The right glands served as controls. At various time intervals after the injection the animals were killed and the amount of ³H-metaraminol in the salivary glands was determined. The symbols are the means of 6—18 (controls), 3—4 (denervated), 4—8 (decentralized), and 2—5 (atrophied) experiments. Symbols without indication of the s. e. m. denote single values (*Almgren* and *Waldeck*, 1967).

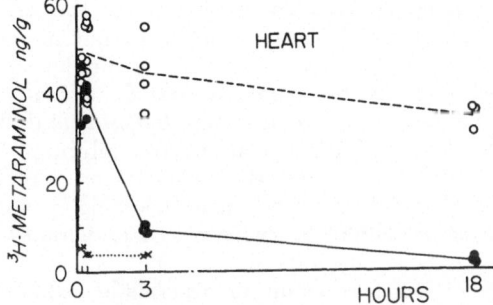

Fig. 4. A. Effects of reserpine and desmethylimipramine (DMI) on ³H-metaraminol uptake and disappearance in mouse hearts. o - - - - - o Control; ●————● Reserpine; x . . . . . x DMI. Reserpine, 10 mg/kg intraperitoneally, was given 6 hrs, or DMI 10 mg/kg intravenously 5 min before administering 0.02 mg/kg ³H-metaraminol (*Carlsson* and *Waldeck*, 1967).

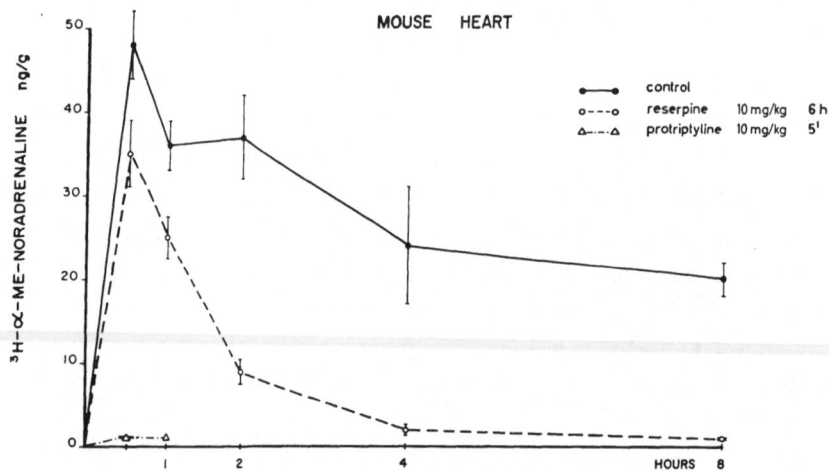

Fig. 4 B. Corresponding ³H-α-methyl noradrenaline data. In these experiments
protriptyline was used instead of DMI *(Carlsson, Lundborg, Stitzel* and *Waldeck,*
1967).

Fig. 4 *(Carlsson* and *Waldeck,* 1965; *Carlsson* and coll., 1967) shows
that after blockade of the intraneuronal storage granules by reserpine, the
amines can still be taken up to an approximately normal extent. However,
they cannot be retained and are lost in the course of a few hours, in
contrast to the prolonged retention normally. Imipramine-like drugs, such
as desipramine and protriptyline, block the uptake almost completely. These
drugs presumably block an uptake mechanism located at the level of the
cell membrane. The last mentioned group of drugs also block the uptake
of NA. After reserpine treatment, however, NA does not behave like its
methylated analogues. Very little uptake can be demonstrated. This is
probably due to the fact that unlike the methylated analogues the NA
taken up by the reserpine resistant mechanism at the cell membrane of
the nerve fiber can be metabolized by intraneuronal monoamine oxidase.
After blockade of this enzyme by nialamide, accumulation of ³H-NA can
be demonstrated even in the reserpine-treated animal. However, this
accumulation is not as efficient as normally *(Carlsson* and *Waldeck,* 1967).
Furthermore, the dosage of monoamine oxidase inhibitor is critical. The
reason for these difficulties is not clear. One factor may be that monoamine oxidase inhibitors have some amine-releasing properties, at least
partly owing to accumulation of endogenous monoamines *(Carlsson* and
Waldeck, 1966). At any rate, the α-methylated NA analogues offer the
great advantage of being resistant to monoamine oxidase. This greatly
facilitates the investigation of amine concentrating and releasing mechanisms.

Studies on the subcellular distribution of the amines gives further
support to the concept of a dual uptake mechanism for amines and gives

important additional insight into the processes involved. In the hearts of mice the distribution of NA and α-methyl NA between particulate and supernatant fractions was rather similar, the latter amine showing a slightly lower affinity to the particles. Metaraminol showed a considerably lower affinity to the particles, especially at short intervals after administration of the amine (*Lundborg* and *Stitzel*, 1967 a). This is surprising, since the NA depleting action of metaraminol is considerable and appears to be quite comparable to that of α-methyl NA (*Shore*, 1966). In the reserpine-treated animal the uptake of all amines by the particulate fraction was considerably reduced. A low but significant uptake still persisted, however, indicating the existence of a reserpine-resistant uptake mechanism in the particles. In the case of metaraminol, the difference in subcellular distribution between normal and reserpine-treated animals was, in fact, hardly detectable at the shortest interval studied (30 min). Nevertheless, a reserpine-sensitive uptake mechanism appears to be important also for metaraminol even at short intervals, as judged by the dramatic release induced by the combined treatment with protriptyline and reserpine, by far exceeding the effect of either drug alone (*Carlsson* and *Waldeck*, 1966; *Lundborg* and *Stitzel*, 1967 b).

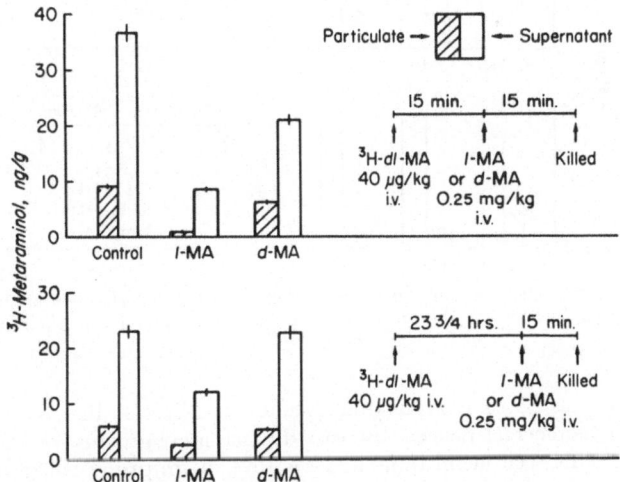

Fig. 5. Influence of *d*- and *l*-metaraminol on the ³H-*dl*-metaraminol content of subcellular fractions of the mouse heart (data compiled from *Lundborg* and *Stitzel*, 1967).

Additional evidence for a movement of metaraminol into a more stable part of the particulate fraction was obtained by experiments with different releasing agents. Unlabelled *l*-metaraminol was found to be much more efficient in releasing ³H-metaraminol, preferentially from the particulate fraction, at a short interval (30 min) than at a long interval (24 hr) after administration of the tritiated amine. *d*-Metaraminol which has been shown to be retained much less efficiently than the *l*-form (*Shore*, 1966) was but slightly effective at the shorter interval and had no significant effect at the longer interval (*Lundborg* and *Stitzel*, 1967 c; fig 5). In agreement with the hypothesis presented above, protriptyline was capable of releasing only supernatant metaraminol, whereas reserpine acted preferentially on the particulate fraction (*Stitzel* and *Lundborg*, 1967; fig. 6). The former drug was active only at a short interval after ^{3}H-metaraminol, the latter preferentially at a long interval. — Nerve stimulation was found to act preferentially on particle-bound metaraminol and was relatively inactive on metaraminol taken up after pretreatment with reserpine (*Almgren* and coll., unpublished data).

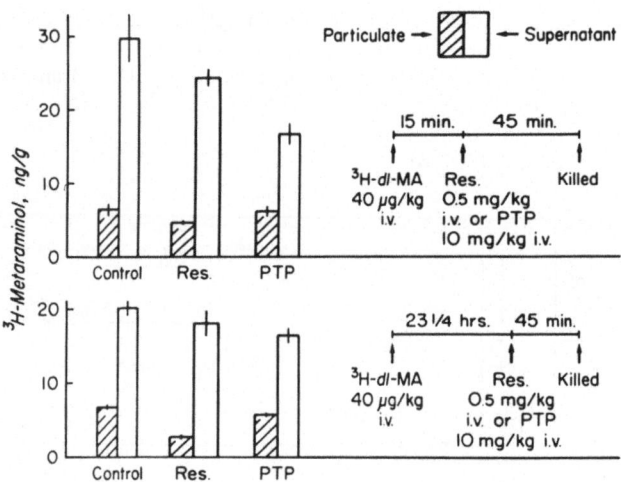

Fig. 6. Influence of reserpine and protiriptyline on the ^{3}H-*dl*-metaraminol content of subcellular fractions of the mouse heart (data compiled from *Stitzel* and *Lundborg*, 1967).

It may thus be concluded that pseudo-transmitters, like NA, are taken up in the adrenergic nerves by two distinct mechanisms, one located at the level of the cell membrane and sensitive to imipramine-like drugs, the other located in the particles and sensitive to reserpine (although some reserpine-resistant uptake by particles also exists). Similar conclusions have been reached on the basis of histochemical observations (*Malmfors*, 1965). Furthermore, evidence has been presented for the movement of an amine from a labile to a stabile part of the particulate fraction. Whether these parts occur in different particles or form different fractions within one and the same particle cannot be decided at present. Some evidence has been presented that the incorporation into the labile fraction is partly sensitive, partly insensitive to reserpine. The release by the nerve impulse appears to occur preferentially from a labile part of the particulate fraction.

Concluding Remarks

It may be concluded that the data available thus far are insufficent to establish with any higher degree of accuracy the validity of the pseudo-transmitter concept. More accurate information on several points is required. This is true both for the storage and release of the pseudo-transmitters from the adrenergic nerve fibers and for their activity at receptor sites. As to the intraneuronal mechanisms, it seems possible that further insight into the movements of amines between different pools may throw light on certain problems. If the concept of two reserpine-sensitive particulate fractions proves valid, some obscure points may be clarified. For example, the lack of a time correlation between NA depletion and loss of function will be explained if we assume that the NA of a small labile fraction, immediately available for release by the nerve impulse and which might be called "the essential pool", is displaced by the pseudo-transmitter only at an early stage, when the concentration of free pseudo-transmitter in the cytoplasm is sufficiently high to compete with newly synthesized NA. At a later stage, when less pseudo-transmitter is available for such competition, newly synthesized NA will again occupy the essential pool. This will be masked by the relative inertia of the large "reserve pool", in which the pseudo-transmitter still predominates.

The concept of the two pools may possibly also explain why two highly effective NA-displacing agents, such as α-methyldopa and α-MMT, may differ in hypotensive activity. The possibility exists that the metabolites of α-MMT are not so efficient as those of α-methyldopa in displacing NA from the small essential pool, although they are both active in displacing NA in the large reserve pool. In support of this, the marked difference between α-methyl NA and metaraminol in subcellular distribution may be quoted. Another difference between α-methyldopa and α-MMT may be important: the former is a more potent inhibitor of NA synthesis. This may contribute to the functional interference caused by the pseudo-transmitter but is probably not the main factor, since other more potent inhibitors of catecholamine synthesis are not so efficient in reducing blood pressure (for review, see *Muscholl*, 1966).

Concerning the activity of pseudo-transmitters at receptor sites more accurate information is required. For example, in the rat α-methyl NA equals NA as a pressor agent according to *Muscholl* and *Weber* (1965). If this is true, it would certainly invalidate the pseudo-transmitter hypothesis, as formulated by *Day* and *Rand* (1964), since α-methyldopa is clearly hypotensive in this species. However, *Day* and *Rand* find that NA is 2.9 times more potent than α-methyl NA in this respect. But even if the latter is true it may be questioned if an activity factor of 3 is enough to explain the pronounced hypotensive action of α-methyldopa — in the rat it appears to be comparable to that seen after almost complete adrenergic neuron blockade — especially if the marked efficiency of blood pressure homeostasis is taken into account. A moderate increase in impulse flow should easily compensate for a loss of transmitter potency of this magnitude.

It does not seem unlikely, that we will finally have to look for a site of action of α-methyldopa in the central nervous system in order to clarify its blood-pressure effect. Certainly, α-methyldopa is active centrally. It produces sedation, abolishes the central stimulating action of amphetamine in the reserpine-treated animal (*Hanson* and *Henning*, 1967) and causes depletion of central monoamines, especially NA. As a matter of fact, α-methyldopa injected into the vertebral artery has been found amazingly potent in reducing blood pressure in the chloralosed cat (*Henning* and *van Zwieten*, 1967). More accurate knowledge of the activity of α-methylated NA analogues at central catecholamine receptor sites is thus highly desirable. At present very little is known on this point. — The lack of hypotensive activity of α-methyldopamine would of course be readily explained if the main site of attack of α-methyldopa proved to be in the central nervous system: unlike the amino acid, its decarboxylation products do not readily pass the blood-brain barrier.

References

Almgren, O., and *B. Waldeck:* On the disposition of ³H-metaraminol in the rat salivary gland. J. Pharm. Pharmacol., London, *19,* 705—708 (1967).

Andén, N.-E., and *T. Magnusson:* Functional significance of noradrenaline depletion by α-methyl metatyrosine, metaraminol and dextro-adrenaline. In: Pharmacology of cholinergic and adrenergic transmission. *G. B. Koelle, W. W. Douglas* and *A. Carlsson* eds. (Proceedings of the Second Internat. Pharmacol. Meeting, Prague 1963, Vol. *3*). 319—328. Oxford 1965.

Brunner, H., P. R. Hedwall, L. Maître, and *M. Meier:* Antihypertensive effects of α-methylated catecholamine analogues in the rat. Brit. J. Pharmacol. *30,* 108—122 (1967).

Carlsson, A., and *M. Lindqvist:* In vivo decarboxylation of α-methyl DOPA and α-methyl metatyrosine. Acta. physiol. Scand. *54,* 87—94 (1962).

Carlsson, A., P. Lundborg, R. E. Stizel, and *B. Waldeck:* Uptake, storage and release of ³H-α-methyl noradrenaline. J. Pharmacol. Exper. Therap., Baltimore, *158,* 175—182 (1967).

Carlsson, A., and *B. Waldeck:* Mechanism of amine transport in the cell membranes of the adrenergic nerves. Acta Pharmac. Tox., K'hvn. *22,* 293—300 (1965).

Carlsson, A., and *B. Waldeck:* Release of ³H-metaraminol by different mechanism. Acta physiol. Scand. *67,* 471—480 (1966).

Carlsson, A., and *B. Waldeck:* The accumulation of [³H] noradrenaline in the adrenergic nerve fibres of reserpine-treated mice. J. Pharm. Pharmacol. *19,* 182—190 (1967).

Crout, J. R., H. S. Alpers, E. L. Tatum, and *P. A. Shore:* Release of metaraminol (Aramine) from the heart by sympathetic nerve stimulation. Science *145,* 828—829 (1964).

Crout, J. R., R. R. Johnston, W. R. Webb, and *P. A. Shore:* The antihypertensive action of metaraminol in man. Clin. Res. *13,* 204 (1965).

Day, M. D., and *M. J. Rand:* Some observations on the pharmacology of α-methyldopa. Brit. J. Pharmacol. *22,* 72—86 (1964).

Haefely, W., H. Thoenen, and *A. Hürlimann:* Qualitativ und quantitativ unterschiedliche Wirkungen der α-Methyldopa-Vorbehandlung an Katzen je nach der Dosierungsschema. Naunyn-Schmiedebergs Arch. exper. Path. *257,* 25 (1966).

Hanson, L. C. F., and *M. Henning:* Effects of α-methyl-dopa on conditioned behaviour in the cat. Psychopharmacologia, Berlin, *11,* 1—7 (1967).

Henning, M., and *P. A. van Zwieten:* Central hypotensive effect of α-methyl-dopa. J. Pharm. Pharmacol., London, *19*, 403—405 (1967).

Horwitz, D., and *A. Sjoerdsma:* Effects of alpha-methyl-meta-tyrosine intravenously in man. Life Sci. *3*, 41—48 (1964).

Lundborg, P., and *R. E. Stitzel:* Uptake of biogenic amines by two different mechanisms present in adrenergic granules. Brit. J. Pharmacol. *29*, 342—349 (1967 a).

Lundborg, P., and *R. E. Stitzel:* Studies on the dual action of guanethidine in sympathetic nerves. Acta physiol. Scand. *72*, 100—107 (1967 b).

Lundborg, P., and *R. E. Stitzel:* Stereospecificity and intracellular binding of metaraminol. Acta physiol. Scand. (1967 c; in press).

Maître, L., and *H. Brunner:* Antihypertensive und Katecholamin-entleerende Wirkungen von α-Methyl-tyramin und α-Methyl-octopamin. Naunyn-Schmiedebergs Arch. exper. Path. *257*, 40—41 (1966).

Malmfors, T.: Studies on adrenergic nerves. Acta physiol. Scand. *64* (suppl. 248), 1—93 (1965).

Muscholl, E.: Autonomic nervous system: nerve mechanism of adrenergic blockade. Ann. Rev. Pharmacol. *6*, 107—128 (1966).

Muscholl, E., and *L. Maître:* Release by sympathetic stimulation of α-methyl-noradrenaline stored in the heart after administration of α-methyldopa. Experientia, Basel, *19*, 658—659 (1963).

Muscholl, E., and *K. H. Rahn:* Die Ausscheidung von α-Methylnoradrenalin im Harn von Hypertonikern während einer Behandlung mit α-Methyldopa. Naunyn-Schmiedebergs Arch. exper. Path. *257*, 44—45 (1966).

Muscholl, E., and *E. Weber:* Die Hemmung der Aufnahme von α-Methylnor-adrenaline in das Herz durch sympathomimetische Amine. Naunyn-Schmiedebergs Arch. exper. Path. *252*, 134—143 (1965).

Shore, P. A.: The mechanism of norepinephrine depletion by reserpine, metaraminol and related agents. Pharmacol. Rev., Baltimore, *18*, 561—568 (1966).

Stitzel, R. E., and *P. Lundborg:* Effect of reserpine and protriptyline on the subcellular distribution of ^{3}H-metaraminol in the mouse heart. Brit. J. Pharmacol. *30*, 379—384 (1967).

Author's address: Prof. Dr. *Arvid Carlsson*, Department of Pharmacology, University of Göteborg, Göteborg, Sweden.

Discussion

Van Riezen: Dr. *Carlsson*, you showed evidence of inhibition of uptake of catecholamines into the sympathetic nervous system of the heart by protriptyline. Can you offer us similar evidence for uptake blockage in the catecholamine systems of the brain stem?

Carlsson: Yes, such evidence is available (see *Carlsson, Fuxe, Hamberger* and *Lindqvist*, Acta Physiol. Scand. *67*, 481—497 (1966); *Hamberger, B.* Reserpine-resistant uptake of catecholamines. A histochemical and biochemical study, Stockholm 1967).

Saxena: Dr. *Carlsson* mentioned that α-methyl-Dopa-induced hypotension cannot be fully accounted for by pseudotransmitter substance and added that, in his laboratory, the observation has been made that α-methyl-Dopa causes hypotension when injected by the intraventricular route. Similar observations have been made in Prof. *Bhargava's* laboratory in Lucknow (India). Intra-cerebroventricular injection of α-methyl-Dopa caused hypotension together with a blockade of pressor responses of carotid occlusion, central vagal stimulation and vasomotor (medullary) center stimulation. Thus, it appears that α-methyl-Dopa has certainly a component of central vasomotor inhibitory action together with pseudotransmitter-formation effect.

Carlsson: Thank you for this interesting piece of information. Perhaps it should be made clear that the evidence of a central point of attack of α-methyl-Dopa does not argue against the pseudo-transmitter concept, since the central nervous system contains catecholamine-storing nerve fibers, whose function might be interfered with by a pseudo-transmitter.

Schümann: Muscholl and coll. found that in patients treated with α-methyl-Dopa the sum of the urinary excretion of α-methyl-noradrenaline and noradrenaline was about of the same order of magnitude as the amount of noradrenaline excreted in normal, untreated patients. This result shows that α-methyl-noradrenaline has also been released as a false transmitter in man and furthermore that the antihypertensive activity of α-methyl-Dopa is at least partly due to the displacement of noradrenaline, also in the periphery. I agree with Prof. *Carlsson* that the displacement of noradrenaline in the brain leads very likely to a reduced activity of the vasomotor centers and lowers, in this way, the bloodpressure.

Carlsson: I agree with Prof. *Schümann* that a combined peripheral and central point of attack may be the most likely alternative to explain the antihypertensive action of α-methyl-Dopa. In fact, the finding of *Muscholl* et al. that the sum of noradrenaline and α-methyl-noradrenaline was not above normal can be taken as evidence to support a central point of action. With a purely peripheral effect this sum should be above normal, owing to a compensatory increase in sympathetic impulse flow, induced by lowered blood pressure.

Schönbaum: Prof. *Carlsson,* your studies of the effect of temperature on noradrenaline uptake in the reserpine-treated rat are extremely interesting, also from the point of view of temperature regulation. Would you like to elaborate on these experiments? Your finding of an increased uptake in the hypothermic animal is particularly intriguing.

Carlsson: The mechanism of increased retention of ^{3}H-NA in hypothermic animals treated with reserpine is not clear. Two possibilities suggest themselves. 1. If the amine-releasing action of nialamide is due to accumulation of endogenous amines replacing the labelled analogues, hypothermia might prevent this process by inhibiting monoamine synthesis. 2. The optimum dose of nialamide (about 10 mg/kg) is certainly insufficient to cause complete inhibition of monoamine oxidase. Hypothermia might inhibit deamination by the remaining enzyme activity.

Journal of Neuro-Visceral Relations, Suppl. IX, 261—276 (1969)

Subcellular Distribution of Neurohumors and Chemical Receptors in the Central Nervous System *

E. De Robertis

Instituto de Anatomía General y Embriología, Facultad de Medicina — Universidad de Buenos Aires, Buenos Aires, Argentina

With 4 Figures

Summary

Electron microscopic studies have revealed that both, neurohumors and neurohormones, are stored within membrane-bound vesicles, which may have different size, shape and electron density. Knowledge about these different types of vesicles may be obtained by the separation of nerve endings from brain and of the synaptic vesicles themselves. Two large groups of nerve endings, one aminergic and containing the various biogenic amines, *i. e.* acetylcholine, noradrenaline, dopamine, 5-hydroxytryptamine and histamine, and the other, a non-aminergic group, were separated from the brain.

The non-aminergic group contains glutamic acid decarboxylase, the enzyme that irreversibly synthesizes γ-aminobutyric acid (GABA). The possibility that GABA may be present in such nerve endings and in the synaptic vesicles, playing a synaptic inhibitory role, is discussed.

The isolation of the synaptic vesicles has provided definite evidence that the various neurohumors are contained in quantal units within such compartments of the nerve endings.

The separation of the nerve ending membranes and of the junctional complex has provided a direct way to the study of the receptor properties and to the isolation of the chemical receptors for the neurohumors.

At the third International Symposium on Neurosecretion held in 1961, I took the rather heretical position that there are no essential differences between synaptic processes in which neurohumors are synthesized and secreted by the nerve endings acting rapidly and at a short distance on specific chemical receptors, and the neurosecretory processes in which the neurohormones are delivered by the nerve endings acting more slowly by way of diffusion or the circulation upon more distant receptors (*De Robertis,* 1962). This unitary concept of neurosecretory mechanisms was later developed in monograph form (*De Robertis,* 1964). I am most pleased that the present symposium will allow us to discuss the neurohormones

* The original research of which this article gives an account was supported by Grants from the National Institutes of Health NB-06953-02 and Consejo Nacional de Investigaciones Científicas y Técnicas, Argentina.

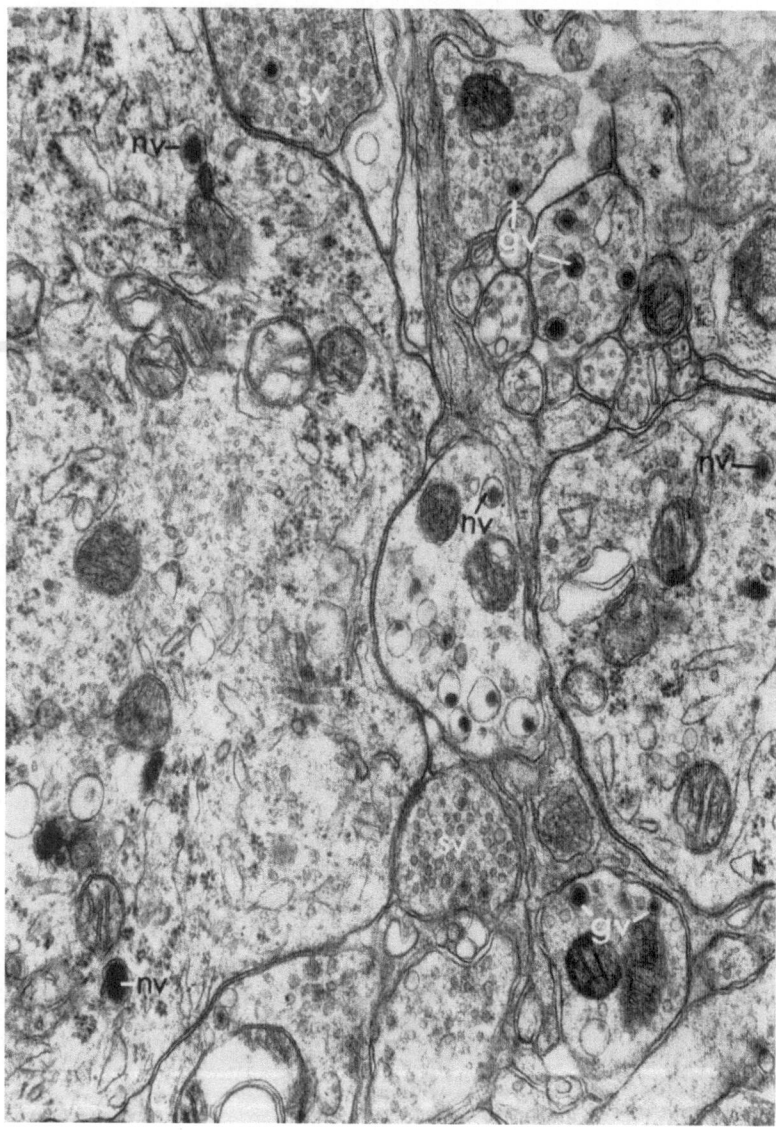

Fig. 1. Electronmicrograph of the supraoptic nucleus of the rat. On the left, the perikaryon of a neurosecretory neuron containing two neurosecretory vesicles (nv). On the right, a part of another neuron with a neurosecretory vesicle. In between, nerve endings containing clear synaptic vesicles (sv) and some with granulated vesicles (gv), probably containing catecholamines. A neurosecretory axon is observed in the center. Notice the variation in size of the different vesicular elements but the basic similarity in ultrastructural organization. X 45,000 (courtesy of Dr. Zambrano).

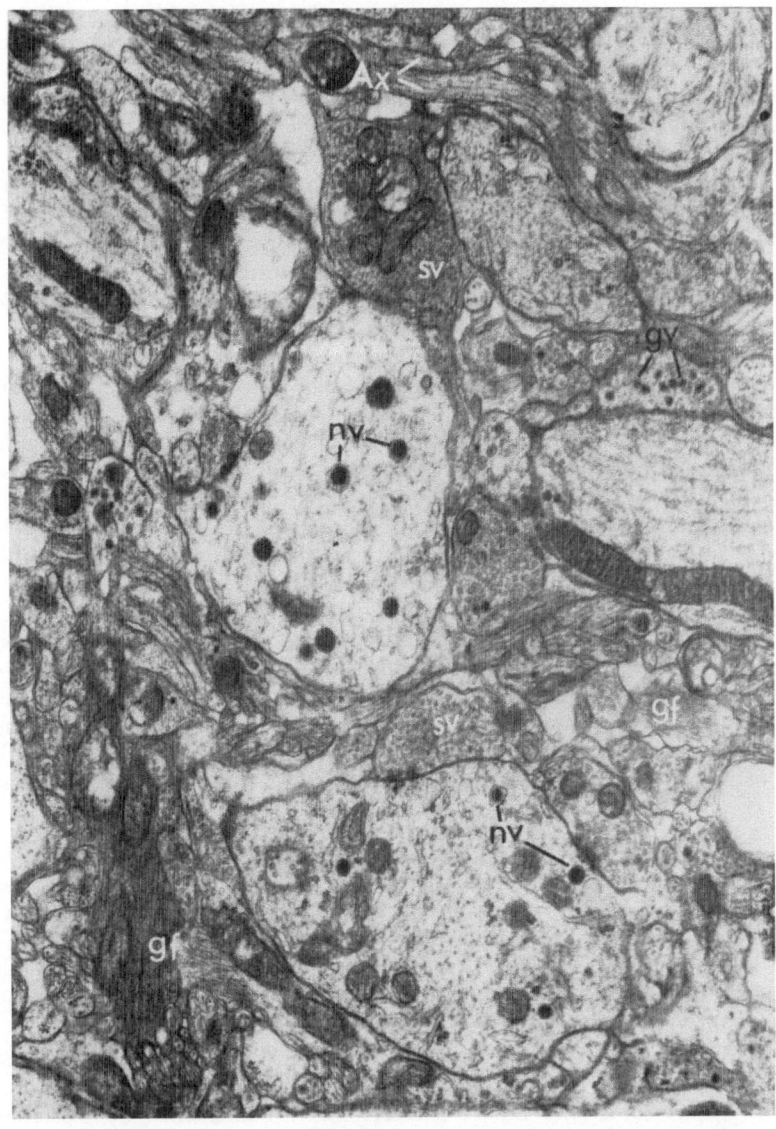

Fig. 2. The same as Fig. 1, but showing a neuropile region. Two large axons containing neurosecretory vesicles (nv) are observed. On both of them small nerve endings terminate showing clear synaptic vesicles (sv) and a few granulated ones. Observe the axons (Ax) containing neurotubules and the small axons and endings with granulated vesicles (gv). A few glial processes with gliofilaments (gf) are intermingled with the other components of the neuropile. The ultrastructural similarities between the different vesicular elements is evident (courtesy of Dr. *Zambrano*). × 24,000.

and the neurohumors together and to see the basic similarities that exist in the mechanism of formation, storage and release of these products.

The electron microscope has revealed that in every case the active substance is accumulated within a membrane which isolates it from the surrounding cytoplasm. These secretory vesicles reach a rather constant size. From this it may be followed that they contain a quantal number of active molecules. Such vesicular units may be formed in the perikaryon, along the axon or at the nerve ending proper. Finally they accumulate near the surface membrane at strategic points where they can be released upon arrival of the nerve impulse.

The ultrastructural distinction between neurohumoral and neurohormonal processes has become more and more difficult with the discovery of a wide variety of products secreted by neurons such as biogenic amines, aminoacids, polypeptides and so forth, and also of vesicular elements having different size and morphology. In axons and nerve endings, in addition to the typical synaptic vesicles of about 500 Å in diameter and clear content (*De Robertis* and *Bennett*, 1954, 1955), a series of granulated vesicles with a dense core was described. With Pellegrino de Iraldi we found special granulated vesicles in adrenergic peripheral endings (*De Robertis* and *Pellegrino de Iraldi*, 1961) and in the anterior hypothalamus, a region particularly rich in catecholamines. In the latter we also found elliptical and more complex vesicles (*Pellegrino de Iraldi* and coll., 1963).

The morphological similarities that at the ultrastructural level exist between clear synaptic vasicles, granulated vesicles in axons and nerve endings and neurosecretory vesicles (granules) in axons and perikarya are illustrated in Figs. 1 and 2 corresponding to the supraoptic nucleus of the rat. Here, neurosecretory neurons and axons are seen side by side with axons and nerve endings containing granulated and typical synaptic vesicles.

Distribution of Neurohumors in Nerve Endings

An understanding of the true significance of the various types of vesicles may only be reached by separating them in a rather pure form. Fortunately, the cell fractionation techniques applied to the CNS have permitted the use of quantitative methods for the assay of neurohumors and of the related enzymes and to correlate the results with the submicroscopic organization of the synaptic region (*Whittaker*, 1959).

Independently, *Gray* and *Whittaker* (1960) and our group (*De Robertis* and coll., 1960, 1961) demonstrated that the mitochondrial fraction of the brain contained isolated nerve endings which could then be separated by gradient centrifugation.

In our laboratory three methods of cell fractionation have been developed to produce: (1) the separation of several types of nerve endings, (2) the isolation of synaptic vesicles, and (3) the separation of different types of nerve ending membranes. More recently a method for the separation of the junctional complex, formed by the two synaptic membranes and associated structures, has also been developed.

Table 1. *Distribution of the biogenic amines and enzymes of the γ-aminobutyric acid system. A-E submitochondrial fractions isolated by gradient centrifugation. Results are expressed as relative specific concentration, which is the percentage of amine or enzyme recovered, divided by the percentage of protein recovered.*

Structure	Submitochondrial fraction				
	A Myelin	B Small nerve endings	C Nerve endings	D Nerve endings	E Mitochondria
Biogenic amines					
Acetylcholine	0,15	2,24	2,99	0,94	0,58
5-Hydroxytryptamine	0,61	0,78	2,17	0,76	0,48
Noradrenaline	0,32	2,05	1,66	0,77	0,72
Dopamine	0,79	1,85	1,13	0,91	0,71
Histamine	0,72	2,70	1,56	0,44	0,70
Enzymes					
Glutamic acid decarboxylase	0,02	0,49	1,22	2,00	0,40
γ-aminobutiric acid aminotransferase	0,15	0,11	0,29	1,10	8,00
Succinic dehydrogenase	—	—	0,52	2,10	7,60

Table 1 shows a summary of the distribution of the various biogenic amines, acetylcholine, 5-hydroxytryptamine, noradrenaline, dopamine and histamine, in the submitochondrial fractions of the brain. These results, together with others on the enzymes of the cholinergic system (i. e. acetyl-cholinesterase and cholineacetylase, *De Robertis* and coll., 1962, 1963), 5-hydroxytryptophane decarboxylase (*Rodriguez de Lores Arnaiz* and *De Robertis,* 1964) and catechol-0-methyltransferase (*Alberici* and coll., 1965), demonstrate that two groups of nerve endings may be differentiated. One of them, represented by subfractions B and C, is essentially aminergic (that is, they contain biogenic amines), while the other (subfraction D) is non-aminergic and lacks these active neurohumors. Recently *Kataoka* and *De Robertis* (1967) have shown that in the cerebral cortex there are tiny nerve endings rich in histamine which sediment in the heavy microsomal fraction.

Isolation of Inhibitory Nerve Endings and the Localization of GABA

Recent studies have stressed that certain aminoacids may have an important physiological role in the CNS. Glutamic acid and aspartic acids have a powerful excitatory action while γ-aminobutiric acid (GABA) has a general depressant effect on neurons. This aminoacid, which is irreversibly formed by decarboxylation from glutamic acid, is of particular interest because, together with the synthesizing enzyme glutamic acid decarboxylase (GAD), it is only found in brain and particularly in gray matter. While it is impossible to review here the vast literature on GABA, the amount of evidence that was given at the recent symposium on Neuronal Inhibitory Mechanisms in Stockholm (1966) indicated that this aminoacid plays an essential role in inhibitory synapses of the mammalian brain.

In 1963, *Salganicoff* and *De Robertis* demonstrated that the two main enzymes of the GABA system, i. e. GAD and GABA-aminotransferase, had a very different localization in the submitochondrial fractions of the brain. GAD was found concentrated in sub-fraction D of non-aminergic nerve endings while GABA-aminotransferase is preferentially located in mito-chondria and has a distribution similar to succinic dehydrogenase (Table 1). Because of the strict correlation existing between GAD and GABA (*Sisken* and coll., 1960), we postulated that the nerve endings of subfraction D also contain GABA and that this aminoacid would play a synaptic role at the junction. Evidence was also given that the enzyme GAD could be attached to the outer surface of the synaptic vesicles by Ca^{2+} and it was postulated that GABA could be taken by the synaptic vesicle thus participating as a true neurotransmitter at the synaptic junction.

Table 2. *Percent variation of the cholinergic system and of various enzymes related to the glutamine-glutamate-GABA system during MSO convulsions in the rat brain cortex (from De Robertis and coll., 1967 a).*

Cholinergic system	ChAc —3	AChE —8	ACh —3
Glutamate glutamine and GABA systems	GDH +2	Asp-AT —9	GABA-AT +8
	GAD —25	Alan-AT —60	GS —66

Abbreviations: ACh acetylcholine, AChE acetylcholinesterase, Alan-AT ala-nine aminotransferase, Asp-AT asparate aminotransferase, ChAc choline-acetylase, GABA-AT gamma-aminobutyric acid aminotransferase, GAD L-glutamic acid decarboxylase, GDH glutamic dehydrogenase, GS glutamine synthetase.

Also related to the inhibitory nerve endings are the recent observations on the action of the convulsant drug methionine sulfoximine (MSO) on the cerebral cortex (*De Robertis* and coll., 1967 a). Giving $C^{14}MSO$ sub-arachnoidally, it was observed that at the time of convulsions the drug was mainly fixed to the nerve endings. MSO had no effect on the components of the ACh system but it produced considerable inhibition of the enzymes glutamine synthetase and alanine aminotransferase while GAD was only slightly inhibited (Table 2). Probably the most interesting results came from

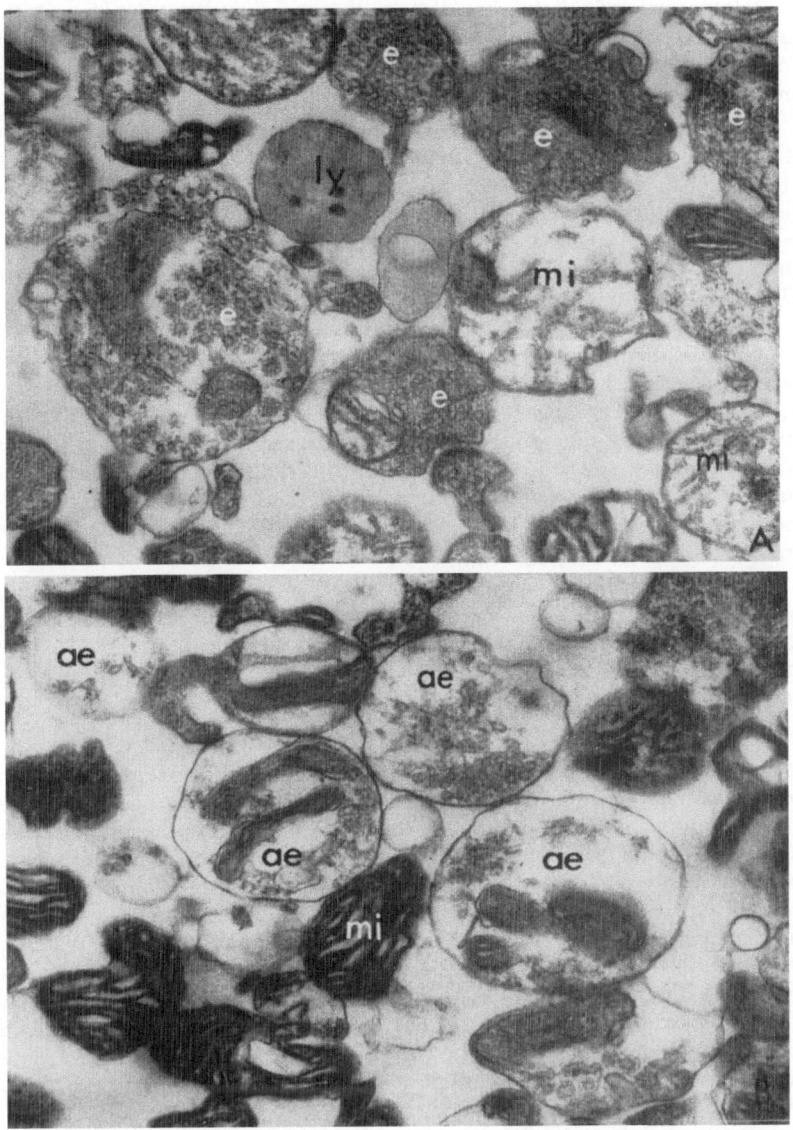

Fig. 3. Electronmicrographs of the mitochondrial fraction isolated from the cerebral cortex of the rat. A, from a normal control rat. Several nerve endings (e) filled with synaptic vesicles are observed. There are also a few mitochondria (mi) and two lysosomes (ly), × 60,000. B, from a rat injected with 400 mg/kg of MSO intraperitoneally and killed while in convulsions. Notice that most of the nerve endings are altered (ae) showing swelling and a decreased amount of synaptic vesicles. mi, free mitochondria, × 60,000.

the electronmicroscopic observation of the nerve endings isolated at the time of convulsions. Numerous nerve endings, particularly these non-aminergic ones, separated in subfraction D, showed considerable swelling and loss of synaptic vesicles (Fig. 3).

At the present time it is difficult to establish which is the initial cause of convulsions and the exact chain of events by which MSO causes the disequilibrium in the aminoacid metabolism. The strong inhibition of glutamine synthetase and alanine-aminotransferase and, to a lesser extend, of GAD, may alter many of the interconnected reactions of the glutamine-glutamate and GABA metabolism. The resulting effect may be very complex. The observation that most of the non-aminergic nerve endings of subfraction D are profoundly changed in the convulsant animal is in line with the previous finding that such endings contain the GABA-system and are probably involved in inhibitory functions in the brain. This conclusion is also supported by the fact that the cholinergic system is not changed by MSO and that structurally the aminergic nerve endings of subfractions B + C are less affected by this convulsant drug. In this respect it is interesting that in convulsions induced by thiosemicarbazide, in which there is a decrease in GABA content, the acetylcholine content of the brain also remains unchanged (*Roa* and coll., 1964).

Basing us mainly on the above findings on the distribution of GAD and the action of MSO, we have recently suggested that the non-aminergic nerve endings, separated in subfraction D, contain the bulk of the inhibitory synapses of the cerebral cortex (*De Robertis*, 1966 b).

Isolation of the Synaptic Vesicles

In 1952, *Fatt* and *Katz* discovered that acetylcholine is released in multimolecular packets or quanta of uniform size. This quantal release was evidenced by the so-called spontaneous miniature end-plate potentials at the myoneural junction. With the arrival of the nerve impulse, there would be a synchronous discharge of a certain number of quantal units, the release of which produce the much larger synaptic potential. With the discovery of the synaptic vesicles (*De Robertis* and *Bennett*, 1954, 1955), the correlation with the quantal release was made. Several early experiments suggested that synaptic vesicles were implicated in neural transmission (*De Robertis* and *Vaz Ferreira*, 1957). However, the isolation of such a subcellular component would be the most direct approach to elucidate the chemical nature and physiological significance of the synaptic vesicles.

In 1962, our laboratory developed an isolation method which is now widely used (*De Robertis* and coll., 1962). It consists of the treatment of the crude mitochondrial fraction or the isolated nerve endings with a hypotonic solution. This causes swelling of the isolated nerve endings with rupture of the limiting membrane and release of the enclosed vesicles, mitochondria and axoplasm. The vesicles may then be separated by differential centrifugation into a fraction M_2 while the membranous portion of the endings and the mitochondria goes into the much bulkier fraction

M_1. More recently, using a gradient technique, a more purified fraction of synaptic vesicles was obtained (*Lapetina* and coll., 1967).

Table 3 shows some of the main results obtained on the localization of neurohumors. In the synaptic vesicles the highest concentration of bound acetylcholine (*De Robertis* and coll., 1963), noradrenaline, dopamine (*Zieher* and *De Robertis*, 1964) and histamine (*Kataoka* and *De Robertis*, 1967) was found. 5-hydroxytryptamine has also been found concentrated in the vesicular fraction by *Maynert* and coll. (1964).

Table 3. *Content of biogenic amines in the bulk fraction (M₁), synaptic vesicles (M₂), and in the soluble fraction (M₃) (De Robertis and coll., 1963). The results are expressed in relative specific concentration as defined in Table 1.*

	Fraction		
	M_1	M_2	M_3
5-Hydroxytryptamine	0,48	1,90	2,30 [*]
Acetylcholine	0,55	2,85	1,20
Noradrenaline	0,40	2,56	1,93
Dopamine	0,46	2,46	1,72
Histamine	0,39	2,24	2,27

[*] Calculated from *Maynert* et al., 1964.

Recently, synaptic vesicles rich in histamine were separated from the nerve endings of the heavy microsomal fraction of the cerebral cortex (*Kataoka* and *De Robertis*, 1967). Vesicles, isolated from the anterior hypothalamus, were found to contain five to six times more noradrenaline than a similar fraction from total hemispheres. In this preparation, many granulated vesicles were observed (*De Robertis* and coll., 1965). This fact and other indirect information from pharmacological changes of these vesicles led us to conclude that they contain the adrenergic transmitter (see *De Robertis*, 1966 a).

These studies, confirmed by others (*Whittaker* and coll., 1964, *Whittaker* and *Sheridan*, 1965), strongly support the postulate that the synaptic vesicles are the quantal units for storage and release of the neurohumors in neurons and that, probably, they function in synaptic transmission.

Isolation of the Junctional Complex and the Chemical Receptor

A basic postulate of the chemical theory of synaptic transmission is that the subsynaptic membrane should contain the chemical receptors for the various neurohumors secreted by the nerve ending. Since the junctional complex, containing the subsynaptic membrane and other structures, remains, after the osmotic shock, attached to the isolated nerve ending and to the nerve-ending membrane, it was expected that such membranes would also contain the chemical receptors. This property, so far, has been investigated with the use of radioactive cholinergic blocking agents such as dimethyl C^{14}-d-tubocurarine or methyl C^{14}hexamethonium. C^{14}5HT and radioactive adrenergic blocking drugs are being also used in our laboratory.

The use of a gradient centrifugation of the bulk fraction M_1 led to the separation of different types of nerve-ending membranes (*De Robertis* and coll., 1966). It was demonstrated that those membranes which were richer in AChE had also the highest binding capacity for the cholinergic blocking agents (Table 4). Further studies using the non-ionic detergent Triton X-100 led to the solubilization of most of the AChE, but the binding capacity of the sediment remained intact (*Fiszer* and *De Robertis*, 1967). The electron microscope revealed that, in this solution, most of the nerve ending membranes had disintegrated while the junctional complex with the subsynaptic membrane remained intact (*De Robertis* and coll., 1967 b) (Fig. 4).

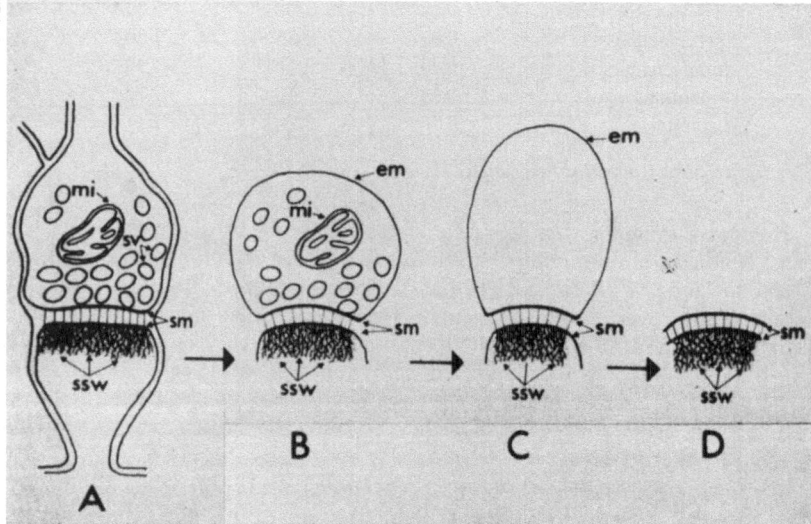

Fig. 4. Diagram of the systematic dissection of a synaptic ending from the cerebral cortex. Synaptic ending as observed (A) in situ, (B) after isolation, and (C) after the osmotic shock. D the junctional complex left after treatment with a detergent. Em Ending membrane, mi mitochondria, sv synaptic vesicles, sm synaptic membranes joined by the intersynaptic filaments, ssw subsynaptic web.

These findings opened the possibility of isolating the chemical receptor. Studies, now in progress, demonstrate that the binding capacity of the junctional complex toward dimethyl C^{14}-d-tubocurarine is probably due to a specific proteolipid protein which can be isolated from the nerve-ending membranes (*De Robertis* and coll., 1967 c).

Table 4. *Distribution of acetylcholinesterase (AChE), uptake of dimethyl-C^{14}-d-tubocurarine ($C^{14}DMTC$), methyl-C^{14}-hexamethonium $C^{14}MHM$), and H^3-alloferin in subfractions of M_1 on a density gradient Acetylcholinesterase is expressed in its relative specific activity and uptake of radioactive blocking agents as the specific binding ratio (counts per milligram of protein in fraction divided by counts per milligram of protein in the total particulate fraction).*

Sub-fraction	Structure	AChE	C^{14} DMTC	C^{14} MHM	H^3-Alloferin
M_1 0,8	Myelin	1,64	2,14	2,92	3,86
M_1 0,9	nerve-ending membranes	3,40	4,16	4,44	4,04
M_1 1,0	nerve-ending membranes	3,45	6,88	4,76	4,37
M_1 1,2	nerve-ending membranes	1,44	3,00	2,52	2,89
M_1 p	Mitochondria	0,38	1,60	0,72	1,87

References

Alberici, M., G. Rodriguez de Lores Arnaiz, and *E. De Robertis:* Catechol-O-methyltransferase in nerve endings of rat brain. Life Sci. *4,* 1951—1960 (1965).

De Robertis, E.: Ultrastructure and function in some neurosecretory systems. In: Neurosecretion (Proceedings of the Third Internat. Conf. on Neurosecretion, Bristol 1961), 3—20. London and New York: Academic Press, 1962.

De Robertis, E.: Histophysiology of synapses and neurosecretion. Oxford, London, Pergamon Press, 1964.

De Robertis, E.: Adrenergic endings and vesicles isolated from brain. (Second Symposium on Catecholamines. Milano, Italy 1965). Pharmacol. Rev. *18,* 413—424 (1966 a).

De Robertis, E.: Isolation of inhibitory nerve endings. In: Structure and Functions of Inhibitory Neuronal Mechanisms. Oxford-New York: Pergamon Press, 1968.

De Robertis, E., M. Alberici, G. Rodríguez de Lores Arnaiz, and *J. M. Azcurra:* Isolation of different types of synaptic membranes from the brain cortex. Life Sci. *5,* 577—582 (1966).

De Robertis, E., J. M. Azcurra, and *S. Fiszer:* Ultrastructure and cholinergic binding capacity or junctional complexes isolated from rat brain. Brain Res. *5,* 45—56 (1967 b).

De Robertis, E., and *H. S. Bennett:* Submicroscopic vesicular component in synapses. Fed. Proc. *13,* 35 (1954).

De Robertis, E., and *H. S. Bennett:* Some features of the submicroscopic morphology of synapses in frog and earthworm. J. biophys. biochem. Cytol. *1,* 47—58 (1955).

De Robertis, E., S. Fiszer, and *E. F. Soto:* Cholinergic binding capacity of proteolipids from isolated nerve-ending membranes. Science *158,* 928—929 (1967 c).

De Robertis, E., and *A. Pellegrino De Iraldi:* A plurivesicular component in adrenergic nerve endings. Anat. Rec., Philadelphia, *139,* 299 (1961).

De Robertis, E., and *A. Pellegrino De Iraldi:* A plurivesicular component in *C. J. Gómez:* Aislamiento de terminaciones nerviosas y vesículas sinápticas. (Sesiones de la Soc. Argentina de Biología, Mendoza 1960) 24—25 (1960).

De Robertis, E., A. Pellegrino De Iraldi, G. Rodriguez de Lores Arnaiz, and On the isolation of the nerve endings and synaptic vesicles. J. biophys. biochem. Cytol, *9,* 229—235 (1961).

De Robertis, E., A. Pellegrino De Iraldi, and *G. Rodriguez de Lores Arnaiz: L. Salganicoff:* Cholinergic and non-cholinergic nerve endings in rat brain. I. Isolation and subcellular distribution of acetylcholine and acetylcholinesterase. J. Neurochem. *9,* 23—25 (1962).

De Robertis, E., A. Pellegrino De Iraldi, G. Rodriguez de Lores Arnaiz, and *L. M. Zieher:* Synaptic vesicles from the rat hypothalamus. Isolation and norepinephrine content. Life Sci. *4,* 193—201 (1965).

De Robertis, E., G. Rodríguez de Lores Arnaiz, L. Salganicoff, A. Pellegrino De Iraldi, and *L. M. Zieher:* Isolation of synaptic vesicles and structural organization of the acetylcholine system within brain nerve endings. J. Neurochem. *10,* 225—235 (1963).

De Robertis, E., O. Z. Sellinger, G. Rodriguez de Lores Arnaiz, M. Alberici, and *L. M. Zieher:* Nerve Endings in methionine sulfoximine convulsant rats. A neurochemical and ultrastructural study. J. Neurochem. *14,* 81—89 (1967 a).

De Robertis, E., and *A. Vaz Ferreira:* Submicroscopic changes of the nerve endings in the adrenal medulla after stimulation of the splanchnic nerve. J. biophys. biochem. Cytol. *3,* 611—614 (1957).

Fatt, P., and *B. Katz:* Spontaneous subthreshold activity at motor nerve endings. J. Physiol. *117,* 109—128 (1952).

Fiszer, S., and *E. De Robertis:* Action of Triton X-100 on ultrastructure and membranes-bound enzymes of isolated nerve endings from rat brain. Brain Res. *5,* 31—44 (1967).

Gray, E. G., and *V. P. Whittaker:* The isolation of synaptic vesicles from the central nervous system. J. Physiol. *153,* 35—37 P (1960).

Kataoka, K., and *E. De Robertis:* Histamine in isolated small nerve endings and synaptic vesicles or rat brain cortex. J. Pharmacol. Exper. Therap., Baltimore, *156,* 114—125 (1967).

Lapetina, E. G., E. F. Soto, and *E. De Robertis:* Ganglioside and acetylcholinesterase in isolated membranes of the rat brain cortex. Biochem. biophysica acta, N. Y., *135,* 33—43 (1967).

Maynert, E. W., R. Levi, and *A. J. D. De Lorenzo:* The presence of norepinephrine and 5-hydroxytryptamine in vesicles from disrupted nerve endings particles. J. Pharmacol. Exper. Therap., Baltimore, *144,* 385—392 (1964).

Pellegrino De Iraldi, A., H. J. Farini Duggan, and *E. De Robertis:* Adrenergic synaptic vesicles in the anterior hypothalamus of the rat. Anat. Rec., Philadelphia, *145,* 521—531 (1963).

Roa, P. D., J. K. Tews, and *W. E. Stone:* A neurochemical study of thiosemicarbazide seizures and their inhibition by amino-oxacetic acid. Biochem. Pharmacol. *13,* 477—487 (1964).

Rodriguez de Lores Arnaiz, G., and *E. De Robertis:* 5-Hydroxytryptophanedecarboxylase activity in nerve endings of rat brain. J. Neurochem. *11,* 213—219 (1964).

Salganicoff, L., and *E. De Robertis:* Subcellular distribution of glutamic decarboxylase and gamma-aminobutyric alphaketoglutaric transminase. Life Sci. *2,* 85—91 (1963).

Sisken, B., E. Roberts, and *C. F. Baxter:* "γ-Aminobutyric acid and glutamic decarboxylase activity in the brain of the chick. In: Inhibition in the Nervous System and GABA. Roberts, E. ed. New York. Pergamon Press, 1960.

Whittaker, V. P.: The isolation and characterization of acetylcholine containing particles from brain. Biochem. J., London, 72, 694—706 (1959).

Whittaker, V. P., I. A. Michaelson, and *R. J. Kirkland:* The separation of synaptic vesicles from nerve endings particles (synaptosomes). Biochem. J., London, 90, 293—305 (1964).

Whittaker, V. P., and *M. N. Sheridan:* The morphology and acetylcholine content of isolated cerebral cortical synaptic vesicles. J. Neurochem. 12, 363—372 (1965).

Zieher, L. M., and *E. De Robertis:* Distribución subcelular de noradrenalina y dopamina en el cerebro de rata. (Congreso Latinoamericano de Ciencias Fisiológicas, Viña del Mar, Chile), November 23—28, 1964.

Author's address: Prof. Dr. *E. De Robertis,* Instituto de Anatomía General y Embriología, Facultad de Medicina, Universidad de Buenos Aires, Buenos Aires, Argentine.

Discussion

Aalberse: Dr. *De Robertis,* what was the percentage of dense-core vesicles that you found in your vesicle fraction? Did you try *in vitro* incubation of the vesicles with acetylcholine?

De Robertis: In the vesicle fraction isolated from the anterior hypothalamus of the rat there were between 10—20% of granulated vesicles. In a similar fraction from the cortex only very few granulated vesicles may be seen among the clear type of vesicles. Changes in the percentage of granulated vesicles were found by *Pellegrino De Iraldi* and myself in *in vivo* experiments on peripheral adrenergic nerve endings using the action of different precursors for noradrenaline, MAO inhibitors and releasing drugs. We did not incubate our vesicular fraction with acetylcholine.

Philippu (to the question of Dr. *Aalberse*): We isolated noradrenaline containing vesicles from the hypothalamus of the pig and then incubated at 37° C. We found that acetylcholine (55 m moles/ml) caused a significant increase of the spontaneous release of noradrenaline.

Streefkerk: After potassium permanganate fixation catecholaminergic nerve fibers contain small dense-core vesicles, which possibly represent catecholamines (*Hökfelt, T.:* Acta Physiol. Scand. 69, 119—120 [1967]). This may agree with *Fuxe's* opinion that catecholamines are stored in small vesicles which are agranular after routine fixation (*Fuxe, K.,* and coll.: Amer. J. Anat. 117, 33—46 [1965]). However, it is accepted that catecholamines are stored in large (± 800 Å) dense-core granules. *Rinne* and *Arstila* (Med. Pharmacol. Exper. 15, 357—369 [1966]) found, following reserpine treatment, a decrease in number of such granules. I wonder whether small as well as large dense-core vesicles may contain catecholamines. The small vesicles then may be derivatives of the larger granules. May I ask two questions to Prof. *De Robertis:* 1. Do you think it possible that part of the small (so-called synaptic-like) vesicles in axons and nerve endings are remnants of larger (catecholamine-containing or neurosecretory) granules? 2. Is anything known on gradient centrifugation of neurosecretory nerve endings, especially with regard to the properties of the fraction containing the small (synaptic-like) vesicles?

De Robertis: Ad 1. I think there exist at least three types of catecholamine containing vesicles or granules, all of them dense-cored: a) small (400 Å) vesicles, in peripheral nerve endings; b) larger (± 800 Å) granules, in nerve endings of the central nervous system; c) yet larger granules, *e. g.* in the cells of the adrenal medulla. I would stress again the necessity of a combined morphological and physiological approach to this sort of problems. Ad 2. No answer can be given.

Picard: I should like to give some additional information on the aminergic fibers and their vesicles and granules:

1. Large granules of unknown nature are to be found in such fibers in the peripheral nervous system; *Taxi* terms them "neurosecretory-like" on the basis of their morphology, density and size; but he thinks no decisive argument exists as to their functional significance. Comparison with recent findings of *Hökfelt* on aminergic neurons in the central nervous system seems to demonstrate similar granules in these fibers. *Hökfelt* suggests that these granules originate in the pericaryon, namely in the Golgi apparatus. They might give rise to smaller ones, *i. e.* dense-core vesicles usually accepted as characteristic of aminergic terminals.

2. During the recent meeting of the Association des Anatomistes, Orsay, March 1967, two observations were presented, one from *Filogamo's* laboratory, demonstrating by the fluorescence method numerous noradrenergic fibers in the myenteric plexus of the small intestine, the other from *Taxi* who showed that fibers in that same plexus, though containing electron-lucent microvesicles only, are able to incorporate labelled noradrenaline injected *in vivo*.

3. *Richardson,* confirmed by *Fuxe* and *Hökfelt,* and also *Tranger* and *Thoener,* demonstrated that fixation procedures (respectively permanganate or glutaraldehyde) may increase the number of granular vesicles seen in adrenergic fibers.

Finally, a number of consistent findings appear to confirm the view expressed by *Fuxe* and *Hökfelt* that aminergic fibers contain only one kind of vesicles which might be either "empty" or "granular", according to various factors including technical processing, so that these fibers might contain an apparently heterogeneous, though actually homogeneous population of vesicles. This is why, in the particular case of neurosecretory systems, it should be suggested to re-investigate completely the significance of the term "synaptic-like" microvesicles. In the present state of our knowledge, morphological features of electron-lucent microvesicles might correspond either to acetylcholine-bearing or to monoamine-bearing structures, which can only be decided by convergent techniques.

Csillik: I think that Prof. *Picard* is stressing the importance of fixation in questions related to the morphology of synaptic vesicles. One has to bear in mind the fact that we are dealing with biological material, *i. e.* with living tissue, the very *in vivo* structure of which we do not know *a priori.* Then we introduce some fixative and start the electron microscopic procedure. Accordingly, the pattern we finally observe in the electron microscope is partly characteristic of the living tissue, but partly characteristic also of the fixative. Therefore, with objects in or below the range of 200 Å one must be very careful; electron microscopy, in my opinion, is not a direct but an indirect method of investigation, in spite of the optical patterns this method results in. Another point might be relevant for the dense core in the granulated vesicles. Recently, my coworker Miss *Knyihár* and I undertook a study of the secondary degeneration of sympathetic nerve fibers in the rat pineal body. As is well known, this organ contains a lot of adrenergic fibers producing a strong fluorescence with the *Falck*-technique. After bilateral removal of the cervical superius ganglion, the catecholamine content of these fibers disappears as may be judged from the complete absence of fluorescence 48 hours after ganglionectomy. In spite of this, however, one finds a fair number of dense-

core vesicles in the degenerating fragments of nerve fibers, even 3 or 4 days after denervation. Therefore, let me propose to regard the dense core as a subcellular machinery, capable of storing catecholamines, but not as identical with the catecholamines themselves.

Hauswirth: I wonder not to have heard the name of an author whom I judge to be very important, namely *Zondek.* His findings are some more then 20 years old and perhaps obsolete. But, in fact, the sympathetic nerves cannot function without calcium and the parasympathetic ones not without potassium. It seems to me, that both these ions are quite forgotten in the investigations of neurohumoral and neurohormonal functions. Only Prof. *De Robertis* has mentioned calcium. In my physiotherapeutic praxis I always give, if other therapies do not work, calcium. Then, stress therapy becomes effective again.

De Robertis: Ca^{2+} is an important ion for the preservation of membrane structures at the ultrastructural level and is usually added to fixatives. Furthermore it is well known that in cholinergic synaptic transmission there is an entrance of Ca^{++} into the nerve endings which is related to the release of the transmitter. This ion is also important in the release of catecholamines from the adrenal medulla as well as from other neurohumoral and neurohormonal secreting organs.

Carlsson: It would be of great help if the electron microscopists would care to clarify the relationsship between vesicles of various densities and various sizes and the sites of catecholamine storage. Some investigators appear to look upon the large dense-core vesicles as the main storage sites, others (*e. g. Fuxe* and *Hökfelt*) claim that the small vesicles are mainly responsible. These vesicles are usually empty in electron micrographs, but with improved techniques they are said to possess a density which can be correlated with the presence of catecholamines.

De Robertis: The intermediate type of dense-core vesicles (800—1200 Å) described by *Pellegrino De Iraldi* and coll. (Anat. Rec. *145*, 521 [1963]) was interpreted as containing catecholamines in view of the high content of monoamines in the anterior hypothalamus. Separation of the synaptic vesicles of this region showed a 5—6 fold increase in norepinephrine as compared with the same fraction of the whole brain. At the same time 10—20% intermediate dense-core vesicles were found in such a fraction (*De Robertis* and coll., Life Sci., *4*, 193 [1965]). Pharmacological evidence in favor of this interpretation was provided by the work of *Masuoka* and coll. (Experentia *31*, 121 [1965]; *Hashimoto* and coll., Jap. J. Pharmacol. *15*, 395 [1965]; *Ishii* and coll., Biochem. Pharmacol. *14*, 183 [1965]).

Recently, *Pellegrino De Iraldi* and *Etcheverry* have demonstrated a 49% increase in intermediate dense-core vesicles in the nerve endings of the median eminence of the rat treated with nialamide-l-Dopa. These findings, although highly suggestive, do not discard the possibility that smaller vesicles could contain catecholamines as is true for peripheral adrenergic nerve endings (*De Robertis* and *Pellegrino De Iraldi*, Anat. Rec. *139*, 299 [1931]). Undoubtedly, preparatory techniques influence the preservation of the vesicular content observed under the electron microscope.

Blümcke: Dr. *Carlsson*, have you any idea about the nature of the osmiophilic core (the so-called dense core) of the catecholamine vesicles? Do you believe that the dense core could be the transmitter substance itself or would you rather believe that the osmiophilic material consists of some compounds of proteins and lipids?

Carlsson: Since protein and other material (*e. g.* nucleotides) may be released along with the catecholamines this question may be difficult to answer at this moment.

Feldberg: I should like to ask Prof. *De Robertis* a question, but before doing so I must congratulate him on the beautiful demonstration of D-tubocurarine uptake by the subsynaptic fraction, an observation whith has a special appeal to the pharmacologist because thas compound acts postsynaptically. The question concerns your demonstration of a high histamine content in one of the nerve ending fractions. As there is no convincing evidence that histamine has a transmitter function, the presence of a pharmacologically active substance in the nerve ending fraction therefore does not necessarily indicate that it acts as a trinsmitter. Thus the question arises whether the presence of γ-aminobutyric acid (GABA) is evidence for its transmitter function. I should like to have your opinion on this problem.

De Robertis: I am very pleased with Prof. *Feldberg*'s remarks on our work on the binding capacity of the subsynaptic membrane. Regarding histamine, recently *Kataoka* and *De Robertis* (J. Pharmacol. and Exper. Therap. *156*, 114—125 [1967]) have shown that this biogenic amine is in the small nerve endings present in the mitochondria and in the heavy microsomal fractions. This fact, together with the high concentration of the amine in the synaptic vesicle suggests a possible synaptic role of histamine in the cerebral cortex. The evidences for GABA being a transmitter are generally accepted for Crustaceans and other invertebrates. In vertebrates the neurophysiological studies with intracellular recordings from several cell types, and particularly the Purkinje cell, have demonstrated that GABA is able to produce typical inhibitory synaptic potentials with hyperpolarization of the subsynaptic membrane. The conclusion reached at the Symposion of Stockholm (1966) on "Inhibitory Mechanisms" was that GABA could indeed be an inhibitory transmitter in vertebrates

Journal of Neuro-Visceral Relations, Suppl. IX, 277—282 (1969)

Role of Hypothalamic Monoamines in the Control of Pituitary Secretions

P. G. Smelik

Department of Pharmacology, University of Utrecht, Utrecht, The Netherlands

Summary

Apart from the role which may be played by hypothalamic polypeptides in the control of pituitary function, the possible function of hypothalamic monoamines should also be considered in this respect.

A dopaminergic tubero-infundibular tract has been described. The implantation of reserpine into the hypothalamus of rats causes a disappearance of the fluorescent material representing dopamine. In such animals pituitary-adrenocortical activity is normal, suggesting that the presence of monoamines in the median eminence is not obligatory for the control of ACTH secretion. However, hypothalamic reserpine implants appear to induce pseudopregnancy, indicating that prolactin secretion is increased. It is concluded that dopamine may act as an inhibitory neurotransmitter for prolactin release.

Introduction

Transmission of signals within the CNS may occur through release of substances which belong to two different groups, neurohormones or neurohumors, i. e. they are either of polypeptide nature, or they are much smaller in molecular size and generally called biogenic amines. There are some apparent functional and morphological differences between these two groups concerning their mode of synthesis, release, destination and action, although the differences are now not so clear as it seemed to be some years ago. In fact, the question whether and how the functions of polypeptides and of biogenic amines of central origin are correlated, can be considered as one of utmost importance and interest. It is quite remarkable that recently an increasing amount of data suggests that for the action of peptides the presence of biogenic amines may be necessary and vice versa. This relationship also seems to exist in the field of neuro-endocrinology, where the high degree of integration between the endocrine system and the autonomic nervous system is apparent in such a clear fashion, that it almost seems appropriate to speak of only one neuroendo-crine system. Since we know that the function of the pituitary gland as the endocrine mastergland is in fact controlled by the hypothalamus, this integration should not be astonishing. The hypothalamus has been recognized not only as the regulatory center for the pituitary gland, but also (and even much earlier) as the integrative center of the autonomic nervous system.

It is, however, only since about 15 years that endocrinologists started to discover the hypothalamus as a kind of super-master gland, producing and delivering a number of neurohormones through which the control of the pituitary hormones was thought to be achieved. The search for the identity of these peptides, commonly designated as releasing factors (RF) and thought to be derived from neurosecretory material, involved a vast amount of biochemical work with relatively little success, since not a single RF has been identified up to now. It is perhaps also because of the difficulties encountered in this field, that neuroendocrinologists increasingly became aware of the fact that the hypothalamus still is the integrative center for many vegetative functions in the first place, and that control of pituitary function may in fact be part and parcel of metabolic and emotional regulations which are located in this area of the brain stem.

The fact that the hypothalamus contains very high amounts of known neurotransmitter substances, like acetylcholine, serotonin, noradrenaline and histamine, may suggest a role of these compounds in the control of this pituitary, apart from the releasing factors. Whether such neurotransmitters would influence the pituitary directly or via the release of the peptide factors, is still an open question, but there are several indications that a mutual influence may exist.

Hypothalamic Monoamines

Some information is available on one group of neurotransmitters, since it is possible to visualize monoamines in histological sections with the histochemical fluorescence procedure, worked out by Drs *Carlsson, Falck* and coll. (see *Falck* and *Owman*, 1965). If one applies this method to the hypothalamus of the rat, an intense fluorescence representing monoamines can be seen in the median eminence. The location is such that the highest accumulation of monoamines is found in the outer zone of the median eminence, especially around the capillary loops of the portal vessel system. It is believed that this accumulation is due to the production of monoamines by neurons which have their cell bodies in the region of the arcuate nucleus (*Fuxe*, 1965). A few more aspects of this system deserve special attention.

1. It can be shown by suitable pharmacological pretreatment, that most if not all of the fluorescent material is derived from dopamine. Therefore, this neural system is considered to be dopaminergic (*Fuxe* and *Hökfelt*, 1966; *Smelik*, 1966).

2. Although the highest accumulation is in the median eminence, we were able to show that dopaminergic nerves also traverse the hypophysial stalk and seem to end in the intermediate lobe, suggesting a dopaminergic innervation of the pars intermedia (*Smelik*, 1966).

3. The location so close to the capillary loops strongly suggests that dopamine can be released into the portal vessel system, and thereby act directly upon the pituitary cells instead of via release of peptide factors.

4. The dopaminergic system appears to function in several respects, independently of the well-known neurosecretory tract. It has been shown by *Streefkerk* (1967) that under conditions in which the neurosecretory material is depleted, *e. g.* in dehydration and severe stress, the fluorescent material in the median eminence remains unchanged.

Local Depletion of Hypothalamic Monoamines

In order to know whether these monoamines in the median eminence are really involved in the control of pituitary secretions, we have tried to find out what happens with the secretion of pituitary hormones in the absence of these monoamines. It is well known that a drug like reserpine is able to deplete serotonin and catecholamines, and it has been shown that reserpine treatment completely abolishes the fluorescence derived from these compounds. This is interesting because reserpine also elicits a long-lasting hypersecretion of ACTH from the anterior pituitary, and it could be asked whether this ACTH release may actually be due to the central depletion of monoamines. In fact, this has been suggested by *Westermann,* and he has argued that not only ACTH hypersecretion but also the sedation following treatment with reserpine has a common cause, namely the depletion of monoamines, especially serotonin (*Westermann* and coll., 1962). To establish whether the monoamines in the median eminence would play a role in the control of ACTH secretion, systemic reserpine injections would not be useful, since one cannot differentiate between effects on the hypothalamus and effects on other brain structures, or even on peripheral targets. Therefore, it is necessary to obtain a local depletion of these substances in the hypothalamus. We could achieve this by implantation of $\pm 2\,\mu g$ reserpine into the hypothalamus. Depletion could be verified by the histochemical fluorescence method. It appeared that reserpine implants in the ventral part of the hypothalamus, just rostral to the median eminence, completely depleted the fluorescent material. The animals, treated in this way, were otherwise normal; there was no sedation and no diarrhoea, indicating that the implants did not have systemic effects.

Hypothalamic Monoamines and ACTH Secretion

The effect of systemic reserpine injections on ACTH secretion was studied first. As judged by the *in vitro* corticosteroid production by excised adrenals, ACTH hypersecretion after intravenous reserpine was identical in reserpine-implanted and sham-implanted rats. This would rule out the possibility that reserpine stimulates ACTH release by the depletion of hypothalamic monoamines.

Of course, it is still possible that reserpine would cause sedation and ACTH hypersecretion by acting on other brain structures. Therefore, we implanted reserpine into several subcortical sites, among which the amygdala complex, the hippocampus, caudate nucleus, septal area and reticular formation. However, we could not find an effect on either ACTH secretion or emotional behaviour (*Smelik* and *Van der Bilt*, 1967).

Since there are many stressful stimuli which can be expected to activate central adrenergic systems, we also studied the effect of traumatic stimuli, ether anesthesia and emotional disturbance on ACTH secretion in rats bearing reserpine implants. In these animals, there was also no indication of a change in stress-induced ACTH release, neither in intensity nor in duration, when hypothalamic monoamines were depleted (*Smelik*, 1967).

In conclusion, these experiments indicate that the presence of mono-amines is not obligatory for the control of ACTH secretion. This would appear in line with the observation mentioned earlier, that there is no apparent relation between the activity of the neurosecretory system and the dopaminergic system.

Hypothalamic Monoamines and Prolactin Secretion

In contrast to these findings, it appears that monoamines in the median eminence do play a role in the regulation of other pituitary hormones. This has been worked out now for prolactin. It has been known since long that reserpine and other drugs may cause pseudopregnancy, which is indicative for an increased release of prolactin (*Barraclough* and *Sawyer*, 1959; *Ratner* and coll., 1965; *Coppola* and coll., 1965). We found that local depletion of hypothalamic monoamines by reserpine implantation indeed causes pseudopregnancy as judged by the disappearance of the vaginal cycle and the formation of deciduomata after traumatization of one uterus horn.

That this is really due to the depletion of monoamines and not to other effects of reserpine in this region could be shown by treatment with a MAO-inhibitor. Inhibition of the breakdown of monoamines by monoamine oxydase abolishes the effect of reserpine, since the depleted monoamines are then still capable of reaching the receptor sites. The injection of a MAO-inhibitor before and after reserpine implantation completely prevented pseudopregnancy iduction (*Van Maanen* and *Smelik*, 1968).

These experiments indicate that monoamines (probably dopamine) would act as an inhibitor of prolactin release. Its removal enables the anterior pituitary to secrete prolactin in higher amounts. It remains to be seen whether dopamine is identical to the so-called PIF (Prolactin Inhibiting Factor), which is generally supposed to be a peptide, or whether dopamine would be essential for the availability of PIF. This aspect, and also the role of hypothalamic monoamines in the control of some other pituitary hormones, is under present investigation.

Concluding Remarks

The experimental evidence presented suggests that chemical mediators, not being polypeptides, in the brain stem are linked up in some fashion with pituitary function. This is certainly not only true for monoamines, but for other mediators as well. For instance, we have been able to show that the anticholinergic drug atropine blocks ACTH secretion when implanted into the hypothalamus (*Hedge* and *Smelik*, 1968). It could well be, therefore, that not monoaminergic but cholinergic transmission is involved in the control of ACTH secretion.

It is still a long way to go before a certain pattern will emerge of the interaction between neurohormones and neurohumors in controlling vegetative and endocrine regulation. For progress in this field, further integration of neuropharmacological, neuroendocrinological and morphological techniques will be essential.

References

Barraclough, C. A., and *C. H. Sawyer:* Induction of pseudopregnancy in the rat by reserpine and chlorpromazine. Endocrinology 65, 563—571 (1959).

Coppola, J. A., R. G. Leonardi, W. Lippmann, J. W. Perrine, and *I. Ringler:* Induction of pseudopregnancy in rats by depletors of endogenous catecholamines. Endocrinology 77, 485—490 (1965).

Falck, B., and *Chr. Owman:* A detailed methodological description of the fluorescence method for the cellular demonstration of biogenic monoamines. Acta Univ. Lund., Sectio II, 7, 5—23 (1965).

Fuxe, K.: Cellular localization of monoamines in the median eminence and the infundibular stem of some mammals. Zschr. Zellforsch. 61, 710—724 (1965).

Fuxe, K., and *T. Hökfelt:* Further evidence for the existence of tubero-infundibular dopamine neurons. Acta physiol. Scand. 66, 245—246 (1966).

Hedge, G. A., and *P. G. Smelik:* Corticotropin release: Inhibition by intrahypothalamic implantation of atropine. Science 159, 891—892 (1968).

Ratner, A., P. K. Talwalker, and *J. Meites:* Effect of reserpine on prolactin-inhibiting activity of rat hypothalamus. Endocrinology 77, 315—319 (1965).

Smelik, P. G.: A dopaminergic innervation of the intermediate lobe of the pituitary? Acta physiol. pharmacol. Neerl. 14, 92—93 (1966).

Smelik, P. G.: ACTH secretion after depletion of hypothalamic monoamines by reserpine implants. Neuroendocrinology 2, 247—254 (1967).

Smelik, P. G., and *F. van der Bilt:* The effect of reserpine implantation in subcortical structures on ACTH secretion and on locomotor activity. Acta physiol. pharmacol. Neerl. 14, 523 (1967).

Streefkerk, J. G.: Functional changes in the morphological appearance of the hypothalamo-hypophyseal neurosecretory and catecholaminergic neural system, and in the adenohypophysis of the rat. Thesis, Free University of Amsterdam, Department of Histology, 110 pp (1967).

Van Maanen, J. H., and *P. G. Smelik:* Depletion of monoamines in the hypothalamus and prolactin secretion. Acta physiol. pharmacol. Neerl. 14, 519 (1967).

Van Maanen, J. H., and *P. G. Smelik:* Induction of pseudopregnancy in rats following local depletion of monoamines in the median eminence of the hypothalamus. Neuroendocrinology 3, 177—186 (1968).

Westermann, E. O., R. P. Maickel, and *B. B. Brodie:* On the mechanism of pituitary-adrenal stimulation by reserpine. J. Pharmacol. Exper. Therap., Baltimore, 138, 208—217 (1962).

Discussion

Feldberg: Dr. *Smelik,* could you tell us if the implants were made on one side or bilaterally?

Smelik: Hypothalamic implants were made in the midline, just in front of the median eminence; others, like in the amygdala or hippocampus, were made bilaterally.

Westermann: Concerning the prolactin release you made the statement that dopamine exerts an inhibitory action. I should like to know which data led to this assumption.

Smelik: I have said that my impression is that dopamine in the CNS may play an inhibitory role. This is not only indicated by the fact that its depletion facilitates prolactine release, but it is also suggested by its action on other pituitary hormones and in other CNS structures, *e. g.* the extrapyramidal system.

Author's address: Dr. *P. G. Smelik,* Department of Pharmacology, Medical Faculty, University of Utrecht, Vondellaan 6, Utrecht, The Netherlands.

Journal of Neuro-Visceral Relations, Suppl. IX, 283—296 (1969)

The Autonomic Nervous System and Energy Metabolism

E. Westermann and **K. Stock**

Department of Pharmacology, Medical School of Hannover, Hannover, Germany

With 8 Figures

Summary

The mechanism by which fat stores are mobilized is of central importance to an understanding of energy metabolism in general. Apart from the fact that the dynamic state of adipose tissue can be regulated by the influx of glucose into the fat cell, a number of hormones are able to stimulate lipolysis as well as glycogenolysis by a direct interaction with a "Metabolic Receptor".

In this context it is difficult to overestimate the importance of the work of *Sutherland* and coworkers who developed the concept that cyclic 3', 5'-AMP is a "Second Messenger" in metabolic processes. This nucleotide, formed by adenyl cyclase in all nucleated cells, is the trigger through which a physiologic event — the release of a hormone — is translated to a metabolic event — lipolysis or glycogenolysis. Thus, it appears that the metabolic receptor is identical with the enzyme adenylcyclase which converts ATP to cyclic 3', 5'-AMP. In this reaction catecholamines exert a catalytic function by interaction with the enzyme and substrate as well.

Our experiments with various stimulants and inhibitors of lipolysis indicate that the adenyl cyclase — ATP — complex in adipose tissue behaves like a sympathetic β-receptor. Although both, α- and β-adrenolytics, are able to block the action of noradrenaline and adrenaline, only β-adrenolytics produce this effect already in very low concentrations, stero-specifically and in a competitive manner. In addition, β-adrenolytics have a second point of attack in the lipolytic systems, which is probably located beyond the adenyl cyclase step and possibly identical with that of α-adrenolytics. This suggestion, however, remains speculative as long as nothing is known about the interconversion of inactive and active triglyceride lipase.

Behaviour in unicellular organisms is manifested by a direct adaption of biochemical reactions to the stimuli of environmental changes. In the highly organized mammal, the information about the outside world must be relayed from sensory organs to the brain. Accordingly, the brain must coordinate the mechanisms which change the activity of biochemical reactions in response to sensory input. In this manner the brain controls not only psychological and physiological behaviour, but biochemical behaviour as well.

Some 20 years ago, W. R. Hess (1948) developed his concept of an "ergotropic" and an opposing "trophotropic" system. By electrical stimulation of well-defined areas in the diencephalon he was able to differentiate between "dynamogenic" and "endophylactic" zones in the central nervous system. The effects following a stimulation of the "trophotropic" system are mediated predominantly by the parasympathetic division of the autonomic nervous system, producing *inter alia* sleepiness and decreased responsiveness. Conversely, the effects of stimulation of "ergotropic" centers in the hypothalamus are mediated by the sympathetic nervous system, producing wakefulness, alertness and — last not least — a mobilization and combustion of metabolic fuel.

It has been known for a long time that metabolic fuel is stored as glycogen in various organs. In situations of increased energy requirements — to respond to environmental stimuli — the organism is in the position to rapidly enhance the mobilization and combustion of glycogen. However, the body's reserves of carbohydrates are very limited and would hardly be sufficient to cover the energy requirements during a longer period of starvation.

Research work of the past decade has disclosed that the organism has an additional large reservoir of matabolic fuel at its disposal. In contrast to the limited ability to store carbohydrate, the body has — perhaps unfortunately — an almost infinite capacity to store fat, and it is well known, that the energy density of triglycerides (9.3 kcal/g) is more than twice as high as that of glycogen (4.1 kcal/g). The actual difference is even greater, since fat is deposited already in droplets which can be easily displaced and accomodated in spaces of odd shape, while the necessary hydration dilutes the energy density of glycogen to about one quarter of its original value. Thus, more than 90% of the total amount of metabolic fuel is stored in fat depots.

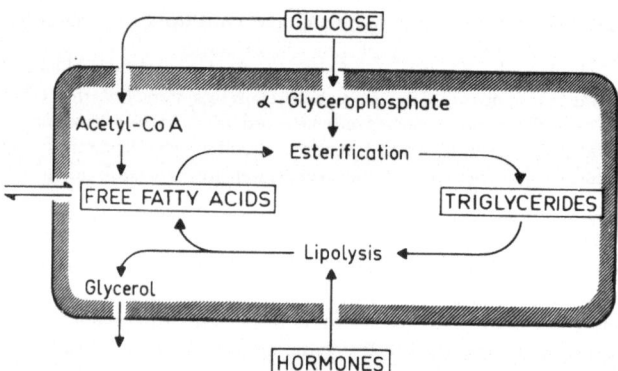

Fig. 1. The dynamic state of adipose tissue. Schematic representation of an adipose tissue cell, indicating some major pathways of fat deposition and mobilization. For explanation see text.

An understanding of the mechanisms by which fat stores are laid down and mobilized is of central importance to an understanding of energy metabolism in general. As can be seen from Fig. 1, glucose is normally the major precursor of adipose tissue triglycerides. An increased influx of glucose can give rise to both, the fatty acid moieties by way of acetyl CoA, and the glycerol moiety by way of α-glycerophosphate, thus promoting the esterification process in adipose tissue. The breakdown of triglycerides — lipolysis — is catalyzed by a hormone-sensitive lipase, also present in adipose tissue. While glycerol is leaving the fat cell, most of the formed free fatty acids are re-esterified to triglycerides, starting another cycle of fat deposition and mobilization. From this simplified scheme it becomes clear, that depot fat is in a true dynamic state.

In situations of increased energy requirements, the steady state in adipose tissue is disturbed: free fatty acids are released into the blood, serving as metabolic fuel in various organs. Therefore, the question arises: what are the mechanisms that regulate the output of free fatty acids from adipose tissue? Obviously, glucose is one of the controlling gears. If, for example, the influx of glucose is decreased, the esterification process is also reduced and a higher percentage of the formed free fatty acids are released into the blood. Conversely, an increased influx of glucose enhances the esterification process, and most of the formed free fatty acids are instantaneously deposited as triglycerides. This metabolic "self-regulation" can be demonstrated *in vivo* as well as *in vitro*.

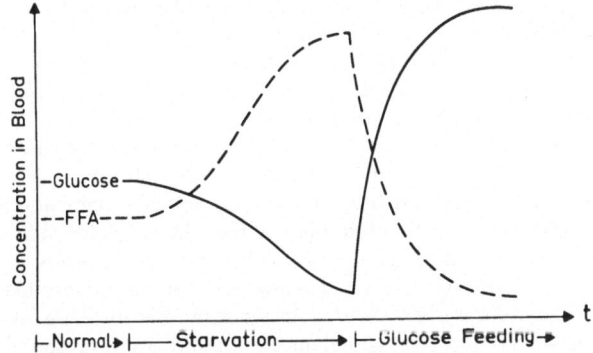

Fig. 2. Mirror-like changes of blood glucose and plasma free fatty acids (FFA). Schematic representation of experiments in rats. For details see text.

In normal rats on standard diet the level of free fatty acids is about 0.2 µEq/ml of plasma, and blood glucose amounts to 6 µmoles per milliliter. During starvation, as shown in Fig. 2, blood sugar decreases to 3.5 µmoles/ml within 24 hours. However, this reduction in carbohydrate fuel is compensated by an increased mobilization of fat, the level of plasma free fatty acids rising to 0.7 µEq/ml, thus providing the organism with a steady supply of a different metabolic fuel. Conversely, feeding the animals with glucose not only produces hyperglycemia but also decreases the level of plasma free fatty acids below the normal value. This mirror-like behaviour of the two metabolic fuels is mainly due to different rates of esterification in adipose tissue and controlled by the influx of glucose into the fat cells.

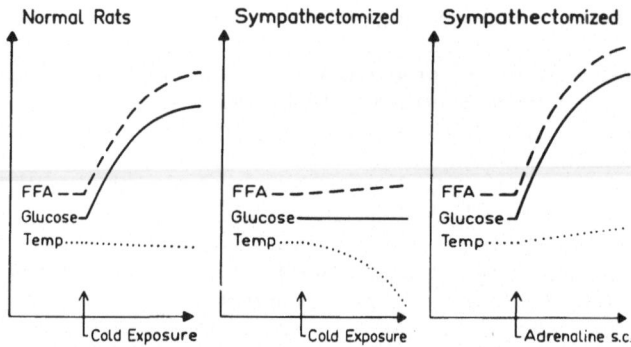

Fig. 3. Changes blood glucose, plasma free fatty acids (FFA) and body temperature (Temp). Schematic representation of experiments in rats. For details see *Stock* and *Westermann* (1965), and text.

However, there are situations in which the mobilization of free fatty acids is not regulated simply by the level of blood glucose. From the graphs shown in Fig. 3 it can be seen, that exposure to cold — by putting rats into the cold room at 4^0 C — increases the level of blood glucose. However, in this case hyperglycemia does not inhibit the mobilization of free fatty acids. The level of plasma free fatty acids is, on the contrary, increased by more than 100% after 4 hours of cold exposure.

A clue to the mechanism of this process is given by experiments with sympathectomized rats the results of which are also shown in Fig. 3. After adrenal demedullation and additional blockade of the sympathetic nerves, exposure to cold neither elevated plasma free fatty acids nor produced hyperglycemia. The inability of the sympathectomized animals to mobilize metabolic fuel is obviously the main reason for the rapid decline of body temperature which does not occur in intact rats. From these experiments it can be concluded that the simultaneous breakdown of stored fat and glycogen during cold exposure is governed by the sympathetic nervous system and is mediated by the release of catecholamines. Therefore, the injection of adrenaline should be able to mimic this behaviour. As can be seen from the right-hand graph in Fig. 3, in sympathectomized rats the subcutaneous injection of 0.5 mg/kg of adrenaline also produced hyperglycemia and an increase of plasma free fatty acids. In addition, a slight but definite increase in body temperature was observed in these animals, demonstrating the well known calorigenic effect of catecholamines.

Now the following questions arise: What is the exact mechanism of this rapid mobilization of metabolic fuel during sympathetic stimulation?

How can catecholamines enhance the breakdown of glycogen and triglycerides? Where is the link between adrenergic receptors and energy
metabolism?

The observation that adrenaline enhances the breakdown of glycogen
in isolated frog muscle was already made in 1920 by *Lesser*. In 1934.
Hegnauer and *Cori* were able to demonstrate that this effect is accompanied by a large increase in hexosephosphate and a corresponding fall
in inorganic phosphorus. However, it took another quarter of a century
until the intimate mechanism of this action was disclosed. In 1951,
Sutherland and *Cori* found that liver slices, pre-incubated with adrenaline,
contained more active phosphorylase than those not exposed to the hormone.
This work led to the most important discovery by *Sutherland* and *Rall*
(1900) that the increase in phosphorylase activity after exposure to adrenaline is mediated by cyclic adenosine-3',5'-monophosphate (3,5-AMP), a
metabolite of ATP. Research work of the past years proved that, besides
phosphorylase, an increasing number of other biochemical reactions is also
mediated by this cyclic nucleotide (Table 1).

Table 1. *Some Metabolic Processes known to be Influenced by Cyclic 3', 5'-AMP*

(+) = Increased;   (—) = Decreased

Phosphorylase (+)	*Sutherland* and *Rall* (1960)
Glycogen Synthetase (—)	*Rosell-Perez* and *Larner* (1964)
Triglyceride Lipase (+)	*Rizack* (1964)
Acetate → Fatty Acids (—)	*Berthet* (1960)
Tryptophane — Pyrrolase (+)	*Chytil* and *Skrivanova* (1963)
Amino Acids → Protein (—)	*Pryor* and *Berthet* (1960)
Phosphofructokinase (+)	*Mansour* (1965)
Steroidogenesis (+)	*Haynes* and coll. (1960)
Ketogenesis (+)	*Berthet* (1960)

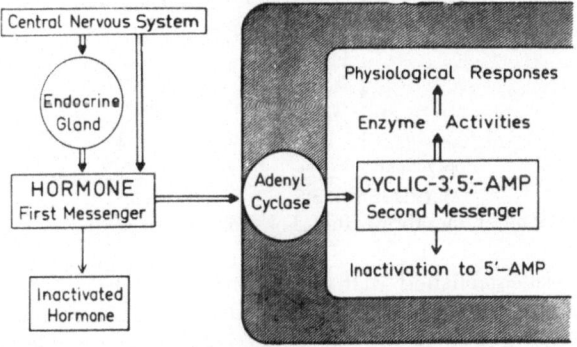

Fig. 4. The second messenger concept according to *Sutherland* and coll. (1965).
For explanation see text.

According to the concept of *Sutherland* and coworkers (1965), shown in Fig. 4, hormones like adrenaline only serve as "First Messengers" in biological reactions. Directed by the central nervous system, hormones are released from endocrine glands or autonomic nerves, thus getting in contact with cell membranes. It has been shown that in all nucleated mammalian cells adenyl cyclase is present, an enzyme which converts ATP to cyclic 3,5-AMP. This nucleotide, defined as "Second Messenger" by *Sutherland* and coll., serves as an immediate regulator of enzyme activities and physiological responses. Because of its poor lipid solubility, cyclic 3,5-AMP can hardly penetrate cell membranes, its effects therefore being restricted to the cell where it is formed. If, however, cyclic 3,5-AMP stimulates the biosynthesis of another hormone, this hormone once formed may leave the cell, thus acting as "Third Messenger" on other systems. In the pituitary-adrenal-axis, for example, the first messenger ACTH stimulates adenyl cyclase in the adrenal cortex, producing increasing amounts of the second messenger, 3,5-AMP. In this organ, the second messenger stimulates the synthesis of other hormones (corticoids), which can easily penetrate the cell membrane, thus serving as third messenger and affecting the function of other tissues.

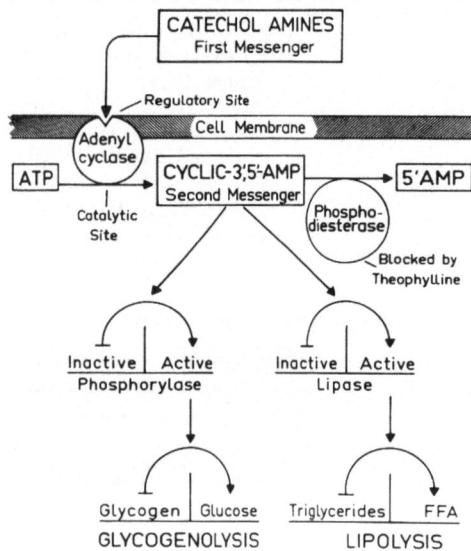

Fig. 5. Mobilization of metabolic fuel. For explanation of the diagram see text.

It is now well-established that the mobilization of fuel — the activation of lipolysis and glycogenolysis — is also mediated by the second messenger, 3,5-AMP. As can be seen from Fig. 5, catecholamines are able to stimulate the enzyme adenyl cyclase in various organs, thus converting increasing amounts of ATP to cyclic 3,5-AMP. From the work of *Sutherland* and his group it is known, that 3,5-AMP converts an inactive phosphorylase into an active form which catalyzes the breakdown of glycogen to glucose. Very recently published data strongly indicate that a similar mechanism holds also true for the activation of lipolysis (for reviews see *Butcher*, 1966; *Stock* and *Westermann*, 1965 a, b; 1966 a, b; *Westermann*, 1966, 1967 a, b). The second messenger, 3,5-AMP, converts an inactive triglyceride lipase into an active form, which catalyzes the breakdown of triglycerides to free fatty acids and glycerol. Eventually, glucose and free fatty acids are released into the blood, thus serving as a metabolic fuel in various organs (Fig. 5).

The question, whether catecholamines produce either a mobilization of free fatty acids or of glucose merely depends upon the enzymatic profile of the target organ and the availability of metabolic fuel to be metabolized. If, for example, liver is exposed to adrenaline, the 3,5-AMP formed will stimulate predominantly the breakdown of glycogen by an activation of phosphorylase. If, on the other hand, adipose tissue is exposed to adrenaline, the 3,5-AMP formed will stimulate predominantly the breakdown of triglycerides by an activation of the hormone-sensitive lipase.

However, there are hormones which stimulate adenyl cyclase in one organ only but not in another. A number of pituitary hormones are known to activate adenyl cyclase in adipose tissue without affecting this enzyme in liver. Probably, the molecular configuration of adenyl cyclase varies from one tissue to another, a circumstance which is not very surprising, since there are numerous examples of enzyme variation from tissue to tissue, and even within a single tissue. Possibly, there is also a variation in the components which are intimately associated with the enzyme in the particulate complex. Either type of variation could account for the specifity of hormone action in various organs.

In contrast to the rather specific action of hormones on the enzyme adenyl cyclase, an accumulation of 3,5-AMP is also observed after blockade of phosphodiesterase. This enzyme, which normally inactivates 3,5-AMP, can be inhibited in all organs by xanthine derivates such as theophylline. Therefore, it is to expect that after treatment with theophylline 3,5-AMP accumulates in all tissues, thus stimulating lipolysis as well as glycogenolysis.

Finally, exposure of the fat pads to the second messenger, 3,5-AMP itself, should also stimulate lipolysis and glycogenolysis. However, because of its poor lipid solubility, this nucleotide penetrates cell membranes only with difficulty and is rapidly metabolized by the enzyme phosphodiesterase. Therefore, only high concentrations of 3,5-AMP — if any at all — are able to affect enzyme systems within the intact cell. It has been shown that acyl derivatives of 3,5-AMP, as first described by *Posternak* and coll. in 1962, apparently penetrate cell membranes much easier than the parent compound and resist breakdown by phosphodiesterase. In studying the biological significance of the second messenger, the dibutyryl derivative of 3,5-AMP proved to be an excellent tool since its lipid solubility is approximately as high as its water solubility.

In summary it can be stated that various compounds of extremely different chemical structure produce an accumulation of 3,5-AMP, although by different mechanisms: catecholamines and other hormones by activating the enzyme adenyl cyclase, xanthine derivatives by inhibiting the enzyme phosphodiesterase and dibutyryl-AMP by mimicking the action of the second messenger.

Since adenyl cyclase is embedded in the cell membrane, two different sites of this enzyme can be distinguished: a regulatory site facing the extracellular fluid and a catalytic site which is in contact with the interior of the cell. According to this model, catecholamines and other hormones are able to interact with the regulatory site of adenyl cyclase and to influence the configuration of the catalytic site, thus affecting the enzyme activity and the rate of 3,5-AMP formation. Therefore, it is reasonable to assume that the regulatory site of adenyl cyclase is identical with a metabolic receptor.

Now the question arises: what are the pharmacological properties of this "metabolic receptor" which can be stimulated by catecholamines? According to *Ahlquist* (1948), two different types of adrenergic receptors can be distinguished: α-receptors which mediate excitatory actions of catecholamines, for example vasoconstriction, and β-receptors which mediate inhibitory actions of catecholamines. However, there are exceptions to this rule. Therefore, *Furchgott* (1959) introduced two additional types of adrenergic receptors: the so called gamma and delta receptors. This classification of adrenergic receptors is based on experiments with various sympathomimetic agents and sympatholytic drugs.

In order to get some information on the type of the metabolic receptor, we performed experiments with various stimulants and inhibitors of lipolysis. To avoid interferences with other hormones we choose a very simple *in vitro* system. Pieces of epididymal fat of rats were incubated for 60 minutes with various lipolytic stimulants. The formed free fatty acids were extracted with an organic solvent, determined quantitatively by titration and expressed as microequivalents of free fatty acids per gram tissue and hour.

As can be seen from Fig. 6, increasing concentrations of noradrenaline produced a dose-dependent lipolytic effect which was maximal at a molar concentration of less than 10^{-6}. A half-maximal effect — the formation of about 7 μEq of free fatty acids per gram and hour — was observed already at concentrations of less than 10^{-7} molar, which roughly corresponds to only 0.01 μg/ml of noradrenaline.

Incubation of the fat pads with theophylline also produced a lipolytic effect. In order to obtain a half-maximal effect, molar concentrations between 10^{-3} and 10^{-2} of this inhibitor of phosphodiesterase were needed. Exposure of the fat pads to cyclic 3,5-AMP also enhanced lipolysis. However, the concentration which was necessary to induce a half-maximal effect was approximately 6 orders of magnitude higher than that of noradrenaline. The dibutyryl derivative of 3,5-AMP proved to be about

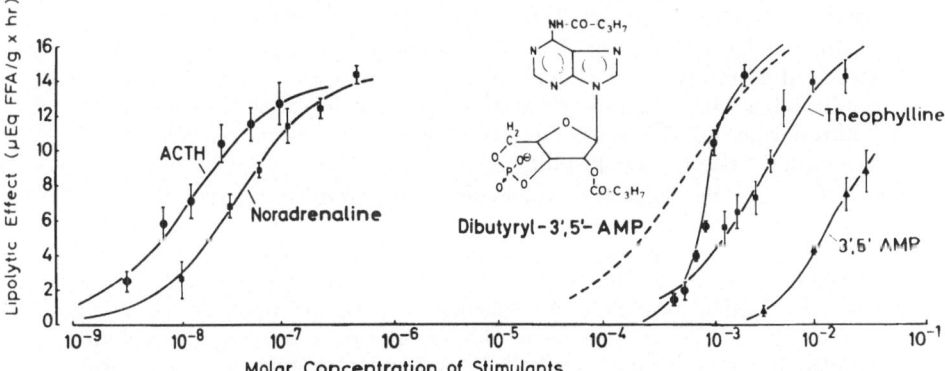

Fig. 6. Activity of various lipolytic stimulants on isolated epididymal fat pads of rats. For details see *Stock* and *Westermann* (1966), *Aulich, Stock* and *Westermann* (1967), and text.

100 times more active in promoting lipolysis than the unsubstituted nucleotide, but was still 4 orders of magnitude less active than noradrenaline. The polypeptide ACTH was the most active substance in this system: molar concentrations of about 10^{-8} already produced a half-maximal lipolytic effect.

Further studies were carried out with some adrenolytic agents (Fig. 7). In experiments shown in the upper graph of Fig. 7, lipolysis was stimulated by norepinephrine. The effect of 0.1 μg/ml of noradrenaline was set at 100%. Pre-incubation of the fat pads with increasing concentrations of the α-adrenolytic phentolamine — shown on the abscissa — did not affect lipolysis up to 10^{-5} M. However, higher concentrations of phentolamine gradually decreased the lipolytic response to noradrenaline: a 50%

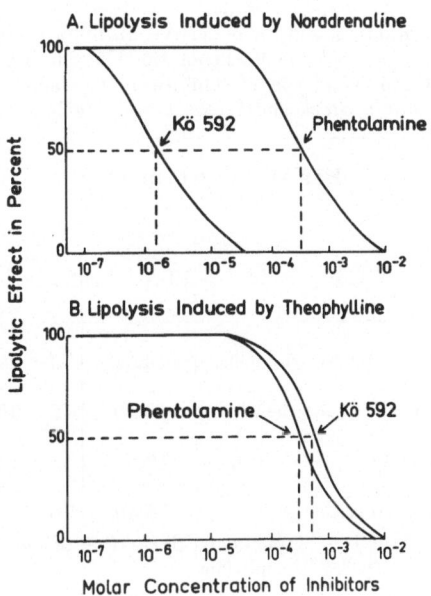

Fig. 7. Inhibition of lipolysis by the β-adrenolytic Kö 592 and the α-adrenolytic phentolamine. Chemical structure of the inhibitors are shown in Fig. 8. For details see *Stock* and *Westermann* (1965), and text.

inhibition was observed at a concentration of 4×10^{-4} molar, as indicated by the broken lines. The β-adrenolytic Kö 592, however, proved to be about 200 times more potent in inhibiting the norepinephrine-induced lipolysis than phentolamine. As indicated by the broken lines, 50% inhibition was observed already at a concentration of 2×10^{-6} M Kö 592.

In the lower graph of Fig. 7, norepinephrine was substituted by a standard concentration of theophylline as the lipolytic stimulant. While the α-adrenolytic phentolamine retained its potency in these experiments, the β-adrenolytic Kö 592 was now even less active than phentolamine in blocking the effect of theophylline.

Of course, the numerical values for a 50% inhibition, as indicated by the vertical dotted lines in this rather schematic presentation, are of limited significance since nothing can be concluded about the type of inhibition and the affinity of the inhibitors for their respective sites of action. However, by further analyzing the kinetics involved in lipolysis and its inhibition by adrenolytics we were able to determine the type of inhibition and the inhibitor constants of both Kö 592 and phentolamine as a measure of their affinities. The results are summarized in Table 2.

Table 2. *Profile of some Stimulants and Inhibitors of Lipolysis*

Determination of intrinsic activities (E_{max}) and apparent affinities (K_A) of lipolytic stimulants, and the respective inhibitor constants (K_i) of the α-adrenolytic phentolamine and the β-adrenolytic Kö 592. Chemical structure of the inhibitors see Figure 8. Type of inhibition: C competitive, NC non-competitive. Data from *Aulich, Stock* and *Westermann* (1967).

	Lipolytic Stimulants		Lipolytic Inhibitors	
Compounds	E_{max} (μEq FFA/g/hr)	K_A (moles/liter)	Inhibitor Constants (K_i=moles/liter) Phentolamine	Kö 592
D (—)—Noradrenaline	14.61	7.7×10^{-8}	2.8×10^{-4} NC	1.9×10^{-7} NC
ACTH (Synacthen®)	13.99	1.2×10^{-8}	3.3×10^{-4} NC	5.2×10^{-4} NC
Cyclic 3', 5' — AMP °	17.33	5.3×10^{-4}	4.8×10^{-4} NC	5.2×10^{-4} NC
Theophylline	16.80	9.3×10^{-3}	4.0×10^{-4} NC	7.4×10^{-4} NC

° The N^6-2'-O-dibutyuyl derivative of the nucleotide was used in the presence of 1×10^{-3}M Theophylline

The α-adrenolytic phentolamine inhibited non-competitively not only the effects of norepinephrine and theophylline but also those of dibutyryl-3,5-AMP and of ACTH, thus indicating that its site of action must be different from that of all stimulants applied. From these data the conclusion can be drawn that phentolamine acts beyond the adenyl cyclase step, *i. e.*, it blocks the action of 3,5-AMP formed in or added to the tissues.

On the other hand, the β-adrenolytic Kö 592 inhibited the effect of norepinephrine competitively. Its high affinity is emphasized by its low inhibitor constant of 2×10^{-7} M at which concentration a 50% saturation of the receptor sites is achieved. This type of inhibition and the structural resemblance between norepinephrine and Kö 592 leave little doubt that both compete for the same site at adenyl cyclase.

In contrast to norepinephrine, the effect of ACTH was blocked non-competitively by Kö 592 at considerably higher concentrations, although both hormones act on adenyl cyclase. This could mean that Kö 592, by occupying the "norepinephrine site" of adenyl cyclase, also prevents ACTH bound at its own receptor from activating the enzyme. However, since theophylline as well as dibutyryl-3,5-AMP responded in exactly the same manner to Kö 592 as ACTH did, this explanation is rather unlikely. Alternatively, it is more likely that the same type of inhibition and the practically identical inhibitor constants found with ACTH, theophylline and the nucleotide, reflect a second site of action of Kö 592 for which this β-adrenolytic agent has a considerably lower affinity. It should be added that identical results were obtained with other β-adrenolytics, such as propranolol and the newly developed Inpea (p-nitrophenyl-isopropyl-aminoethanol). Similar to phentolamine, this second site of action of β-adrenolytics must be localized beyond the adenyl cyclase step in the lipolytic system.

Further evidence for two different points of attack of β-adrenolytics is brought by experiments with the optical isomers of Inpea. We observed that the primary site of action of Inpea at the adenyl cyclase shows a high degree of stereo-specificity; in other words, l-Inpea was about 100 times more potent than d-Inpea in blocking competitively the effect of norepinephrine. In contrast, no stereospecificity was observed in the non-competitive inhibition of the lipolytic effect of ACTH, theophylline and dibutyryl-3,5-AMP.

The diagram in Fig. 8 is meant to summarize these rather complex data in a rather simplified way. According to the molecular concept of β-agonism developed by *Belleau* (1966) and by *Bloom* and *Goldman* (1966), the substrate ATP is attached to the surface of adenyl cyclase by weak interactions such as hydrogen bonding. Catecholamines are visualized to be bound to the enzyme by *Van der Waals* forces and hydrogen bonds as well as to ATP by a metal chelating mechanism requiring Mg^{++}. The steric configuration of this enzym-ATP-Mg^{++}-amine complex allows for a charge neutralization between the positively charged nitrogen of the amine and a negatively charged oxygen of the 5'-phosphate moiety of ATP. This ion-pair formation facilitates a phosphoryl group transfer or nucleophilic displacement reaction between the 3'-ribosyl-hydroxy group and the innermost phosphorus atom of ATP leading to the formation of cyclic 3,5-AMP. In turn, the nucleotide then promotes activation of the lipase.

Our experiments have shown that the action of noradrenaline was blocked in a competitive manner by the β-adrenolytic Kö 592. This effect is now easy to understand. Because of the similarity in chemical structure, Kö 592 should be able to displace noradrenaline in the enzyme-ATP-complex, but is obviously unable to take over its catalytic function.

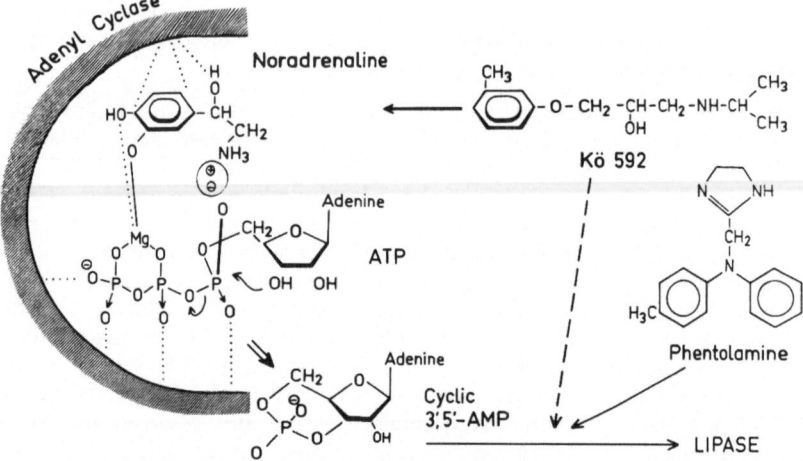

Fig. 8. Catalytic action of noradrenaline in the formation of cyclic 3', 5'-AMP according to the concept of *Belleau* (1966). Sites of antilipolytic action of the β-adrenolytic Kö 592 and of the α-adrenolytic phentolamine according to data of *Aulich, Stock* and *Westermann* (1967). For explanation see text.

The chemical structure of the α-adrenolytic phentolamine alone renders the possibility of a direct competition with catecholamines unlikely, and, indeed, this compound proved to be a non-competitive inhibitor. Since phentolamine not only inhibited the action of noradrenaline, but the lipolytic action of theophylline as well, it is reasonable to assume that α-adrenolytics do not interfere with the formation but block the action on lipase of the formed 3,5-AMP. High concentrations of the β-adrenolytic Kö 592 also inhibited the action of theophylline, *i. e.* the action of endogenous, 3,5-AMP. Therefore, a second site of attack must be assumed, which is possibly identical with that of α-adrenolytics.

References

Ahlquist, R. P.: A study of the adrenotropic receptors. Amer. J. Physiol. *153*, 586—600 (1948).

Aulich, A., Stock, and *E. Westermann:* Lipolytic effects of cyclic adenosine-3', 5'-monophosphate and its butyryl derivatives, and their inhibition by α- and β-adrenolytics. Life Sci. *6*, 929—938 (1967).

Belleau, B.: Steric effects in catecholamine interactions with enzymes and receptors. Pharmacol. Rev. Baltimore, *18*, 131—140 (1966).

Berthet, J.: Action du glucagon et de l'adrénaline sur le métabolisme des lipides dans le tissu hépatique. 4. Internat. Congr. Biochem. *17*, 107 (1960).

Bloom, B. M., and *I. M. Goldman:* The nature of catecholamine-adenine mononucleotide interactions in adrenergic mechanisms. Advances in Drug Res. Vol. 3, 121—169. London: Academic Press, 1966.

Butcher, R. W.: Cyclic 3', 5'-AMP and the lipolytic effects of hormones on adipose tissue. Pharmacol. Rev., Baltimore, *18*, 237—241 (1966).

Chytil, F., and *J. Skrivanova:* Reactivation of cortisone-induced liver tryptophan pyrrolase by boiled liver cell sap and by cyclic adenosine-3', 5'-phosphate. Biochim. Biophys. Acta *67*, N. Y. 164—166 (1963).

Furchgott, R. F.: The receptors for epinephrine and norepinephrine (adrenergic receptors). Pharmacol. Rev., Baltimore, *11*, 429—441 (1959).

Haynes, R. C., E. W. Sutherland and *T. W. Rall:* The role of cyclic adenylic acid in hormone action. Recent Progr. Hormone Res., N. Y., *16*, 121—133 (1960)

Hegnauer, A. H., and *G. T. Cori:* The influence of epinephrine on chemical changes in the isolated frog muscle. J. Biol. Chem., Baltimore, *105*, 691—703 (1934).

Hess, W. R.: Die funktionelle Organisation des Vegetativen Nervensystems. Basel. Benno Schwabe, 1948.

Lesser, E. J.: Der Mechanismus der Zuckermobilisierung durch das Adrenalin. Biochem. Zschr. *102*, 304—319 (1920).

Mansour, T. E.: Studies on heart phosphofructokinase; active and inactive forms of the enzyme. J. Biol. Chem., Baltimore, *240*, 2165—2172 (1965).

Posternak, Th., E. W. Sutherland, and *W. F. Henion:* Derivatives of cyclic 3', 5'-adenosine monophosphate. Biochim. Biophys. Acta, N. Y., *65*, 558—560 (1962).

Pryor, J., and *J. Berthet:* The action of adenosine-3', 5'-monophosphate on the incorporation of leucine into liver proteins. Biochim. Biophys. Acta *43*, 556—557 (1960).

Rizack, M. A.: Activation of an epinephrine-sensitive lipolytic activity from adipose tissue by adenosine-3', 5'-monophosphate. J. Biol. Chem., Baltimore, *239*, 392—395 (1964).

Rosell-Perez, M., and *J. Larner:* Studies on UDPG-α-glucan-transglucosylase. V. Two forms of the enzyme in dog skeletal muscle and their interconversion. Biochemistry *3*, 81—88 (1964).

Stock, K., and *E. Westermann:* Effects of adrenergic blockade and nicotinic acid on the mobilization of free fatty acids. Life Sci. *4*, 1115—1124 (1965 a).

Stock, K., and *E. Westermann:* Über die Bedeutung des Noradrenalingehalts im Fettgewebe für die Mobilisierung unveresterter Fettsäuren. Naunyn-Schmiedebergs Arch. exper. Path. *251*, 465—487 (1965 b).

Stock, K., and *E. Westermann:* Hemmung der Lipolyse durch α- und β-Sympathicolytica, Nicotinsäure und Prostaglolyin E₁. Naunyn-Schmiedebergs Arch. Pharmak. exper. Path. *254*, 334—354 (1966 a).

Stock, K., and *E. Westermann:* Competitive and noncompetitive inhibition of lipolysis by α- and β-adrenergic blocking agents, methoxamine derivatives and prostaglandine E₁. Life Sci. *5*, 1667—1678 (1966 b).

Sutherland, E. W., and *C. F. Cori:* Effect of hyperglycemic-glycogenolytic factors and epinephrine on liver phosphorylase. J. Biol. Chem., Baltimore, *188*, 531—543 (1951).

Sutherland, E. W., and *T. W. Rall:* Fractionation and characterization of a cyclic adenine ribonucleotide formed by tissue particles. J. Biol. Chem., Baltimore, *232*, 1077—1091 (1958).

Sutherland, E. W., and *T. W. Rall:* The relation of adenosine-3', 5'-phosphate and phosphorylase to the actions of catecholamines and other hormones. Pharmacol. Rev., Baltimore, *12*, 265—299 (1960).

Sutherland, E. W., and *G. A. Robison:* The role of cyclic 3', 5'-AMP in responses to catecholamines and other hormones. Pharmacol. Rev., Baltimore, *18,* 145—161 (1966).

Sutherland, E. W., I. Oye, and *R. W. Butcher:* The action of epinephrine and the role of the adenyl cyclase system in hormone action. Rec. Progr. Horm. Res. *21,* 623—646 (1965).

Westermann, E.: Drugs affecting the mobilization of free fatty acids. Pathophysiological and clinical aspects of lipid metabolism, pp. 38—48. Stuttgart: Georg Thieme 1966.

Westermann, E.: Sympathicus und Fettstoffwechsel. Acta neuroveget, Wien, *30,* 19—29 (1967).

Westermann, E.: Mechanismus und pharmakologische Beeinflussung der endokrinen Lipolyse. 12. Symp. Dtsch. Ges. Endokrinologie, pp. 154—173. Heidelberg/ New York: Springer, 1967b.

Westermann, E., K. Stock, and *P. Bieck:* False transmitter substances in mammalian adipose tissue. Progr. biochem. Pharmacol. *3,* 233—247 (1967).

Authors' address: Prof. Dr. med. *Erik Westermann* and P. D. Dr. med. *Klaus Stock,* Department of Pharmacology, Medical School of Hannover, Bissendorfer Straße 9, 3000 Hannover-Kleefeld, Germany.

Discussion

De Robertis: It is well known that adenyl cyclase has the highest concentration in brain. Prof. *Sutherland,* Dr. *Butcher* and our group working together in Buenos Aires (J. Biol. Chem., in press) have been able to show that this enzyme, as well as part of the cyclic phosphodiesterase, are preferentially localized in the nerve ending membranes. Furthermore, by using media with different sucrose concentrations on isolated nerve endings, it was demonstrated that adenyl cyclase has the active groups situated inside the membrane, so that it is only fully active when the nerve ending is disrupted osmotically. Now I would like to ask Prof. *Westermann* what is known about the possible function of adenyl cyclase in brain?

Westermann: I am afraid that I am not in a position to answer your question precisely, since little is known about central actions of cyclic 3,5-AMP. The high concentration of adenyl cyclase in brain at least indicates that a great number of biochemical processes in the central nervous system are regulated via a formation of cyclic 3,5-AMP. Recently, it hast been reported that the activity of adenyl cyclase in brain slices is affected not only by catecholamines but also by histamine — an interesting observation which indicates that this "local hormone" may play a role in brain function. Until today, only about one dozen of enzymatic reactions are known to be influenced by cyclic 3,5-AMP. but I am pretty sure that research work in the near future will show that in addition a large number of other enzymatic activities and physiological responses are also controlled by this cyclic nucleotide. Probably, numerous negative attempts to demonstrate an effect of cyclic 3,5-AMP on biochemical reactions are due to the fact that this nucleotide penetrates cell membranes only with difficulty. Therefore, studies in this field should be made with acyl derivatives (e. g. the dibutyril derivate) of cyclic 3,5-AMP which penetrate cell membranes much better and are more resistant towards breakdown by phosphodiesterase.

Journal of Neuro-Visceral Relations, Suppl. IX, 297—308 (1969)

The Importance of Monoamine Metabolism for the Pathology of the Extrapyramidal System

W. Birkmayer

Neurological Department of the Geriatric Hospital of the City of Vienna, Lainz

With 7 Figures

Summary

1. It has been made possible to objectify the kinetic effect of L-Dopa in Parkinson's disease by recording the kinetic energy with the help of the physiological acceleration transducer (Philips).

2. P-Tyrosine + NADH in the Parkinson patient lead to an increased active mobility similar to the Dopa-effect. P-Tyrosine without additives in choreatic persons leads to an increased choreatic commotion.

3. Alpha-methyl-P-Tyrosine leads to an increased akinesia in the Parkinson patient, having a sedative effect on the commotion of the choreatic.

4. An aminohydrazone derivative (Ro. 4-4602, Hoffmann-La Roche), being an effective counter-agent of the decarboxylase in the Parkinson patient, leads to a protracted and more intense kinetic L-Dopa-effect.

5. On the basis of these results the possibility is discussed that the defective Dopamine-synthesis in Parkinson's disease depends on an insufficient activity of tyrosinase.

In Parkinson's disease, particularly in the post-encephalitic form, in addition to the standard symptoms of rigor, tremors and akinesia, we observe crises in the form of Ewald's spastic vision, attacks of profuse perspiration, a type of flush syndrome, sebaceous secretion, flow of saliva, oedema in the leg, periods of sleep lasting for days, periods of gluttonous eating, diarrhoea hyperthermia and depressive indispositions. The phasic nature of these vegetative and affective disorders led us to believe that they might be due to an uncontrolled release effect of biogenic transmitter substances. Because of the vagomimetic reactions we first assumed a serotonin effect and believed a deficiency of noradrenaline to be responsible for the akinesia. During the phase of the first biochemical analyses the communication of *Carlsson* and coll. (1958) on the accumulation of dopamine in the extrapyramidal ganglia of the brain stem was published. Supplementary analyses of dopamine carried out subsequently showed considerable deficiencies. Table 1 shows the values registered. Thus it was proved that in Parkinson's disease there are deficiencies — in part considerable — of biogenic amines in the caudate nucleus, the putamen, the pallidum, the substantia nigra and in the hypothalamus (*Ehringer* and

Hornykiewicz, 1960; *Bernheimer* and coll., 1961). Since the biogenic amines do not pass the blood-brain barrier we tried to relieve the deficit in the extrapyramidal structures by supplying precursors such as L-Dopa,

Table 1

(In brackets the corresponding values in Parkinson patients)

	5 HT µg/g Tissue		Noradrenaline		Dopamine	
Striatum	0.33	(0.12)	0.1	(0.02)	3.5	(0.1)
Pallidum	0.23	(0.14)	0.02	(0.01)	0.1	(0.07)
Thalamus	0.26	(0.13)	0.05	(0.05)	0.01	(0.01)
Hypothalamus	0.29	(0.12)	1.33	(0.47)	0.02	—
Subst. nigra	0.55	(0.26)	0.04	(0.02)	0.46	(0.07)
Floor of the 4th ventricle	0.60	(0.55)	0.35	—	0.6	—

5 HTP and dops. Only in the case of L-Dopa at first clinically observable effect was produced, *i. e.* an improvement of the akinesia (*Birkmayer* and *Hornykiewicz,* 1961).

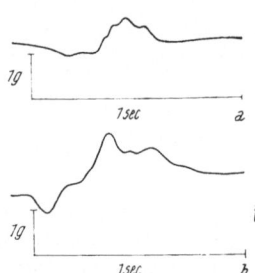

Fig. 1. Parkinson syndrome, straight pushing, registered by a "physiological acceleration transducer" (Fa. Philips).
a: kinetic energy 0.8 G. (G=9.81), time 0.25 sec.
b: 30 min. after 50 mg. l-Dopa i. v., kinetic energy 1.7 G/0.13 sec.

Fig. 1 shows such a kinetic effect measured with the physiological acceleration transducer (Philips). A pick-up registers changes in the physical state and transforms them so that they can be demonstrated on the oscillograph. In graph a the acceleration of a push straight forward was measured. This acceleration is 0.8 G (G = 9.81) and is reached within 0.25 seconds. In graph b, 30 minutes after the intravenous administration of 50 mg L-Dopa, an acceleration of 1.7 G is reached after 0.13 seconds. Fig. 2 (a, b, c) shows in graph a a push straight forward with an acceleration of 1.6 G/O, 32 seconds. Graph b represents the values reached 30 minutes after intravenous administration of 50 mg. L-Dopa, *i. e.* 4 G/O, 24 seconds. Graph c shows the values recorded after a repeated injection of 50 mg L-Dopa after another 60 minutes; now the acceleration is 3.4 G/O, 44 seconds. The primary kinetic effect cannot be increased by additional doses of L-Dopa. In accordance with these results we observe clinically increased active mobility, a cessation of propulsion and retropulsion and increased motoric agility. This kinetic effect lasts for 12 to 72 hours and may be reproduced time and again in every patient. The question now arose whether this dopamine deficiency is the result of accelerated reduction

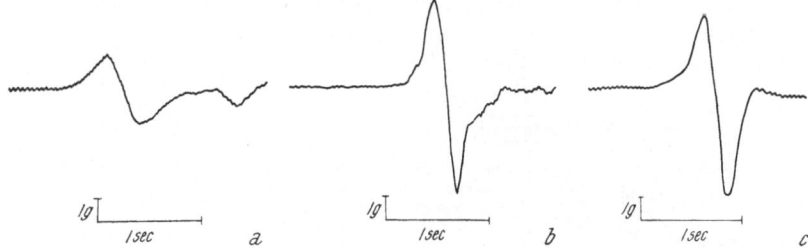

Fig. 2. Parkinson syndrome, straight pushing.
a: kinetic energy 1.6 G/0.32 sec.
b: 30 min. after 50 mg l-Dopa i.v.
   kinetic energy 4 G/24 sec
c: 60 min. later a 2nd injection of l-Dopa (50 mg i.v.),
   kinetic energy 3.4 G/0.44 sec.

or of insufficient synthesis. The activity of monoaminoxydase was found to be normal in the corresponding regions of the brain in patients suffering from Parkinson's disease (*Bernheimer* and *Hornykiewitz*, 1962). At the same time considerably reduced values of homovanillic acid in the cerebrospinal fluid of patients with Parkinson's disease indicated that increased reduction could not be the cause of the deficiency. The average values in 17 patients with Parkinson's disease amounted to 15 mg/ml cerebrospinal fluid, whereas the value was 60 ng/ml in 14 normal test persons (*Bernheimer* and coll., 1966). Fig. 3 shows the increase in biogenic amines following monoaminoxydase inhibitor therapy, whereas the values remained low, a fact that also indicates a blockage of the synthesis (*Bernheimer* and coll., 1962).

Now we had to clarify at which stage of the catecholamine synthesis the effect is produced (Fig. 4). In 50 patients with Parkinson's disease phenylalanine (administered in doses of 50—100 mg. intravenously) produced no kinetic effect. In some patients with Parkinson's disease M-tyrosine (50 mg

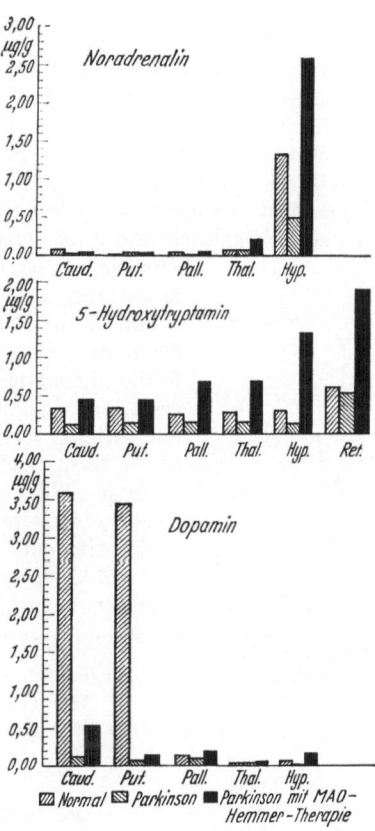

Fig. 3.

intravenously) produced considerable increase in blood pressure but no
kinetic effects (*Birkmayer*, 1966). P-tyrosine (200 mg.) plus 20 mg.
NADH (nicotinamide-adenindinucleotide-diphosphate) administered intra-

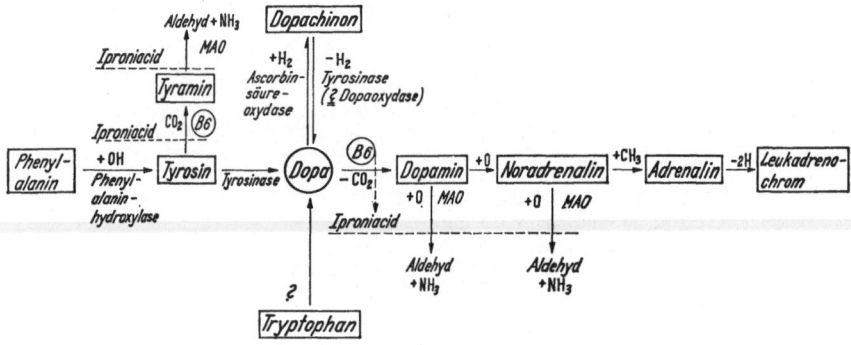

Fig. 4.

venously produced the same kinetic effects as did L-Dopa (*Birkmayer*
and *Mentasti*, 1967). Fig. 5 shows a push straight forward carried out by
a patient suffering from Parkinson's disease at a rate of 0.5 G/O, 6 seconds.
60 minutes after the administration of 200 mg of P-tyrosine plus 20 mg.
NADH the kinetic energy produced a reading of 1.5 G/O, 2 seconds. Table 2
shows that in patients suffering from Huntington's chorea P-tyrosine leads
to a considerable increase in the choreatic commotion.

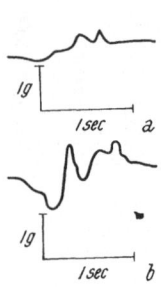

Administration of alpha-methyl-p-tyrosine (200 mg. administered intravenously) caused an additional aggra-
vation of the Parkinson akinesia and resulted in a slowing
down of motions and an incapacity for active mobility.
In the choreatic patients this treatment caused a complete
cessation of jactitation (Fig. 2). In further tests the Parkinson patients were given an aminoacid hydrazine derivate
(Ro 4-4602, Hoffmann-La Roche, Fig. 6) in addition to
L-Dopa. This resulted in a protracted and intensified
kinetic effect; in chorea patients we observed an increase
in the choreatic commotion (Table 2). As *Pletscher* and
coll. (1963) have shown, this hydrazine derivate is a very
effective decarboxylase inhibitor. In animal tests we observed a marked accumulation of L-Dopa in the brain

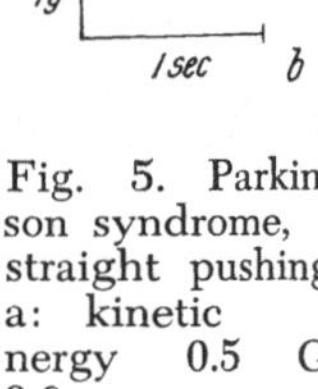

Fig. 5. Parkin-
son syndrome,
straight pushing.
a: kinetic
nergy 0.5 G/
0.6 sec.
b: 60 min.
after 200 mg. P-
Tyrosin+20 mg.
NADH i. v.,
kinetic energy
1.5 G/0.2 sec.

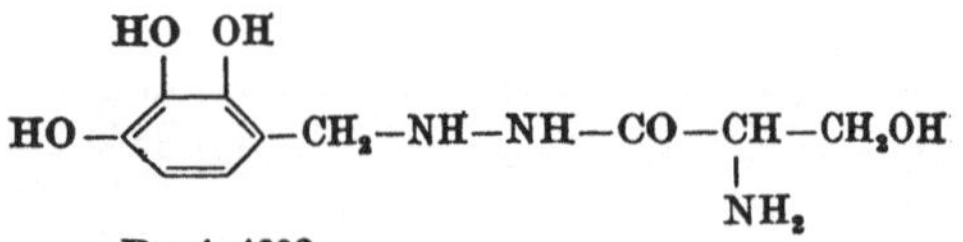

HO OH

HO—⟨◯⟩—CH₂—NH—NH—CO—CH—CH₂OH
$$HO-\bigcirc-CH_2-NH-NH-CO-CH-CH_2OH$$
$$NH_2$$

Ro 4-4602

Fig. 6.

(Bartholini and coll. 1967, **Fig. 7**). 25 Parkinson patients and 4 persons suffering from chorea Huntington were closely tested in a joint research effort with *Seitelberger* (neuropathology) and *Hornykiewicz* (biochemistry).

Table 3 presents a summary of the values recorded. Cases 1 to 6 were definitely postencephalitic Parkinson cases from the histological and clinical points of view, and a positive kinetic effect was observed after administration of L-Dopa. Biochemically, tremendously decreased values of dopamine were observed in the caudatum and in the putamen. The values for homovanillic acid were increased by one power on a base of 10; however, compared to other groups of Parkinson patients they were lower by half approximately. From a histological and clinical point of view cases 7 to 14 corresponded to degenerative morbus Parkinson. In these cases L-Dopa injections produced no objectifiable kinetic effect. Compared to postencephalitic Parkinson cases, the dopamine values were, on the average, considerably higher in the caudatum and equally low in the putamen; on the average the homovanillic acid was higher by half compared to the postencephalitic cases. In the caudatum and putamen the chorea cases showed far higher dopamine values and the values of homovanillic acid were also considerably increased.

Table 4 shows a comparison of the average values; this presentation does not claim statistical validity, but shows clearly the characteristic differences. Particularly striking is the biochemical differentiation between the postencephalitic Parkinson cases and morbus Parkinson in the dopamine content of the caudatum. The values for homovanillic acid in morbus Parkinson are twice those of the postencephalitic cases.

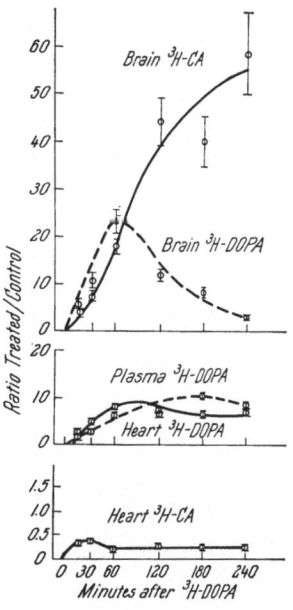

Fig. 7. Ratio of ^{3}H-Dopa and ^{3}H-catecholamines in treated versus control rats. The controls were injected with 375 γ/rat ^{3}H-Dopa i. p. alone, the treated were administered 50 mg./kg. Ro 4-4602 i. p. followed by 375 γ ^{3}H-Dopa i. p. after 30 min. The points represent averages of 3—5 experiments ± S. E. after A. *Pletscher* and coll.

The ratio between dopamine and its principal product of decomposition, homovanillic acid, may be regarded as a measure of the metabolic activity. In the caudatum of postencephalitic cases this ratio is one third of that in morbus Parkinson cases. However, in the putamen the value of this ratio for morbus Parkinson cases is cut in half. In chorea patients the dopamine and homovanillic acid readings in the caudatum and putamen are considerably higher and the value of the ratio of dopamine and homovanillic acid is almost ten times higher.

Table 2. *Effect of activity and sedation after different drugs*

Parkinson-Syndrome	Effect after 200 mg. P-Tyro-sine + NADH i. v.	Alpha-methyl P-Tyrosine 200 mg i. v.	Ro. 4-4602 + L-Dopa
H. M., 65 years, female	+ + +	— — —	+ + +
R. A., 60 years, female	+ + +	— — —	+ + +
D. E., 77 years, female	+ +	— —	+ +
K. M., 52 years, female	+ +	— —	+ +
Ch. A., 55 years, female	+	— —	+ +
T. A., 70 years, female	+ + +	— — —	+ + +
F. M., 71 years, female	+	— —	+
K. F., 72 years, male	±	±	±
R. F., 68 years, male	+ +	— —	+ +
St. R., 49 years, male	+ + +	— —	+ +
H. J., 71 years, male	+	—	+
K. L., 54 years, female	+ +	— —	+ +
E. M., 53 years, female	+ + +	—	+ +
F. K., 69 years, female	+	±	+
W. M., 72 years, female	±	±	±

Chorea syndrome	Effect after 200 mg. P-Tyro-sine + NADH i. v.	Alpha-methyl P-Tyrosine 200 mg i. v.	Ro. 4-4602 without L-Dopa
S. M., 53 years, female	+ + +	— — —	+ +
F. K., 73 years, female	+ +	— —	+ + +
K. M., 66 years, female	+ + +	— —	+ + +
S. M., 48 years, female	+ +	— —	+ +
H. M., 54 years, female	+	— —	+
D. J., 61 years, male	+ + +	— — —	+ + +

+ = Increase of activity or choreatic agitation.
— = Increase of akinesis or sedation of choreatic agitation.

Table 3

	Kinetic effect after L-Dopa	Dopamine μg./g.		Homovanillic acid μg/g		
		Caud.	Put.	Caud.	Put.	Pall.
1. R. F., 68 years	+ +	0.01	0.01	0.01	0.09	0.24
2. P. M., 49 years	+ + +	0.01	0.01	0.25	0.34	0.33
3. P. H., 78 years	+ +	0.07	0.03	0.31	0.51	0.31
4. S. H., 72 years		0.14	0.14	1.38	1.38	1.41
5. Sch. J., 62 years	+ +	0.04	0.04	0.06	0.11	0.14
6. Sch. J., 75 years	+ +	0.01	0.01	0.01	0.01	0.05
7. K. K., 69 years		0.15	0.01	1.8	1.7	0.92
8. V. N., 82 years	+ —	0.47	0.01	1.09	1.24	0.36
9. P. M., 74 years	+ —	0.66	0.01	0.85	0.49	0.53
10. R. H., 67 years	+ —	0.41	0.04	1.02	0.75	0.70
11. S. O., 65 years	—	0.01	0.09	0.01	0.64	0.95
12. Sch. A., 64 years	+ —	0.09	0.01	1.02	0.67	0.53
13. P. P., 75 years	+ —	0.31	0.06	0.76	1.18	n. i.
14. Sch. N., 75 years	—	0.57	0.13	1.54	0.66	1.92
15. G. M., 50 years	Chorea Huntington	0.64	2.65	0.55	2.55	1.04
16. H. H., 59 years	Chorea Huntington	1.19	2.03	1.55	2.87	1.42
17. S. M., 50 years	Chorea Huntington	1.34	2.15	1.61	1.85	1.42
18. K. M., 60 years	Chorea Huntington	1.56	2.43	2.11	3.75	1.75

Table 4. *Comparison of Dopamine and Homovanillic acid values and the ratio between Dopamine and Homovanillic acid*

	Dopamine		Homovanillic acid		Ratio Dopamin/ Homovanillic acid	
Postencephalit. Parkinson.	Caudatum	0.04	Caudatum	0.33	Caudatum	0.12
	Putamen	0.04	Putamen	0.4	Putamen	0.1
Morbus Parkinson	Caudatum	0.3	Caudatum	0.96	Caudatum	0.3
	Putamen	0.03	Putamen	0.91	Putamen	0.03
Chorea patient	Caudatum	1.18	Caudatum	1.3	Caudatum	0.9
	Putamen	2.31	Putamen	2.75	Putamen	0.85

Discussion

Which are the conclusions to be drawn on the basis of these findings for the pathogenesis of Parkinson's disease and chorea Huntington?

Alpha-methyl-P-tyrosine definitely aggravates the akinesia in Parkinson's disease. This substance blocks the synthesis of P-tyrosine into L-Dopa (*Spector* and coll., 1965). In animal experiments *Porter* and coll. (1966) succeeded in a long-term lowering of the dopamine content of mouse brains by application of alpha-methyl-P-tyrosine. The application of P-tyrosine plus NADH, a co-ferment of tyrosinase, causes a kinetic effect analogous to that obtained after administration of L-Dopa and, thus, is able to counteract temporarily the tyrosine deficiency. The administration of Ro 4-4602, which caused an increase of Dopa in the brain in animal experiments (*Bartholini* and coll., 1967), improves the akinesia in Parkinson's disease.

Thus, the decisive block in the catecholamine synthesis in Parkinson's syndrome occurs during the step leading from tyrosine to Dopa, a fact that may be regarded as the consequence of insufficient tyrosinase activity. This interpretation also corresponds to the morphological findings in Parkinson's syndrome, which show depigmentation of the substantia nigra as the only constant phenomenon. Since in the organism of animals the cells containing melanin are the main sources for the formation of tyrosinase, the insufficiency of this ferment in Parkinson's syndrome must, in the last analysis, be regarded as the causative factor for the lowered dopamine content. *Hornykiewicz* and coll. (1967) have shown that under the influence of alpha-methyl-P-tyrosine the dopamine content in the substantia nigra drops much more markedly than in the striatum, a fact confirming the above conclusions. The tyrosinase insufficiency inhibits the synthesis of dopamine at the site where it is mainly produced, *i. e.* in the substantia nigra. This agrees with our clinical experiments and closes the pathogenetic causal chain for Parkinson akinesia. The destruction of the ganglion cells containing melanin, particularly in the substantia nigra, leads to insufficient tyrosinase activity resulting, in turn, in blocking the step from tyrosine to Dopa in the synthesis. The immediate consequences are the dopamine deficiency and the akinesia caused by it. *Falck* was able to demonstrate in impressive pictures the transportation of dopamine from the substantia nigra by way of dopaminergic fibers into the cells of the pallidum, putamen and caudatum. It is, therefore, not surprising that in Parkinson's syndrome, as shown in tables 3 and 4, the content of dopamine and homovanillic acid is particularly decreased in these nuclear areas. Even if there is no histological damage (putamen, caudatum), the deficiency in the transmitter substance dopamine is responsible for reduced metabolism, which, in turn, causes the deficiency symptom of akinesia. The specially lowered ratio between dopamine and homovanillic acid as a measure of reduced dopamine metabolism is merely a biochemical consequence of insufficient nigral synthesis. However, we cannot yet explain, why in cases of morbus Parkinson the dopamine content of the caudatum should be considerably higher that that of the putamen, and why application of Dopa does not result in a kinetic effect.

In chorea of Huntington the dopamine and homovanillic acid values in the substantia nigra, the pallidum, the putamen and the caudatum are within the normal range. However, compared to Parkinson's syndrome the value for the ratio between dopamine and homovanillic acid shows an approximately ten-fold increase in the metabolism of the transmitter substance.

It is natural to regard this increased dopamine metabolism as the cause of choreatic hyperkinesis [*]. The application of P-tyrosine or L-Dopa increases the choreatic hyperkinesis (*Birkmayer*, 1966; *Gerstenbrand* and *Pateisky*, 1962). The administration of the hydrazine derivate Ro 4-4602, which leads to an increase in Dopa in the brain, also increases choreatic commotion. Based on these clinical experiments we may term a disinhibited dopamine metabolism the cause of choreatic hyperkinesis. At present, we do not know at all which morphological lesion causes this metabolic disorder. The only morphological basis of Huntington's chorea that is definitely established is the destruction of the small striatum cells. Since, however, data on the function of these cells are lacking, we can only point to this correlation without drawing any final conclusions.

References

Bartholini, G., W. P. Burkhard, A. Pletscher, and *H. M. Bates:* Increase of cerebral catecholamines by 3,4-dihydroxyphenylalanine after inhibition of peripheral decarboxylase. Nature *215,* 852—853 (1967).

Bernheimer, H., and *O. Hornykiewicz:* Das Verhalten einiger Enzyme im Gehirn Normaler und Parkinson Menschen. Naunyn-Schmiedebergs Arch. exper. Path. *243,* 295 (1962).

Bernheimer, H., W. Birkmayer, and *O. Hornykiewicz:* Verteilung des 5-Hydroxytryptamin im Gehirn des Menschen und sein Verhalten bei Patienten mit Parkinson Syndrom. Klin. Wschr. *39,* 1056—59 (1961).

Bernheimer, H., W. Birkmayer, and *O. Hornykiewicz:* Verhalten der Monoaminoxydase im Gehirn des Menschen. Wien. klin. Wschr. *74,* 558—59 (1962).

Bernheimer, H., W. Birkmayer, and *O. Hornykiewicz:* Homovanillinsäure im Liquor cerebrospinalis. Untersuchungen beim Parkinson Syndrom und anderen Erkrankungen des Zentralnervensystems. Wien. klin. Wschr. *78,* 417—419 (1966).

Birkmayer, W.: Experimentelle Befunde und neue Aspekte bei extrapyramidalen Erkrankungen. Wien. Zschr. Nervenhk. *23,* 128—139 (1966).

Birkmayer, W., and *O. Hornykiewicz:* Der L-3.4.Dioxyphenylalanin (Dopa) Effekt bei der Parkinson-Akinese. Wien. klin. Wschr. *73,* 787—788 (1961).

Birkmayer, W., and *O. Hornykiewicz:* Weitere experimentelle Untersuchungen beim Parkinson Syndrom und Reserpin Parkinsonismus. Arch. Psychiatr. *206,* 367—381 (1964).

Birkmayer, W., and *M. Mentasti:* Weitere experimentelle Untersuchungen über den Catecholaminstoffwechsel bei extrapyramidalen Erkrankungen (Parkinson- und Chorea Syndrom). Arch. Psychiatr. *210,* 29—35 (1967).

---

[*] Alpha-methyl-P-tyrosine has a sedative effect upon the choreatic commotion.

Carlsson, A., M. Lindquist, T. Magnusson, and *B. Waldeck:* On the presence of 3-hydroxytyramine in brain. Science *125,* 471 (1958).

Ehringer, H., and *O. Hornykiewicz:* Verteilung von Noradrenalin und Dopamin im Gehirn des Menschen und ihr Verhalten bei Erkrankungen des extrapyramidalen Systems. Klin. Wschr. *38,* 1236—1239 (1960).

Gerstenbrand, F., and *K. Pateisky:* Über die Wirkung von L-Dopa auf die motorischen Störungen beim Parkinson Syndrom. Wien. Zschr. Nervenhk. *20,* 90—100 (1962).

Hornykiewicz, O., J. Lisch, and *H. Denk:* Wirkung von Chlorpromazin und Alpha-methyl P-Tyrosin auf den Dopamingehalt des nigrostriären Systems. Naunyn-Schmiedebergs Arch. exper. Path., im Druck (1967).

Pletscher, A., K. F. Gey, and *W. P. Burkhard:* Beeinflussung des cerebralen Stoffwechsels von 5-Hydroxytryptamin durch Decarboxylase-Hemmung. Helvet. physiol. pharmacol. Acta *21,* 46—50 (1963).

Porter, C. C., J. A. Totaro, A. Burcin, and *E. R. Wynosky:* The effect of the optical isomeres of alpha-methyl-P-tyrosine upon brain and heart catecholamines in the mouse. Biochem. Pharmacol. *15,* 583—590 (1966).

Spector, S., A. Sjoerdsma, and *S. Udenfried:* Blockade of endogenous norepinephrine synthesis by alpha-methyl-tyrosine, an inhibitor of tyrosinehydroxylase. J. Pharmacol. Exper. Therap. Baltimore *147,* 86—95 (1965).

Author's address: Prof. Dr. W. *Birkmayer,* Neurologische Abteilung des Altersheimes der Stadt Wien-Lainz, Versorgungsheimplatz 1, A-1130 Wien.

Discussion

Feldberg: You might be interested, Prof. *Birkmayer,* in an observation recently reported by *Martha Vogt.* She perfused the anterior horn of the cerebral ventricles in cats and found in the effluent an increased release of homovanillic acid from the caudate nucleus on stimulation of the substantia nigra.

Birkmayer: This finding suggests that our opinion about the way of the transmitter substance dopamine is probably right.

Ratzenhofer: In the large obduction material of the "Landeskrankenhaus" in Graz, comprising also many deceased of advanced age, cases of Parkinsonism have become very rare. This is partly due to the favorable result of operation in cases in which this disease is not in a much advanced stage, but also partly to the accumulation of patients in Parkinson-centers as that of Prof. *Birkmayer.* I can remember only a few cases of postencephalitic Parkinsonism showing unequal paling of the zona nigra and signs of degeneration in the basal ganglia. Furthermore, I should like to ask whether the preparation Ro 4-4602 is identical with the antibiotic and cytostatic drug of the same name.

Birkmayer: No, Ro 4-4602 is not identical with a cytostatic drug. For the neurohistological differentiation between degenerative and postencephalitic Parkinsonism especially the inclusion bodies and Alzheimer's fibrils are important. It is interesting that Alzheimer's fibrils are localized where biogenic amines have been demonstrated and that they exclusively occur in postencephalitic Parkinsonism. In cases of Morbus Parkinson Alzheimer's fibrils are absent, but inclusion bodies can be observed. Therefore, by means of neurohistopathology a differential diagnosis is quite possible. It is not right to base the diagnosis exclusively on the findings in the substantia nigra.

Pedersen: In animals, neuroaminergic synapses have been demonstrated in the spinal cord. These are terminals of descending inhibitory pathways from the brain stem, and they influence afferent inflow to the spinal cord and interneuronal activity. It would be interesting to know whether increase in the catecholamine level in man gives rise to changes in the stretch reflex, especially the phasic part, and motoneuron excitability tested by the H-reflex of *Hoffmann*. Were such alterations demonstrated in the patients?

Birkmayer: The rigor after l-Dopa is diminished. The T-reflex (tendon reflex) is inhibited and the H-reflex is without alteration.

Schümann: I should like to know how long the favorable effect lasts of an intravenous injection of l-Dopa given in combination with the Ro compound?

Birkmayer: From 2—6 days, varying individually. The kinetic effect is, indeed, striking but after the second day the patient starts to vomit. Vomitting may occur during several days. This is a disadvantage.

Carlsson: Yes, Dopa has, of course, an emetic effect that is very strong. Why you get this, in using this particular combination, is hard to tell. Of course there is first the problem why you get this potentiation of the Dopa response by this special compound. I understand that the Ro compound is a decarboxylase-inhibitor. The dosis used in these patients is sufficient to block the peripheral Dopa-decarboxylase in the kidney and so on. It is not sufficient to block the Dopa-decarboxylase in the brain, because the compound does not penetrate so readily into the brain and, therefore, the Dopa that is preserved in the periphery will get into the brain. This is probably the main mechanism, but of course there could be another compound involved in this action. We have found in animals that in very large doses some monoamine-oxidase inhibition probably occurs. So, this could be an additional action. The reason why I bring this up is that in Göteborg a doctor gave a monoamine-oxidase inhibitor and than Dopa to a Parkinson patient. This patient started to vomit. This vomitting, after treatment for one or two days with Dopa followed by an interruption, went on for several days. It became really very troublesome and the doctor stopped trying. Could it be, Dr. *Birkmayer,* that you gave too much Dopa? Possibly, if you give a little bit too much than you get this vomitting.

The Parkinson patient may be also in an unfavorable position, because his dopaminergic neurons which start in the nigra and end in the putamen and caudate nucleus have probably degenerated, as is shown by biochemical evidence. Dopamine is no longer present. I think the pathologists have to look very carefully whether the cells did not also die. Anyway, the enzyme is not to be found in any considerable amount, so how could Dopa act? That is the problem. If really you have a 100% Parkinson patient, there are no dopaminergic fibers left and no Dopa-decarboxylase. I do not think there will be any action at all. The reason why you have an effect probably is that there are a few fibers left.

The ideal drug in this case, I think, would not be anything like Dopa, which acts presynaptically that is it has to be converted presynaptically in those nerve endings that have more or less disappeared. Such a drug would have to act like Dopamine in activating the dopamine receptors in the postsynaptic cell which is probably present, just waiting for the missing transmitter. Prof. *Ernst* has made the very interesting observation that apomorphine, which incidentally has a structure that is very similar to Dopa, is able to activate dopamine receptors even in the absence of these neurons. Now we are again in a dilemma of course, because, just like Dopa, apomorphine is an emetic. But from the theoretical point of view, I think, it would possibly be more favorable. Actually there are some very early reports that apomorphine is acting clinically. Prof. *Ernst* also has initiated some clinical studies which seem to confirm this.

Birkmayer: Does apomorphine pass the blood-brain barrier?

Carlsson: It does. It is a dopamine-analogue that is able to pass the blood-brain barrier and to activate dopamine receptors. In Göteborg we have made studies that confirm Prof. *Ernst's* observations and indicate very strongly that we are dealing with a direct activation of dopamine receptors and a selective activation, so that noradrenaline receptors are apparently not activated. It might be interesting for neurologists to take this up again. Of course there is the problem of the emetic effect, but Prof. *Ernst* has a dose that he can recommend. How much is it? A couple of milligrams perhaps?

Ernst: It is a dose of 2 milligrams per day.

Carlsson: Divided in several doses?

Ernst: No, just one dose. Vomitting does not occur.

Kappers: Is it known whether Dopa is also stored in the area postrema or in adjacent regions? I ask this because this area has by some authors been regarded as an emetic center.

Carlsson: The histochemists in Stockholm have been looking for dopamine in this region. So far, they have not been successful. Apparently only noradrenergic fibers are present.

Birkmayer: Can I ask again something? We have seen with the combination of L-Dopa and MAO-inhibitors paranoid-schizophrenic reactions in 10% of 350 Parkinson-diseased. Hallucinations frequently occur. If we discontinue the injection with MAO-inhibitor the schizophrenic symptoms disappear, and if we give the MAO-inhibitor again the same symptoms reappear after 10 days. This effect is reproducable. Do you have any explanation for this?

Carlsson: We know, of course, that amphetamine is one of the most potent drugs to produce paranoid reactions in patients. Misuse of amphetamine or metaphetamine is very dangerous from that point of view. It is also known that amphetamine is a drug which acts centrally indirectly by releasing catecholamines in the brain. Whether it acts by releasing either dopamine or noradrenaline I do not know. It could be noradrenaline and it would also be quite possible that the balance between the noradrenaline-like effect of Dopa and the dopamine-like effect of Dopa is shifted by giving a MAO-inhibitor so that you get more of the noradrenaline-like reaction. In favor of this is that the noradrenergic fibers are much more sensitive to release by amphetamine than the dopaminergic fibers. This comment, therefore, is somewhat in favor of noradrenaline being involved in the schizophrenic reaction after amphetamine.

This is only a tentative explanation. I think, you should publish this very interesting observation Dr. *Birkmayer*, because I do not think you have done so yet.

Neurohormonal and Neurohumoral
Regulatory Mechanisms

Journal of Neuro-Visceral Relations, Supp. IX, 311—328 (1969)

The Neuronal Connexions of the Hypothalamic Neurosecretory Nuclei in Mammals

A G e n e r a l S u r v e y

H. J. Lammers

Department of Anatomy and Embryology, University of Nijmegen,
The Netherlands

With 5 Figures

Summary

Regarding the hypothalamic magnocellular and parvocellular neurosecretory nuclei as constituting a final common path *(Sherrington)*, influencing the activity of neuro- and adenohypophysis respectively, a survey is given of the intrinsic and extrinsic afferent neuronal relations of these nuclei as revealed by Golgi and Nauta-Gygax studies.

The parvocellular nuclei receive their most important afferent input from the medial hypothalamus, *viz.* the ventro-medial nucleus as well as from the periventricular hypothalamic area. The ventro-medial nucleus serves as a relay center between the lateral hypothalamus forming part of the limbic system-midbrain circuit on one hand and the parvocellular nuclei, *viz.* the arcuate or infundibular nucleus on the other.

Data on the afferent relations of the magnocellular nuclei are still scarce. It seems justified however, to assume that these nuclei receive their main afferent input from the limbic system-midbrain circuit, *viz.* from the septal and lateral preoptic area. Attention is paid to recent findings on aminergic and cholinergic afferent projections to the neurosecretory nuclei of the hypothalamus. As a preliminary conclusion one may presume a rather dominant aminergic influence on the activities of the parvocellular nuclei, the magnocellular nuclei being influenced by both aminergic and cholinergic afferents.

Introduction

Even if a number of details about the mechanism of neurosecretion and the nuclei concerned with this mechanism are still to be elucidated, there is substantial evidence to designate as neurosecretory nuclei not only the magnocellular supraoptic and paraventricular nucleus, but also the parvocellular arcuate or infundibular nucleus of the pars tuberalis of the hypothalamus and the ventral part of the anterior periventricular nucleus. The parvocellular as well as the magnocellular nuclei form the ultimate efferent link between the central nervous system and the pituitary gland. In this respect they may be considered (*Scharrer*, 1965) as constituting a *'final common path'*, a term originally coined by *Sherrington* (1947) to designate the function of the motor neurons of the spinal cord. Thus the central nervous system, by an affluence of nervous stimuli converging upon the neurosecretory cells, controls the production of specific substances to be released at the termination of their axons in the neural lobe or in the median eminence.

Doubtless the activities of the neurosecretory cells are influenced not only by nervous stimuli but also by humoral stimuli from the circulating blood and from the cerebrospinal fluid. Although this gives rise to a number of interesting questions, in particular concerning the interrelation of the nervous and humoral stimuli converging upon these cells, the highly complex problems underlying such questions are beyond our scope here. I intend only to deal with the afferent neuronal relations of these nuclei and those areas of the central nervous system in which they originate.

In spite of the large number of studies that have appeared about the hypothalamus and its connexions, reliable data on the afferents to the neurosecretory cells must still be said to be scarce, at least if strict criteria for their evaluation are applied.

Most of the older data at our disposal derive either from studies of normal material of the hypothalamus or from studies of experimental material, based on histological methods which apparently were apt to give misleading information. Since the last decade, however, studies have appeared that are based on the more reliable Nauta-Gygax-technique (*Nauta* and *Gygax*, 1954) for demonstrating the course of degenerated fibers after experimental lesions. Nevertheless, a drawback still inherent in this technique is that the degenerated terminal fibers, and degenerated very fine fibers in general, escape impregnation. So even this technique may cause some discrepancies concerning the determination of the final termination of fiber projections. In this respect much might be expected of a recent modification of the Nauta-method as described by *Fink* and *Heimer* (1967).

Furthermore, when trying to investigate experimentally the interrelations of the secretory and non-secretory hypothalamic nuclei we are faced with the double problem posed by the very small dimensions of the hypothalamus and by the presence of numerous fiber systems passing through this area, owing to which our interpretations may be easily misguided.

The only method capable of supplying us with reliable if not complete information about the intrinsic relationships of neuronal complexes is the Golgi-technique, for this reason in recent years once more a focus of interest. However, up to the present only one study on the hypothalamus has been carried out by means of this technique (*Szentágothai* and coll., 1962). This study has yielded some very interesting informations about the intrinsic structure of the hypothalamus to which I shall refer at greater length in the next section dealing with the intra-hypothalamic relations of the neurosecretory nuclei. In the subsequent sections, the extra-hypothalamic afferent relations will be discussed. As it would carry us too far to have a full discussion of all the afferent projections to the hypothalamus described in the literature (for extensive references see *Diepen*, 1962, and *Knook*, 1965), I shall restrict myself to a survey of the most important ones, as revealed by means of the Nauta-Gygax-technique as well as by the recent histochemical methods for the demonstration of cholinergic and aminergic systems respectively.

1. The Intrinsic Neuronal Relations of the Hypothalamus

Szentágothai and coll. in their study already mentioned arrive at the important conclusion that the basic structure of the hypothalamus is one that almost ideally gives shape to an n-dimensional neural network. *Harris* already in 1958 drew attention to the remarkable similarity in structure of the hypothalamus and the reticular formation of the brainstem. *Nauta* (1963) suggested bringing the paleothalamic cell groups, the subthalamus, the hypothalamus, the preoptic area, and the septum under one common denominator as the *"forebrain reticular formation"*.

The principal nodal point of the hypothalamus appears to be the ventro-medial nucleus. The axons of this nucleus, although finally running in a dorsal direction, at first take a highly tortuous course, with numerous initial collaterals at random. These initial collaterals "secure connexions of each cell with probably the majority of its immediate neighbours situated in whatever direction" (*Szentágothai* and coll., 1962). Thus the ventro-medial nucleus has contact on one side with the lateral hypothalamic zone, where most of the afferent extrinsic connexions arrive, and on the other with the dorso-medial hypothalamic nucleus as well as with the nucleus arcuatus (nc. infundibularis, *Diepen*, 1962) the axons of which constitute the tubero-infundibular tract (tubero-hypophyseal tract of Spatz) terminating around the capillary loops of the primary plexus of the hypothalamo-hypophyseal portal system. In this way, the ventro-medial nucleus forms a vital link between the extrinsic afferent systems on the one side and the nucleus arcuatus on the other.

Another important associative system in the hypothalamus is made up of the periventricular area. We have to do here with an extensive complex of small cells, for the greater part vertically or obliquely orientated, the main dendrites of which take their ramifications parallel to the ventricle's surface and run generally from dorso-lateral to ventro-medial. The axons leave the cells in a lateral direction and make contact with the dorso- and ventro-medial nuclei.

It is evident that the periventricular area receiving a large number of afferents from extra-hypothalamic areas (see section 2.6) may subserve an important relay function for the parvocellular neurosecretory nuclei too. Unfortunately, the Golgi-technique has failed to inform us about the synaptic field of the magnocellular nuclei, because their cells appear to be absolutely refractory to this technique.

2. The Extrinsic Afferent Relations of the Neurosecretory Nuclei

By means of the Nauta-Gygax-technique, the following afferent relations
have been revealed:

1. afferents from the basal olfactory area and the amygdala,
2. afferents from the limbic area and the septum,
3. afferents from the neocortical areas,
4. afferents from the strio-pallidum,
5. afferents from the dorsal thalamus and the epithalamus,
6. afferents from the mid-brain.

For the topography of the regions involved I refer to Fig. 1.

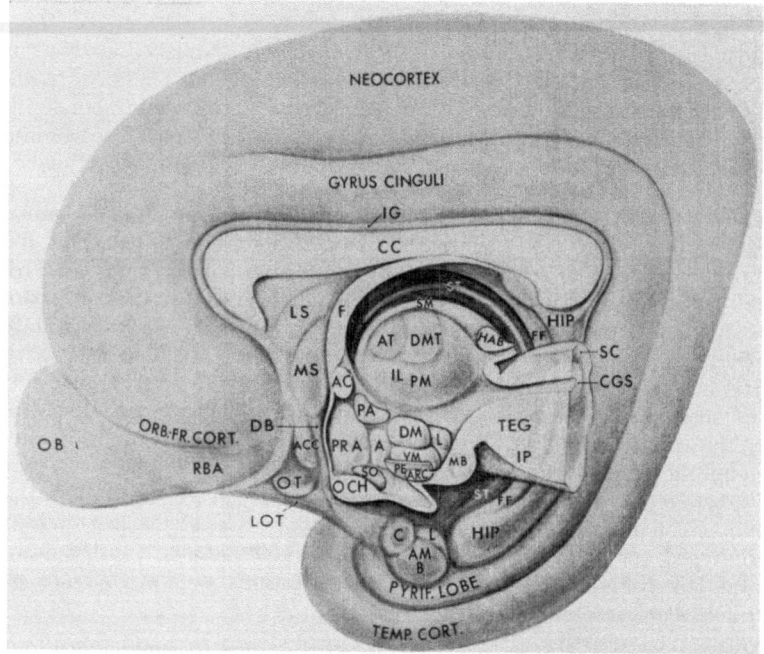

Fig. 1. Topography of the hypothalamus. Medial view. A anterior hypothalamus,
AC anterior commissure, ACC nucleus accumbens, AMB amygdala, basal nucleus,
AMC amygdala, cortico-medial nucleus, AML amygdala, lateral nucleus, ARC
arcuate nucleus, AT anterior thalamic nuclei, CC corpus callosum, CGS central
grey midbrain substance, DB diagonal band of Broca, DM dorso-medial hypothalamic nucleus, DMT dorso-medial thalamic nucleus, F fornix, FF fimbriae
fornicis, HAB habenula, HIP hippocampus, IL intralaminar thalamic nuclei,
IG indusium griseum, L lateral hypothalamic nucleus, LOT lateral olfactory
tract, LS lateral septal area, MB mamillary body, MS medial septal area,
OB olfactory bulb, OCH optic chiasma, ORB-Fr. Cort. orbito-frontal cortex,
OT olfactory tubercle, PA paraventricular nucleus, PE periventricular nucleus
(ventral part), PM paramedian thalamic nuclei, PRA preoptic area, PYRIF.
LOBE pyriform lobe, RBA retrobulbar area, SC superior colliculus, SM stria
medullaris, SO supraoptic nucleus, ST stria terminalis, TEG midbrain tegmentum,
          TEMP. CORT. temporal cortex, VM ventro-medial hypothalamic nucleus.

2.1. Afferents from the basal olfactory area and the amygdala (Fig. 2)

This area consists of the olfactory bulb, the retrobulbar area (anterior olfactory nucleus), the olfactory tubercle, the prepyriform cortex, the pyriform lobe, in so far as its periamygdaloid part is concerned, and the amygdala (*Pribram* and *Krüger*, 1954; *Gastaut* and *Lammers*, 1961).

All these areas are either primarily or secondarily related to the projection of olfactory impulses from the olfactory bulb (*Lohman*, 1963; *Lohman* and *Lammers*, 1967).

Incidentally, it should be noted that these areas not only receive olfactory afferents but also afferents from surrounding neocortical areas as well as

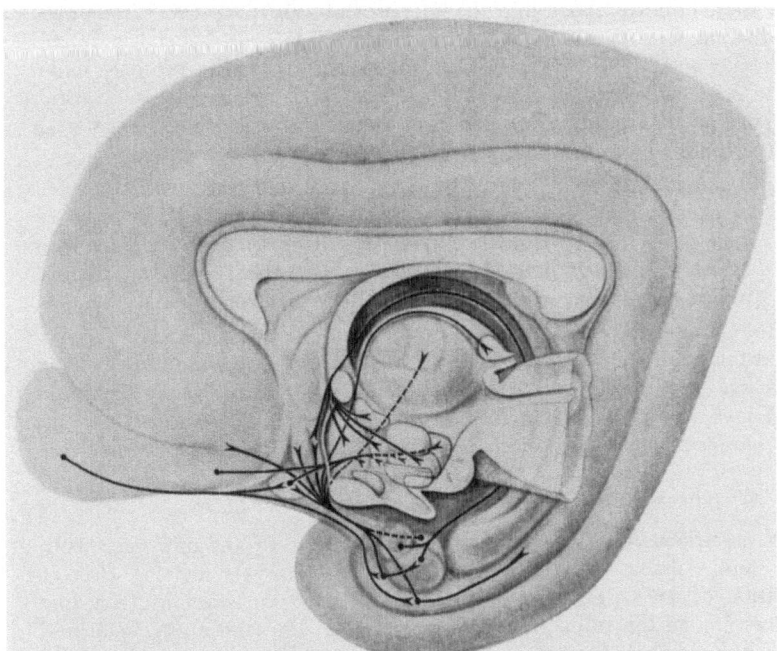

Fig. 2. Afferents from the basal olfactory area and the amygdala. Fibers running most laterally are indicated by interrupted lines.

from lower levels of the cerebral organization (*Whitlock* and *Nauta*, 1956; *Powell* and coll., 1965).

From this basal area the efferent projection directs itself towards the septum, the preoptic area, the lateral hypothalamus and the dorsal thalamus, its course taking various pathways (*Nauta*, 1961, 1962; *Sanders-Woudstra*, 1961).

The retrobulbar area makes an important contribution to the medial forebrain bundle, which constitutes the main fibrous component of the lateral hypothalamic zone and continues in a caudal direction as far as the midbrain tegmentum.

From the cortico-medial and basal nuclei of the amygdala, efferent fibers run via the stria terminalis to the bed nucleus of the stria, Broca's diagonal band, the medial preoptic area, the anterior hypothalamus and to the ventro-medial and the paraventricular nucleus of the hypothalamus.

From the pyriform cortex, whether or not joined by fibers from the lateral and basal amygdaloid nuclei (*Lundberg*, 1960; *Cowan* and coll., 1965; *Powell* and coll., 1965), efferent fibers constitute a ventral amygdalo-fugal system, which, fanning out from the amygdala, terminates in an area that extends, from before backwards, from the medio-frontal cortex, the nucleus accumbens, the olfactory tubercle and Broca's diagonal band, along the lateral hypothalamus to the caudal lateral hypothalamic area. Part of these efferent fibers join the inferior thalamic peduncle to terminate bilaterally in the medio-dorsal thalamic and lateral habenular nuclei (*Powell* and coll., 1965). The caudalward extending fibers not only make synaptic contacts with the cells situated in the lateral hypothalamic zone, but according to *Szentágothai* and coll. (1962) they partly continue in a medial direction to terminate directly in the ventro-medial nucleus.

Apart from these direct routes the basal area, *viz.* the pyriform lobe, is also indirectly connected with the hypothalamus, namely by way of a projection either with (*Powell* and coll., 1965) or without (*Lammers* and *Lohman*, 1957; *Cragg*, 1961) a synaptic relay from the periamygdaloid cortex towards the hippocampus and from there to the hypothalamus via the fornix.

From this basal high-order olfactory area, afferent fibers to the supraoptic nucleus are very rarely reported in studies using the Nauta-technique. Only *Knook* (1965) has described in the rat some fibers running from the nucleus accumbens and the olfactory tubercle to the supraoptic nucleus.

2.2. Afferents from the limbic area and the septum (Fig. 3)

It is superfluous to give a full explanation here of the now very well-known term "limbic", as coined by *Maclean* (1952), reviving *Broca's* designation of 1878. This area is built up by two concentric ring-like structures around the hilus of the telencephalon. The inner ring is formed by the hippocampal formation, with pre-, supra- and retrocommissural parts (*Gastaut* and *Lammers*, 1961; *Stephan*, 1964). The outer ring is constituted by the fornicate gyrus consisting of the parolfactory area (area subcallosa), the cingular gyrus, the retrosplenial area and the parahippo-campal gyrus. This outer ring should be seen as an anatomical as well as a functional zone of transition between the neopallium on the one side and the inner ring's archipallium on the other.

For this limbic area the septum is a very important nodal point. Afferent impulses from the midbrain and the hypothalamus are relayed here by way of the fornix to the hippocampal formation, while in turn efferent impulses from the hippocampal formation are caudally relayed to lower levels, *viz.* to the habenula and the lateral hypothalamus (*Raisman*, 1966 a). Moreover, the septum is one of the two sites where the limbic area and the basal olfactory area join each other. The other site is situated at the transition of the entorhinal cortex of the pyriform lobe into the parahippocampal gyrus.

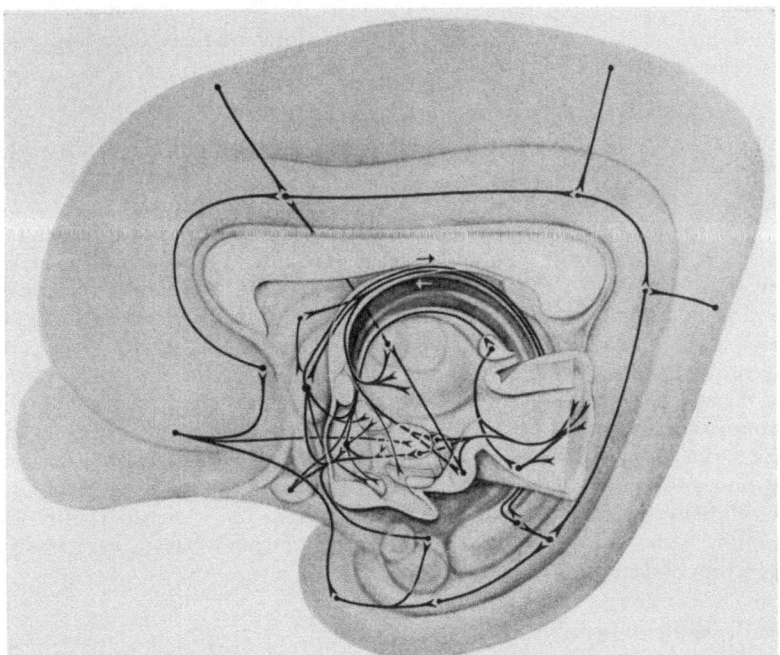

Fig. 3. Afferents from the limbic area and the septum. Interrupted lines indicate course of fibers in the lateral hypothalamus.

The efferent projection from the hippocampal formation occurs along the fornix. The fornix fibers terminate in the septum, in the mamillary body, in the anterior thalamus and in the rostral intralaminar and paramedian thalamic nuclei; some fibers may also end in the periventricular as well as in the lateral hypothalamus (*Guillery*, 1956; *Nauta*, 1956). In some mammals (guinea pig, *Valenstein* and *Nauta*, 1959; elephant, *Diepen* and coll., 1956), part of the fibers of the fornix have been found to continue into the ventral mesencephalic tegmentum (limbic midbrain area, *Nauta*, 1958); in other species the fibers do not run caudalward so far as this. However, this difference is of little weight since it may be assumed that, likewise in the latter animals, the hippocampal formation has still close connexions with the reticular formation of the midbrain, be it by means of one or more synapses in the septum and lateral hypothalamus.

Direct connexions from the limbic area and septum to the supraoptic and paraventricular nucleus are not, or only tentatively, reported in studies using the Nauta-technique. *Knook* (1965) reported on the occurrence of a few fibers from the septum to the supraoptic nucleus in the rat.

Szentágothai and coll. (1962) observed a small number of degenerated fibers running into the supraoptic nucleus after septal lesions in the cat. However, *Powell* and *Rorie* (1967) have recently reported on a more distinct septal projection in the squirrel monkey, not only to the supraoptic nucleus but also to the paraventricular nucleus. From the mid-septal area they found fibers running mainly to the supraoptic nucleus, from the dorso-posterior septum chiefly to the paraventricular nucleus.

2.3. Afferents from the neocortical areas

On the basis of a number of physiological and anatomical studies it was until recently assumed (for references see *Brutkowski*, 1965; *Knook*, 1965) that from the frontal cortex, in particular from the orbito-frontal cortex, as well as from the temporal cortex direct projections exist to the hypo-thalamus, in particular to the ventro-medial nucleus and the arcuate nucleus. However, *Lundberg* (1960) and after him *Szentágothai* and coll. (1962), as a result of their findings in the rabbit and the cat respectively. reject such direct neocortico-hypothalamic connexions. *Nauta* (1962), on the other hand, postulates a direct connexion between the orbito-frontal and lateral prefrontal cortex and the lateral hypothalamus, at least in the monkey and in man. Nevertheless, irrespective of the existence of any direct similar neocortico-hypothalamic connexions, we can safely assume the neocortex to influence the activity of the hypothalamus, both through its relations with the limbic area and through its connexions with the reticular formation of the midbrain, the two of them having an extensive projection to the hypothalamus.

2.4. Afferents from the striato-pallidum

It is generally assumed (for references see *Knook*, 1965) that part of the efferent projection of the pallidum runs to the hypothalamus, especially to the ventro-medial nucleus. *Nauta* and *Mehler* (1961, 1966), however, have raised doubts about this. According to them the pallido-hypothalamic fibers (fasciculus pallido-hypothalamicus) described by previous workers in this field are, in fact, mainly aberrant fibers of a pallido-subthalamic system.

From the caudate nucleus hypothalamic afferent fibers have been reported, running to the homolateral (*Knook*, 1965) as well as to the hetero-lateral hypothalamus, the latter by way of the supraoptic commissures (*Glees*, 1945). However, the existence of a direct striato-hypothalamic connexion is still an open question. At present it is thought more likely that the subthalamus and the midbrain tegmentum subserve a relay function between the basal ganglia and the hypothalamus (*Nauta*, 1963).

2.5 Afferents from the dorsal thalamus and epithalamus

The thalamo-hypothalamic connexions are still a controversial subject. In the pertinent literature, connexions have been described between the dorso-medial nucleus of the thalamus and various hypothalamic areas, *inter alia* the paraventricular nucleus and the dorso- and ventro-medial nucleus

the basal olfactory area join each other. The other site is situated at the transition of the entorhinal cortex of the pyriform lobe into the parahippocampal gyrus.

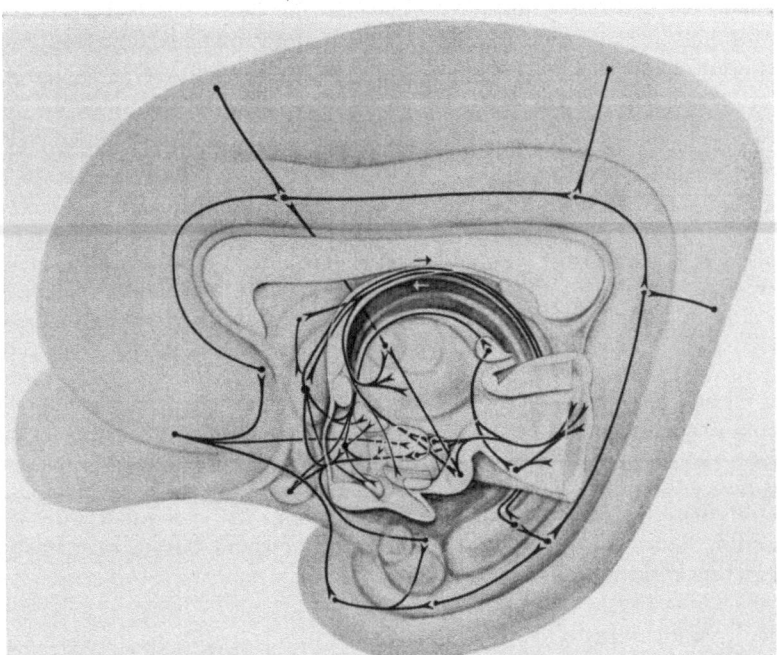

Fig. 3. Afferents from the limbic area and the septum. Interrupted lines indicate course of fibers in the lateral hypothalamus.

The efferent projection from the hippocampal formation occurs along the fornix. The fornix fibers terminate in the septum, in the mamillary body, in the anterior thalamus and in the rostral intralaminar and paramedian thalamic nuclei; some fibers may also end in the periventricular as well as in the lateral hypothalamus (*Guillery*, 1956; *Nauta*, 1956). In some mammals (guinea pig, *Valenstein* and *Nauta*, 1959; elephant, *Diepen* and coll., 1956), part of the fibers of the fornix have been found to continue into the ventral mesencephalic tegmentum (limbic midbrain area, *Nauta*, 1958); in other species the fibers do not run caudalward so far as this. However, this difference is of little weight since it may be assumed that, likewise in the latter animals, the hippocampal formation has still close connexions with the reticular formation of the midbrain, be it by means of one or more synapses in the septum and lateral hypothalamus.

Direct connexions from the limbic area and septum to the supraoptic and paraventricular nucleus are not, or only tentatively, reported in studies using the Nauta-technique. *Knook* (1965) reported on the occurrence of a few fibers from the septum to the supraoptic nucleus in the rat.

(for references see *Diepen*, 1962). But as a result of observations by *Nauta* and by *Szentágothai* and coll. the reality of a good number of these connexions must now be considered doubtful, and we may assume that, instead, the dorsal thalamus is directly connected only with the dorso-medial nucleus and the lateral hypothalamic and pre-optic area of the hypothalamus, *viz.* along two pathways: the periventricular system and the inferior thalamic peduncle. Via the periventricular system a relation is brought about between the paramedian thalamic nuclei and the dorso-medial hypothalamic nucleus. Via the inferior thalamic peduncle fibers run from the magnocellular medial part of the dorso-medial thalamic nucleus to the lateral preoptic and hypothalamic area, as well as to Broca's diagonal band, the olfactory tubercle, the innominate substance and the orbito-frontal cortex (*Nauta* and *Whitlock*, 1954; *Nauta*, 1962).

The habenular nuclei of the epithalamus are only indirectly related to the hypothalamus. On one side they are related to the anterior hypothalamic and preoptic areas by way of the stria medullaris, on the other with the midbrain tegmentum by way of the habenulo-peduncular and the habenulo-tegmental tracts. The stria medullaris is a highly complex fiber-system, the composition of which continues to be a matter of dispute. Broadly speaking, we may say that the fibers of this bundle have their origin in the basal olfactory area, the septum and the preoptic area. A contribution from the fornix, as occasionally reported, has not been confirmed by studies using the Nauta-technique (*Guillery*, 1956; *Nauta*, 1956). Thus the limbic and basal olfactory areas are related along two routes to the ventral midbrain tegmentum, *viz.* a hypothalamic and an epithalamic route (Fig. 3).

2.6. Afferents from the midbrain (Fig. 4)

The ventral part of the central grey substance and the ventral tegmentum of the midbrain constitute a source of afferent fibers vitally important to the hypothalamus.

On broad lines, we may distinguish between two systems: a periventricular and a tegmental system, even if they are not entirely separable because there is evident overlapping.

The *periventricular system* originates from the ventral part of the periaquaductal grey substance, and, taking its course in a rostral direction, forms two components, a posterior and an anterior one (*Szentágothai* and coll., 1962). Via the posterior component there is an extensive projection not only towards the supra- and premamillary and lateral hypothalamic areas, but also towards the ventro- and dorso-medial nuclei. The anterior component continues as far as the anterior hypothalamic area and the anterior periventricular nucleus.

The *tegmental system* consists on the one hand of a more or less separate projection towards the mamillary nuclei (mamillary peduncle), on the other hand of a projection which links up with the medial forebrain bundle in the lateral hypothalamic area and runs along with this into the preoptic area, the septal area and the amygdala.

Guillery (1957), *Nauta* (1963) and *Raisman* (1966 b) are of the opinion that this ascending tegmental projection terminates exclusively in the lateral hypothalamic and preoptic areas. On the other hand, *Szentágothai* and

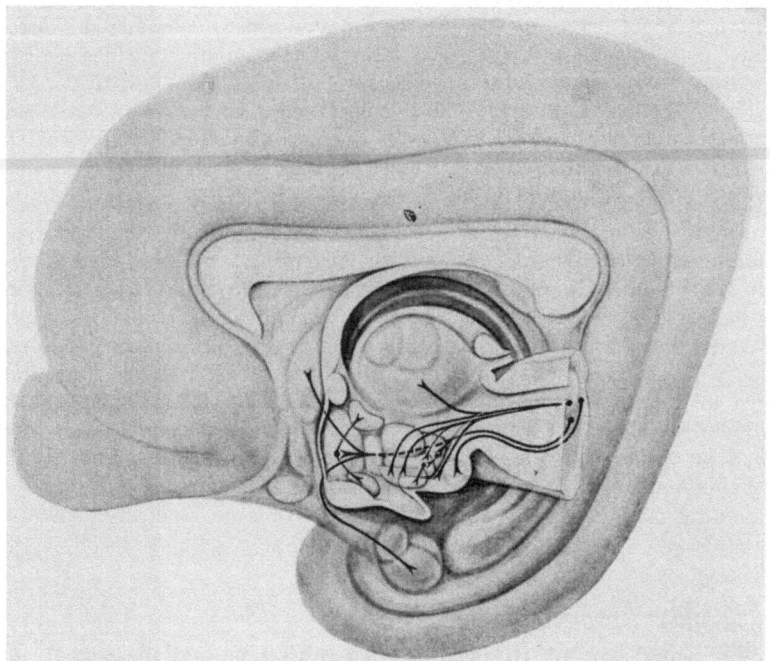

Fig. 4. Afferents from the midbrain. Interrupted line indicates course of fibers in the lateral hypothalamus.

coll. (1962) state that this tegmental projection continues at least partly in a medial direction to terminate directly in the ventro-medial nucleus. We shall have to wait for a new technique to determine with greater certainty the actual termination of this projection.

However that may be, it is evident that the basal olfactory area, the limbic area, the lateral hypothalamus and the midbrain tegmentum are closely interrelated in a system of both ascending and descending fibers: *Nauta's* (1963) limbic system — midbrain circuit (Fig. 5). From this circuit a very important afferent input derives to the neurosecretory nuclei. It might be suggested that, whereas the ventro-medial hypothalamic nucleus is serving as the main relay centre for afferents from this circuit to the parvocellular nuclei, the preoptic area and the septal area are subserving a similar function for the magnocellular nuclei.

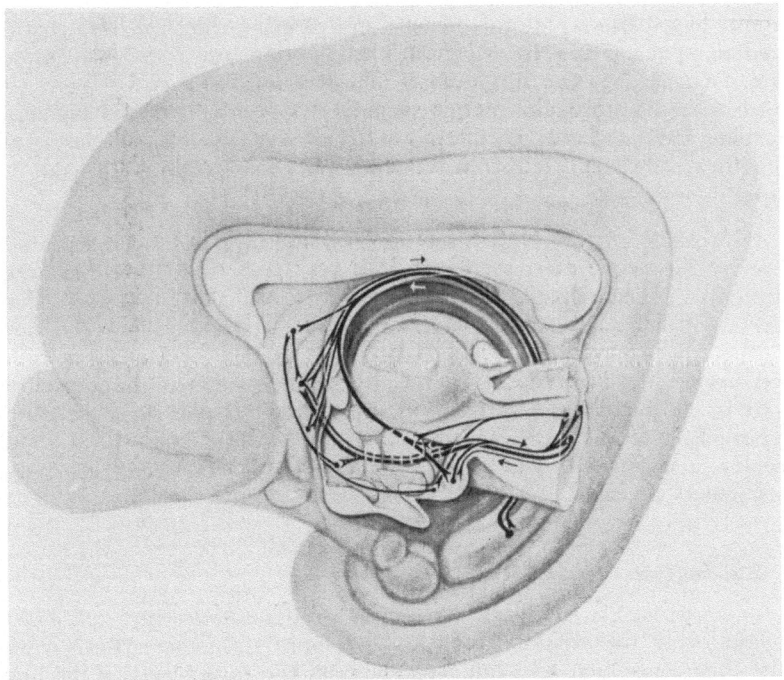

Fig. 5. Limbic system — mid-brain circuit. Interrupted lines indicate course of fibers in the lateral hypothalamus.

3. The Cholinergic and Aminergic Afferents to the Hypothalamus

Recently, our knowledge of the afferent hypothalamic connexions has been amplified by the studies of *Shute* and *Lewis* (1961, 1966) and of the well-known Swedish group of workers (*Carlsson* and coll., 1961, 1962; *Falck*, 1962; *Falck* and coll., 1962; *Fuxe*, 1965; *Dahlström* and *Fuxe*, 1965; *Andén* and coll., 1966) revealing by means of sensitive histochemical methods the existence of cholinergic and aminergic systems in the central nervous system.

3.1. The cholinergic afferents

Shute and *Lewis* (1966) hold that from the tegmentum of the midbrain two cholinergic systems take their origin: a dorsal and a ventral tegmental system.

The *dorsal system* arises from the cuneiform nucleus in the dorso-lateral part of the tegmentum, spreading towards the tectum, pretectum, geniculate bodies and the specific as well as the non-specific thalamic nuclei.

The *ventral system* arises from the substantia nigra and the ventral tegmental area, fanning out from there in a rostral direction, *i. e.* towards the mamillary nuclei, anterior thalamic nuclei, globus pallidus, striatum, lateral hypothalamus, lateral preoptic area, Broca's diagonal band and the medial septum. From the diagonal band and septum a secondary projection continues into the hippocampal formation, while from the lateral preoptic area there are cholinergic connexions not only with the supraoptic nucleus and the basal olfactory area (olfactory tubercle, olfactory bulb, olfactory cortex and entorhinal cortex), but also with the lateral and superior neocortex.

Of special interest is the fact that the supraoptic nucleus appears to receive cholinergic afferent fibers. It is very likely that these cholinergic afferents originate from the lateral preoptic area which is, as we have seen, a nodal point in the whole complex of hypothalamic connexions. Neither the paraventricular nor the arcuate nucleus are found by *Shute* and *Lewis* to be involved in the cholinergic projection from the tegmentum.

It is also a striking fact that in the ventro-medial nucleus of the medial hypothalamus, which according to *Szentágothai* and coll. (1962) should receive a massive direct projection from the midbrain, no cholinergic terminals were found.

3.2. The aminergic afferents

Summarizing the findings of previous studies, *Andén* and coll. (1966) concluded to the existence of four ascending aminergic systems which take their course from the midbrain and from the lower levels of the brain stem towards different areas of the diencephalon and the telencephalon linking up with the medial forebrain bundle.

Besides a nigro-striatal dopamine(DA)-system they distinguish a DA-system running to the nucleus accumbens, the interstitial nucleus of the stria terminalis and the olfactory tubercle; a noradrenaline(NA)-system to the hypothalamus, the limbic brain area and the neocortex; and, finally, a 5-hydroxytryptamine(5-HT)-system, likewise to the hypothalamus and the limbic area. The DA-system to the nucleus accumbens is found to have its origin from catecholamine (CA)-cells in the interpeduncular nucleus of the midbrain. In addition to this, a very high density of DA-terminals has been demonstrated in the external layer of the median eminence. According to *Fuxe* and *Hökfelt* (1966) these terminals belong to a tubero-infundibular DA-system, the cellbodies of which are found in the arcuate nucleus and in the ventral part of the anterior periventricular nucleus. The NA-system to the hypothalamus originates from CA-cells of the pons and the lateral reticular nucleus of the medulla oblongata, and terminates in the hypothalamus, *viz.* in the dorso-medial, paraventricular and periventricular nuclei, and next in the preoptic area, the interstitial nucleus of the stria terminalis, the septal area, the amygdala, the hippocampal formation, the cingular gyrus and the neocortex. The 5-HT-system, for the greater part, has its origin in the raphe nuclei of the midbrain and finds its termination in the globus pallidus, hypothalamus, septal area, amygdala, cingular gyrus and neocortex.

With regard to the magnocellular as well as the parvocellular neuro-secretory nuclei it is worth noting that all of them present a high density of NA-terminals.

4. Conclusions

From the foregoing the following conclusions regarding the afferent neuronal relations of the neurosecretory nuclei would seem to be justified:
1. The *parvocellular nuclei* receive their main afferent input from the medial hypothalamus as well as from the periventricular hypothalamic area.

The *medial hypothalamus,* in particular the ventro-medial nucleus, serves as a relay center between the lateral hypothalamus on one side and the parvocellular nuclei on the other. The lateral hypothalamus is the most important afferent area of the hypothalamus. It forms part of the limbic system — midbrain circuit (Nauta), by means of which circuit the basal olfactory and limbic areas of the brain are closely interrelated to the lateral hypothalamus and the midbrain tegmentum. Afferent stimuli may also reach the ventro-medial nucleus of the hypothalamus without a synaptic relay in the lateral hypothalamus, in particular such as arise from the amygdala and, according to *Szentágothai* and coll., from the midbrain tegmentum as well. The latter direct tegmental projection to the ventro-medial nucleus is not reported by other authors using the Nauta-Gygax-technique.

In this respect it is worth noting that neither cholinergic nor aminergic (or only a very few of the latter) terminals of fibers originating in the midbrain tegmentum have been observed in the ventro-medial nucleus of the hypothalamus. However, as has been revealed by the Golgi-study of *Szentágothai* and coll. (1962), the ventro-medial nucleus has connexions with probably the majority of its immediate neighbours. Most of these show a very high density of cholinergic as well as aminergic terminals. Hence it may be assumed that the ventro-medial nucleus will still be directly influenced by specific (and probably also by non-specific) tegmental afferents, be it rather by axodendritic than by axosomatic synapses.

The *periventricular area* constitutes a relay center to the parvocellular nuclei for afferents originating mainly in the central grey substance of the midbrain and in the paramedian nuclei of the thalamus. In addition, a small number of afferents may reach the periventricular area by way of the stria terminalis and the fornix.

In the parvocellular nuclei no cholinergic terminals could be revealed by the cholinesterase technique. On the other hand, the density of NA-terminals, which apparently belong to an aminergic system originating in CA-cells in the pons and the medulla oblongata, is very high here. This may suggest a rather dominant aminergic influence on the activities of the parvocellular neurosecretory cells.

2. As regards *the magnocellular nuclei,* it may be assumed that they, too, receive an important afferent input from the limbic system — midbrain circuit, in particular from the septal area (*Szentágothai* and coll.,

21*

1962; *Knook*, 1965; *Powell* and *Rorie*, 1967) and the lateral preoptic area (*Nauta*, 1958). According to *Shute* and *Lewis* (1966), the afferent projection from the lateral preoptic area should be a cholinergic one. These authors also hold that the magnocellular nuclei are cholinoceptive rather than cholinergic. This would be in accordance with the physiological findings that the supraoptic nucleus is under cholinergic afferent control (*Akert*, 1959).

However, we may also presume an aminergic influence to be at work upon the magnocellular nuclei, because they contain a very high concentration of NA-terminals.

In addition to the afferent relations just mentioned there are some others, more or less problematical. In older studies, lemniscal as well as non-lemniscal afferents to the hypothalamus have been held to originate from lower levels of the brainstem. However, recent studies using the Nauta-Gygax-technique did not reveal any such afferents (*Nauta* and *Kuypers*, 1958). On the other hand, the fluorescence technique has demonstrated aminergic hypothalamic afferents to originate in the tegmentum of the pons and the medulla oblongata (*Andén* and coll., 1966). This difference of opinion as to the existence of hypothalamic afferents from lower levels of the brainstem might be due to the great sensitiveness of the fluorescence technique in visualizing fibers and terminals of a caliber too small to be demonstrated by the silver technique.

In the literature we also find references to an influence of the visual system on the activities of the hypothalamus, in particular of the magnocellular nuclei, by way of a direct retino-hypothalamic projection (*Frey*, 1950; *Knoche*, 1957, 1960; *Blümcke*, 1961; *Feldman*, 1964). I cannot enter into a detailed discussion of this very controversial projection. I can only state here that current research at our laboratory in different mammals (guinea pig, cat, rabbit and monkey) with the Nauta-Gygax-technique have made us side with those authors (*Hayhow*, 1959; *Hayhow* and coll., 1960; *Giolli*, 1963; *Singleton* and *Peele*, 1965) who think that from the retina no fibres run either to the supraoptic nucleus or to any other hypothalamic area.

Acknowledgements

I wish to express my gratitude to Mr. *H. E. van der Aa* (med. cand.) and Mr. *D. C. Busman* (med. cand.) for their valuable assistance in the preparation of this survey, to Mr. *van Huyzen* for the illustrations, and to Mr. *L. Grooten* for the translation.

References

Akert, K.: Physiology and pathophysiology of the hypothalamus. In: *Schaltenbrand, G.,* and *P. Bailey:* Introduction to stereotaxis with an atlas of the human brain, 152—225. Stuttgart: G. Thieme Verlag, 1959.

Andén, N.-E., A. Dahlström, K. Fuxe, K. Larsson, L. Olson, and *U. Ungerstedt:* Ascending monoamine neurons to the telencephalon and diencephalon. Acta physiol. Scand. *67,* 313—326 (1966).

Blümcke, S.: Vergleichend experimentell-morphologische Untersuchungen zur Frage einer retino-hypothalamischen Bahn bei Hund, Meerschweinchen und Katze. Zschr. mikrosk.-anat. Forsch., Leipzig, *67,* 469—513 (1961).

Broca, P.: Anatomie comparée des circonvolutions cérébrales: le grand lobe limbique et la scissure limbique dans la série des mammifères. Rev. anthrop., Paris, 2e sér., *1,* 385—498 (1878).

Brutkowski, St.: Functions of prefrontal cortex in animals. Physiol. Rev., Baltimore, *45,* 721—746 (1965).

Carlsson, A., B. Falck, N.-Å. Hillarp, G. Thieme, and *A. Torp:* A new histochemical method for visualization of tissue catecholamines. Med. Exper. *4,* 123—125 (1961).

Carlsson, A., B. Falck, N.-Å. Hillarp, and *A. Torp:* Histochemical localization at the cellular level of hypothalamic noradrenaline. Acta physiol. Scand. *54,* 385—386 (1962).

Cowan, W. M., G. Raisman, and *T. P. S. Powell:* The connexions of the amygdala. J. Neurol., London, *28,* 137—151 (1965).

Cragg, B. G.: Olfactory and other afferent connexions of the hippocampus in the rabbit, rat and cat. Exper. Neurol. *3,* 588—600 (1961).

Dahlström, A., and *K. Fuxe:* Evidence for the existence of monoamine containing neurons in the central nervous system. I. Demonstration of monoamines in the cell bodies of brain stem neurons. Acta physiol. Scand. *62,* Suppl. 232 (1965).

Diepen, R.: Der Hypothalamus. In: Handb. d. mikr. Anatomie des Menschen, von *Möllendorf,* ed., *Bd. 4,* Berlin: Springer Verlag (1962).

Diepen, R., P. Jansen, Fr. Engelhardt, et *H. Spatz:* Recherches sur le cerveau de l'éléphant d'Afrique *(Loxodonta africana Blum).* Données sur l'hypothalamus. Acta neurol. psychiatr. Belg. *11,* 759—788 (1956).

Falck, B.: Observations on the possibilities of the cellular localizations of monoamines by a fluorescence method. Acta physiol. Scand. *56,* Suppl. 197, 1—26 (1962).

Falck, B., N.-Å. Hillarp, G. Thieme, and *A. Torp:* Fluorescence of catecholamines and related compounds condensed with formaldehyde. J. Histochem. Cytochem. *10,* 348—354 (1962).

Feldman, S.: Visual projections to the hypothalamus and preoptic area. Ann. N. Y. Ac. Sci. *117,* 53—68 (1964).

Fink, R. P., and *L. Heimer:* Two methods for selective silver impregnation of degenerating axons and their synaptic endings in the central nervous system. Brain Res. *4,* 369—374 (1967).

Frey, E.: Neue anatomische und experimentelle Ergebnisse über das optische Gebiet im Hypothalamus. Schweiz. Arch. Neurol. Psych. *66,* 67—86 (1950).

Fuxe, K.: Evidence for the existence of monoamine neurons in the central nervous system. IV. The distribution of monoamine terminals in the central nervous system. Acta physiol. Scand. *64,* Suppl. 247, 37—85 (1965).

Fuxe, K., and *T. Hökfelt:* Further evidence for the existence of tubero-infundibular dopamine neurons. Acta physiol. Scand. *66,* 245—246 (1966).

Gastaut, H., et *H. J. Lammers:* Anatomie du Rhinencéphale. Les grandes activités du Rhinencéphale, vol. *I.* Paris: Masson & Cie. 1961.

Giolli, R. A.: An experimental study of the accessory optic system in the cynomolgus monkey. J. Comp. Neur. *121,* 89—108 (1963).

Glees, P.: The interrelation of the strio-pallidum and the thalamus in the macaque monkey. Brain, *68,* 331—346 (1945).

Guillery, R. W.: Degeneration in the post-commissural fornix and the mamillary peduncle of the rat. J. Anat. *90,* 350—371 (1956).

Guillery, R. W.: Degeneration in the hypothalamic connexions of the albino rat. J. Anat. *91,* 91—115 (1957).

Harris, G. W.: The reticular formation, stress, and endocrine activity. In: Reticular Formation of the Brain (Henry Ford Hospital International Symposium), 207—221 (1958). Boston: Little, Brown and Co.

Hayhow, W. R.: An experimental study of the accessory optic fiber system in the cat. J. Comp. Neur. *113,* 281—314 (1959).

Hayhow, W. R., C. Webb, and *A. Jervie:* The accessory optic fiber system in the rat. J. Comp. Neur. *115,* 187—215 (1960).

Knoche, H.: Die retino-hypothalamische Bahn von Mensch, Hund und Kaninchen. Zschr. mikrosk.-anat. Forsch. *63,* 461—486 (1957).

Knoche, H.: Ursprung, Verlauf und Endigung der retino-hypothalamischen Bahn. Z. Zellforsch. *51,* 658—704 (1960).

Knook, H. L., The fibre-connections of the forebrain. Med. Thesis. Assen: v. Gorcum & Cie, 1965.

Lammers, H. J., and *A. H. M. Lohman:* Experimenteel anatomisch onderzoek naar de verbindingen van piriforme cortex en amygdala kernen bij de kat. Ned. T. Geneesk. *101,* 602—603 (1957).

Lohman, A. H. M.: The anterior olfactory lobe of the guinea pig. Acta Anat. *53,* Suppl. 49 (1963).

Lohman, A. H. M., and *H. J. Lammers:* On the structure and fibre connexions of the olfactory centres in mammals. In: Progress in Brain Research, *23.* Sensory mechanisms, 65—82 (1967). Amsterdam: Elsevier Publ. Cie.

Lundberg, P. O.: Cortico-hypothalamic connexions in the rabbit. Acta physiol. Scand. *49,* Suppl. 171 (1960).

MacLean, P. D.: Some psychiatric implications of physiological studies on fronto-temporal portion of limbic system (visceral brain). EEG Clin. Neurophysiol. *4,* 407—418 (1952).

Nauta, W. J. H.: An experimental study of the fornix system in the rat. J. Comp. Neurol. *104,* 247—271 (1956).

Nauta, W. J. H.: Hippocampal projections and related neural pathways to the midbrain in the cat. Brain *82,* 319—340 (1958).

Nauta, W. J. H.: Fibre degeneration following lesions of the amygdaloid complex in the monkey. J. Anat. *95,* 515—531 (1961).

Nauta, W. J. H.: Neural associations of the amygdaloid complex in the monkey. Brain *85,* 505—519 (1962).

Nauta, W. J. H.: Central nervous organization and the endocrine motor system. In: Advances in neuroendocrinology. A. V. Nalbandov ed., Urbana: University of Illinois Press, 5—21 (1963).

Nauta, W. J. H., and *P. A. Gygax:* Silver impregnation of degenerating axons in the central nervous system. A modified technique. Stain Technol. *92,* 91—93 (1954).

Nauta, W. J. H., and *H. Kuypers:* Some ascending pathways in the brain stem reticular formation. In: Reticular Formation of the Brain (Int. Symp. Henry Ford Hospital, Detroit) 3—30 (1958). Boston: Little, Brown and Cy.

Nauta, W. J. H., and *W. R. Mehler:* Some efferent connections of the lentiform nucleus in monkey and cat. Anat. Rec. *139,* 260 (1961).

Nauta, W. J. H., and *W. R. Mehler:* Projections of the lentiform nucleus in the monkey. Brain Res. *1,* 3—42 (1966).

Nauta, W. J. H., and *D. G. Whitlock:* An anatomical analysis of the non-specific thalamic projection system. In: Brain mechanism and conciousness (Symposium by the Counc. Internat. Organ. Med. Sci.) 81—116 (1954). Oxford: Blackwell.

Powell, E. W., and *D. K. Rorie:* Septal projections to nuclei functioning in oxytocin release. Am. J. Anat. *120,* 605—610 (1967).

Powell, T. P. S., W. M. Cowan, and *G. Raisman:* The central olfactory connexions. J. Anat. *99,* 791—813 (1965).

Pribram, K. H., and *L. Krüger:* Functions of the 'olfactory brain'. Ann. N. Y. Ac. Sci. *58,* 109—138 (1954).

Raisman, G.: The connexions of the septum. Brain *89,* 317—348 (1966 a).

Raisman, G.: Neural connexions of the hypothalamus. In: Recent Studies on the Hypothalamus. Brit. Med. Bull. *22,* 197—201 (1966 b).

Sanders-Woudstra, J. A. R.: Experimenteel anatomisch onderzoek over de verbindingen van enkele basale telencephale hersengebieden bij de albino rat. Thesis, Groningen 1961.

Scharrer, E.: The final common path in neuroendocrine integration. Arch. Anat. micr. Morph. exp. *54,* 359—370 (1965).

Sherrington, Ch.: The integrative action of the nervous system. Cambridge: Univ. Press 1947.

Shute, C. C. D., and *P. R. Lewis:* The use of cholinesterase techniques combined with operative procedures to follow nervous pathways in the brain. Bibl. anat. *2,* 34—49 (1961). Basle: Karger.

Shute, C. C. D., and *P. R. Lewis:* Cholinergic and monoaminergic pathways in the hypothalamus. In: Recent Studies of the Hypothalamus. Brit. Med. Bull. *22,* 221—226 (1966).

Singleton, M. C., and *T. L. Peele:* Distribution of optic fibers in the cat. J. Comp. Neur. *125,* 303—328 (1965).

Stephan, W.: Die kortikalen Anteile des limbischen System. Der Nervenarzt *35,* 396—401 (1964).

Szentágothai, J., B. Flerkó, B. Mess, and *B. Halász:* Hypothalamic Control of the Anterior Pituitary. Budapest: Akadémiai Kiadó, 1962.

Valenstein, E. S., and *W. J. H. Nauta:* A comparison of the distribution of the fornix system in the rat, guinea pig, cat and monkey. J. Comp. Neur. *113,* 337—364 (1959).

Whitlock, D. G., and *W. J. H. Nauta:* Subcortical projections from the temporal neocortex in *Macaca mulatta.* J. Comp. Neur. *106,* 183—212 (1956).

Author's address: Prof. Dr. *H. J. Lammers,* Department of Anatomy and Embryology, Medical Faculty, Catholic University, Nijmegen (Netherlands).

Discussion

Feldberg: I should like to ask Prof. *Lammers* what is the difference between fornicate gyrus and cingulum, and when do we use the one and when the other term?

Lammers: The gyrus cinguli and cingulum are to be considered parts of the fornicate gyrus. By this latter term, more or less forgotten in the textbooks but still very appropriate in my opinion, is indicated the outer ring of the limbic area. It is constituted by the subcallosal area (parolfactory area of Broca), the gyrus cinguli, the retrosplenial area and the parahippocampal area.

Oksche: With respect to the optico-hypothalamic connections which seem to be missing in your material, I would like to add, that our own investigations in a very light-sensitive avian species, *Zonotrichia leucophrys gambelii,* were also negative. A direct retino-hypothalamic root or bundle could neither be demonstrated by means of the Nauta-technique nor by other *(Bielschowsky, Bodian)* silver methods. In *Passer domesticus* axo-somatic synapses were observed on the perikarya of the supraoptic nucleus. In our material some of these endings and also many of the fine neuropile profiles of this region contain dense-core granules of 800—1000 Å diameter. It is possible that these fibers belong to one of the shown aminergic pathways originating in the mesencephalon. In addition I would like to ask if there are aminergic neurons in the preoptic area?

Lammers: As regards this question you better ask Dr. *Carlsson* to answer and I should like to join you with a question myself. It has puzzled me, that the supra-optic nucleus containing a large amount of NA-terminals was not mentioned by *Andén* et al. (1966) as one of the sites of termination of the NA-system described by them. Might this suggest that these terminals originate from other still unknown groups of NA-cells, maybe situated in the hypothalamus itself?

Carlsson: I think there can be little doubt that the cell bodies belonging to the nerve terminals containing noradrenaline in the supra-optic nucleus are located in the lower brain stem. This is clearly indicated by lesion experiments.

Kappers: You mentioned that the dorsal longitudinal fascicle of *Schütz* would contain ascending fibers exclusively. Formerly, it was generally agreed that this bundle contains also descending fibers. Did opinions change in the mean time?

Lammers: It was not my intention to state expressly that there are no descending fibers at all in the fascicle of *Schütz.* I only wanted to stress the point that in recent studies using the Nauta-technique the ascending fibers have come to the fore as constituting the most important component of this fascicle, whereas in older studies this fascicle is described as containing mainly descending fibers.

Bargmann: Dr. *Lammers,* you was a little reluctant, it seemed to me, concerning the idea of a final common pathway. Would you be so kind to comment on this opinion?

Lammers: This is, indeed, one of the main problems I have met in thinking about this subject. You may look, when you are talking about this final common pathway, for direct afferent relations to the nuclei in discussion. As far as that is concerned we know more about the arcuate and the periventricular nucleus than about the magnocellular nuclei. You may evade this problem by stressing the reticular structure of the hypothalamus, saying that each incoming impulse will spread over the whole complex and also to the magnocellular nuclei, even if we do not know the synaptic field of those magnocellular nuclei. Then, however, you may fall into another pit. How do you explain in that case the differentiated reaction of the hypothalamus to the various kinds of stimuli coming from extrahypothalamic regions? So you can understand why I am rather reluctant to enter into this problem.

Journal of Neuro Visceral Relations, Suppl. IX, 320 361 (1060)

Neurohormonal Mechanisms Controlling Trophic Hormone Secretion of the Anterior Pituitary

Béla Halász

Department of Anatomy, University Medical School, Pécs, Hungary

With 18 Figures

Summary

Anterior pituitary function is primarily controlled by neurohormonal mechanisms. The immediate control of trophic hormone secretion seems to be exercised by the medial basal hypothalamus, the hypophysiotrophic area, producing the hypothalamic releasing and inhibiting factors carried by the portal circulation to the adenohypophysis. This system is probably responsible for the maintenance of "basal" secretion of pituitary hormones. Normal function of the anterior lobe is controlled by neural elements outside the medial basal hypothalamus. This second control level presumably acts on the pituitary via the hypophysiotrophic area by modifying the synthesis and release of the hypothalamic releasing and inhibiting substances. The feedback action of hormones on the adenohypophysis appears to be mediated, at least partly, by way of the hypothalamic control mechanism.

Zusammenfassung

Hypophysenvorderlappenfunktion ist durch neurohormonale Mechanismen reguliert. Es wird angenommen, daß der unmittelbare Regler der Sekretion der Trophhormone der mediale basale Hypothalamus, das „hypophysiotrophe" Gebiet, ist, der die mittels des Portalkreislaufes zu den Hypophysenzellen gelieferte sog. „releasing" und „inhibiting factors" produziert. Dieses System ist wahrscheinlich verantwortlich für die Erhaltung der „basalen" Sekretion der Trophhormone. Die normale Funktion der Hypophyse wird von neuralen Elementen gesteuert, die außer dem medialen basalen Hypothalamus liegen. Diese zweite Regelungsstufe wirkt auf den Vorderlappen, wahrscheinlich mittels des „hypophysiotrophen" Gebietes, durch Modifizierung der Produktion und Abgabe der hypothalamischen „releasing" und „inhibiting" Faktoren. An der Rückwirkung von Hormonen auf die Hypophysenfunktion ist der hypothalamische Steuerungsmechanismus mitbeteiligt.

Introduction

Anterior pituitary function is controlled by two main factors: 1. the hypothalamus; 2. the feedback action of hormones. In the light of the data available, it may be assumed that the hypothalamus plays the key role in the control of the anterior lobe. In the absence of this control, trophic hormone secretion becomes markedly reduced leading to target organ atrophy. The hormonal feedback action as well as the influence of various other factors seem to be mediated, at least partly, by the hypothalamic control mechanism.

1. Hypothalamic Control Mechanism

The three main questions to be discussed in this chapter are:

1. The neurovascular link between the hypothalamus and the anterior pituitary in general,

2. Site of production of the hypothalamic "hypophysiotrophic" substances.

3. Two levels in the hypothalamic control of the anterior pituitary.

1. The neurovascular link between the hypothalamus and the anterior pituitary in general

Harris (1947) and *Green* and *Harris* (1947) first recognized that the link between the hypothalamus and the adenohypophysis is neurovascular. They assumed that the hypothalamus produces some substances to be carried by the portal circulation to the pituitary where they act on the cells of the anterior lobe. This suggestion was originally based on the observations (1) that the principal blood supply of the anterior pituitary is the portal system, which is characterized by the intimate contact of the blood vessels with the nervous tissue of the hypothalamus; (2) that there are no nerve endings of hypothalamic origin — or at least occurring only exceptionally — in the pars distalis of the hypophysis; (3) that hormone secretion of the anterior lobe is normal in pituitary stalk sectioned animals, if the vascular connections between the hypothalamus and the pituitary have regenerated while, conversely, pituitary and target organ atrophy develops if, due to some reason, regeneration does not occur. (For details see *Harris*, 1955; *Szentágothai* and coll., 1962; *Daniel*, 1966).

The neurovascular hypothesis of *Harris* and *Green* gained support from a great number of observations. Perhaps the most convincing evidence for this view is provided by pituitary transplantation experiments. Several authors have observed that, if the pituitary is disconnected from the hypothalamus and transplanted at more remote sites (in the anterior ocular chamber, beneath the renal capsule, or in the temporal lobe), its histological structure is dedifferentiated and trophic hormone secretion is markedly reduced (*Cutuly*, 1941; *Cheng* and coll., 1949; *Fortier*, 1951; *Harris* and *Jacobsohn*, 1952; *Greer* and coll., 1953; *Siperstein* and *Greer*, 1956; *Goldberg* and *Knobil*, 1957; *Nikitovitch-Winer* and *Everett*, 1958, 1959, etc.). In contrast, structure and function of the anterior lobe is well maintained if the pituitary is grafted under the median eminence of the hypothalamus and its vascular connections with the portal system are re-established (*Harris* and *Jacobsohn*, 1952). *Nikitovitch-Winer and*

Everett (1958, 1959) have even shown that normal histology as well as hormone secretion of the dedifferentiated pituitary graft, after having been transplanted under the renal capsule, can be restored by retransplanting the anterior lobe under the median eminence.

It has also been shown that hypothalamic extracts have a direct effect on pituitary hormone secretion. Such an action was first demonstrated on adrenocorticotrophic hormone (ACTH) secretion (*Saffran* and *Schally*, 1955; *Guillemin* and *Rosenberg*, 1955; *Royce* and *Sayers*, 1958; *Rumsfeld* and *Porter*, 1959). The active principle has been called Corticotrophic Hormone Releasing Factor (CRF). Further studies revealed that, besides CRF, hypothalamic extracts contain also additional substances which stimulate the secretion of other pituitary hormones. Evidence has been presented that hypothalamic extracts stimulate the release by the hypophysis of thyrotrophic hormone (TSH) (*Shibusawa* and coll., 1956; *Schreiber* and coll., 1961; *Guillemin* and coll., 1962), luteinizing hormone (LH) (*McCann* and coll., 1960; *Campbell* and coll., 1961; *Nikitovitch-Winer*, 1962), follicle stimulating hormone (FSH) (*Igarashi* and *McCann*, 1964) and growth hormone (GH) (*Franz* and coll., 1962; *Deuben* and *Meites*, 1964). These hypothalamic substances have been named TSH releasing factor (TRF), LH releasing factor (LRF), FSH releasing factor (FRF) and GH releasing factor (GRF), respectively. Beside the factors causing hormone release from the pituitary, there is evidence that the hypothalamus produces a substance which exerts a tonic inhibitory effect on prolactin release (prolactin inhibiting factor, PIF) (*Pasteels*, 1961 a, b, 1962; *Meites* and coll., 1962; *Talwalker* and coll., 1963). The first observations on the existence of hypothalamic releasing — and inhibiting — factors have been confirmed in several laboratories by *in vivo* as well as *in vitro* studies. In addition it has been shown that the action of the releasing substances is not limited to the release of the hormones, but they might stimulate also their synthesis (for details on the hypothalamic releasing factors the reader is referred to the excellent work of *McCann* and *Dhariwal*, 1966).

Evans and *Nikitovitch-Winer* (1965) have demonstrated that ovine median eminence extracts reactivate functionally and cytologically 8-67 day old renal pituitary autografts if the extracts are infused continuously into the renal artery for periods of 7-36 days (1-3 median eminence fragments/24 hrs). The assumption that the hypothalamic substances act on the pituitary via the portal vessels is supported by the findings of *Porter* and *Rumsfeld* (1956, 1959) and of *Averill* and coll. (1966) who demonstrated the presence of hypothalamic releasing factors in the hypophysial portal blood.

The findings mentioned strongly suggest that the mechanism by which the hypothalamus exerts its control action on the pituitary might be neurovascular. The question, however, arises where the hypothalamic "hypophysiotrophic" substances are produced. Is the whole hypothalamus involved in their production or is this function restricted to a certain hypothalamic region?

2. Site of production of the hypothalamic "hypophysiotrophic" substances

a) The existence of a hypophysiotrophic area in the hypothalamus. Location and extent of the area

By implanting anterior pituitaries into different hypothalamic and extra-hypothalamic regions of rat brains it was observed that the structure and function of the anterior lobe depends entirely on the location of the graft. If the graft was situated in the medial basal hypothalamus, numerous PAS-positive basophile cells were present in the implant whereas in grafts located outside this hypothalamic region, either in the hypothalamus, in other regions of the brain or under the kidney capsule, PAS-positive cells could not be found in the pituitary transplant (Fig. 1). Since the basophile cells seemed to be most sensitive to hypothalamic connections — the presence or absence of eosinophile cells depending on blood thyroxine levels rather than on direct hypothalamic connections (*Halász* and coll., 1963) —, the occurrence of this cell type was used as criterion to map the extension of the hypothalamic region being capable to maintain normal pituitary structure. In Fig. 2 five pituitary grafts, the outlines of which are indicated by heavy lines, are projected onto the midsagittal plane of a rat hypothalamus. Part of the grafts containing PAS-positive basophile cells are indicated by small circles. The area capable of maintaining PAS-positive cells reaches ventralward from the region of the paraventricular nuclei to the optic chiasm and extends caudalward as far as the mamillary nuclei. The lateral extent of this area is at about 0.5 mm from the midline. This hypothalamic region has been called the hypophysiotrophic area (HTA). The area includes the arcuate nuclei, the ventral part of the anterior periventricular nuclei, the medial part of the retrochiasmatic region and the median eminence. Our findings concerning the location and extent of the HTA (*Halász* and coll., 1962) have been confirmed by *Flament-Durand* (1965).

The HTA is, however, not only capable of maintaining the differentiated structure of the pituitary, but also to maintain fairly well trophic hormone secretion of the transplanted anterior lobe (*Halász* and coll., 1962, 1965). It was found that normal gonadal function returns in the hypophysectomized female rat if a pituitary graft is placed into the HTA. The vaginal smears

---

Fig. 1. Position and histology of pituitary grafts situated in different parts of the brain or under the kidney capsule. — 1. Anterior pituitary graft in the HTA. There is no contact with capillary loops of the median eminence. Mann's method, reduced from $\times$ 62. — 2. Same grafts as in 1. Numerous PAS-positive basophiles are present. PAS, reduced from $\times$ 320. — 3. Anterior pituitary graft beneath renal capsule. Note the absence of PAS-positive cells in the graft. PAS, reduced from $\times$ 320. — 4. Anterior pituitary graft in lateral ventricle in contact with choroid plexus. PAS-positive cells are completely lacking. PAS, reduced from $\times$ 320. — 5. Anterior pituitary graft placed outside the HTA of the hypothalamus. Mann's method, reduced from $\times$ 62. — 6. Same graft as in 5. No PAS-positive cells are present. PAS, reduced from $\times$ 320. (*Halász* and coll., 1962)

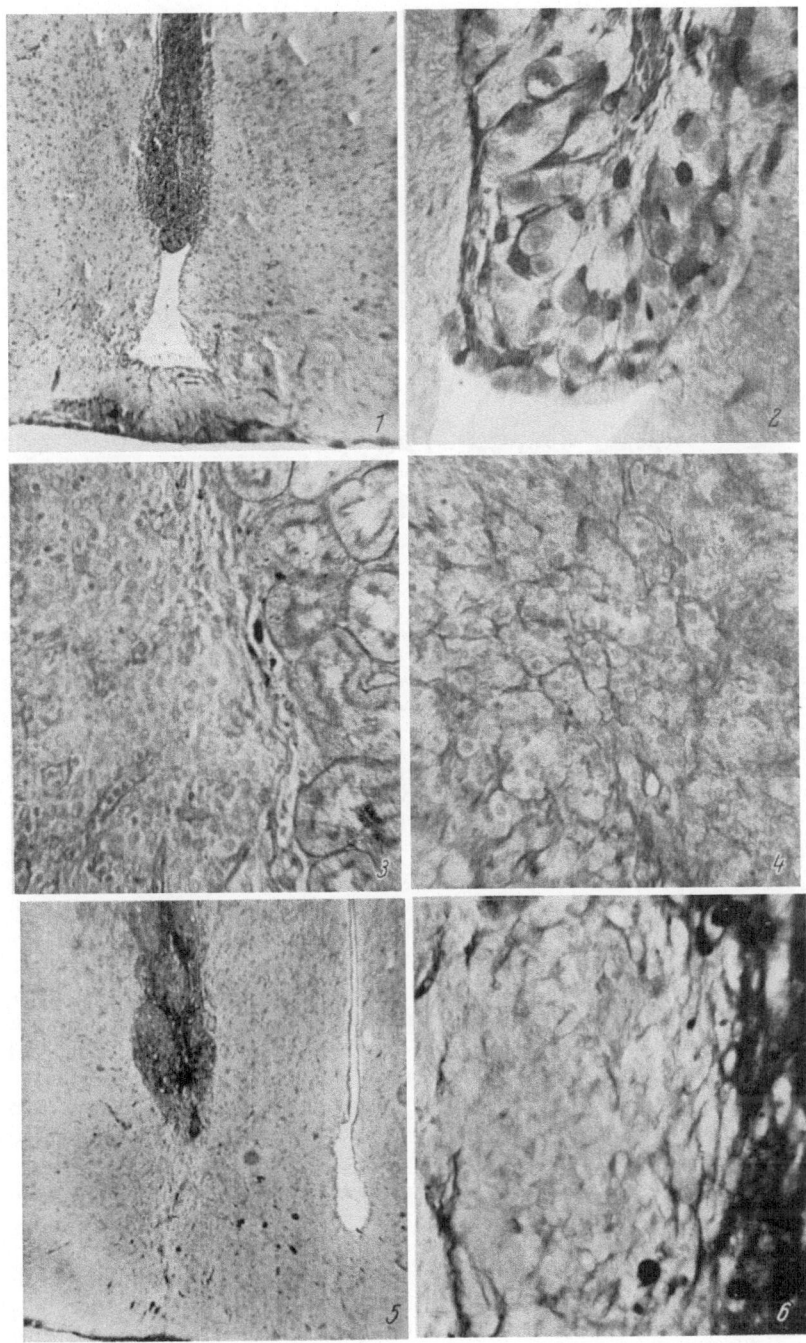

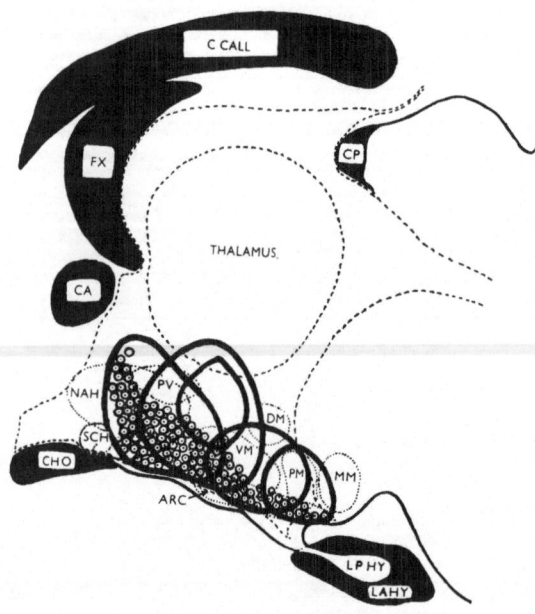

Fig. 2. The location of the hypophysiotrophic area of the hypothalamus.
- - - Outline of third ventricle, ... mid-sagittal projection of main hypothalamic nuclei, —— outlines of five relatively midline pituitary grafts, ⊙⊙
PAS-positive basophile cells. ARC arcuate nucleus, CA anterior commissure,
C CALL corpus callosum, CHO optic chiasm, CP posterior commissure, DM
dorsomedial nucleus, FX fornix, LA HY anterior lobe of hypophysis, LP HY
posterior lobe of hypophysis, MM medial mamillary nucleus, NAH anterior
hypothalamic nucleus, PM premamillary nucleus, PV paraventricular nucleus,
SCH suprachiasmatic nucleus, VM ventromedial nucleus (*Halász* and coll., 1962).

of these animals indicated that nearly normal cycles reappear after a
shorter or longer time interval following implantation (Fig. 3). In accordance, the ovaries of these rats contain ripe follicles and fresh corpora
lutea (Fig. 4). On the other hand, if the pituitary was either implanted
into the hypothalamus but outside the HTA, in other parts of the brain,
or under the kidney capsule, the animals remained in di-oestrus and their
ovaries were atrophic (Figs. 3 and 4). Concerning the maintenance of
pituitary GTH secretion, similar observations were made in male rats.
It was observed that pituitary grafts situated in the HTA maintain normal
spermiogenesis and spermiohistogenesis whereas, if the grafts are located
outside the HTA, the testes are strongly atrophic (Fig. 4).

TSH secretion by the graft is also significantly better preserved if the
graft is placed into the HTA than outside this region. The difference
is demonstrated in figure 5. Thyroid histology is nearly normal in animals
bearing an anterior lobe in the HTA. These rats are even capable of responding to methylthiouracil treatment with an increased TSH secretion. This
response is significantly stronger if the graft is directly connected with

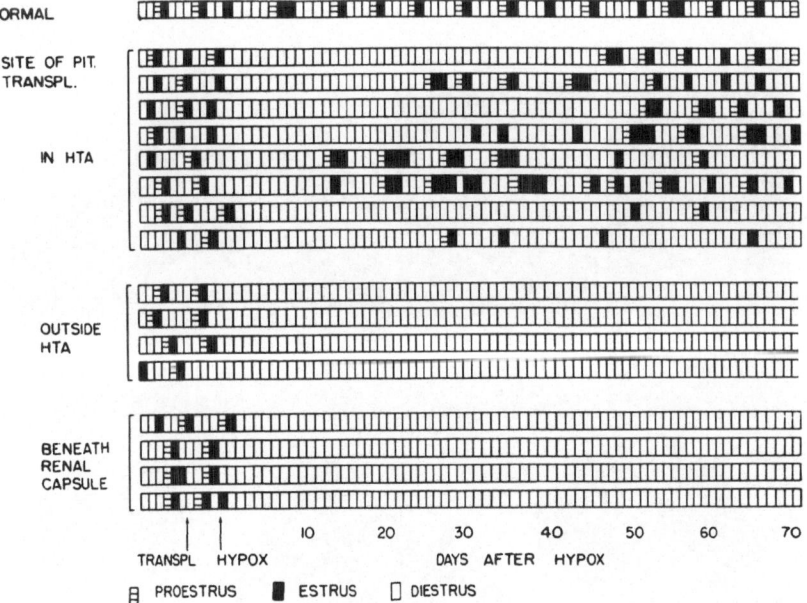

Fig. 3. Daily vaginal smears of intact and of hypophysectomized rats bearing a pituitary transplant in the hypothalamus or under the renal capsule (*Halász* and coll., 1965).

the capillary loop system of the median eminence than when this connection is lacking. The thyroid is highly inactive if the graft is located outside the HTA (Fig. 5 c). In that case, no signs of thyroid hyperactivity are found following methylthiouracil treatment (Fig. 5 h).

ACTH secretion also depends on the location of the pituitary graft. Fig. 6 shows that corticosterone in the adrenal venous blood is only detectable in those animals which have a pituitary graft implanted into the HTA. The zona fasciculata and zona reticularis of the adrenal cortex are atrophic and the so-called sudanophobe zone between the zona glomerulosa and the zona fasciculata is wide if the graft is implanted outside the HTA (Fig. 4 c). Similar changes do not occur if the anterior lobe is present within the HTA (Fig. 4 b).

Growth hormone (GH) secretion showed similar features as other trophic hormones do. If young rats of 60—70 g body weight are hypophysectomized and an anterior pituitary homograft is implanted into different parts of the brain or under the kidney capsule, the growth rate appeared to be significantly better in those animals in which the graft was situated in the HTA than in all other grafted groups (Fig. 7).

The data obtained in animals bearing intrahypothalamic pituitary implants strongly suggest that the hypothalamic substances essential for the maintenance of the structure and function of the anterior lobe might be produced and released by the HTA.

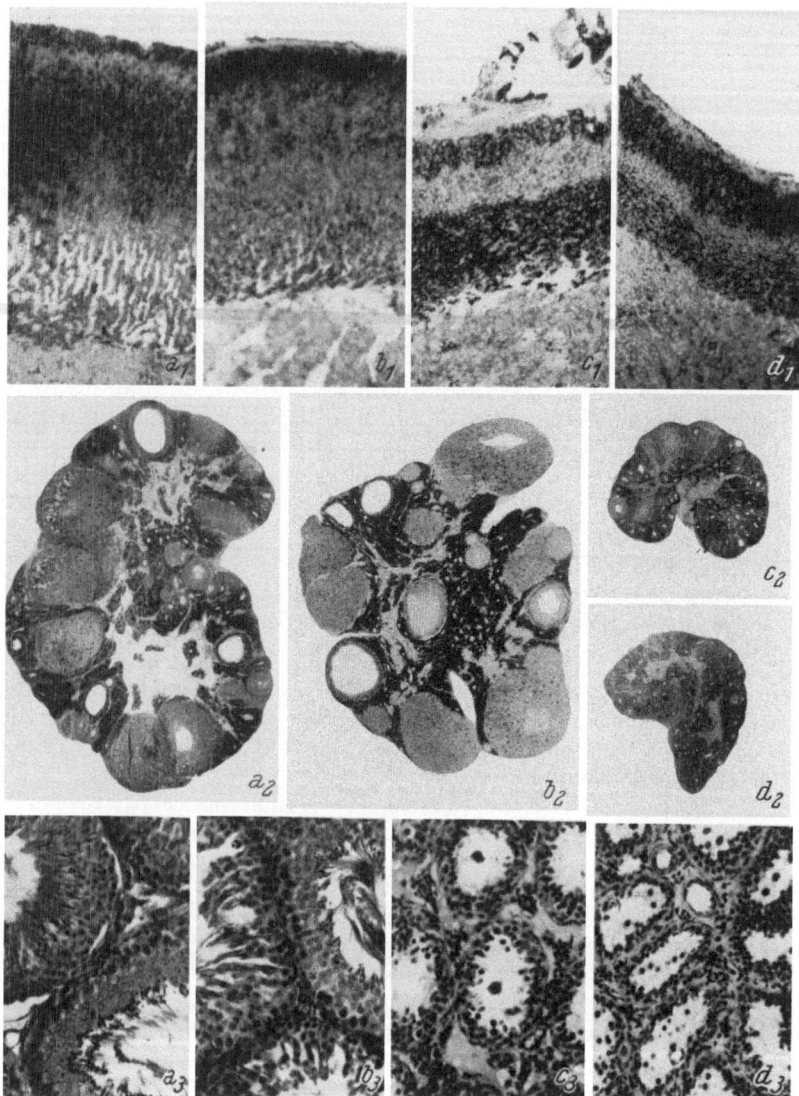

Fig. 4. Lipid picture of the adrenal cortex (row 1), the histological structure of the ovaries (row 2), and the testes (row 3) in normal (a), in hypophysectomized (d), and in hypophysectomized animals bearing a pituitary transplant in the hypophysiotrophic area (b), and outside the area in the hypothalamus (c). Row 1: Sudan black, reduced from × 80; row 2: Scharlach-R-hematoxylin, reduced from × 16; row 3: hematoxylin-eosin, reduced from × 280 (*Halász* and coll., 1965).

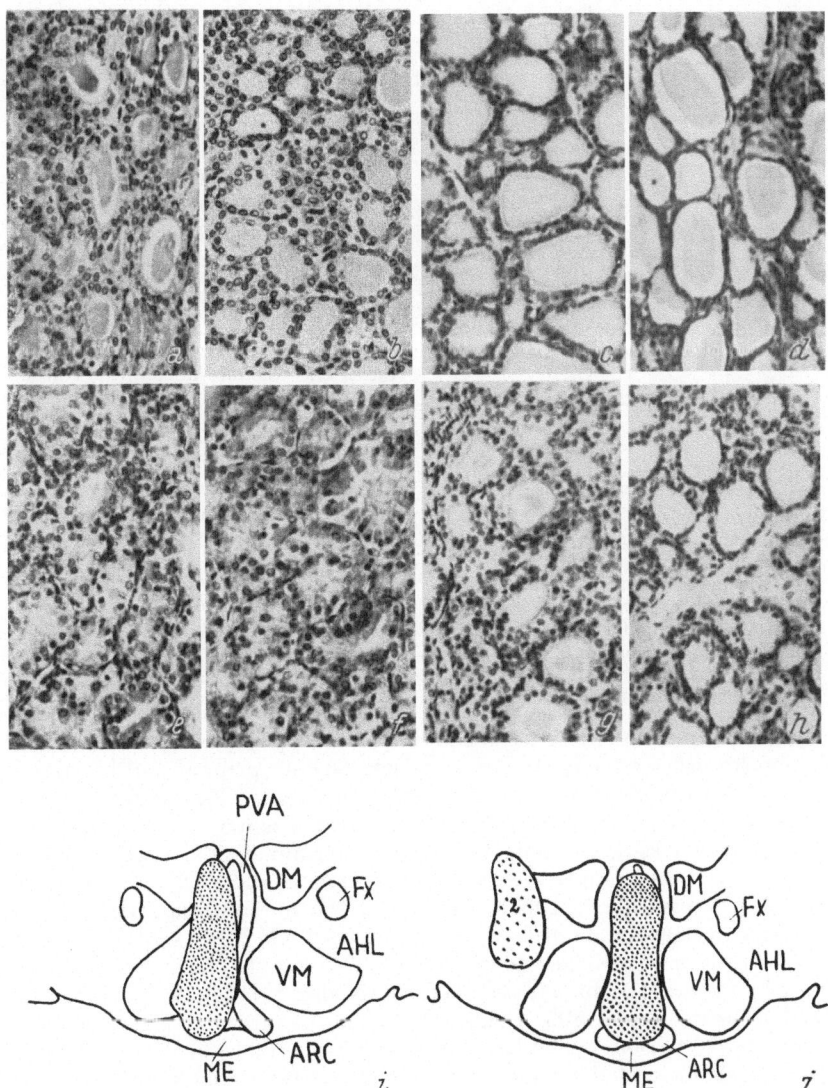

Fig. 5. Histological structure of the thyroid gland of untreated and methylthiouracil-treated normal, hypophysectomized, and hypophysectomized animals bearing a pituitary transplant in the hypothalamus. Untreated: a) normal, b) pituitary in the hypophysiotrophic area, c) pituitary outside the hypophysiotrophic area but in the hypothalamus, d) hypophysectomized control; methylthiouracil-treated: e) intact, f) pituitary in the hypophysiotrophic area in close connection with the median eminence. The location of the transplant is indicated

Continued on p. 338.

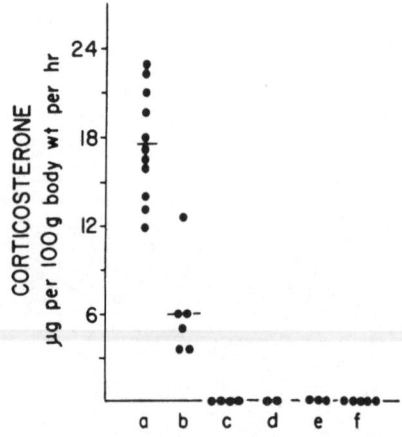

Before analyzing the role of the HTA in the control of the anterior pituitary, some data about the morphology of this area may be mentioned.

b) Some morphological data about the hypophysiotrophic area. The tubero-infundibular tract as the site of hypothalamic trophic and/or releasing factor production

Fig. 6. Corticosterone content of the adrenal venous blood in normal (a), hypophysectomized (f), hypophysectomized animals bearing a pituitary graft in the hypophysiotrophic area (b), outside the hypophysiotrophic area but in the hypothalamus (c), in other parts of the brain (d), or under the renal capsule (e). Each point represents one animal. Group averages indicated by horizontal line *(Halász* and coll., 1965).

As mentioned above the HTA includes the arcuate nuclei, the ventral part of the anterior periventricular nuclei, the medial part of the retrochiasmatic region and the median eminence. The axons of these nuclei turn in a ventral or ventromedial direction and form the fine-calibered tubero-hypophysial tract of *Laruelle* (1934), *Spatz* (1951), *Nowakowski* (1951), *Martinez* (1960), or tubero-infundibular tract of *Szentágothai* (1962, 1964). The fibers of this tract terminate mainly in the superficial layer of the median eminence and in the proximal part of the pituitary stalk. A much smaller number of nerve endings of this pathway can be found around the capillary loops of the median eminence (Figs. 8 and 9). The superficial layer of the median eminence, which was termed zona palisadica by *Martinez* (1960), is composed of nerve terminals of the tubero-infundibular tract intermingled with some ependymal and glial processes and end feet (*Szentágothai,* 1962). Under the electron microscope, many axon endings can be seen in this layer (Figs. 10 and 11). They contain a large number of small vesicles, about 200—700 Å sized, resembling synaptic vesicles, as well as bigger so-called dense-core vesicles the diameter of which varies between 500 Å and 1300 Å, the majority showing a diameter between 700 Å—900 Å. The large neurosecretory granules of 1000 Å—3000 Å in diameter, most of them measuring 1500 Å—2100 Å, which are characteristic of the supraoptico- and paraventriculo-hypophysial tracts, do not occur

Continuation of p. 337.

in i. g) Pituitary in the hypophysiotrophic area but not in direct contact with the portal system. The location of this midline graft is indicated in j no. 1. h) Pituitary transplant outside the hypophysiotrophic area. The location of this graft is indicated in j no. 2. AHL lateral hypothalamic area, ARC arcuate nucleus, DM dorsomedial nucleus, FX fornix, ME median eminence, PVA periventricular nucleus, VM ventromedial nucleus. Hematoxylin-eosin, reduced from × 320 *(Halász* and coll., 1965).

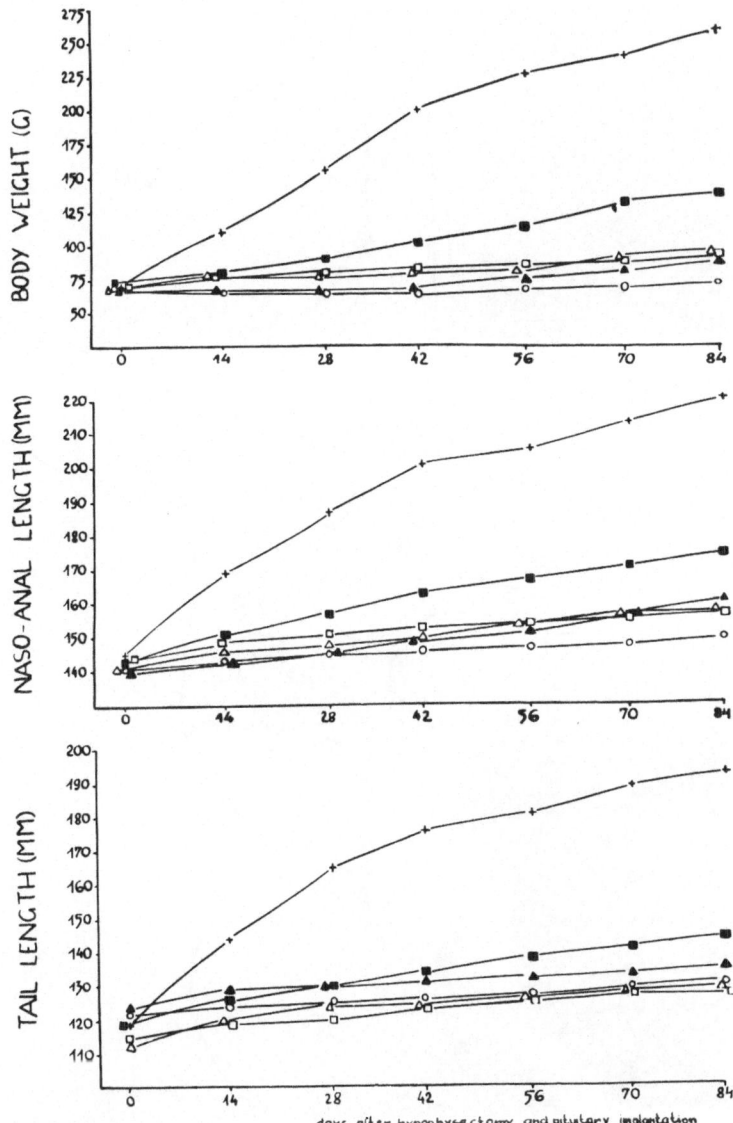

Fig. 7. Mean body weight gain and the increase in naso-anal and tail length in normal, hypophysectomized and hypophysectomized and anterior-pituitary grafted groups. +—+ Normal, ■—■ Hypophysectomized + pituitary graft in the hypophysiotrophic area of the hypothalamus, □—□ Hypophysectomized + pituitary graft outside the hypophysiotrophic area of the hypothalamus, △—△ Hypophysectomized + pituitary graft under the renal capsule, ▲—▲ The same + thyranon and cortisone treatment, o—o Hypophysectomized (Halász, 1964).

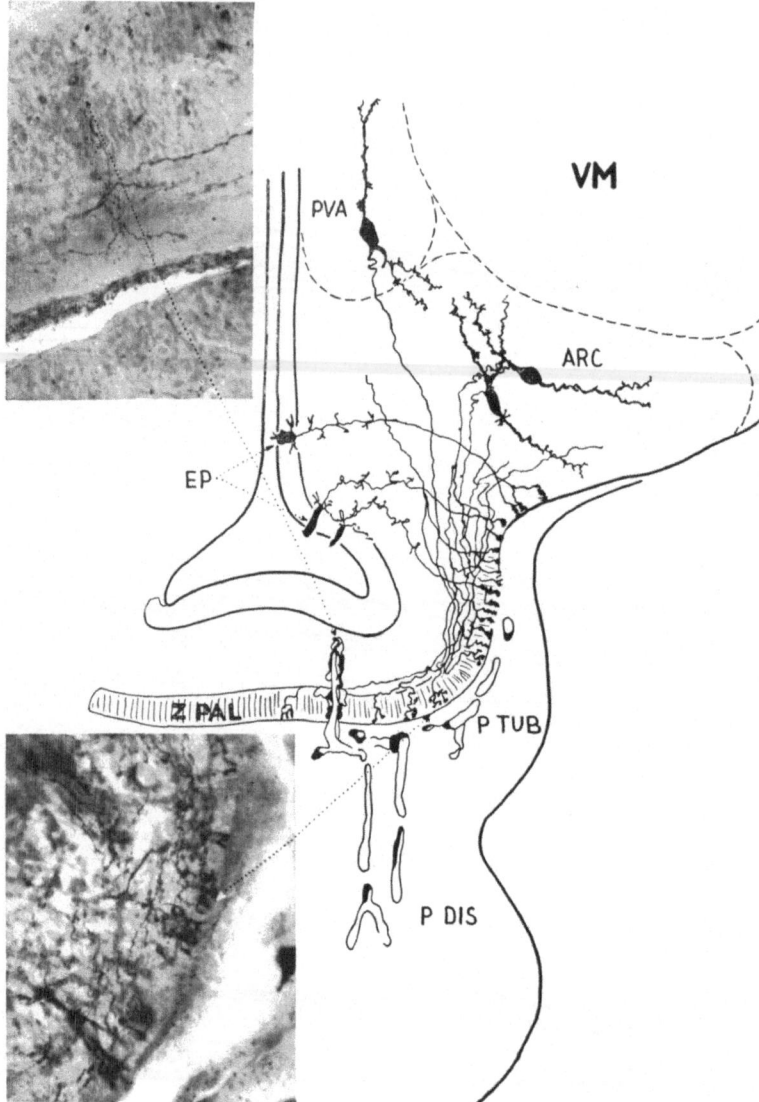

Fig. 8. Semi-diagrammatic representation of the tubero-infundibular neuron system. Fine axons of small nerve cells situated in the arcuate nucleus (ARC) and in the ventral part of the anterior periventricular nucleus (PVA) enter the stalk through the lateral lip of the transition zone between the median eminence and the infundibular process. The fibers terminate partly around capillary loops of the median eminence and the proximal stalk (see upper inset). Their larger part terminates in the outer zone (zona palisadica) of the median eminence and the proximal part, *i. e.* on the surface having immediate contact with the pars tuberalis (see endings in lower inset). EP ependyma, P DIS pars distalis, P TUB pars tuberalis, VM ventromedial nucleus, Z PAL zona palisadica (by courtesy of *Szentágothai*, 1962).

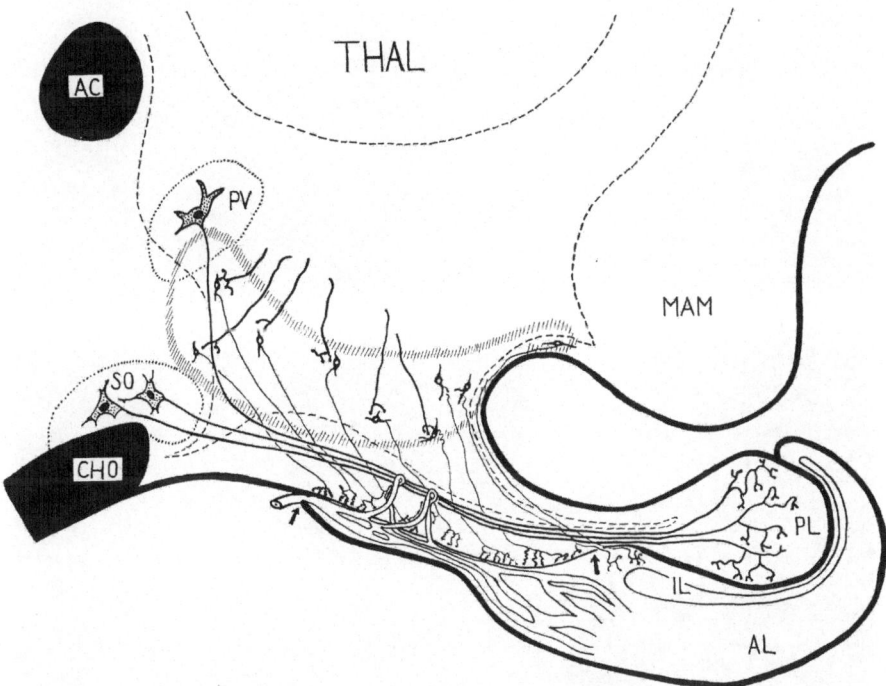

Fig. 9. Diagram illustrating the tubero-infundibular and the supraoptico- and paraventriculo-hypophysial tract. Short arrows indicate proximal and distal ends of the zona palisadica containing the nerve terminals of the tubero-infundibular tract. Larger nerve fibers are the fibers of the supraoptico- and paraventriculo-hypophysial tract, which can be traced to the neurohypophysis. Hatched zone in the diagram indicates the HTA which corresponds to the localization of the nerve cells giving rise to the tubero-infundibular tract. AC anterior commissure, AL anterior lobe, CHO optic chiasm, IL intermediate lobe, MAM mamillary body, PL posterior lobe, PV paraventricular nucleus, SO supraoptic nucleus, THAL thalamus. Dashed line indicates outlines of 3rd ventricle (by courtesy of Szentágothai, 1964).

in this layer (Szentágothai and Halász, 1964; Monroe, 1967). These latter granules are exclusively present in the deep or fibrous layer of the median eminence where the fibers of the tracts mentioned do pass. The nervous tissue of the median eminence is separated from the blood vessels of the portal system by a thin tissue interspace. Between an extremely delicate outer membrane of the median eminence and the basement membrane of the capillaries there is a connective tissue space of 1—2 µ in width. The capillaries of the portal system are lined by a fenestrated endothelium.

When taking into account that (1) the hypothalamic substances essential for the maintenance of normal structure and function of the anterior pituitary are exclusively present in the HTA, as indicated by the intra-

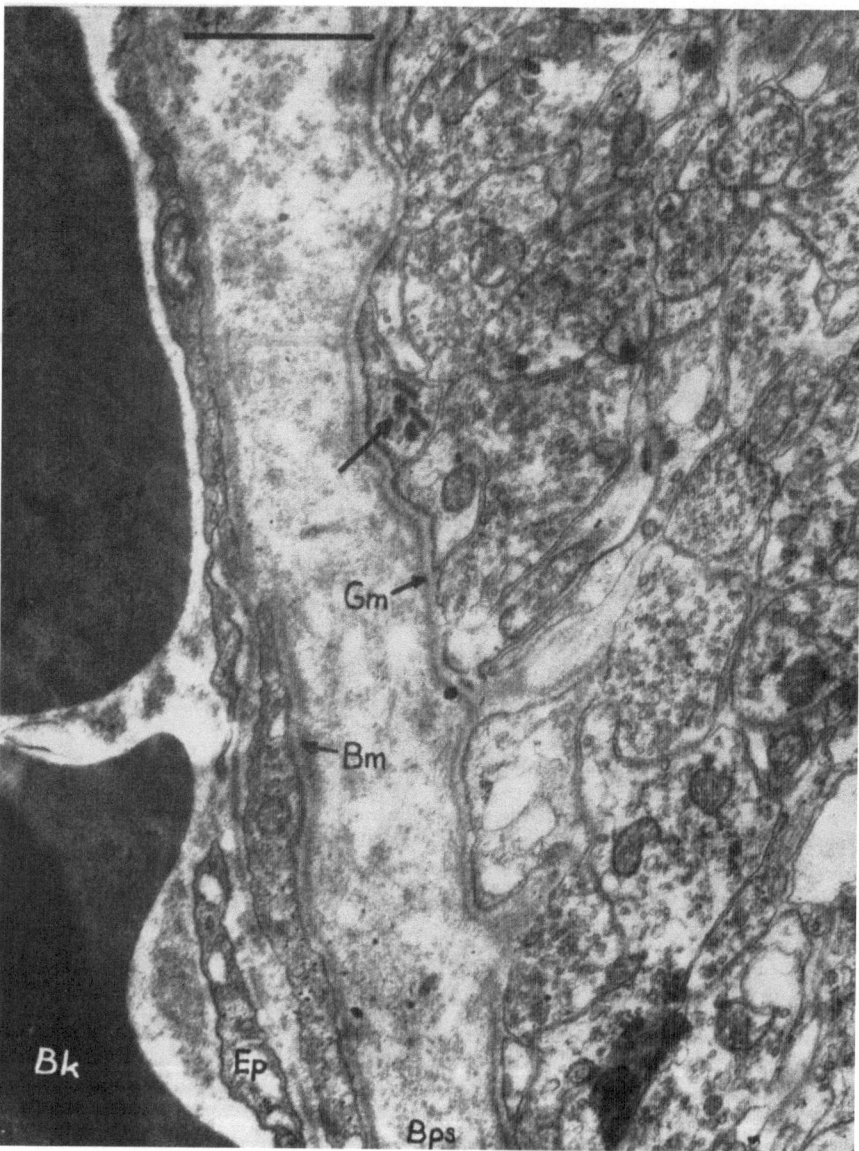

Fig. 10. Superficial zone (zona palisadica) of the median eminence showing to the right closely packed profiles of nerve terminals filled with synaptic vesicles and some dense-core vesicles. Arrow shows larger electron-dense bodies. Bps tissue space with connective tissue filaments, Gm border membrane of the brain tissue, Bm basement membrane of the capillaries belonging to the portal system ("Mantelplexus"-capillaries), Ep porous endothelium of the capillaries, Bk red blood cell. Scale line at top = 1 μ (*Szentágothai* and *Halász*, 1964).

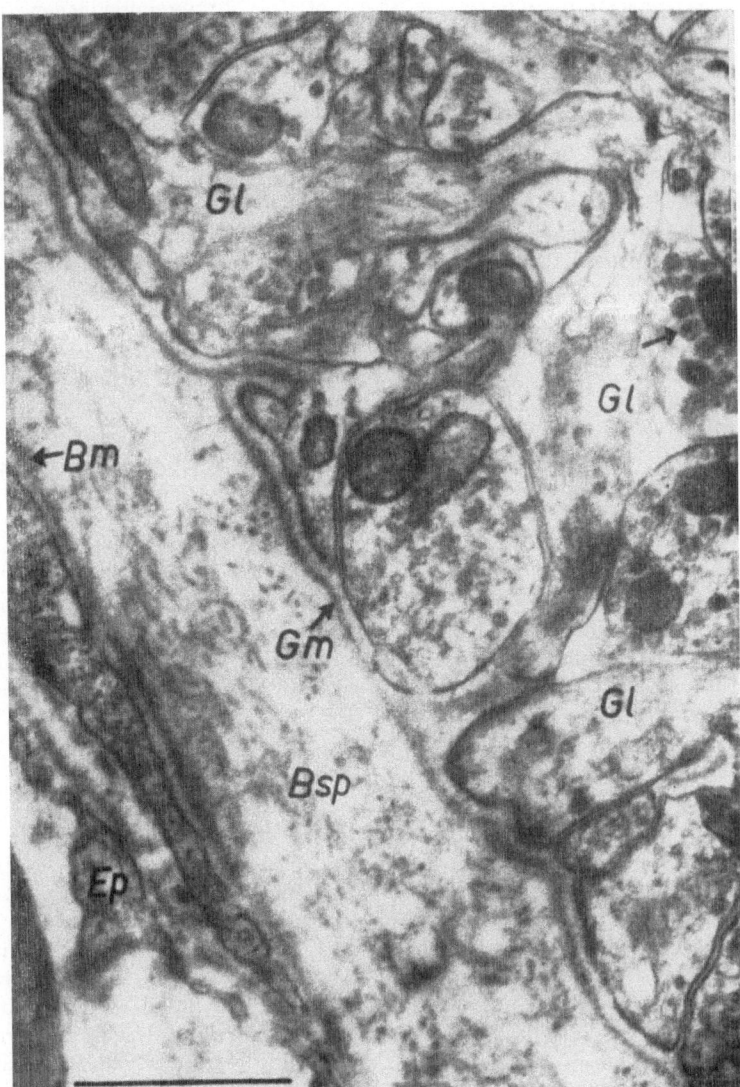

Fig. 11. Higher power view of zona palisadica, ca. $\times$ 31,700. Ep "Mantel-plexus" capillary with porous endothelium, BSP connective tissue space, Bm capillary basement membrane, Gm border membrane of the median eminence, Gl glial pillars with endfeet reaching the surface. Between the glial pillars there are nerve terminals with synaptic vesicles. The arrow on the right points to "dense-core" vesicles. Scale line at bottom = 1 μ (*Szentágothai* and *Halász,* 1964).

hypothalamic pituitary implantation studies; that (2) the HTA corresponds to the region from where the fibers of the tubero-infundibular tract originate; that (3) the nerve endings of this tract are in close connection with the blood vessels of the portal system; and that (4) trophic hormone releasing activity could be demonstrated only in extracts made of the median eminence region, *i. e.* of the hypophysiotrophic area (*McCann*, 1962; *Vernikos-Danellis*, 1964), it is close at hand to assume that the neurons of the tubero-infundibular tract produce and release the hypothalamic "hypophysiotrophic" substances. Although there is no clear-cut morphological evidence as yet, it may be proposed, as it has already been done (*Szentágothai*, 1964; *Monroe*, 1967), that the tubero-infundibular tract is a distinct neurosecretory system which is entirely separated from the supraoptico- and paraventriculo-hypophysial system.

There is some indication that an adrenergic mechanism might be involved in the release of the hypothalamic hypophysiotrophic substances. *Fuxe* (1964) demonstrated great amounts of monoamines — probably dopamine — in the nerve cells of the HTA as well as in the superficial zone of the median eminence, *i. e.* at the terminations of the tubero-infundibular tract. The role of monoamines in the control of the anterior lobe has been dealt with in detail by Dr. *Smelik* (see this volume).

By accepting the theory that the hypothalamic substances, essential for the structural and functional maintenance of the anterior lobe, are produced by the HTA, one is faced with the question: what is the functional significance of the HTA? Does this area merely convert and convey the influence of other hypothalamic structures to the pituitary or does it also exert a regulatory influence on the adenohypophysis by itself? This question is dealt with in the next section.

3. Two levels in the hypothalamic control of the anterior pituitary

a) The hypophysiotrophic area as the first level

To study the exact role and functional capacity of the HTA in the control of the anterior lobe, a special knife assembly was designed (Fig. 12 b; *Halász* and *Pupp*, 1965). One end of a stainless steel wire was bent into a bayonet shape and ground to form a double-edged knife. The wire itself was inserted into a piece of steel tubing. This knife assembly was fixed on the holder of a stereotaxic instrument. By means of this knife, all neural connections of the HTA could be interrupted leaving the area only in contact with the pituitary by the uninterrupted pituitary stalk (Fig. 12 a, c). After this operation it was found that, in the absence of all neural afferents, the HTA is capable to maintain the normal structure and "basal" secretion of the anterior lobe. The following observations were made in rats having a neurally isolated HTA.

1. An appreciable change in anterior pituitary histology did not occur (*Halász* and *Pupp*, 1965).

2. Gonadotrophic hormone secretion was fairly well maintained in male rats (Fig. 13 d), but seriously altered in females. It was found that such

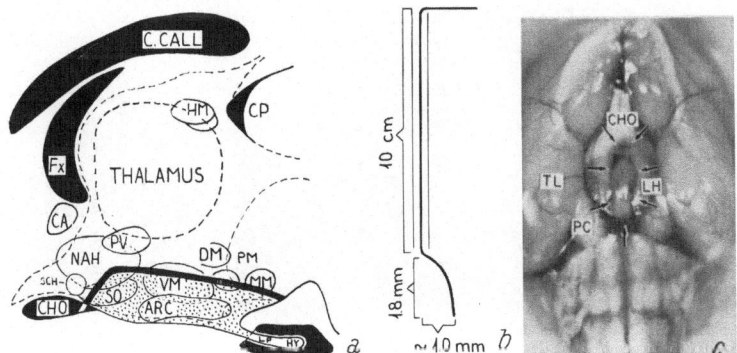

Fig. 12. a) Schematic drawing of the experimental situation in the animals with complete deafferentation of the HTA. The area cut around (hatched region) was in contact with the pituitary but was disconnected from other parts of the brain as indicated by the heavy line. ARC arcuate nucleus, FX fornix, HL lateral habenular nucleus, HM medial habenular nucleus, LPHY posterior lobe of hypophysis, MM medial mamillary nucleus, NAH anterior hypothalamic nucleus, PM premamillary nucleus, PV paraventricular nucleus, SCH supra-chiasmatic nucleus, SO supraoptic nucleus, VM ventromedial nucleus. — b) A schematic drawing of the knife assembly used for deafferentation. — c) Demonstration of the knife cut indicated by arrows as seen from the base of the brain. The picture was taken right after deafferentation. CHO optic chiasm, LH lateral hypothalamus, PC cerebral peduncle, TL temporal lobe (*Halász* and *Pupp*, 1965).

animals are not able to ovulate (Fig. 14). Their ovaries do not contain fresh corpora lutea (Fig. 13 f, g), and unilateral ovariectomy is not followed by the compensatory hypertrophy of the remaining ovary (Fig. 14). There is, however, a pituitary response to castration, *i. e.* pituitary LH secretion increases after ovariectomy (Fig. 14), and castration cells develop in the anterior lobe of spayed rats having a deafferented HTA, although this response is far from normal (*Halász* and *Gorski*, 1967).

3. Basal TSH secretion was slightly depressed (Fig. 15). In spite of this, propylthiouracil treatment (0.15% in food) produced thyroid hyper-trophy (Figs. 15 and 16; *Halász* and coll., 1967).

4. As indicated by the high plasma corticosterone values (Fig. 17) and by the lipid distribution of the adrenal cortex (Fig. 13 b), basal ACTH secretion was enhanced after the operation. In spite of the increased resting levels, surgical stress (ether anaesthesia + unilateral adrenalectomy) caused further increase in blood corticoids (Fig. 17), and unilateral adrenalectomy was followed by compensatory hypertrophy of the remaining adrenal (Fig. 17). Diurnal variations in ACTH secretion, characteristic of normal animals, were not evident (Fig. 17; *Halász* and coll., 1967).

5. Only a slight retardation in body growth was observed indicating that GH secretion is fairly well maintained under these circumstances (*Halász, Schalch* and *Gorski*, unpublished).

From these findings it may be concluded that the role of the HTA is not merely to transform neural signals, coming from various parts of

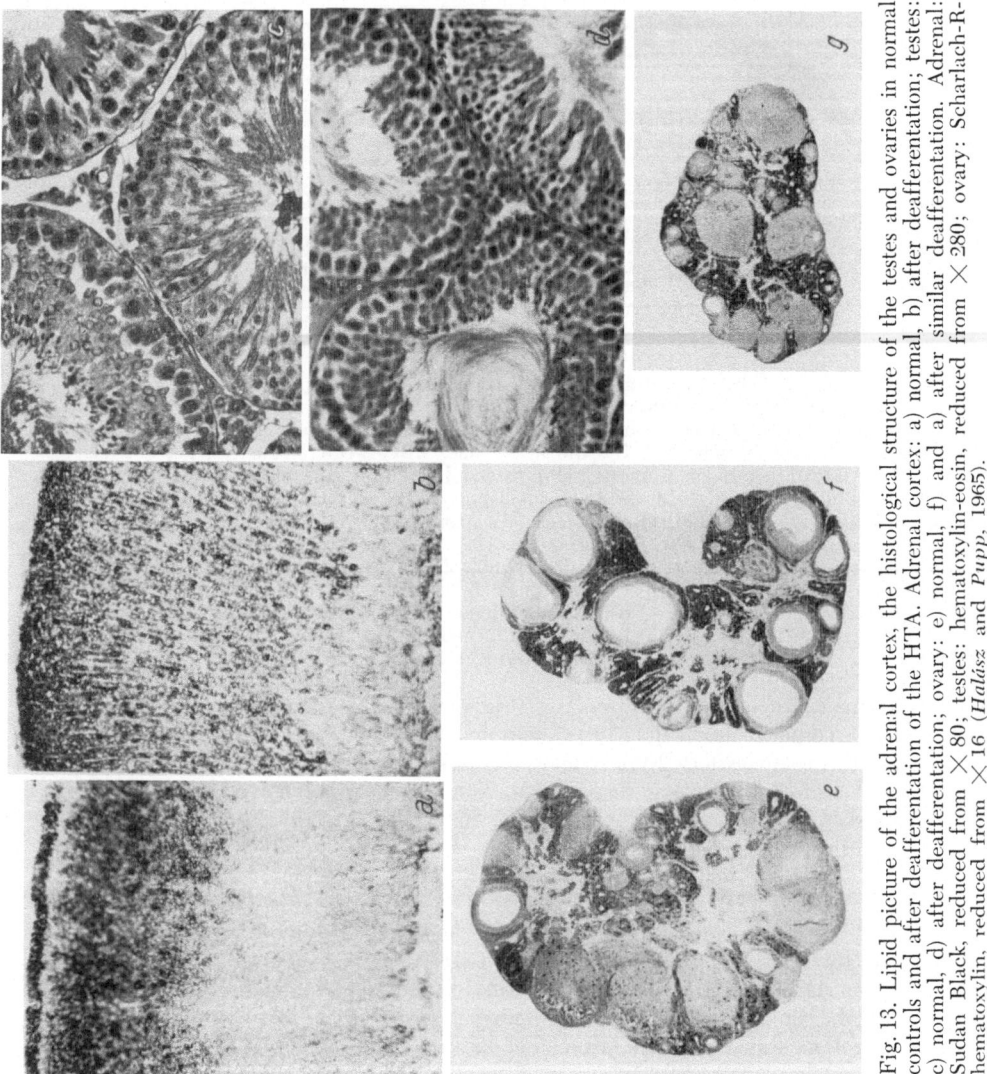

Fig. 13. Lipid picture of the adrenal cortex, the histological structure of the testes and ovaries in normal controls and after deafferentation of the HTA. Adrenal cortex: a) normal, b) after deafferentation; testes: c) normal, d) after deafferentation; ovary: e) normal, f) and a) after similar deafferentation. Adrenal: Sudan Black, reduced from ×80; testes: hematoxylin-eosin, reduced from ×280; ovary: Scharlach-R-hematoxylin, reduced from ×16 (*Halász* and *Pupp*, 1965).

the CNS to the area, into humoral ones but that the area represents a distinct level in the pituitary control which, generally speaking, seems to be responsible for the maintenance of "basal" secretion of the adenohypophysis.

b) The second level in the hypothalamic control mechanism

The findings that ovulation does not occur and the diurnal ACTH rhythm is disturbed if the neural connections of the HTA are interrupted, indicate clearly that the HTA by itself is not able to ensure normal function of the anterior lobe, neural afferents being required for this. This assumption is in good agreement with all observations pointing to the fact that neural structures outside the HTA, such as various other hypothalamic regions, the limbic system, mesencephalic structures, the cortex, etc., are deeply involved in these mechanisms.

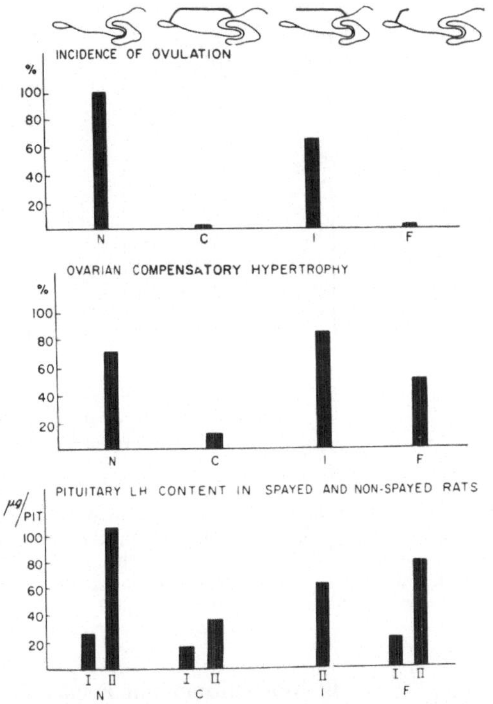

The neural elements outside the HTA may be considered as a second level in the control of the anterior pituitary. Taking all the data available about the morphology and function of the HTA into account, it can be evidently assumed that this second level does not act directly on the pituitary, but might influence it via the HTA by modulating the release, and perhaps also the production, of the "hypophysiotrophic" substance.

There is a vast body of evidence indicating that this second level consists of stimulatory as well as of

Fig. 14. Data about GTH secretion in female rats with different kinds of deafferentation of the HTA. N: normal; C: complete; I: incomplete deafferentation of the HTA; F: frontal cut. I: pituitary LH content in non-spayed; II: pituitary LH content in spayed rats. On the top, the sagittal schematic drawings of the hypothalamic-pituitary complex show the type of deafferentation (heavy line) performed in the different groups (Halász and Gorski, 1967).

inhibitory pathways and that the actual hormone release depends always on the balance of these two influences (for details see Martini and Ganong, 1966).

Studying the functional significance of the various afferent pathways to the HTA we (Halász and Gorski, 1967) could demonstrate that the fibers, indispensable for ovulation, reach the HTA from an anterior direction. The interruption of the lateral, dorsal and posterior connections of the area did not interfere with ovulation, whereas the severance of the anterior afferents leaving all other connections intact blocked ovulation (Fig. 14). The same seems to be true for the afferents, essential for the diurnal variations in ACTH secretion. It was found (Halász and coll., 1967) that, if the anterior connections of the HTA are intact, the diurnal ACTH rhythm is present while, if these pathways are severed, the rhythm is abolished (Fig. 17).

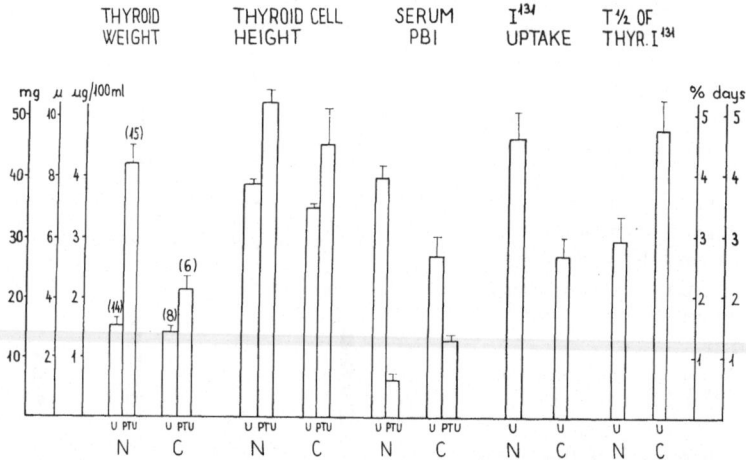

Fig. 15. The columns show thyroid weight, thyroid cell height, serum PBI, thyroidal I[131] uptake and the biological half life of thyroidal I[131] in animals with complete deafferentation of the HTA (C) compared to normal animals (N). U: untreated; PTU: prophylthiouracil-treated group. Horizontal bars above the columns indicate standard error, the numbers in brackets the animal number (*Halász* and coll., 1967, modified).

II. Neurohormonal Feedback Mechanisms

External feedback. It is well known that an automatic self-control exists between the anterior pituitary and the target glands. An excess of corticoids, thyroxine, estrogens or androgens shows an inhibitory effect while a deficiency of these substances exerts a stimulating effect on the secretion and/or on the release of their respective trophic substances by/from the pituitary (external feedback).

Until recently, it was generally believed that the peripheral hormonal milieu influences production and release of anterior pituitary hormones exclusively by direct feedback actions on the adenohypophysis. Although there is no doubt about the existence of such a mechanism, there is increasing evidence indicating that hormone sensitive nervous elements might exist in the hypothalamus and that these are directly involved in the feedback action of target gland hormones on the pituitary. This idea was first raised by *Hohlweg* and *Junkmann* in 1932. These authors did not find castration cells in the pituitaries transplanted under the kidney capsule. Therefore, they suggested that the gonadal hormones influence gonadotrophic functions by way of a hypothetical sexual center located somewhere in the brain. Since then, extensive studies have been performed along these lines. It has been shown for instance that thyroidectomy (*Sinha* and *Meites*, 1965/66), adrenalectomy or administration of adrenal corticoids (*Vernikos-Danellis*, 1965), gonadectomy or treatment with sexual steroids (*Kobayashi* and coll., 1963; *Dávid* and coll., 1965; *Martini* and coll., 1966;

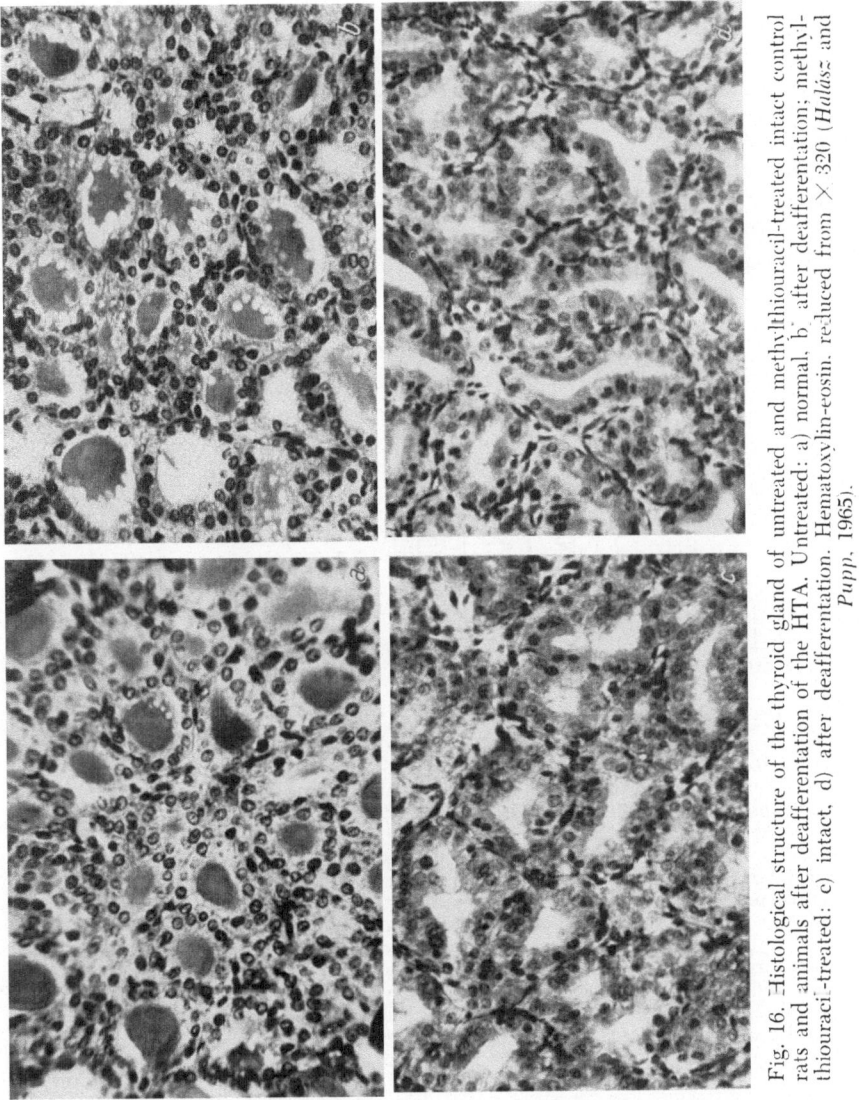

Fig. 16. Histological structure of the thyroid gland of untreated and methylthiouracil-treated intact control rats and animals after deafferentation of the HTA. Untreated: a) normal. b) after deafferentation; methyl-thiouracil-treated: c) intact, d) after deafferentation. Hematoxylin-eosin. reduced from × 320 (*Halász* and *Pupp*. 1965).

Meites and coll., 1966) alters the hypothalamic content of TRF, CRF, LRF and FRF, respectively.

In the light of the data available it seems very probable that there are at least two neural levels at which the hormonal feedback on the pituitary might be mediated. It appears that one of these levels would be the HTA. This assumption is also supported by the following findings.

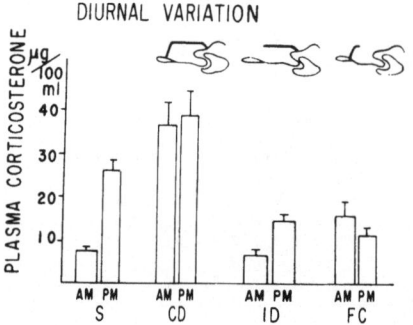

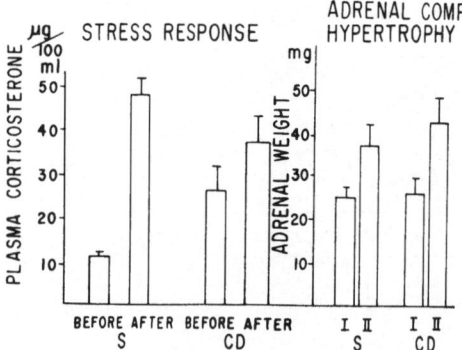

Fig. 17. Plasma corticosterone values and adrenal weight changes following unilateral adrenalectomy in normal animals and in rats with deafferentation of the HTA. Horizontal bars above the columns indicate standard error. AM values at 9 o'clock in the morning; PM at 4 o'clock in the afternoon. S sham operated, CD complete deafferentation, ID incomplete deafferentation, FC frontal cut, *i. e.* only the anterior connections of the HTA were interrupted. On the top, the sagittal schematic drawings of the hypothalamo-pituitary complex show the type of deafferentation (heavy line) performed in the different groups. I. The weight of the adrenal removed first. II. The weight of the "second" adrenal (*Halász* and coll., 1967).

1. Small amounts of estrogens, androgens or corticoids implanted into this region cause marked reduction in pituitary gonadotrophic or adrenocorticotrophic function. This, however, does not seem to be true for thyroxine (see *Reichlin*, 1966). (For references see *Flerkó*, 1966; *Davidson*, 1966; *Mangili* and coll., 1966).

2. Estrogen appears to be taken up selectively by the hypothalamus and the uptake site is, among others, the HTA (*Michael*, 1964).

3. There is a pituitary LH response to castration and adrenal compensatory hypertrophy develops in animals with a neurally isolated HTA (*Halász* and *Gorski*, 1967; *Halász* and coll., 1967).

The second level appears to be represented by hormone sensitive nervous structures outside the HTA. The existence of such elements has been proposed first by *Flerkó* and *Szentágothai* (1957). This theory is supported by the following observations.

1. Electrolytic lesions of the anterior hypothalamic region, *i. e.* of a region outside the HTA (*Flerkó* and *Bárdos*, 1961), as well as the interruption of all neural connections of the HTA (*Halász* and *Gorski*, 1967) interferes with ovarian compensatory hypertrophy.

2. Besides the HTA, the septal and preoptic regions take up estrogen selectively (*Michael*, 1964).

3. Corticoids implanted into the midbrain are effective in inhibiting

ACTH secretion (*Endröczi* and coll., 1961; *Davidson* and *Feldman*, 1963; *Corbin* and coll., 1965).

It seems very likely that the hormone-sensitive nervous elements are sensitive to local increase as well as to local decrease of the hormonal level. In the case of thyroxine, however, the neurons are probably only sensitive to a local decrease of the hormonal level and not to its increase (*Reichlin*, 1966). The observations that the pituitaries of rats with a neurally isolated HTA respond to castration but not to unilateral ovariectomy, favour the idea that the hormone sensitivity of the various nervous structures may be different. It may be assumed that the nervous elements outside the HTA are more sensitive than those within this area. These latter ones seem to respond only to major changes in blood hormonal levels, at least as far as estrogens are concerned. It appears that under physiological circumstances the hormonal feedback is mediated mainly by nervous structures located outside the HTA, but it is very probable that these elements exert their influence on the pituitary via the HTA.

In most cases the neurohormonal feedback is negative or "degenerative". There is, however, evidence that also a positive feedback might exist, at least in the control of gonadotrophic function. Several years ago it has been demonstrated that small amounts of estrogen increase LH output from the pituitary (*Hohlweg* and *Chamorro*, 1937; *Hellbaum* and *Greep*, 1946). *Sawyer* and coll. (1949) and *Everett* and *Sawyer* (1949) have shown that the effects of estrogen or progesterone in producing spontaneous LH release might be eliminated by neural blocking agents indicating that this action would be mediated by way of a neural link (for a detailed discussion of this problem see *Flerkó*, 1966).

There is evidence that some hormones acting at a critical stage of life are able to alter basic patterns of pituitary hormone secretion for the whole life. *Pfeiffer* (1936) was first to observe that, if testes are implanted into newborn female rats, the ovaries of these animals when adult contain only follicles and no corpora lutea while constant vaginal cornification ensued in these rats. In contrast, ovaries grafted to adult male rats which had been castrated at birth showed normal follicular development and formation of corpora lutea. *Barraclough* and *Leathem* (1954) and *Barraclough* (1955, 1961) demonstrated that a single injection of testosterone propionate, administered during the first few days of life to female mice or rats, induces lasting sterility. By the experiments of *Segal* and *Johnson* (1959) the possibility of a direct action of androgens on the pituitary is ruled out. These authors observed that pituitaries of androgen-treated anovulatory female rats are capable of securing cyclic reproductive functions when transplanted in contiguity with the hypothalamus of normal females.

Thus it is close at hand to assume that it is the neural mechanism controlling gonadotrophic hormone secretion that is impaired by androgen. In the light of the observations mentioned as well as of many other it may be inferred that androgens can cause permanent alterations in the cyclic pattern of gonadotrophins, characteristic of females. Under physiological circumstances this action might be exerted by the androgens produced by the testes of the newborn rat. It appears that the brain of the newborn rat of either sex has the inherent ability to maintain a cyclic release of gonadotrophins. It is only during the first few postnatal days that normal male animals under the influence of testicular androgens or female rats that are given testosterone, lose the ability to release periodically gonadotrophins by which ability ovulation is secured.

The mechanism by which androgen may elicit its effect on the neural structures responsible for the cyclic release of gonadotrophins is unknown. It is very interesting, however, that progesterone, administered during 10 days to newborn male rats, prevents the "masculinizing" effect of testicular androgens (*Flerkó* and coll., 1967).

Internal feedback. Apart from the feedback action of target gland hormones, there is some evidence that trophic hormones themselves would be fed back to the hypothalamus (internal feedback), thus influencing anterior pituitary function. The anatomical basis for this assumption has been delivered by *Török* (1954, 1962) and *Szentágothai* and coll. (1957). These authors observed that a small fraction of the blood which has passed the pars tuberalis and, under certain circumstances, even blood which irrigates some part of the pars distalis by internal vessels of the stalk, previously unknown, is drained towards the hypothalamus.

The existence of an internal feedback mechanism has been first proposed in connection to ACTH secretion (*Halász* and *Szentágothai*, 1958, 1960). It has been observed that small pieces of pituitary tissue implanted into the infundibular recess of the hypothalamus produce a depression of ACTH secretion of the animal's own hypophysis. Other tissues, such as muscle or liver, implanted into the same site, or pituitary tissue implanted into other regions of the hypothalamus, did not cause similar changes. These findings have been corroborated by *Motta* and coll. (1965), who showed that ACTH implants situated in the median eminence significantly depress blood corticosterone levels in the rat whereas the same implants in the pituitary are ineffective.

It has been suggested by *Sawyer* and *Kawakami* (1959, 1961) and by *Kawakami* and *Sawyer* (1959) that an internal feedback mechanism might also exist as gonadotrophic functions are concerned. This assumption has been supported by several authors. *Szontágh* and coll. (1962) showed a striking increase of the number of PAS-positive gonadotrophic cells in the pituitaries of both, intact and castrated rats treated with placental gonadotrophins. According to *Corbin* (1966 a), FSH implanted into the median eminence causes a significant reduction in pituitary FSH and the FRF activity of the stalk-median eminence extracts. Implants of LH into the median eminence of normal (*Corbin* and *Cohen*, 1966) or of spayed female (*Corbin*, 1966 b) and of normal and castrated male rats (*Dávid* and coll., 1966) deplete pituitary LH stores significantly.

There is some evidence that such a mechanism would operate also in the control of growth hormone secretion (*Krulich* and *McCann*, 1966; *Müller* and *Pecile*, 1966).

Many examples show that, besides the hormonal feedback action controlling anterior pituitary function, neural afferent connections primarily those conducting stimuli from the reproductive organs, play an important role in the regulation of gonadotrophic hormone secretion. Copulation in the rabbit, cat and ferret causes an almost immediate release of LH from the pituitary with the result that ripe follicles rupture and are replaced by functional corpora lutea. *Long* and *Evans* (1922) observed that the development of corpora lutea and their function could be induced in rat by inserting glass rods into the cervix uteri. Mechanical and electrical stimulation of the cervix (*Shelesnyak*, 1931) are standard methods for inducing pseudopregnancy in the rat.

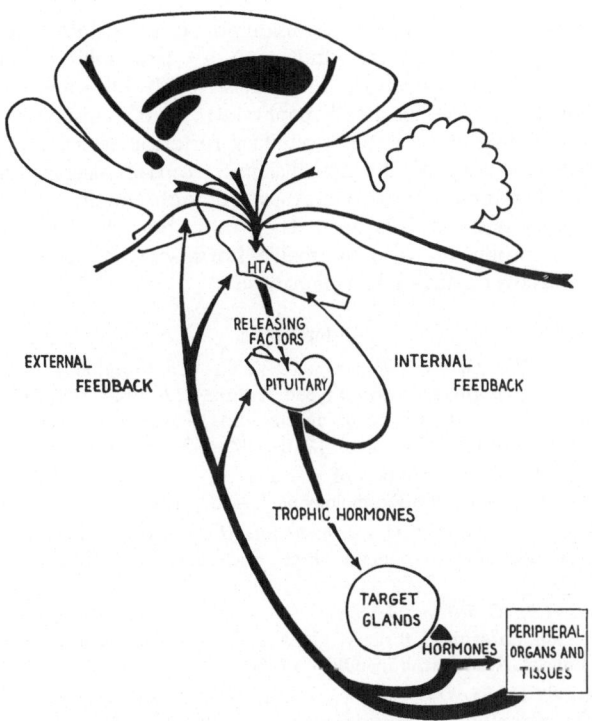

Fig. 18. A summary of the neurohormonal mechanisms controlling anterior pituitary function (*Szentágothai*, 1962, modified).

Conclusions

Anterior pituitary function is primarily controlled by neurohormonal mechanisms. The immediate control of trophic hormone secretion seems to be exercised by the hypophysiotrophic area (HTA) of the hypothalamus. It appears that this area produces and releases the substances, called hypothalamic releasing and inhibiting factors, which are essential for the structural and functional maintenance of the anterior lobe. Anatomically, the area corresponds fairly well with the region from which the fibres of the tubero-infundibular tract originate. It can be assumed that the "hypophysiotrophic" substances are produced by the neurons of this tract. After being released in the median eminence or in the proximal part of the pituitary stalk they enter the blood vessels of the portal system thus reaching the cells of the anterior lobe by blood transport. The tubero-infundibular tract appears to be a distinct neurosecretory system. There is evidence that this system represents a separate level in the neural control of the anterior lobe, which may be responsible for the control of "basal" secretion of pituitary trophic hormones. For the maintenance of normal function of the adenohypophysis, however, neural elements outside the HTA are indispensable. They are considered to constitute a second level in the neural regulation of the pituitary. Very probably they do not act directly on the adenohypophysis, but exercise their influence via the HTA by modifying the release of the "hypophysiotrophic" substances. The feedback action of hormones on anterior pituitary function seems to be mediated, at least partly, by way of the hypothalamic control mechanism. There is evidence that hormone sensitive nervous elements might exist within the HTA as well as outside this area. They may be sensitive to peripheral hormones and, probably, also to trophic hormones, at least to some. In Fig. 18 the concepts outlined are summarized.

References

Averill, R. L. W., D. F. Salaman, and W. C. *Worthington, Jr.:* Thyrotrophin releasing factor in hypophyseal portal blood. Nature *211,* 144—145 (1966).

Barraclough, C. A.: Influence of age on the response of preweaning female mice to testosterone propionate. Amer. J. Anat. *97,* 493—522 (1955).

Barraclough, C. A.: Production of anovulatory, sterile rats by single injections of testosterone propionate. Endocrinology *68,* 62—67 (1961).

Barraclough, C. A., and *J. H. Leathem:* Infertility induced in mice by a single injection of testosterone propionate. Proc. Soc. Exper. Biol. Med., N. Y., *85,* 673—674 (1954).

Campbell, H. J., G. Feuer, J. Garcia, and *G. W. Harris:* The infusion of brain extracts into the anterior pituitary gland and the secretion of gonadotrophic hormone. J. Physiol., London, *157,* 30P—31P (1961).

Cheng, C. P., G. Sayers, L. S. Goodman, and *C. A. Swinyard:* Discharge of ACTH from transplanted pituitary tissue. Amer. J. Physiol. *159,* 426—432 (1949).

Corbin, A.: The „internal" feedback mechanism: effect of median eminence (ME) implants of FSH on pituitary FSH and on stalk-median eminence (SME) FSH-RF. 2nd Int. Cong. Horm. Ster. Milan, Excerpta Med. Found., Int. Cong. Ser. *111,* 194, Amsterdam, New York, London, Milan, Tokyo and Buenos Aires, 1966 a.

Corbin, A.: Pituitary and plasma LH of ovariectomized rats with median eminence implants of LH. Endocrinology *78,* 893—896 (1966 b).

Corbin, A., and *A. I. Cohen:* Effect of median eminence implants of LH on pituitary LH of female rats. Endocrinology *78,* 41—46 (1966).

Corbin, A., G. Mangili, M. Motta, and *L. Martini:* Effect of hypothalamic and mesencephalic steroid implantations on ACTH feedback mechanism. Endocrinology 76, 811—818 (1965).

Cutuly, E.: Autoplastic grafting of anterior pituitary in male rats. Anat. Rec., Philadelphia, 80, 83—97 (1941).

Daniel, P. M.: The anatomy of the hypothalamus and pituitary gland. In *L. Martini* and *W. F. Ganong* (eds.): Neuroendocrinology I. 2, Academic Press, New York and London, 1966.

Dávid, M. A., F. Fraschini, et *L. Martini:* Parallélisme entre le contenu hypophysaire en FSH et le contenu hypothalamique en FSH-RF (FSH-Releasing Factor). C. R. Acad. Sci., Paris, 261, 2249—2251 (1965).

Dávid, M. A., F. Fraschini, and *L. Martini:* Control of LH secretion: role of a "short" feedback mechanism. Endocrinology 78, 55—60 (1966).

Davidson, J. M.: Control of gonadotropin secretion in the male. In *L. Martini* and *W. F. Ganong* (eds.): Neuroendocrinology I. 14. Academic Press, New York and London, 1966.

Davidson, J. M., and *S. Feldman:* Cerebral involvement in the inhibition of ACTH secretion by hydrocortisone. Endocrinology 72, 936—946 (1963).

Deuben, R. R., and *J. Meites:* Stimulation of pituitary growth hormone release by a hypothalamic extract in vitro. Endocrinology 74, 408—414 (1964).

Endröczi, E., K. Lissák, and *M. Tekeres:* Hormonal feed-back regulation of pituitary-adrenocortical activity. Acta physiol. Hung. 18, 291—299 (1961).

Evans, J. S., and *M. B. Nikitovitch-Winer:* Reactivation of hypophyseal grafts by continuous perfusion with median eminence extracts (MEE). Fed. Proc. 24, 190 (1965).

Everett, J. W., and *C. H. Sawyer:* A neural timing factor in the mechanism by which progesterone advances ovulation in the cyclic rat. Endocrinology 45, 581—595 (1949).

Flament-Durand, J.: Observations on pituitary transplants into the hypothalamus of the rat. Endocrinology 77, 446—454 (1965).

Flerkó, B.: Control of gonadotropin secretion in the female. In *L. Martini* and *W. F. Ganong* (eds.): Neuroendocrinology I. 15, Academic Press, New York and London, 1966.

Flerkó, B., and *V. Bárdos:* Absence of compensatory ovarian hypertrophy in rats with anterior hypothalamic lesions. Acta endocr., K'hvn, 36, 180—184 (1961).

Flerkó, B., P. Petrusz, and *L. Tima:* On the mechanism of sexual differentiation of the hypothalamus. Factors influencing the "critical period" of the rat. Acta biol. Hung. 18, 27—36 (1967).

Flerkó, B., and *J. Szentágothai:* Oestrogen sensitive nervous structures in the hypothalamus. Acta endocr., K'hvn, 26, 121—127 (1957).

Fortier, C.: Dual control of adrenocorticotropin release. Endocrinology 49, 782—788 (1951).

Franz, J., C. H. Haselbach, and *O. Libert:* Studies on the effect of hypothalamic extracts on somatotrophic pituitary function. Acta endocr., K'hvn, 41, 336—350 (1962).

Fuxe, K.: Cellular localization of monoamines in the median eminence and the infundibular stem of some mammals. Zschr. Zellforsch. 61, 710—724 (1964).

Goldberg, R. C., and *E. Knobil:* Structure and funktion of intraocular hypophysial grafts in the hypophysectomized male rat. Endocrinology 61, 742—752 (1957).

Green, J. D., and *G. W. Harris:* The neurovascular link between the neurohypophysis and adenohypophysis. J. Endocr. 5, 136—146 (1947).

Greer, M. A., R. O. Scow, and *C. Grobstein:* Thyroid function in hypophysectomized mice bearing intraocular pituitary implants. Proc. Soc. Exper. Biol. Med., N. Y., *82,* 28—30 (1953).

Guillemin, R., and *B. Rosenberg:* Humoral hypothalamic control of anterior pituitary: A study with combined tissue cultures. Endocrinology 57, 599—607 (1955).

Guillemin, R., E. Yamazaki, M. Jutisz, et *E. Sakiz:* Présence dans un extrait de tissus hypothalamiques d'une substance stimulant la sécrétion de l'hormone hypophysaire thyréotrope (TSH). Première purification par filtration sur le Sephadex. C. R. Acad. Sci., Paris, *255,* 1018—1020 (1962).

Halász, B.: Neural control of growth hormone secretion. Proceedings 2nd Internat. Congr. Endocrinology, London. Excerpta Medica Foundation, International Congress Series 83, 517—521, Amsterdam, New York, London, Milan, Tokyo and Buenos Aires, 1964.

Halász, B., and *R. A. Gorski:* Gonadotrophic hormone secretion in female rats after partial or total interruption of neural afferents to the medial basal hypothalamus. Endocrinology 80, 608—622 (1967).

Halász, B., and *L. Pupp:* Hormone secretion of the anterior pituitary gland after physical interruption of all nervous pathways to the hypophysiotrophic area. Endocrinology 77, 553—562 (1965).

Halász, B., und *J. Szentágothai:* Über die unmittelbare Rückwirkung einer vom Hypophysenvorderlappen erzeugten Substanz auf den Hypothalamus. Acta physiol. Hung. *14,* Suppl. 6 (1958).

Halász, B., and *J. Szentágothai:* Control of adrenocorticotrophic function by direct influence of pituitary substance on the hypothalamus. Acta morph. Hung. 9, 251—261 (1960).

Halász, B., L. Pupp, and *S. Uhlarik:* Hypophysiotrophic area in the hypothalamus. J. Endocr. *25,* 147—154 (1962).

Halász, B., M. Slusher, and *R. A. Gorski:* Adrenocorticotrophic hormone secretion in rats after partial or total deafferentation of the medial basal hypothalamus. Neuroendocrinology 2, 43—55 (1967).

Halász, B., W. H. Florsheim, N. L. Corcorran, and *R. A. Gorski:* Thyrotrophic hormone secretion in rats after partial or total interruption of neural afferents to the medial basal hypothalamus. Endocrinology *80,* 1075—1082 (1967).

Halász, B., L. Pupp, S. Uhlarik, and *L. Tima:* Growth of hypophysectomized rats bearing pituitary transplants in the hypothalamus. Acta physiol. Hung. 23, 287—292 (1963).

Halász, B., L. Pupp, S. Uhlarik, and *L. Tima:* Further studies on the hormone secretion of the anterior pituitary transplanted into the hypophysiotrophic area of the rat hypothalamus. Endocrinology 77, 343—355 (1965).

Harris, G. W.: The blood vessels of the rabbits pituitary gland, and the significance of the pars and zone tuberalis. J. Anat., London, *81,* 343—351 (1947).

Harris, G. W.: Neural Control of the Pituitary Gland. Monographs of the Physiol. Soc. Arnold, London, 1955.

Harris, G. W., and *D. Jacobsohn:* Functional grafts of the anterior pituitary gland. Proc. Roy. Soc. London, Biol. Sc. *139,* 263—276 (1952).

Hellbaum, A. A., and *R. O. Greep:* Action of estrogen on release of hypophyseal luteinizing hormone. Proc. Soc. Exper. Biol. Med., N. Y., *63,* 53—56 (1946).

Hohlweg, W., und *A. Chamorro:* Über die luteinisierende Wirkung des Follikelhormons durch Beeinflussung der luteogenen Hypophysenvorderlappensekretion. Klin. Wschr., Wien, *16,* 196—197 (1937).

Hohlweg, W., und *K. Junkmann:* Die homonal-nervöse Regulierung der Funktion des Hypophysenvorderlappens. Klin. Wschr., Wien, *11,* 321—323 (1932).

Igarashi, M., and *S. M. McCann:* A hypothalamic follicle stimulating hormone-releasing factor. Endocrinology *74,* 446—452 (1964).

Kawakami, M., and *C. H. Sawyer:* Neuroendocrine correlates of changes in brain activity thresholds by sex steroids and pituitary hormones. Endocrinology *65,* 652—668 (1959).

Kobayashi, T., T. Kobayashi, T. Kigawa, M. Mizuno, and *Y. Amenomori:* Influence of rat hypothalamic extract on gonadotropic activity of cultivated anterior pituitary cells. Endocrinol. Japon. *10,* 16—24 (1963).

Krulich, L., and *S. M. McCann:* Influence of growth hormone (GH) on content of GH in the pituitaries of normal rats. Proc. Soc. Exper. Biol. Med. N. Y. *121,* 1114—1117 (1966).

Laruelle, L.: Le système végétatif mésodiencéphalique, partie anatomique. Rev. neurol., Paris, *61,* 808—842 (1934).

Long, J. A., and *H. M. Evans:* The oestrous cycle in the rat and its associated phenomena. Mem. Univ. Calif. *6,* 1—148 (1922).

Mangili, G., M. Motta, and *L. Martini:* Control of adenocorticotropic hormone secretion. In *L. Martini* and *W. F. Ganong* (eds.): Neuroendocrinology I. 9, Academic Press, New York-London 1966.

Martinez, P. M.: The structure of the pituitary stalk and the innervation of the neurohypophysis in the cat. Dissertation, Leiden, Luctor et Emergo, 1960.

Martini, L., and *W. F. Ganong* (eds.): Neuroendocrinology I, Academic Press, New York-London 1966.

Martini, L., F. Fraschini, and *M. Motta:* New data on the nervous control of the pituitary gland. 2nd Int. Congr. Horm. Ster., Milan, Excerpta Med. Found., Int. Cong. Ser. *111,* 94, Amsterdam-New York-London-Milan-Tokyo-Buenos Aires 1966.

McCann, S. M.: A hypothalamic luteinizing hormone-releasing factor. Amer. J. Physiol. *202,* 395—400 (1962).

McCann, S. M., and *A. P. S. Dhariwal:* Hypothalamic releasing factors and the neurovascular link between the brain and the anterior pituitary. In *L. Martini* and *W. F. Ganong* (eds.): Neuroendocrinology I. 8, Academic Press, New York-London 1966.

McCann, S. M., S. Taleisnik, and *H. M. Friedman:* LH-releasing activity in hypothalamic extracts. Proc. Soc. Exper. Biol. Med., N. Y., *104,* 432—434 (1960).

Meites, J., B. E. Piacsek, and *J. C. Mittler:* Effects of castration and gonadal steroids on hypothalamic content of FSH-RF and LRF in rats. 2nd Int. Congr. Horm. Ster., Milan, Excerpta Med. Found., Int. Cong. Ser. *111,* 91, Amsterdam, New York, London, Milan, Tokyo, Buenos Aires 1966.

Meites, J., P. K. Talwalker, and *A. Ratner:* Evidence for a prolactin inhibiting factor in rat hypothalamic tissue. Program 44th Meeting Endocrine Soc., Chicago, *19* (1962).

Michael, R. P.: The selective accumulation of oestrogen in the neural and genital tissues of the cat. Proc. 1st Intern. Congr. Hormonal Steroids. Milan 1962, Academic Press, New York 2, 457—469 (1964).

Monroe, B. G.: A comparative study of the ultrastructure of the median eminence, infundibular stem and neural lobe of the hypophysis of the rat. Zschr. Zellforsch. *76,* 405—432 (1967).

Motta, M., G. Mangili, and *L. Martini:* A short feedback loop in the control of ACTH secretion. Endocrinology *77,* 392—395 (1965).

Müller, E., and *A. Pecile:* Influence of exogenous growth hormone on endogenous growth hormone release. Proc. Soc. Exper. Biol. Med., N. Y. *122*, 1289 to 1291 (1966).

Nikitovitch-Winer, M. B.: Induction of ovulation in rats by direct intrapituitary infusion of median eminence extracts. Endocrinology *70*, 350—358 (1962).

Nikitovitch-Winer, M. B., and *J. W. Everett:* Functional restitution of pituitary grafts re-transplanted from kidney to median eminence. Endocrinology *63*, 916—930 (1958).

Nikitovitch-Winer, M. B., and *J. W. Everett:* Histologic changes in grafts of rat pituitary on the kidney and upon re-transplantation under the diencephalon. Endocrinology *65*, 357—368 (1959).

Nowakowski, H.: Infundibulum and Tuber cinereum der Katze. Dtsch. Zschr. Nervenhk. *165*, 201—239 (1951).

Pasteels, J. L.: Sécrétion de prolactine par l'hypophyse en culture de tissus. C. R. Soc. Biol., Paris, *253*, 2140—2142 (1961 a).

Pasteels, J. L.: Premiers résultats de culture combinée *in vitro* d'hypophyse et d'hypothalamus, dans le but d'en apprécier la sécrétion de prolactine. C. R. Soc. Biol., Paris, *253*, 3074—3075 (1961 b).

Pasteels, J. L.: Élaboration par l'hypophyse humaine en culture de tissus d'une substance stimulant le jabot de Pigeon. C. R. Acad. Sci., Paris, *254*, 4083—4085 (1962).

Pfeiffer, C. A.: Sexual differences of the hypophyses and their determination by the gonads. Amer. J. Anat. *58*, 195—225 (1936).

Porter, J. C., and *H. W. Rumsfeld jr.:* Effect of lyophylized plasma and plasma fraction from hypophyseal portal vessel blood on adrenal ascorbic acid. Endocrinology *58*, 359—364 (1956).

Porter, J. C., and *H. W. Rumsfeld, jr.:* Further study of an ACTH-releasing protein from hypophyseal portal vessel plasma. Endocrinology *64*, 948—954 (1959).

Reichlin, S.: Control of thyrotropic hormone secretion. In *L. Martini* and *W. F. Ganong* (eds.): Neuroendocrinology I. 12. Academic Press, New York-London 1966.

Royce, P. C., and *G. Sayers:* Corticotropin-releasing activity of a pepsin labile factor in the hypothalamus. Proc. Soc. Exper. Biol. Med., N. Y., *98*, 677—680 (1958).

Rumsfeld, H. W., and *J. C. Porter:* ACTH-releasing activity in an acetone extract of beef hypothalamus. Arch. Biochem. *82*, 473—475 (1959).

Saffran, M., and *A. V. Schally:* Release of corticotrophin by anterior pituitary tissue *in vitro*. Can. J. Biochem. Physiol. *33*, 408—415 (1955).

Sawyer, C. H., J. W. Everett, and *J. E. Markee:* A neural factor in the mechanism by which estrogen induced the release of luteinizing hormone in the rat. Endocrinology *44*, 218—233 (1949).

Sawyer, C. H., and *M. Kawakami:* Characteristics of behavioral and electroencephalographic after-reactions to copulation and vaginal stimulation in the female rabbit. Endocrinology *65*, 622—630 (1959).

Sawyer, C. H., and *M. Kawakami:* Interactions between the central nervous system and hormones influencing ovulation. In *C. A. Villee* (ed.): Control of Ovulation, Pergamon Press, Oxford, 79—97 (1961).

Schreiber, V., A. Eckertova, Z. Franz, J. Koci, M. Rybák, and *V. Kmentova:* Effect of a fraction of bovine hypothalamic extract on the release of TSH by rat adenophypophysis *in vitro*. Experientia, Basel, *17*, 264—265 (1961).

Segal, S. J., and *D. C. Johnson:* Inductive influence of steroid hormones on the neural system: ovulation controlling mechanisms. Arch. anat. microsc., Paris, 48, 261—274 (1959).

Shelesnyak, M. C.: Induction of pseudopregnancy in the rat by means of electrical stimulation. Anat. Rec., Philadelphia, 49, 179—183 (1931).

Shibusawa, K., S. Saito, K. Nishi, T. Yamamoto, K. Tomizawa, and *C. Abe:* The hypothalamic control of the thyrotroph-thyroidal function. Endocrinol. Japan. 3, 116—124 (1956).

Sinha, D., and *J. Meites:* Effects of thyroidectomy and thyroxine on hypothalamic concentration of "thyrotropin releasing factor" and pituitary content of thyrotropin in rats. Neuroendocrinology 1, 4—14 (1965/66).

Siperstein, E. R., and *M. A. Greer:* Observation on the morphology and histochemistry of the mouse pituitary implanted in the anterior eye chamber. J. Nat. Cancer Inst., Wash. 17, 569—599 (1956).

Spatz, H.: Neues über die Verknüpfung von Hypophyse und Hypothalamus. Acta neuroveget., Wien, 3, 1—49 (1951).

Szentágothai, J.: Anatomical considerations. In *J. Szentágothai, B. Flerkó, B. Mess,* and *B. Halász:* Hypothalamic control of the anterior pituitary, 2, Akadémiai Kiadó, Budapest 1962.

Szentágothai, J.: The parvicellular neurosecretory system. Progress in Brain Research. Elsevier Publishing Company, Amsterdam, 5, 135—146 (1964).

Szentágothai, J., und *B. Halás:* Regulation des endokrinen Systems über den Hypothalamus. Nova Acta Leopoldina N. F., Leipzig, 28, 227—248 (1964).

Szentágothai, J., I. Rozsos, and *J. Kutas:* Posterior lobe and blood circulation of the anterior pituitary. Magyar Tud. Akad. Biol. Orv. Tud. Oszt. Közl., Budapest 8, 104—106 (1957), in Hungarian.

Szentágothai, J., B. Flerkó, B. Mess, and *B. Halász:* Hypothalamic control of the anterior pituitary, Akadémiai Kiadó, Budapest 1962.

Szontágh, F. E., S. Uhlarik, and *A. Jakobovits:* The effect of gonadotrophic hormones on the hypophysis of the rat. Acta endocr., K'hvn, 41, 31—34, (1962).

Talwalker, P. K., A. Ratner, and *J. Meites:* In vitro inhibition of pituitary prolactin synthesis and release by hypothalamic extract. Amer. J. Physiol. 205, 213—218 (1963).

Török, B.: Lebendbeobachtung des Hypophysenkreislaufes an Hunden. Acta morph. Hung. 4, 83—89 (1954).

Török, B.: Neue Angaben zum Blutkreislauf der Hypophyse. Verhandlungen des 1. Europäischen Anatomen-Kongresses, Strasbourg 1960. Anat. Anz., Jena, Erg.-Heft zum 109. Bd, 622—629 (1962).

Vernikos-Danellis, J.: Estimation of corticotropin-releasing activity of rat hypothalamus and neurohypophysis before and after stress. Endocrinology 75, 514—520 (1964).

Vernikos-Danellis, J.: Effect of stress, adrenalectomy, hypophysectomy and hydrocortisone on the corticotropin-releasing activity of rat median eminence. Endocrinology 76, 122—126 (1965).

Author's address: Dr. *Béla Halász,* Department of Anatomy, University Medical School, Pécs, Hungary.

Discussion

De Wied: Dr. *Halász,* I should like to congratulate you on a beautiful presentation of excellent work. The question I should like to ask is whether you have an explanation for the fact that in the animals with complete deafferentiation of the hypophysio-trophic area in which you found an increased secretion of adrenal control hormones, adrenal weights were completely normal.

Halász: Concerning your question in regard of the weight of the adrenal glands of the animals showing high plasma corticosterone values, I would like to say that the adrenal weight of these rats was not entirely normal, although the mean weight did not differ significantly from the weight of the sham-operated animals. Plasma corticosterone levels were higher in all animals with a deafferentiated hypophysio-trophic area. There was a great variation, however, in the actual values. In these rats in which we obtained extremely high corticosterone values the adrenal weight was also far above that in the controls. Adrenal weight was in the normal range only in those operated animals in which the plasma corticosterone levels were only slightly increased. In addition I should like to mention that in our first experiments, published in Endocrinology in 1965, the adrenal weight of the animals with complete deafferentiation of the hypophysio-trophic area was significantly higher than in the controls.

Knowles: You have mentioned in your paper the probability of both, releasing factors and inhibitory factors, being concerned in pituitary control. In two of your diagrams you have shown that the hypophysio-trophic area contains not only fibers of the tubero-infundibular tract but also terminations of ependymal fibers (tanycytes). Our studies on the monkey and the ferret suggest that these ependymal cells might exert an inhibitory effect on the gonadotropic activity of the pituitary. Would you care to comment on this possibility?

Halász: First of all I should like to apologize for not mentioning anything about the special type of ependyma and glia present in the hypophysio-trophic area. This has been due to lack of time. Concerning your question it is true indeed that a peculiar type of ependymal cells can be found in this part of the 3rd ventricle. Several authors (*Löfgren, Leveque* and others) did already suggest that these cells might play some role in the control of the anterior lobe. It might be that they exert an inhibitory effect, but nothing is known about this as yet.

Oksche: I am wondering what kind of functional interpretation you would offer for the avian median eminence. As our neurohistological studies have shown, the location of the tubero-infundibular system in birds is in a good agreement with your findings. In *Zonotrichia leucophrys Gambelii* this system is the hypothalamic area responsible for the light-induced increase in gonadal weight (*Wilsson, Farner*). But we are not allowed to overlook that there is a very distinct contribution of pseudo-isocyanine-positive fluorescent fibers to the fine structure of the anterior (rostral) part of the avian median eminence. This material belongs to the magno-cellular system. Beaded fibers can be traced back from the anterior median eminence to the rostral hypothalamus. In this case we are not dealing with aberrant elements. In the organization of the median eminence there is not a uniform pattern for all vertebrates.

Halász: I am sorry, but I cannot give you an interpretation on the functional significance of the mentioned pseudo-isocyanine-positive fluorescent fibers present in the rostral part of the avian median eminence.

Akert: I should like to mention the work of *Ruf* and *Steiner* who could demonstrate the sensitivity of hypothalamic neurons to locally (micro-electrophoresis) applied cortico-steroids. Both, inhibitory and facilitatory effects were seen, thus confirming the notion of negative and positive feed-back mechanisms directly playing upon the hypothalamus.

Halász: Thank you for your comment.

Westermann: Concerning the feed-back mechanisms, I should like to recall a paper by *Kitay, Holub* and *Aller* who were able to demonstrate that high doses of cortisone not only prevented the secretion of ACTH, but actually reduced the ACTH-content of the pituitary. I think these data indicate that corticoids and/or the corticotrophin releasing factor (CRF) not only regulate the release but also the synthesis of ACTH in the gland.

Halász: Yes, I agree with your assumption.

De Robertis: Are you aware of the work of the mexican physiologist *Alvarez Bulle* who found that implantation of different tissues in the hypothalamic region produces a restoration of pituitary function? What is your opinion about this? Did you perform similar experiments implanting other tissues in the hypophysio-trophic region of the hypothalamus?

Halász: I am not familiar with the work of Dr. *Bulle,* therefore, I cannot comment on his observations. Concerning the second part of your question, I have implanted muscle and liver tissue into the hypophysio-trophic area and found increased ACTH secretion in these rats. This was probably a stress effect, because similar observations were made if the piece of muscle or liver was placed in other parts of the hypothalamus.

Journal of Neuro-Visceral Relations, Suppl. IX, 362—384 (1969)

The Role of Monoamines in the Hypothalamus for Temperature Regulation

W. Feldberg

National Institute for Medical Research, Mill Hill, London, England

With 16 Figures

Summary

Our views about the role of the three monoamines, noradrenaline, adrenaline and 5-hydroxytryptamine (5-HT) in temperature regulation of the hypothalamus is based on the finding that they occur in the hypothalamus as chemical transmitter substances of monoaminergic neurons which innervate the cells of the anterior hypothalamus. Therefore these cells are sensitive to the monoamines, *i. e.* they are monoaminoceptive, in the same way as the cells of sympathetic ganglia innervated by preganglionic cholinergic fibers, respond to acetylcholine, *i. e.* they are cholinoceptive.

To study the temperature responses of the monoamines when acting on the anterior hypothalamus they are either injected into or perfused through the cerebral ventricles or applied by microinfusion directly into the anterior hypothalamus. From the results obtained with these methods it became clear that the temperature responses to he monoamines differ in different species. And if these responses mimic the central transmitter functions of the amines in temperature regulation it would follow that in some species these functions are mediated by the catecholamines and by 5-HT, in others mainly by the catecholamines, and again in others by 5-HT alone and further, that the same monoamine may be used in one species as a central transmitter to raise, in another to lower body temperature.

Evidence for the release of the monoamines in the hypothalamus has so far been obtained for one amine, 5-HT, and in one species only, the cat. When the third cerebral ventricle of anaesthetized cats was perfused with artificial cerebrospinal fluid, 5-HT was detected in the effluent from the aqueduct, but in small amounts only. However, the amounts increased and temperature rose when either 5-hydroxytryptophan, the precursor of 5-HT, or tranylcypromine, an inhibitor of the monoamine oxidase was added to the perfusion fluid.

The fall of temperature in anaesthesia is attributed to an increased release of all three monoamines in the hypothalamus, with the hypothermic effect of the released catecholamines predominating. That the fall is due to an action of the anaesthetics on the hypothalamus was shown by the fact that temperature fell when an anaesthetic — chloralose was used for these experiments — was perfused through the ventral half of the third ventricle, the walls of which contain the hypothalamic nuclei. It was also shown that in anaesthesia the hypothalamus retains its sensitivity to the monoamines. Injeced into the cerebral ventricles during anaesthesia the catecholamines postponed and 5-HT accelerated the return of temperature. Further, it was shown that the fall in temperature did not occur when the destruction of 5-HT was prevented by inhibition of the monoamine oxidase. The intravenous injection of chloralose no longer lowered temperature when it was preceded by an intraperitoneal injection of the monoamine oxidase inhibitor tranylcypromine. In cats only 5-HT, not the catecholamines, appear to be a substrate for the monoamine oxidase of the brain. Finally, in species like rabbit and sheep, which lack an efficient hypothermic monoamine in the hypothalamus, anaesthesia can be produced by intravenous pentobarbitone sodium without an appreciable fall in temperature.

Zusammenfassung

Unsere Vorstellungen über die Rolle der drei Monoamine Noradrenalin, Adrenalin und 5-Hydroxytryptamin (5-HT) für die Temperaturregulierung im Hypothalamus basieren auf Befunden, nach denen die Monoamine im Hypothalamus vorkommen, und zwar als chemische Überträgerstoffe monoaminergischer Neurone, die die Zellen des vorderen Hypothalamus innervieren. Diese Zellen reagieren daher auf die Monoamine, d. h. sie sind monoaminoceptiv, ebenso wie die sympathischen Ganglienzellen, die von preganglionären, cholinergischen Nervenfasern innerviert werden, auf Azetylcholin reagieren und somit cholinoceptiv sind.

Wollen wir untersuchen, wie die Monoamine die Temperatur beeinflussen, wenn sie auf den vorderen Hypothalamus einwirken, so injiziert man sie entweder in die Hirnventrikel, oder die Ventrikel werden durchströmt und man setzt die Monoamine der Durchströmungsflüssigkeit zu, oder die Monoamine werden direkt durch Mikroinjektion in den vorderen Hypothalamus eingebracht. Mit diesen Methoden konnte gezeigt werden, daß die Monoamine die Körpertemperatur bei verschiedenen Tierarten verschiedenartig beeinflussen. Falls diese pharmakologischen Befunde die zentrale Überträgerfunktion der Monoamine in der Temperaturregulierung wiederspiegeln, würde das bedeuten, daß diese Funktionen bei einigen Tierarten von den Katecholaminen und vom 5-HT, bei anderen Tierarten hauptsächlich von den Katecholaminen, und bei wieder anderen Tierarten vom 5-HT alleine ausgeübt werden und weiter, daß ein und dasselbe Monoamin in seiner Eigenschaft als zentraler Überträgerstoff bei einer Tierart dazu dient, die Temperatur zu erhöhen, bei einer anderen Tierart sie zu senken.

Das Freiwerden der Monoamine im Hypothalamus ist bisher nur für ein Monoamin, für das 5-HT, und nur bei einer Tierart, bei der Katze, nachgewiesen worden. Durchströmt man in narkotisierten Katzen den dritten Hirnventrikel mit einer Salzlösung, die in ihrer Zusammensetzung der des Liquors entspricht, so enthält die aus dem Aquedukt abfließende Flüssigkeit 5-HT, aber nur in kleinen Mengen. Setzt man der Durchströmungsflüssigkeit aber entweder 5-Hydroxytryptophan, den Precursor von 5-HT, zu oder Tranylcypromin, einen Hemmstoff der Monoaminoxidase, so nimmt der 5-HT-Gehalt der abfließenden Flüssigkeit zu und die Körpertemperatur steigt an.

Die bekannte Temperatursenkung in der Narkose wird damit erklärt, daß es zu einem vermehrten Freiwerden aller drei Monoamine im Hypothalamus kommt, daß aber die temperatursenkende Wirkung der freigewordenen Katecholamine überwiegt. Als erstes wurde gezeigt, daß die Temperatursenkung wirklich auf einer Einwirkung der Narkotika auf den Hypothalamus beruht, denn wenn die ventrale Hälfte des dritten Ventrikels, dessen Wände die hypothalamischen Kerngebiete enthalten, mit einem Narkotikum, Chloralose, durchströmt wurde, kam es zu einem Temperaturabfall. Dann wurde gezeigt, daß die Ansprechbarkeit des Hypothalamus auf die Monoamine in der Narkose erhalten bleibt, denn wenn diese in der Narkose in die Ventrikel injiziert wurden, bewirkten die Katecholamine eine Verlängerung, das 5-HT eine Verkürzung in der Dauer des Temperaturabfalles. Weiter konnte gezeigt werden, daß die Temperatursenkung ausblieb, wenn die enzymatische Zerstörung von 5-HT durch Hemmung der Monoaminoxidase verhindert wurde. Das wurde dadurch erreicht, daß man der Katze, bevor sie mit Chloralose oder Nembutal narkotisiert wurde, den Hemmstoff Tranylcypromin intraperitoneal injizierte. Bei der Katze ist nur 5-HT Substrat der Hirnmonoaminoxidase; die Katecholamine sind es dagegen nicht. Schließlich wurde noch gezeigt, daß Kaninchen und Schafe, denen ein wirksames hypothermisches Monoamin im Hypothalamus fehlt, durch Nembutal narkotisiert werden können, ohne daß es zu einer Temperatursenkung kommt.

Paul Ehrlich once gave excellent advice on how to give a lecture. Never express more than one idea, he said, because no audience, however intelligent, is able to take in more than one idea in one lecture. I shall follow his advice, and the only idea I shall develop is as follows:

We know body temperature is controlled in the hypothalamus. But we do not know what goes on in this part of the brain to bring about this control. We believe it is the release of adrenaline, noradrenaline and 5-hydroxytryptamine (5-HT), the three monoamines that occur naturally in the hypothalamus. Into this one sentence I have put the whole of my lecture, because what now follows is an attempt to show why we have this belief.

The occurrence of the three monoamines in the hypothalamus was first shown in 1954 by *Marthe Vogt* for adrenaline and noradrenaline, and by *Amin, Crawford* and *Gaddum* for 5-HT. The monoamines can be made visible in the hypothalamus by their fluorescence, and from the beautiful fluorescent microscopic studies of Swedish scientists (*Carlsson, Falck* and *Hillarp*, 1962; *Dahlström* and *Fuxe*, 1964; *Andén, Dahlström, Fuxe* and *Larsson*, 1965) it appears that in the hypothalamus they are mainly located not in nerve cells but in nerve fibers and nerve endings. This recalls the location of acetylcholine in sympathetic ganglia. Here, too, the acetylcholine is located not in the cells, but in the preganglionic cholinergic nerve endings which impinge upon the ganglion cells. In the same way, the high content of the monoamines in the hypothalamus signifies monoaminergic nerve fibers which end in the hypothalamus. Most of them appear to be ascending fibers from the lower part of the brain, from the mesencephalon. Further, the ganglion cells, on which the cholinergic fibers end, are sensitive to acetylcholine or, as we say, are cholinoceptive. In the same way, the cells of the hypothalamus on which the monoaminergic fibers end ought to be sensitive to the monoamines, or to be monoaminoceptive, and so they are.

When *Myers* and I injected small amounts of noradrenaline and adrenaline into the cerebral ventricles of the cat, they penetrated the hypothalamus and body temperature fell, whereas 5-HT similarly injected had the opposite effect and produced fever. Before discussing these results I want to show a picture of the human brain with the cerebral ventricles

(Fig. 1) to illustrate how the injections were made and how the ventricles were perfused, the two main methods used in our experiments.

The figure shows the two lateral ventricles, one in each cerebral hemisphere, and their communication through the foramina of Monro with the third ventricle, which is divided into a dorsal and ventral part by

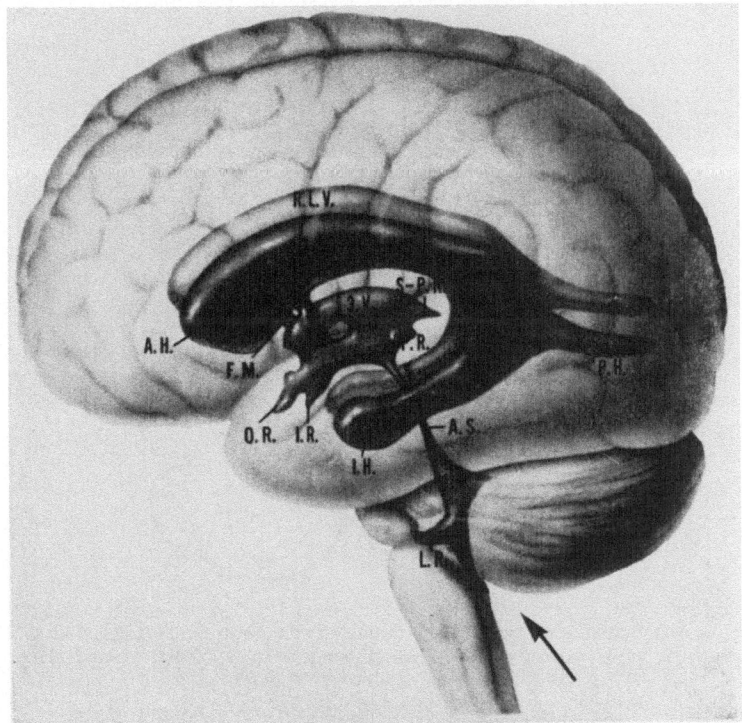

Fig. 1. Human brain with cerebral ventricles. The arrow points to the cisterna magna. (With permission, from the *CIBA Collection of Medical Illustrations,* by *Frank H. Netter,* MD.)

a bridge of nervous tissue, the massa intermedia. In cats and dogs the massa is larger than in man. The third ventricle communicates with the fourth, shown as a triangle in the figure, through the aqueduct. The walls of the third contain the hypothalamic nuclei. The ventricular cavities communicate with the fluid space surrounding the brain, the subarachnoid space, through three openings, all of which are in the walls of the fourth ventricle, two at its lateral recesses, the third at its caudal end. This opening is absent in cats and dogs.

To inject drugs into the cerebral ventricles we implant a Collison cannula into a lateral ventricle, and to prevent drugs from escaping into the subarachnoid space through the openings in the walls of the fourth

ventricle we perfuse the ventricles and exclude the fourth ventricle from
the perfusion. For this purpose an outflow cannula is inserted through
the opened cisterna magna and pushed along the floor of the fourth
ventricle until the tip lies in the middle of the aqueduct. Most of our
experiments were made on cats.

A specially constructed Collison cannula is used for the intraventricular
injections (*Feldberg* and *Sherwood*, 1953). In the diagram of Fig. 2, a

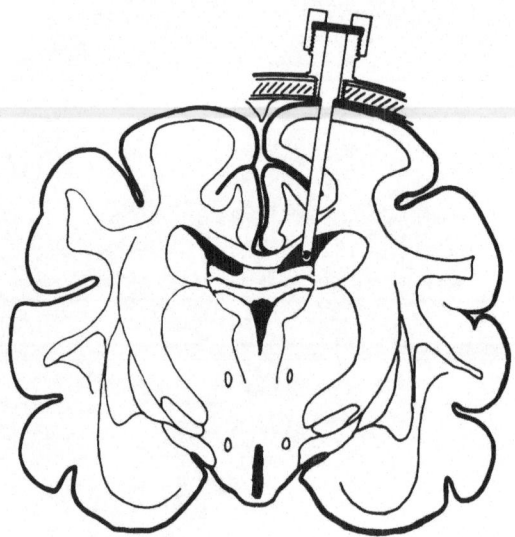

Fig. 2. Diagram of frontal section through cat's brain with implanted Collison
cannula for intraventricular injections. (From a film by *Feldberg* and *Sherwood*,
1953; drawing modified by *Fleischhauer*.)

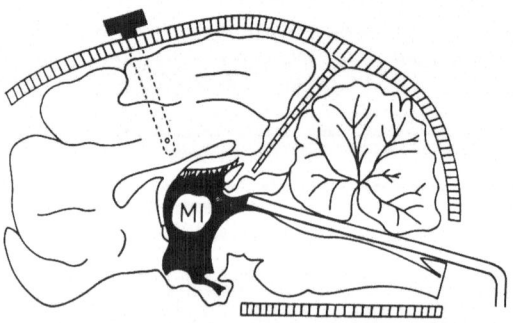

Fig. 3. Diagram of median sagittal section of cat's brain to illustrate perfusion
from Collison cannula in lateral ventricle to cannulated aqueduct. Third ventricle
in black. MI massa intermedia. (Modified drawing from *Bhattacharya* and *Feldberg*, 1958.)

frontal section of the cat's brain, the cannula is shown screwed into the skull with the tip in the lateral ventricle. The protruding end of the cannula is closed with a rubber diaphragm through which the injections are made. They can be made in anaesthetized and unanaesthetized cats.

Perfusions are made in anaesthetized cats only. The procedure is illustrated in Fig. 3. The inflow cannula is the same as for intraventricular injections. As it lies behind the plane of this mid-saggital section, it is given in dotted lines. The tip of the outflow cannula is inserted through the cisterna and lies in the middle of the aqueduct. The perfusion is

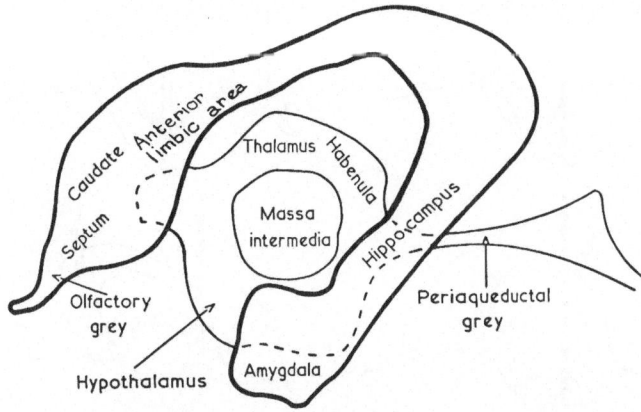

Fig. 4. Diagram of the ventricular system of the cat's brain to illustrate regions of grey matter bordering lateral and third ventricles as well as aqueduct. Only the left lateral ventricle is shown. (Slightly modified from *Carmichael* and coll., 1964).

limited to a lateral ventricle, the third ventricle, and to the upper end of the aqueduct. But this is not enough, because the grey masses reached on perfusion from a lateral ventricle to aqueduct are numerous, as seen from the diagram Fig. 4. We have to restrict the perfusion with drugs, so that they pass only through one lateral ventricle, or only through its anterior or inferior horn, or through the third ventricle, or only through its dorsal or ventral half. The methods we use for this purpose are based on the finding that the perfused drug solution is denied access to any part simultaneously perfused with inert artificial cerebrospinal fluid, and the various modifications are shown in the diagrams of Fig. 5.

Four or five cannulae are inserted, each into a different part. One cannula acts as an outflow; the others attached to separate injectors serve as inflow. Outflow and inflow through the cannulae are marked in these diagrams by correspondingly directed arrows. One cannula, at most two, deliver the drugs, the others artificial cerebrospinal fluid. The cannulae delivering the drug are in solid black and the parts of the ventricular system exposed to the drugs are shaded. The right lateral ventricle is always perfused with artificial cerebrospinal fluid so that the drug cannot enter it.

The arrangement of cannulation is the same whether we perfuse drugs through the left lateral ventricle, or through either its inferior or anterior horn (diagram *a*, *b*, *c*). The same arrangement of cannulation provides a method for perfusing drugs through the third ventricle (diagram *d*). The

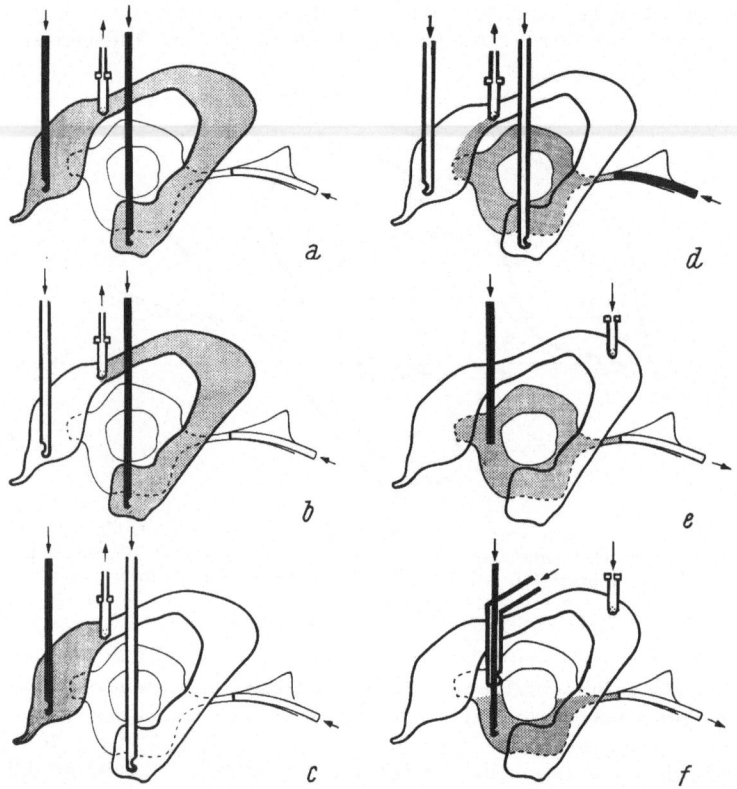

Fig. 5. Diagrams to illustrate the methods of regional perfusion of the cat's cerebral ventricles. For further explanation see text. (Slightly modified from *Carmichael* and coll., 1964.)

drug is delivered through the aqueductal cannula and passes the third ventricle in antidromic direction. Diagram *e* illustrates another method in which the cannula delivering the drug is inserted directly into the third ventricle.

The last diagram, *f*, shows the procedure used for perfusing drugs through either the ventral or dorsal half of the third ventricle. A double-bore cannula is inserted into the third ventricle in such a way that the opening of the inner tube lies ventral, that of the outer tube dorsal to the massa intermedia. The diagram gives the arrangement for the perfusion of drugs through the ventral half. The drug is delivered through the inner, and artificial cerebrospinal fluid through the outer tube. For perfusion through the dorsal half the procedure is reversed.

For our temperature experiments we needed particularly the arrangement for perfusing the third ventricle and its ventral and dorsal half.

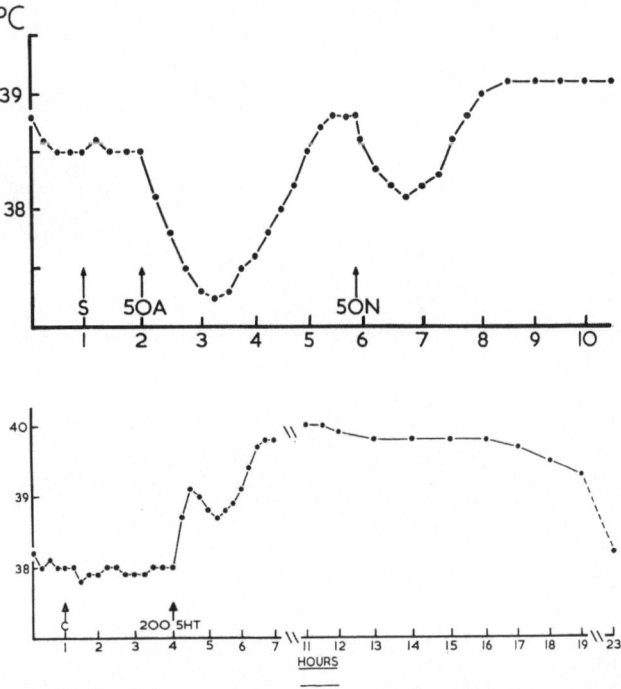

Fig. 6. Records of rectal temperature of two unanaesthetized cats. The arrows indicate injections into the cannulated cerebral ventricles. S: injection of 0.1 ml 0.9% NaCl solution, 50 A : 50 µg adrenaline, 50 N : 50 µg noradrenaline, C: 100 µg creatinine sulphate, 200 5-HT = 200 µg 5-HT. (From *Feldberg* and *Myers*, 1964 a.)

When *Myers* and I injected small amounts of the monoamines into the cerebral ventricles of the cat, so that they penetrated the hypothalamus, we found that adrenaline and noradrenaline lowered body temperature, whereas 5-HT had the opposite effect and produced fever. This is illustrated in Fig. 6.

These hypo- and hyperthermic effects result from an action on the anterior hypothalamus, for the same changes occur with microinjections of the monoamines into this part of the hypothalamus and, as shown in Fig. 7, much smaller amounts were effective.

The most common changes in temperature are the rise produced by bacterial pyrogens, *i. e.* the fever of infectious diseases, and the fall which occurs in anaesthesia. Can these changes be explained in the light of the new concept?

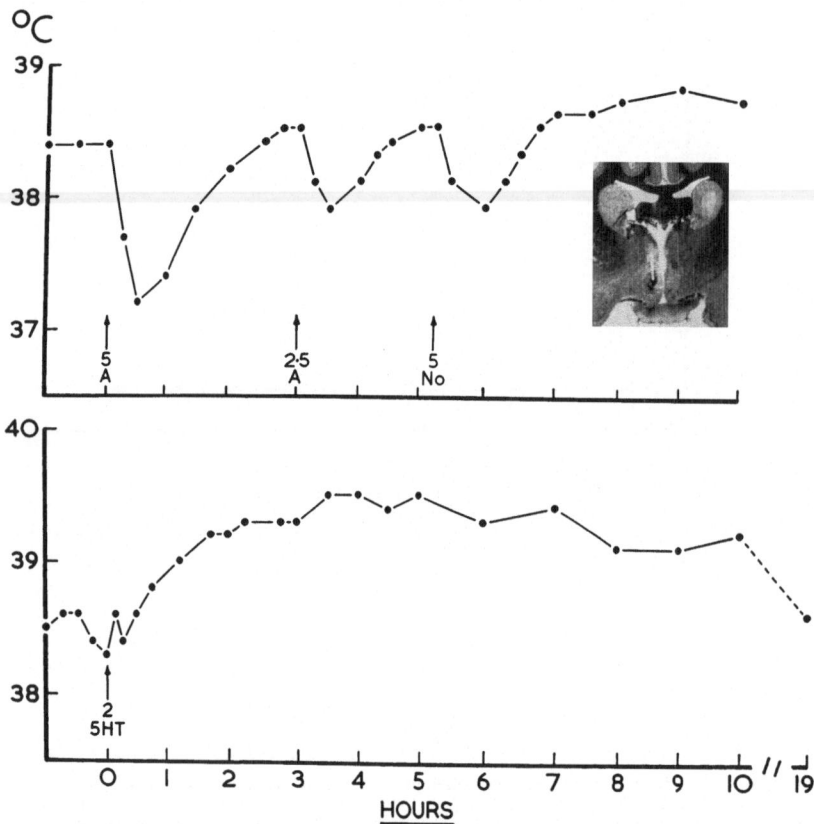

Fig. 7. Records of rectal temperature of two unanaesthetized cats. The arrows indicate microinfusions of 1 µl fluid into the anterior hypothalamus. Upper record, 5 and 0.25 µg. adrenaline and 5 µg noradrenaline; lower record, 2 µg 5-HT. Inset coronal section from the cat's brain with cannula tract ending in anterior hypothalamus. (From *Feldberg* and *Myers*, 1965 a.)

In cats, experiments to elucidate the mechanism of action of the bacterial pyrogens have not yet been performed. They may act by releasing 5-HT, by inhibiting the release of the catecholamines, by rendering the hypothalamus more sensitive to the action of 5-HT, or they may not act through the hypothalamic amines, but may mimic the effect of 5-HT. In rabbits experiments were done which suggest that the febrile response to bacterial pyrogens is due to 5-HT release since it no longer occurred when the

5-HT content in the hypothalamus had been reduced, whereas reduction of the noradrenaline content did not have this effect (*Des Prez, Helman* and *Oates,* 1966). However, as we shall see, in rabbits noradrenaline but not 5-HT causes hyperthermia when injected into the cerebral ventricles or

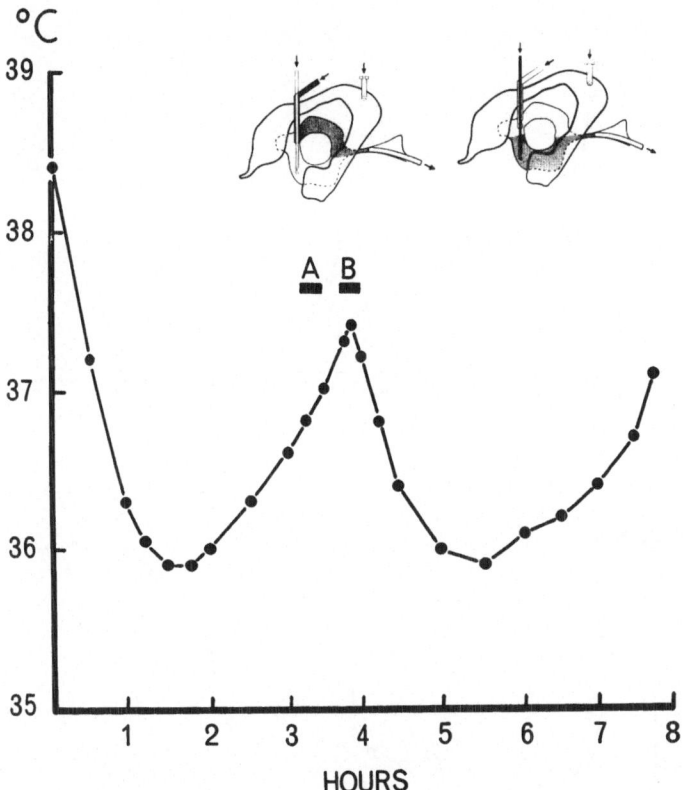

Fig. 8. Record of rectal temperature of a cat during perfusion of cerebral ventricles. The record begins immediately after induction of anaesthesia with intraperitoneal pentobarbitone sodium. At the bars, perfusion for 15 min of chloralose 1/2500 through the dorsal (at *A*) and ventral (at *B*) half of third ventricle as indicated above by the diagrams. (From *Feldberg* and *Myers,* 1965 b.)

directly into the anterior hypothalamus, so the problem is anything but solved.

In anaesthesia, temperature regulation is abolished — the animal has become poikilothermic — temperature falls if no external heat is applied. The curves with the solid circles of Fig. 9 and 13 illustrate the fall for chloralose and pentobarbitone sodium anaesthesia. The fall is attributed to an action of the anaesthetics on the hypothalamus. This has never been

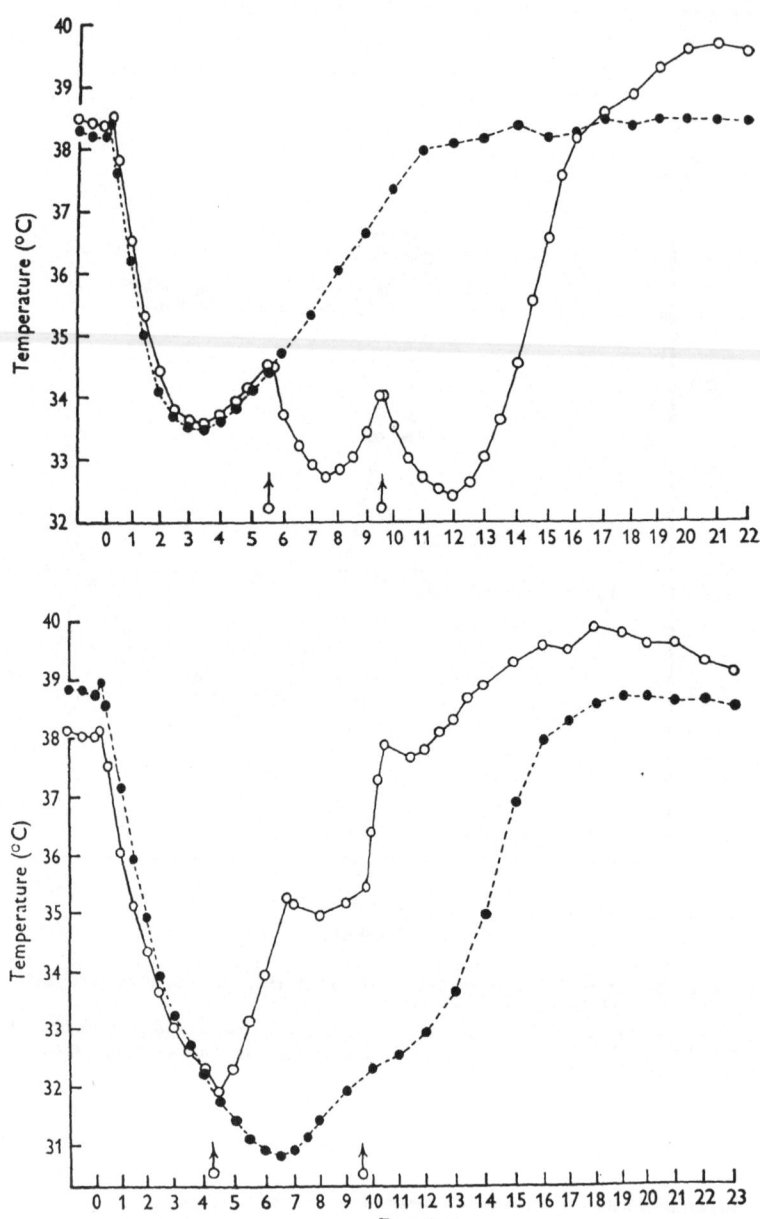

Fig. 9. Upper tracing: two records of rectal temperature obtained from the same cat at an interval of one week. At zero hour intravenous injection of chloralose 45 mg/kg in both instances. Solid circles, choralose alone. Open circles, choralose

proved but the proof can easily be obtained by using the methods of regional perfusion of the cerebral ventricles. If the action is on the hypothalamus, there should be no fall when an anaesthetic is perfused through the ventral half of the third ventricle. Fig. 8 shows such an experiment with chloralose. We anaesthetized the cat with intraperitoneal pentobarbitone sodium and perfused all parts of the cerebral ventricles with artificial cerebrospinal fluid. The anaesthesia produced a fall in temperature, but we waited until anaesthesia lightened and temperature began

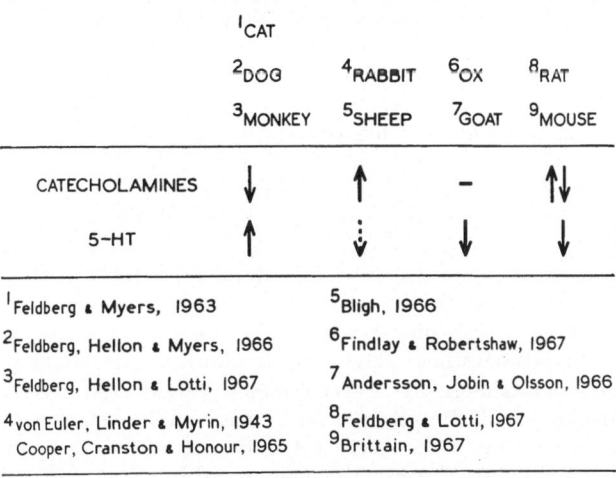

Fig. 10. Diagram to illustrate the species difference in the response to monoamines injected intraventricularly. Direction of the arrows indicates a rise or fall in temperature. (Slightly modified from *Feldberg* and coll., 1967.)

to rise, then we perfused chloralose in a concentration of 1/2500 first through the dorsal, and later through the ventral half of the third ventricle. On its perfusion through the dorsal half for 15 min at the bar *A*, temperature continued to rise. On its perfusion through the ventral half also for 15 min, at the bar *B*, temperature began to fall within a few minutes and continued to fall for over 1 hr by 1.5° C.

What is the cause of this fall in temperature during anaesthesia? Do we anaesthetize the hypothalamus so that it becomes insensitive to the amines? This explanation is only valid for deep anaesthesia. If the anaesthesia was not too deep, we found that the hypothalamus responded

followed by intraventricular injection of 50 µg adrenaline at the first and of 50 µg noradrenaline at the second arrow. Lower tracing: Two records of rectal temperature obtained from another cat at an interval of one week. At zero hour intravenous injection of chloralose. Solid circles, 60 mg/kg choralose alone; open circles, 70 mg/kg chloralose followed at each arrow by intraventricular injection of 200 µg 5-HT. (From *Feldberg* and *Myers*, 1964 b.)

to the amines as illustrated for adrenaline and noradrenaline in the upper, and for 5-HT in the lower half of Fig. 9, in cats anaesthetized by intravenous chloralose. The upper half shows a postponement by 10 hr in the return of temperature by intraventricular injection, first of 50 μg of adrenaline, and then of 50 μg noradrenaline, and the lower half is the opposite effect, the acceleration in the return of temperature by two intraventricular injections of 200 μg 5-HT.

It thus looks as if the anaesthetics modify the release of the amines in the hypothalamus and that the fall in temperature produced by anaesthetics is an active process. The question is, how do they act? Do they increase the release of the catecholamines, or do they inhibit the release of 5-HT or — which seems to be the most likely explanation — do they increase the release of all three amines, with the action of the catecholamines predominating.

It is interesting to note in this connection that shivering is a characteristic feature of some anaesthetics, for instance, halothane in man. The shivering may be the sign of 5-HT release, and if so, it would suggest that various anaesthetics influence the release of the monoamines in a different way.

So far, all experiments discussed were done on cats. Dogs and monkeys were found to respond to the monoamines in the same way as cats, but the hypothalamus of rabbits, sheep, oxen, goats and mice, responded differently to the monoamines. This is shown diagrammatically in Fig. 10.

In rabbits and sheep the catecholamines raised, and 5-HT lowered, body temperature, but the effect of 5-HT was small and was not obtained consistently. Thus the monoamines not only act differently than in cats, dogs and monkeys, but the hypothalamus of rabbits and sheep apparently lacks an efficient hypothermic monoamine. In goats and oxen, on the other hand, the catecholamines had no effect on body temperature when injected into the cerebral ventricles, whereas 5-HT had a strong hypothermic action. Rats and mice responded differently again. 5-HT produced hypothermia, as in oxen and goats, but the catecholamines were not inactive. With the smallest effective doses injected into the cerebral ventricles hyperthermia was produced and larger doses lowered body temperature.

If these pharmacological effects mimic their central transmitter functions in temperature regulation, it would follow that in cats, dogs and monkeys, these functions are mediated by the catecholamines and 5-HT, in rabbits and sheep mainly by the catecholamines, and in oxen and goats by 5-HT alone and, further, that the same monoamine may be used in one species as a central transmitter for raising, in another for lowering body temperature. In rats and mice both the catecholamines and 5-HT would appear to act as central transmitters of temperature regulation, but the question arises of whether the release of the catecholamines raises or lowers temperature, i. e. whether the hyperthermic effect obtained with the smallest effective doses, or the hypothermic effect of the larger ones, or both effects, mimic the physiological effects of the catecholamines. Such great variability of transmitter functions has not been observed elsewhere.

We would naturally like to know how the monoamines act in man, whose hypothalamus also contains the monoamines, as shown by *Bertler* (1961). Although no experimental data are available to tell us how they act in this species, until evidence to the contrary is obtained, we may take the view that they act as in monkeys, dogs and cats, and not as in the other species.

The evidence so far given has consisted entirely of pharmacological results. We require, however, physiological proof, that is, we must show that the amines are released in the hypothalamus and that this release is associated with changes in temperature.

If the released monoamines were to diffuse from the hypothalamus, that is from the walls of the third ventricle into the cerebrospinal fluid, then on perfusion of this ventricle they would also most likely diffuse into the perfusing fluid and could then be detected in the effluent. We had to decide whether to look first for the release of the catecholamines or of 5-HT, and we decided on 5-HT.

The isolated fundus strip of the rat's stomach is very sensitive to 5-HT. Suspended in a 5 ml bath, it contracts to less than a nanogram as shown by *Vane* (1957). A nanogram is a millionth of a milligram. You can imagine how pleased we were when, already in the first experiment, we found that the effluent from the perfused third ventricle contracted the fundus strip (*Feldberg* and *Myers*, 1966). This finding alone did not necessarily mean that the effluent contained 5-HT, because many substances contract the fundus strip, but when we found that the specific antagonist of 5-HT, bromolysergic acid diethylamide (BOL) blocked the effect of the effluent, we could be pretty certain that the contractions were due to 5-HT. The amounts of 5-HT assayed in the effluent, however, were small and it soon became apparent that without some means whereby the output of 5-HT could be increased, it would not be possible to establish a correlation between 5-HT output and changes in temperature.

For demonstrating the release of acetylcholine from cholinergic neurones, the use of inhibitors of cholinesterase like eserine or neostigmine has been of invaluable help in the last 30 years. With regard to 5-HT, two procedures suggested themselves. The use of 5-hydroxytryptophan (5-HTP) the precursor of 5-HT, and the use of inhibitors of monoamine oxidase, the enzyme which destroys 5-HT. Both procedures were successful.

The addition of 5-HTP to the fluid perfusing the third ventricle produces a manifold increase in the 5-HT content of the effluent. But this is not all. This increase is associated with shivering and a rise in rectal temperature, provided the toxic actions of 5-HTP do not supervene. This is illustrated in Fig. 11 which again shows a temperature record from an anaesthetized cat. The temperature has fallen on account of the anaesthesia. The record begins as temperature is returning. When it had levelled off, the third ventricle was first perfused with artificial cerebrospinal fluid, as indicated by the open bar. This did not affect temperature and the effluent contained only small amounts of 5-HT. Then we switched over to perfusion with a solution of 5-HTP 1/10.000. The 5-HT output increased greatly, shivering occurred and temperature rose to fever level.

This rise in 5-HT output, however, is no evidence for an increased release of 5-HT from the hypothalamus, because the 5-HT is derived from the 5-HTP in the perfusion fluid. As this fluid passes along the ventricular walls its 5-HTP comes into contact with an enzyme present in these walls

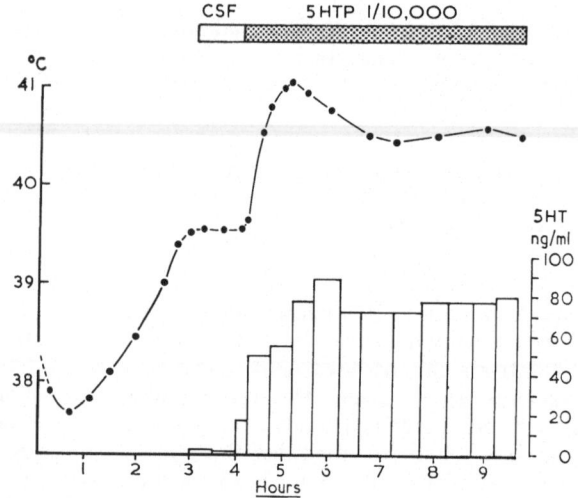

Fig. 11. Record of rectal temperature in a cat anaesthetized by intraperitoneal pentobarbitone sodium. The horizontal bar indicates perfusion of third ventricle at a rate of 0.05 ml/min with artificial cerebrospinal fluid (open bar) and with 5-HTP 1/10.000 (stippled bar). The block diagram below the temperature curve represents 5-HT output in ng/l of effluent. (From *El Hawary* and *Feldberg*, 1966.)

which decarboxylates the 5-HTP to 5-HT. The name of this enzyme is 5-HTP decarboxylase.

The ventricular walls contain not only the enzyme which forms 5-HT, but also the enzyme that destroys it, the monoamine oxidase. This enzyme can be inhibited by a number of substances, the monoamine oxidase inhibitors. If they were to increase the 5-HT output and at the same time to raise body temperature, the result would provide the physiological evidence we have been looking for: a rise in temperature produced by 5-HT released from the hypothalamus.

The monoamine oxidase inhibitor we used was Parnate, or tranyl-cypromine. Injected intraperitoneally, or added to the perfusion fluid, tranylcypromine caused an increase in 5-HT output from the perfused third ventricle associated with shivering and a rise in temperature (*El Hawary, Feldberg* and *Lotti*, 1967). Fig. 12 shows the result of an experiment with an intraperitoneal injection of 5 mg/kg. The arrow indicates the injection. As the tranylcypromine becomes absorbed and reaches the brain, the released 5-HT is no longer destroyed, its output in the perfusing fluid increases as shown in the block diagram. Temperature first falls less steeply and then rises, whilst the cat begins to shiver. Without the injection, the steep fall would have continued for some time.

Once more I want to point out that with these experiments we did in principle the same as was done 30 years ago, in order to demonstrate the role of acetylcholine as a transmitter of cholinergic nerves. At that time

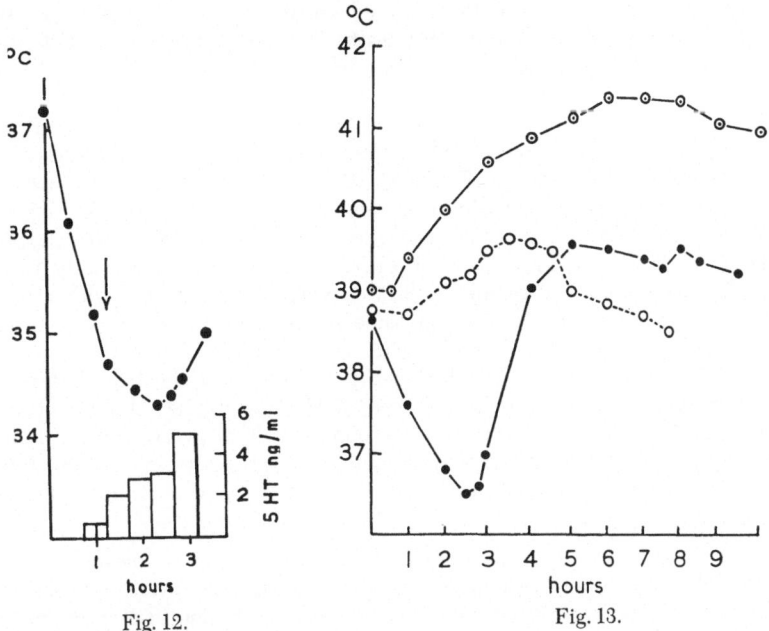

Fig. 12.   Fig. 13.

Fig. 12. Record of rectal temperature from a cat anaesthetized with intravenous chloralose during perfusion of third ventricle. At the arrow, intraperitoneal injection of 5 mg/kg tranylcypromine. The block diagram below the temperature curve represents 5-HT output in ng/ml of effluent. Rate of perfusion 0.05 ml min (From *Feldberg* and *Lotti*, 1967 a.)

Fig. 13. Three records of rectal temperature obtained from the same cat at intervals of at least one week. (o - - - o) intraperitoneal injection of 10 mg/kg tranylcypromine one hour after beginning of record. (●—●) Intraperitoneal injection of 30 mg/kg pentobarbitone sodium 15 min before beginning of record. (☉—☉) Intraperitoneal injection of 10 mg/kg tranylcypromine 30 min, and of 30 mg/kg pentobarbitone sodium 15 min before beginning of record. (From *Feldberg* and *Lotti*, 1967 a.)

the acetylcholine destroying enzyme, cholinesterase, was inhibited using physostigmine as the inhibitor. It then became possible to detect the released acetylcholine and to show that stimulation of cholinergic nerves produced a stronger effect. The enzyme we are inhibiting in the present experiments is monoamine oxidase, the inhibitor of the enzyme is tranylcypromine, and the transmitter is 5-HT instead of acetylcholine. And, instead of the stronger response to nerve stimulation, shivering and a rise in temperature are produced.

Before continuing with the tranylcypromine experiments, I want to mention one other finding because it may explain a phenomenon which has puzzled neurosurgeons for a long time, the brain fever with bouts of shivering, which occurs as a result of brain injuries after accidents.

We thought we must find out how long it takes for the 5-HT to disappear in the effluent from the perfused third ventricle after killing the cat, so we continued perfusion after death. To our surprise, the 5-HT did not decrease, but increased, sometimes up to 20 times, in the effluent collected during the first hour after death, and only then decreased gradually. This rise in 5-HT output occurred whether or not 5-HTP or tranylcypromine had been added to the perfusion fluid (*Feldberg* and *Myers*, 1966; *El Hawary* and *Feldberg*, 1966; *El Hawary* and coll., 1967).

What is the meaning of this post mortem rise, this life in the hypothalamus after death? It means that, when the blood supply to the hypothalamus is disturbed, the release of 5-HT proceeds at an abnormally high rate. This happens after death. If the cat were not dead, it would shiver and its temperature would rise. The same may happen after brain injuries when small vessels in the hypothalamus are torn, or go into spastic contraction. The result would be that in those regions of the hypothalamus in which the blood supply is disturbed, 5-HT would be released at an abnormally high rate, and this in turn would result in pyrexia with shivering. Although this theory is not proved, it at least describes in physiological terms a phenomenon which, so far, has defied any convincing explanation.

To come back to the experiments with tranylcypromine. Injected intraperitoneally into unanaesthetized cats the effect of tranylcypromine on body temperature was surprisingly small. Yet such injections prevented the deep fall in temperature produced by an anaesthetic. This is shown for pentobarbitone sodium in Fig. 13, but the same result was obtained with chloralose. The three records of rectal temperature were obtained from the same cat at weekly intervals. The dotted curve with the open circles shows the very small rise in temperature following an intraperitoneal injection of 10 mg/kg tranylcypromine. The curve with the solid circles shows the typical fall in temperature following an intraperitoneal injection of 30 mg/kg pentobarbitone sodium. The curve with the open circles containing the dots shows the combined effect of both injections, the tranylcypromine followed 15 min later by the pentobarbitone sodium. The anaesthetic no longer produced the fall in temperature, instead temperature rose to fever level; yet the cat became fully anaesthetized.

How to explain this surprizing result, the insignificant effect of tranylcypromine on temperature in the unanaesthetized cat, yet its ability to counteract and even reverse the temperature lowering effect of an anaesthetic?

Anaesthetics appear to increase the release of the three monoamines, noradrenaline, adrenaline and 5-HT, in the hypothalamus, but as long as the monoamine oxidase in the brain is not inhibited, the action of the released catecholamines predominates. Therefore, the fall in temperature produced by anaesthetics.

Now in cats, and also in dogs, 5-HT but not noradrenaline appears to be a substrate for the monoamine oxidase in the brain because, as first shown by *Vogt* (1959), injections of inhibitors of this enzyme, including

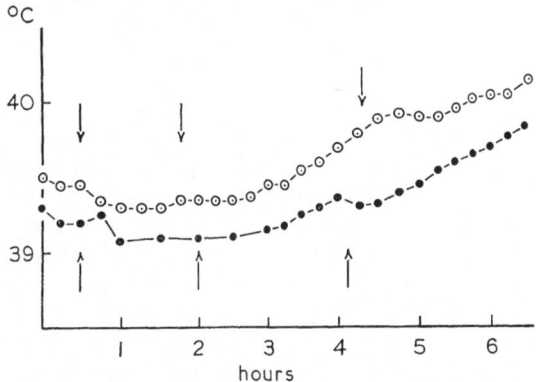

Fig. 14. Records of rectal temperature from one rabbit. The two temperature curves were obtained at an interval of 6 days. At the arrows, intravenous infusions of 25 mg/kg pentobarbitone sodium. ●—● Effect of the pentobarbitone sodium injections alone. o—o Effect of the pentobarbitone sodium injections after an intraperitoneal injection of 10 mg/kg tranylcypromine given at the beginning of the record. (From *Feldberg* and *Lotti*, 1967 a.)

tranylcypromine, increase the brain level of 5-HT, but not that of the catecholamines. Consequently, after an injection of tranylcypromine the 5-HT released by the anaesthetic is no longer destroyed and its hyperthermic effect when acting on the hypothalamus is able to counteract the hypothermic effect of the catecholamines also released by the anaesthetic. Temperature, therefore, no longer falls and may even rise in anaesthesia.

By taking into account the species differences we can test our ideas about the action of monoamine oxidase inhibitors as well as of anaesthetics. Since the hypothalamus of the dog and monkey responds to the monoamines as in cats, anaesthetics like pentobarbitone sodium should produce a fall in temperature. And they do. In dogs we have also found that tranylcypromine prevented the fall (*Feldberg, Hellon* and *Lotti*, 1967). Similar experiments have not yet been performed in the monkey.

Next we come to the rabbit and sheep. It was pointed out that these species lack an efficient hypothermic monoamine in the hypothalamus. Therefore, according to our theory, anaesthetics should not cause a fall in temperature. And this we found as illustrated in Fig. 14. To take the lower record first. At each arrow 25 mg/kg pentobarbitone sodium was infused intravenously resulting in surgical anaesthesia for 1—2 hr, but without producing a fall in temperature. Perhaps we should have expected a rise in anaesthesia because of the temperature-raising effect of the catecholamines in the rabbit. There is at least a tendency for temperature to rise, but the rise is small and very gradual. The upper curve was obtained a week later. The arrows again indicate intravenous infusions of pentobarbitone sodium, but before the first infusion, 10 mg/kg tranylcypromine was injected intraperitoneally. It had no effect. Tranylcypromine was also ineffective when injected into the cerebral ventricles, whereas in cats such injections caused a rise in temperature (Feldberg and Lotti, 1967 a).

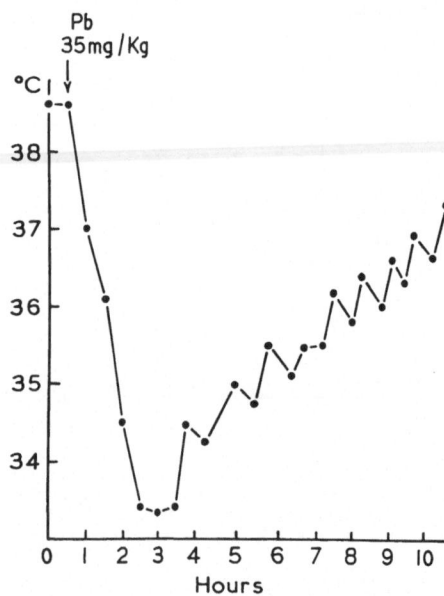

Fig. 15. Record of rectal temperature from a rat. At the arrow intraperitoneal injection of 35 mg/kg pentobarbitone sodium. (From Feldberg and Lotti, 1967 b.)

We have pointed out that in cats only the 5-HT and not the noradrenaline level of the brain rises when the enzyme is inhibited. In rabbits noradrenaline, too, is a substrate for the amine oxidase, but it is a much poorer substrate than 5-HT. Inhibition of the monoamine oxidase produces a much smaller increase in the brain level of noradrenaline than of 5-HT and the increase occurs later (Spector, Shore and Brodie, 1960; Pscheidt, Morpurgo and Himwich, 1964; Spector, 1963). The initial effect of monoamine oxidase inhibition is on the 5-HT level only. Therefore it is perhaps not surprising that in the conditions of our experiments no rise in temperature was observed due to undestroyed noradrenaline. In order to obtain such a result it would probably be necessary to inhibit the other enzyme for the destruction of noradrenaline, the O-methyltransferase.

According to our theory, inhibition of monoamine oxidase should cause a fall in temperature in those species in which intraventricular 5-HT causes a fall. Oxen and goats would have been the ideal test animals, but we have done experiments only in rats (Feldberg and Lotti, 1967 b). In these, anaesthetics produced a fall in temperature, and tranylcypromine, like 5-HT, had a hypothermic effect. The result with tranylcypromine was obtained whether it was injected intraperitoneally or intraventricularly. The last two

figures demonstrate the fall in temperature produced in rats by pentobarbitone sodium (Fig. 15) and by intraventricular tranylcypromine
(Fig. 16).

I am aware that there is still a long way to go before the new theory
is firmly established and before the conditions under which the three monoamines are released are fully elucidated. But just as over 30 years ago
acetylcholine in its newly discovered role as synaptic transmitter in the
peripheral nervous system changed our outlook on practically all problems

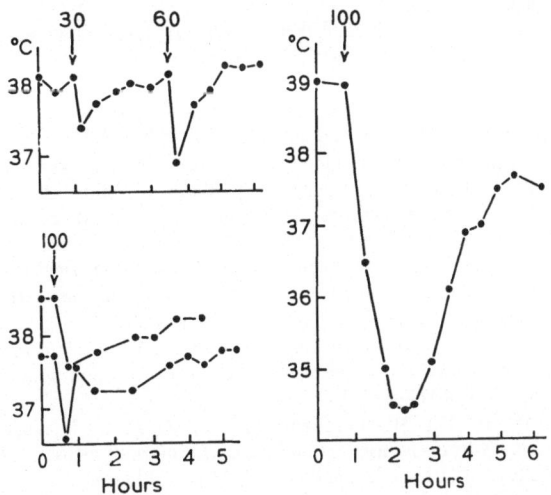

Fig. 16. Records of rectal temperature from four unanaesthetized rats. At the
arrows, injections into the cerebral ventricles of tranylcypromine, 30 and 60 µg
in the one and 100 µg in the other three rats. (From *Feldberg* and *Lotti*, 1967 b.)

of neuromuscular and ganglionic transmission, so also will the role of the
three monoamines in the hypothalamus change our way of thinking when
interpreting the intricate mechanism of temperature regulation within the
hypothalamus, which is only one of its many functions dependent on the
release of its monoamines.

References

Amin, A. N., T. B. B. Crawford, and *J. H. Gaddum:* The distribution of
substance P and 5-hydroxytryptamine in the central nervous system of the dog.
J. Physiol. *126,* 596—618 (1954).

Andén, N. E., A. Dahlström, K. Fuxe, and *K. Larsson:* Mapping out of
catecholamine and 5-hydroxytryptamine neurons innervating the telencephalon and
diencephalon. Life Sciences, *4,* 1275—1279 (1965).

Andersson, B., M. Jobin, and *K. Olsson:* Serotonine and temperature control.
Acta physiol. Scand. *67,* 50—56 (1966).

Bertler, A.: Occurrence and localization of catecholamines in human brain.
Acta physiol. Scand. *51,* 97—101 (1961).

382   W. Feldberg:

Bhattacharya, B. K., and *W. Feldberg:* Perfusion of cerebral ventricles: effects of drugs on outflow from the cisterna and aqueduct. Brit. J. Pharmacol. *13,* 156—162 (1958).

Brittain, R. T., and *S. L. Handley:* Temperature changes produced by the injection of catecholamines and 5-hydroxytryptamine into the cerebral ventricles of the conscious mouse. J. Physiol., London, *192,* 805—813 (1967).

Bligh, J.: Effects on temperature of monoamines injected into the lateral ventricles of sheep. J. Physiol., London, *185,* 46—47 P (1966).

Carmichael, E. A., W. Feldberg, and *K. Fleischhauer:* Methods for perfusing different parts of the cat's cerebral ventricles with drugs. J. Physiol., London, *173,* 354—367 (1964).

Carlsson, A., B. Falck, and *N. Hillarp:* Cellular localization of brain monoamines. Acta physiol. Scand. *56,* Suppl. 196 (1962).

Cooper, K. E., W. I. Cranston, and *A. J. Honour:* Effects of intraventricular and intrahypothalamic injection of noradrenaline and 5-HT on body temperature in conscious rabbits. J. Physiol., London, *181,* 852—864 (1965).

Dahlström, A., and *K. Fuxe:* Evidence for the existence of monoamine-containing neurons in the central nervous system. I. Demonstration of monoamines in the cell bodies of brain stem neurons. Acta physiol. Scand. *62,* Suppl. 232 (1964).

Des Prez, R., R. Helman, and *J. A. Oates:* Inhibition of endotoxin fever by reserpin. Proc. Soc. Exper. Biol. Med., N. Y., *122,* 746—749 (1966).

Euler, U. S. von, E. Linder, und *S. D. Myrin:* Über die fiebererregende Wirkung des Adrenalins. Acta physiol. Scand. *5,* 85—86 (1943).

El Hawary, M. B. E., and *W. Feldberg:* Effect of 5-hydroxytryptophan acting from the cerebral ventricles on 5-hydroxytryptamine output and body temperature. J. Physiol., London, *186,* 401—415 (1966).

El Hawary, M. B. E., W. Feldberg, and *V. J. Lotti:* Monoamine oxidase inhibition: effect on 5-hydroxytryptamine output from the perfused third ventricle and body temperature. J. Physiol., London, *188,* 131—140 (1967).

Feldberg, W., and *V. J. Lotti:* Body temperature responses in cats and rabbits to the monoamine oxidase inhibitor tranylcypromine. J. Physiol., London, *190,* 203—220 (1967 a).

Feldberg, W., and *V. J. Lotti:* Temperature responses to monoamines and an inhibitor of MAO injected into the cerebral ventricles of rats. Br. J. Pharmacol. Chemother. *31,* 152—161 (1967 b).

Feldberg, W., and *R. D. Myers:* A new concept of temperature regulation by amines in the hypothalamus. Nature, London, *200,* 1325 (1963).

Feldberg, W., and *R. D. Myers:* Effects on temperature of amines injected into the cerebral ventricles. A new concept of temperature regulation. J. Physiol., London, *173,* 226—237 (1964 a).

Feldberg, W., and *R. D. Myers:* Temperature changes produced by amines injected into the cerebral ventricles during anaesthesia. J. Physiol., London, *175,* 464—478 (1964 b).

Feldberg, W., and *R. D. Myers:* Changes in temperature produced by microinjections of amines into the anterior hypothalamus of cats. J. Physiol., London, *177,* 239—245 (1965 a).

Feldberg, W., and *R. D. Myers:* Hypothermia produced by chloralose acting on the hypothalamus. J. Physiol., London, *179,* 509—517 (1965 b).

Feldberg, W., and *R. D. Myers:* Appearance of 5-hydroxytryptamine and an unidentified pharmacologically active lipid acid in effluent from perfused cerebral ventricles. J. Physiol., London, *184,* 837—855 (1966).

Feldberg, W., and *S. L. Sherwood:* A permanent cannula for intraventricular injections in cat. J. Physiol. *120,* 3—5P (1953).

Feldberg, W., R. F. Hellon, and *V. J. Lotti:* Temperature effects produced in dogs and monkeys by injections of monoamines and related substances into the third ventricle. J. Physiol., London, *191,* 501—515 (1967).

Feldberg, W., R. F. Hellon, and *R. D. Myers:* Effects on temperature of monoamines injected into the cerebral ventricles of anaesthetized dogs. J. Physiol., London, *186,* 413—423 (1966).

Findlay, J. D., and *D. Robertshaw:* The mechanism of body temperature changes produced by intraventricular injections of monoamines in the ox *(Bos taurus).* J. Physiol., London, *188,* 46P (1967).

Pscheidt, G. R., C. Morpurgo, and *H. E. Himwich:* Nor-epinephrine and 5-hydroxytryptamine in various species. Regional distribution in the brain, response to monoamine oxidase inhibitors, comparison of chemical and biological assay methods for nor-epinephrine. *Comparative Neurochemistry.* ed. *D. Richter,* pp. 401—412 (1964). Oxford: Pergamon Press.

Spector, S.: Monoamine oxidase in control of brain serotonin and norepinephrine content. Ann. N. Y. Acad. Sci *107,* 856—861 (1963).

Spector, S., P. A. Shore, and *B. B. Brodie:* Biochemical and pharmacological effects of the monoamine oxidase inhibitors proniazid, 1-phenyl-2-hydrazinopropane (JB 516) and 1-phenyl-3-hydrazinobutane (JB 835). J. Pharmacol. Exper. Therap., Baltimore, *128,* 15—21 (1960).

Vane, J. R.: A sensitive method for the assay of 5-hydroxytryptamine. Br. J. Pharmacol. *12,* 344—349 (1957).

Vogt, M.: The concentration of sympathin in different parts of the central nervous system under normal conditions and after the administration of drugs. J. Physiol., London, *123,* 451—481 (1954).

Vogt, M.: Catecholamines in brain. Pharmacol. Rev., Baltimore, *11,* 483—489 (1959).

Author's address: Prof. Dr. W. *Feldberg,* National Institute for Medical Research, Mill Hill, London, N.W. 7, England.

Discussion

Saxena: 1. How do you explain the hypothermic effect of catecholamines after its initial hypothermic action? 2. Did you study the effects of specific antagonists of monoamines? 3. In the light of your fascinating theory on temperature regulation and the effects of monoamine, would you care to elaborate on the mechanism of hypothermic action of chlorpromazine and reserpine? Does reserpine resemble the anaesthetics as both release catecholamine? 4. Would you also comment on the potentiation of hyperthermia and antagonism of hypothermia induced by catecholamines and by imipramine-like drugs?

Feldberg: Ad 1. The effect is not always obtained and it is, therefore, difficult to give a definite conclusion; the effect may be due to a compensatory mechanism. Ad 2. We have, so far, only done experiments with ergotamine in cats and found that after an injection of ergotamine into the cerebral ventricles the hypothalamus no longer responds to the monoamines, except that there is actually a "reversal" of the effects of the catecholamines injected intraventricularly. In this condition pentobarbitone sodium no longer causes a fall in temperature. Ad 3. We have not done any experiments with chlorpromazine. Most of the experiments with reserpine have been done in rabbits, and the reserpine was given systemically. It is difficult to interpret the results obtained with this kind of administration, because we do not know when reserpine acts and how it exerts its action on the hypothalamus. The results obtained all point to the role of 5-HT as a hyperthermic monoamine in the hypothalamus of rabbits (*Böing*, 1959; *Des Prez, Holman* and *Oates*, 1966). This, however, requires further investigation. There is one recent observation by *Cooper* and coworker that, injected into the cerebral ventricles of rabbits, reserpine causes a rise in temperature.

Koepchen: I should like to ask Prof. *Feldberg* if he can rule out the possibility that the reactions to intraventricular application of adrenaline and/or noradrenaline on body temperature could be influenced or modified by vascular reactions within the hypothalamus. One could imagine that such effects alter the blood flow and therewith the temperature in those areas of the hypothalamus which contain elements sensitive to temperature changes.

Feldberg: I think that this possibility may certainly exist but I do not believe that it is the cause of these effects. There is not yet any valid evidence for this, except this one experiment which we did on dogs: if you measure the hypothalamic temperature during the injection of the catecholamines you observe a fall in temperature whereas, then, you would have expected the opposite, a rise. I mean, the possibility is not excluded, but I do not think a real pharmacologist thinks the best of it.

De Wied: There is some indication, at least in the rat (*Lommax* and *Jenden*, 1965), that implantation of carbachol into the area preoptica lowers body temperature. Would you care to comment on this finding with respect to the theory on temperature regulation as outlined in your paper?

Feldberg: Acetylcholine and carbachol have as far as I know no effect on temperature when applied into the anterior hypothalamus. The fact that carbachol effects temperature when applied to the preoptic area may suggest that cholinergic fibers impinge on the cells of this area and that the fibers originating from these cells end on anterior hypothalamic nuclei.

Akert: Your findings are of great interest to me because they seem to offer a solution to the difficulties which I have encountered years ago when I was trying to establish a "center" for shivering by localizing the active stimulation points in Prof. *Hess'* material (*K. Akert* and *F. Kesselring*, Helv. Physiol. Acta, 1951). The points were scattered in septal and hypothalamic areas from which shivering could be elicited in unaesthetized animals. There seems to be no focal area, and the only common factor of localization was the wall of the third ventricle. In the light of your observations it would appear that 5-HT was released by electrical stimulation.

Summing up and General Discussion

Scharrer: I would like, more as a start for a more generalized discussion from the floor, to bring up a few points — or bring them into focus — as they appear to me to arise from our preceding invited papers. If you permit me I, of course, have to select those that are somehow close to my heart and therefore seem to be, perhaps, a little more important.

I would like to start out by bringing up again the principal question and it relates to the importance of having such interdisciplinary types of meetings as this one is. It is brought out in particular by the remarks from Dr. *Feldberg* and his colleagues in pharmacology, and biochemists would do the same. They are somewhat surprised and perhaps unhappy about the confidence with which morphologists deal with morphological details or — let me say — look at them as correlates of physiological events. Had Dr. *Feldberg* or some of his colleagues been last year at Strasbourg, at the neurosecretion-meeting, he would be much more fed up with vesicles as a center of discussion. I do want to say, however, that, used in the right way, such physiological-morphological correlations as we attempt them may be useful and that those of us who are concerned with structure like to look at these vesicles and similar things in terms of cytophysiological consideration. Observation of a change in morphology under certain known conditions, either physiologically or experimentally induced, could be useful.

Now, the elongated or round clear vesicle in itself may not be of importance, but if the elongated shaped ones happen to come into appearance in those synaptic endings, which for physiological reasons are considered as inhibitory ones, than one begins to try to correlate. I do want to propose for discussion the possibility, because it has not really come up, that a clear vesicle of small size is after all very indistinctive morphologically. It is a small space closed in by a membrane. It could very well be that this little structural detail serves in one case for harbouring acetylcholine. It might, in another location, where there is no reason to assume that acetylcholine is either present or functional, contain something else or what it contains might serve a purpose different from the classical functional significance that we attributed to acetylcholine, namely that of a transmitter. If we think in such terms, we can perhaps come to some truthful ideas.

A second point is the important new development in neurosecretory neurons as defined formally. As defined today in chemical terms it is no longer to be denied that there may be — or I would say — that there are neurosecretory neurons with substances other than peptides, more specifically catecholamine-type products, which we think serve — or may serve — a special type of neurohormones. It might perhaps also come as a summarizing thought that these are closely concerned with the regulation of the anterior pituitary, in other words, what was discussed beautifully by Dr. *Halász*, with the releasing factors. As a footnote to that, a very much nicer term that I would like to propose is not "releasing factor", but "regulating factor". It preserves the first letters for all abbreviations, and regulation would include inhibition as well as stimulation of release. It would also cover what Dr. *Halász* brought out: the possible effect — or a possible influence — not only on the release or non-release of substances of this kind, but their possible control of the synthesis. So, perhaps this might be something to consider.

Than I come back to the question which we discussed on the first day. The most interesting question concerning the area of overlap between classical, ordinary neurons on one hand and what we like to call neurosecretory neurons on the other. We do have with this added chemical feature and with the feature of contiguity of some bonafide neurosecretory fibers on effector cells, which may either be endocrine cells or muscle cells, at least in invertebrates, a very interesting new area to explore. This is something for future work to consider.

Perhaps another point to consider is the emergence and great interest in ependyma. This can no longer be considered a uniform type of tissue, but a tissue from which a variety of specialized organs or areas — whatever you want to call it — seem to emerge as structures, some of which were very nicely discussed during the meeting. Their function certainly needs further exploration.

A smaller point perhaps, but I was rather pleased that it came up, is the important question of species differences. It came out very beautifully in Dr. *Feldberg's* topic and it came out yesterday in discussions. Some of the controversies that you all get into about differences in results, experimental results, or interpretations may need our thinking in terms of such species differences.

I think these are the major points and I am sure that from now on everybody has a lot more to say than I have. Thank you very much.

Akert: Prof. *Scharrer,* I think you have made several very important points and before they vanish from the short memory store perhaps we should go ahead and discuss them right now. I would like to invite members of the audience to comment on these various aspects.

Kappers: I should like to support strongly the suggestion made by Dr. *Scharrer* to use the term "regulating factors" instead of "releasing factors".

Hommes: A study of 24 human hypothalami showed a strong relation of the size of the nuclei of neurons of the arcuate (infundibular) nucleus with age. After the age of 40—45 years the size increases sharply. From the histoligigal appearance of these neurons this must be interpreted as growth and increased activity. Also in the light of Prof. *Knowles'* observation on the special structure of the ependyma lining the 3rd ventricle at the level of the arcuate nucleus, this change in the histology of the arcuate nucleus during a rather critical age for men seems of importance.

De Robertis: Work done some time ago in Argentine has shown that in animals which were gonadectomized the arcuate nucleus cells, a few months later, showed very big changes, the nuclear size of the neurons being, for instance, much enlarged. This finding may be related with your observations.

Bargmann: The observation that neurosecretory fibers form synaptic contacts with epithelial cells of the intermediate lobe provokes the question, whether octapeptides — if released from the nerve endings — are capable to act on the membrane of the epithelial cells as transmitter substances comparable to the classical transmitters acetylcholine and amines. I should be very grateful for a discussion of this problem from the view of the pharmacologists attending this meeting.

Feldberg: This is, indeed, a very pertinent question. Theoretically, there is this possibility. There are cases in the peripheral nervous system where we know that an effector cell, which is normally impinged by cholinergic neurons, responds well to an octapeptide. Medullary cells of the suprarenal are, for instance, also extremely sensitive to bradikinine. At your special anatomical localization there is as yet no evidence for this condition, but there may possibly be a pharmacological effect.

Kappers: You told us, Dr. *Scharrer,* that neurohumors are characterized by a short range of action. Is this always true? I think, for instance, of the noradrenaline produced by the medulla of the adrenals, which is taken up by the blood and shows a far-range action.

Scharrer: Yes, but is that a neurohumor or a neurohormone in the first place?

Kappers: That is just the question!

Scharrer: I did not, in my deliberations, dare to include that at all. There are some other important questions which, unfortunately, I have had no time to discuss. I think noradrenaline can, even coming from the postganglionic neurons, be either a neurotransmitter in the classical sense, produced in a classical ordinary adrenergic neuron, or it can — and then it belongs to the B-fiber-category — be a hormone. And if it is, than it is acting in the same way as the adrenal hormone of the same chemical composition. We are now in that strange position that we can no longer go by the chemistry or at least by far less than we did. But to come back now to another question which is related: is it — and I asked this to one of our local pharmacologists in New York, Dr. *Douglas* — permitted to extend the neurohumor concept beyond that of the neurotransmitter in synaptic transmission and would a local effect of a peptidergic type with direct contact, as Dr. *Bargmann* brings it into the picture, be permitted to be called a neurohumoral rather than a neurohormonal effect. Dr. *Douglas* said: "Absolutely not." I am, however, not sure that I would follow him. This might be something to discuss.

Knowles: May I remind you of a certain congress on neurosecretion where the adrenal medulla has also been a stumbling-block?

Feldberg: May I just refer to the assumption that peptidergic neurons release the octapeptide at their endings and have an anatomically synaptic electron microscopical appearance? Is there any evidence that these peptides released are really acting on the effector structure? Because we do not know that and I think I would agree with Dr. *Scharrer* in her reply.

Smelik: As a pharmacologist, I would like to add that it is true that many polypeptides, like vasopressin and angiotensin, have effects on different kinds of cells. However, pharmacologists become increasingly aware of the fact, that many of these actions can be blocked by antiadrenergic drugs, suggesting that peptide action is mediated by adrenergic transmission. One good example, however, of an effect of peptides and of catecholamines on the same structure, is their action on the adenylcyclase system, which is situated in the cellular wall, but this of course is not a synaptic structure.

Akert: Prof. *Feldberg*, may I ask you now to make some summing up statements?

Feldberg: Prof. *Scharrer*, you said you were not volunteered to give a talk, with me it was much worse. Prof. *Ariëns Kappers* knows the best way to get a person to do what he wants is to hypnotize him. The moment before I went to give my lecture he said: you are saying a few words after the symposion and summarize it. I thought, why do I need to worry what happens after my paper?

We have had 18 papers and actually, if I use a modern fashion, I can deal with it very briefly, because you all know that, as is said at least in England: A talk should be like a dress of a lady, long enough to cover the whole subject, but short enough to be interesting. I think we must be extremely grateful to Prof. *Ariëns Kappers* to have arranged the symposion in such a logical form.

We got first a review which dealt with the terms "neurohumors" and "neuro hormones" and brought us face to face with the interesting problem that there is a similarity and that there is a dissimilarity and that we have not yet solved the problem, but that we must think about it when we work on it. We had than the two beautiful lectures by Prof. *Picard* and Prof. *Bargmann* on the neurosecretory cells in vertebrates and on the neurosecretory diencephalo-hypophyseal system and its synaptic connections. I think, the most interesting point for physiologists was the attempt by Prof. *Bargmann* to try to get an answer from the physiologists or pharmacologists what they think about peptidergic neurons. The result was, and I could have told him that before, that physiologists and pharmacologists are never able to give the answer. The problem of the second different form of synaptic connections is one about which physiologists have partly thought little. For those who are here as physiologists and pharmacologists it was a very healthy way in being reminded of it.

I am thinking of the papers of Prof. *Akert* on the subfornical organ, of Sir Francis *Knowles* on ependymal secretion, and of Prof. *Oksche* on the subcommissural organ. Again what impressed me most is how little support the anatomists have received in studying these regions from physiologists and pharmacologists. You morphologists know so much now about the detailed structure of these organs and on ependymal secretion that you can even go so far as to accept some evidence that there is secretion into the ventricle and that there may be a feed-back system. The physiologists and the pharmacologists, however, have not given you so far single data, so that we could say and teach the students when you deal with the subfornical organ, ependymal secretion and the subcommissural organ, that and that are the functions. Physiologists and pharmacologists must be grateful to all three of you to have drawn our attention where we lack in doing experiments.

The same does not apply to Prof. *Ariëns Kappers'* communication on the mammalian pineal organ as one of these dark little things of which a normal physiologist does not know enough. He did not only show us the histological and electron microscopical pictures, but he could give us definite physiological evidence about pineal function. That was also a very good transition to the next three papers, which were reviews on acetylcholine, catecholamine, tryptamine and substance P.

Concerning the paper on acetylcholine: It is interesting that you can nowadays give a paper on acetylcholine, a stimulating paper, in which you can follow every word, and do not need anything to say about the actions of acetylcholine. At the moment the emphasis has shifted from the purely physiological and pharmacological point of view to what we may call a biophysical and an electron microscopical point of view. I am also very grateful to Dr. *Quay* about his review on catecholamines and tryptamines, and I am very much looking forward to read it.

Concerning these papers, there is always one thing which I ask myself and I am not giving the answer. I ask you forgiveness if I sound crude but I always admire the people who show these beautiful diagrams. I wonder, and it would be interesting to know, how correct these diagrams will look within the next 5 or 10 years. But, here, the physiologists and pharmacologists have an advantage: if the drug injected into the ventricle causes a fall or a rise in temperature, it will do that in 10 years too.

Coming now to the paper of Prof. *Stern* I think he had in a way one of the most difficult subjects because if we were to give a summary about substance P in a short time you could say: we know it is a polypeptide, we know it occurs in the central nervous system, we know it occurs in the intestinal wall, but we do not know what the physiological function is. I was particularly glad that he drew attention to the fact that there has been the suggestion by *Pernow*, that it may be a central synaptic transmitter of sensory fibers. Although this theory is not proved, it is also not yet disproved. That he brought it again into the foreground is very important.

I come now to Prof. *Carlsson*. Even if he had not given any paper he would have been present by all the other papers. I admired him for his modesty that he took an outside line as the main subject of his paper on the pseudo-transmitters. He has done so much to advance our knowledge on the monoamines that he could have given very well many other papers of similar interest. I personally would like to say that I am rather optimistic about his concept of pseudo-transmitters and their importance as a tool and even as a practical way of trying to find remedies in clinical cases. I am not convinced that hypotension by α-methyldopamine

(DOPA) is not due to what you thought it is due, but probably only in the central nervous system. .

I think we must all be extremely grateful to Prof. *De Robertis*, who succeeded in showing us how far electron microscopy has gone. I was mostly impressed that electron microscopists not only look at their slides, but that they use drugs as tools making conclusions from the action of the drugs. When I made some remarks which looked as if I was not always convinced, I would like to say that this does not mean that the evidence for most of his conclusions about the content of the granules of the vesicles as containing dopamine, catecholamine or 5-HT was not most impressing. For someone who does not work in this field it was particularly important to see how from that side so much evidence has accumulated that has to be taken into account by physiologists and pharmacologists.

Coming now to the afternoon session from yesterday we had three interesting papers, all on different subjects. The role of the hypothalamic monoamines in the control of pituitary secretion was discussed by Dr. *Smelik.* Prof. *Westermann* dealt with the autonomic nervous system and metabolism in such a way that even those who do not know much about biochemistry could understand the great advance which, in recent years, has been made on the influence of the autonomic nervous system on fat and glucose metabolism. It was very good that in this symposion, which is meant also to help the clinicians, we had at least one paper by a pure clinician, Prof. *Birkmayer,* giving us his account on Parkinsonism in showing us that what the non-clinician does will ultimately help the clinician, because the bridge is nearly going to be built between this disease and the results which have been obtained in biochemical, physiological and pharmacological laboratories.

As the last three papers of today are concerned we got some new aspects of neural connections of the hypothalamic nuclei from Prof. *Lammers.* All of his figures are extremely valuable for anyone working on this subject. Then Prof. *Halász* gave a completely new concept for those who do not work on it, on the releasing factors or, as we now have to call them, regulating factors, based on his beautiful experiments. Then in the last paper the importance of species differences was stressed, a fact also emphasized by Prof. *Scharrer.*

Akert: Prof. *Feldberg,* I thank you very much for this masterful analysis. I feel strongly and I hope all of you feel too, that there is no need now of a prolonged discussion. What could be said, has been said. The only thing that has been left for me is to thank those who have contributed to this symposion. In particular we thank our president and host for this wonderful opportunity for exchange of ideas.

Author Index

Page numbers printed in *italics* refer to References

Subject Index